Red Dot Design Yearbook 2015/2016

Edited by Peter Zec

reddot award
product design

About this book

"Living" uses more than 400 design products to demonstrate the state of the art in contemporary homes and lives. All of the products in this book are of outstanding design quality and have been successful in one of the world's largest and most renowned design competitions, the Red Dot Design Award. This book documents the results of the current competition in the field of "Living", also presenting its most important players – the design team of the year, the designers of the best products and the jury members.

Über dieses Buch

„Living" zeigt anhand von mehr als 400 Designprodukten den State of the Art zeitgemäßen Wohnens und Lebens. Alle Produkte in diesem Buch sind von herausragender gestalterischer Qualität, ausgezeichnet in einem der größten und renommiertesten Designwettbewerbe der Welt, dem Red Dot Design Award. Dieses Buch dokumentiert die Ergebnisse des aktuellen Wettbewerbs im Bereich „Living" und stellt zudem seine wichtigsten Akteure vor – das Designteam des Jahres, die Designer der besten Produkte und die Jurymitglieder.

Contents
Inhalt

Professor Dr Peter Zec
Preface of the editor
Vorwort des Herausgebers

Dear Readers,

With the Red Dot Design Yearbook 2015/2016, you are holding a very special book in your hands. In the three volumes, "Living", "Doing" and "Working", it presents to you the winners of this year's Red Dot Award: Product Design, which make up the current state of the art in all areas of product design. At the same time this compendium is a piece of living history, because this year marks the 60th anniversary of the Red Dot Award.

Since 1955, we have set out every year in search of the best that the world of industrial design has to offer. What began 60 years ago as a small industrial show in Essen with a selection of around 100 products, most of which were German, has developed consistently over the past six decades. The competition has grown, ventured out into the world and become cosmopolitan.

A visible sign of this new, international gearing appeared 15 years ago, when the name of the competition was changed to the "Red Dot Design Award" and a new logo, the Red Dot, was introduced. That logo has become synonymous with good design – and the Red Dot Award is now the largest international competition for product design. In this anniversary year alone, 1,994 companies and designers from 56 countries entered 4,928 products in the competition for a Red Dot.

Yet through all the changes, one thing has stayed the same: The heart of the competition is and will always be its independent and fair jury. The following pages will show you this year's selection by our design experts.

I wish you an inspiring read.

Sincerely, Peter Zec

Liebe Leserin, lieber Leser,

mit dem Red Dot Design Yearbook 2015/2016 halten Sie ein besonderes Buch in den Händen: Es präsentiert Ihnen in den drei Bänden „Living", „Doing" und „Working" die Gewinner des diesjährigen Red Dot Award: Product Design und damit den aktuellen State of the Art in allen Bereichen des Produktdesigns. Zugleich ist dieses Kompendium jedoch auch ein Stück Zeitgeschichte, denn der Red Dot Award feiert dieses Jahr sein 60-jähriges Bestehen.

Seit 1955 machen wir uns jedes Jahr aufs Neue auf die Suche nach dem Besten, das die Welt des Industriedesigns zu bieten hat. Was vor 60 Jahren als kleine Industrieschau in Essen mit der Auswahl von rund 100 vornehmlich deutschen Produkten seinen Anfang nahm, hat sich im Laufe der vergangenen sechs Jahrzehnte beständig weiterentwickelt. Der Wettbewerb ist gewachsen, in die Welt hinausgegangen und weltoffen geworden.

Als sichtbares Zeichen für diese neue, internationale Ausrichtung bekam er vor 15 Jahren mit der Änderung zu „Red Dot Design Award" nicht nur einen neuen Namen, sondern auch ein neues Logo, den Red Dot. Dieses Logo ist heute ein Synonym für gutes Design – und der Red Dot Award mittlerweile der größte internationale Wettbewerb für Produktdesign. Alleine in diesem Jubiläumsjahr bewarben sich 1.994 Unternehmen und Designer aus 56 Ländern mit 4.928 Produkten um eine Auszeichnung mit dem Red Dot.

Eine Sache gibt es jedoch, die bei allen Veränderungen gleich geblieben ist: Das Herzstück des Wettbewerbs ist und bleibt seine unabhängige und faire Jury. Welche Auswahl unsere Designexperten in diesem Jahr getroffen haben, das sehen Sie auf den folgenden Seiten.

Ich wünsche Ihnen eine inspirierende Lektüre.

Ihr Peter Zec

The title "Red Dot: Design Team of the Year" is bestowed on a design team that has garnered attention through its outstanding overall design achievements. This year, the title goes to Robert Sachon and the Bosch Home Appliances Design Team. This award is the only one of its kind in the world and is extremely highly regarded even outside of the design scene.

Mit der Auszeichnung „Red Dot: Design Team of the Year" wird ein Designteam geehrt, das durch seine herausragende gestalterische Gesamtleistung auf sich aufmerksam gemacht hat. In diesem Jahr geht sie an Robert Sachon und das Bosch Home Appliances Design Team. Diese Würdigung ist einzigartig auf der Welt und genießt über die Designszene hinaus höchstes Ansehen.

In recognition of its feat, the Red Dot: Design Team of the Year receives the "Radius" trophy. This sculpture was designed and crafted by the Weinstadt-Schnaidt based designer, Simon Peter Eiber.

Als Anerkennung erhält das Red Dot: Design Team of the Year den Wanderpokal „Radius". Die Skulptur wurde entworfen und angefertigt von dem Designer Simon Peter Eiber aus Weinstadt-Schnaidt.

2015	Robert Sachon & Bosch Home Appliances Design Team
2014	Veryday
2013	Lenovo Design & User Experience Team
2012	Michael Mauer & Style Porsche
2011	The Grohe Design Team led by Paul Flowers
2010	Stephan Niehaus & Hilti Design Team
2009	Susan Perkins & Tupperware World Wide Design Team
2008	Michael Laude & Bose Design Team
2007	Chris Bangle & Design Team BMW Group
2006	LG Corporate Design Center
2005	Adidas Design Team
2004	Pininfarina Design Team
2003	Nokia Design Team
2002	Apple Industrial Design Team
2001	Festo Design Team
2000	Sony Design Team
1999	Audi Design Team
1998	Philips Design Team
1997	Michele De Lucchi Design Team
1996	Bill Moggridge & Ideo Design Team
1995	Herbert Schultes & Siemens Design Team
1994	Bruno Sacco & Mercedes-Benz Design Team
1993	Hartmut Esslinger & Frogdesign
1992	Alexander Neumeister & Neumeister Design
1991	Reiner Moll & Partner & Moll Design
1990	Slany Design Team
1989	Braun Design Team
1988	Leybold AG Design Team

Red Dot: Design Team of the Year 2015
Robert Sachon & Bosch Home Appliances Design Team
Technology and Design for a better Quality of Life

For more than 125 years, the name Bosch has stood for pioneering technology and outstanding quality. And for more than 80 years, home appliances made by Bosch have lived up to this name. With the Bosch brand, consumers worldwide associate efficient functionality, reliable quality and internationally acclaimed design. In the Red Dot Award: Product Design alone, the design team led by Robert Sachon has received more than two hundred Red Dots and several Red Dot: Best of the Bests in the past 10 years. To honour this achievement, the design team of Robert Bosch Hausgeräte GmbH is being recognised this year for its consistently high and groundbreaking design performance with the honorary title "Red Dot: Design Team of the Year".

Seit mehr als 125 Jahren steht der Name Bosch für wegweisende Technik und herausragende Qualität. Diesem Anspruch sind auch die Hausgeräte von Bosch seit über 80 Jahren verpflichtet. Mit der Marke Bosch verbinden Konsumenten weltweit effiziente Funktionalität, verlässliche Qualität und ein international ausgezeichnetes Design. Allein im Red Dot Award: Product Design erhielt das Designteam unter der Leitung von Robert Sachon in den vergangenen 10 Jahren mehr als zweihundert Auszeichnungen mit dem Red Dot und darüber hinaus mehrere Auszeichnungen mit dem Red Dot: Best of the Best. In diesem Jahr wird das Designteam der Robert Bosch Hausgeräte GmbH für seine kontinuierlich hohen und wegweisenden Gestaltungsleistungen mit dem Ehrentitel „Red Dot: Design Team of the Year" ausgezeichnet.

Quality of Life

Robert Sachon, Global Design Director of the Bosch brand, summarises the brand values of the company as follows: "Bosch has a long tradition spanning more than 125 years, and has always stood for technical perfection and superior quality." Bosch home appliances always put people first, in keeping with the principles of the company's founder Robert Bosch: technology and quality, credibility and trust, along with a sense of responsibility for people's well-being.

As early as 1932, Robert Bosch wrote that "progress in the development of technology serves, in the broad sense of the term, to afford the largest service to humanity – technology that is intended and able to provide all of humanity with enhanced opportunities and happiness in life." In the wake of the global economic crisis, Robert Bosch anticipated a notion, or value, that was not to enter the public consciousness until the early 1970s: the concept of quality of life.

Bosch develops premium-quality modern home appliances that facilitate people's daily lives and that help them attain a better quality of life. Nevertheless, only few home appliances have a place in the "design hall of fame". Indeed, as a category, home appliances have been largely neglected in the history of design, despite the fact that these products have, due to their technical functionality and everyday use, a high impact on the quality of life.

The Refrigerator

Home appliances from Bosch first saw the light of day at the Leipzig Spring Fair in 1933. The fair visitors must have been impressed when Bosch presented its first refrigerator: a metal barrel with a 60-litre net capacity. At the time, refrigerators were still regarded as "machines" given the obtrusive metal cylinder that gave them more of a tool character. The round form, which we today associate more with a washing machine, served the technical purpose of saving energy and retaining the cold optimally inside the appliance. A refrigerator in the shape of a cabinet, more suitable for daily use, was introduced on the market three years later. Resting on high feet, this newer model looked more like a piece of furniture than a technical machine. In 1949, Bosch then presented the refrigerator that, with its subtle, sober and highly industrialised form, was to become a design classic like no other home appliance.

In 1996, Bosch brought out a new edition of the 1949 design classic, thereby turning the model dating from the time of Germany's post-war economic miracle into a historical icon. Nevertheless, this new edition, introduced in a world that was rapidly changing and becoming faster with the rise of globalisation and the Internet, was designed to remain familiar and recognisable – in other words, to be "cool and calm" at once. That said, the new edition of the classic was much more than a nostalgic reminder of a bygone age. Equipped with modern technology, the Bosch Classic Edition awakened new desires and quickly advanced to become a longtime bestseller on its own terms.

Die Lebensqualität

„Bosch hat eine lange Tradition von mehr als 125 Jahren und steht seit jeher für die Themen ‚Technische Perfektion' und ‚Überlegene Qualität'", fasst Robert Sachon, Global Design Director der Marke Bosch, die Markenwerte des Unternehmens zusammen. Bosch Hausgeräte stellen immer den Nutzen für den Menschen in den Mittelpunkt, getreu den Grundsätzen des Firmengründers Robert Bosch: Technik und Qualität, Glaubwürdigkeit und Vertrauen sowie Verantwortungsbewusstsein zum Wohle der Menschen.

Bereits 1932 schrieb Robert Bosch, „dass die Fortschritte in der Entwicklung der Technik im vollem Umfange des Wortes dazu dienen, der Menschheit die größten Dienste zu leisten. Der Technik, die dazu bestimmt und in der Lage ist, der gesamten Menschheit ein Höchstmaß an Lebensmöglichkeit und Lebensglück zu verschaffen." Unter dem Eindruck der Weltwirtschaftskrise nimmt Robert Bosch einen Gedanken vorweg, der erst zu Beginn der 1970er Jahre in die öffentliche Wahrnehmung treten soll: den Begriff der Lebensqualität.

Bosch entwickelt moderne Haushaltsgeräte von höchster Qualität, die den Alltag der Menschen erleichtern und ihnen zu mehr Lebensqualität verhelfen. Indes ist nur wenigen Hausgeräten ein Platz im Olymp des Designs vorbehalten; wie überhaupt den Hausgeräten nicht immer der gebührende Platz in der Geschichte des Designs eingeräumt wird, obgleich sie aufgrund ihrer technischen Funktionalität und ihres alltäglichen Gebrauchs einen hohen Einfluss auf die Lebensqualität haben, weil sie die Hausarbeit enorm erleichtern und das Leben insgesamt verändern.

Der Kühlschrank

Die Geburtsstunde der Hausgeräte von Bosch schlägt auf der Leipziger Frühjahrsmesse 1933. Die Messebesucher dürften nicht schlecht gestaunt haben, als Bosch seinen ersten Kühlschrank vorstellt: eine Trommel mit 60 Litern Nutzinhalt. Zur damaligen Zeit ist es durchaus üblich, von einer Kältemaschine zu sprechen, da der Metallzylinder unmissverständlich seinen Werkzeugcharakter zum Ausdruck bringt. Die runde Form, die man heute eher mit einer Waschmaschine assoziiert, hat technische Gründe. Es geht darum, Energie zu sparen und die Kälte optimal im Gerät zu halten. Die für den Alltagsgebrauch besser geeignete Schrankform kommt drei Jahre später auf den Markt und erinnert mit ihren hohen Füßen mehr an ein Möbelstück als an eine technische Maschine. 1949 stellt Bosch dann den Kühlschrank vor, der es mit seiner zurückhaltenden, sachlichen und industriell geprägten Form wie kaum ein anderes Hausgerät zum Designklassiker schafft.

1996 bringt Bosch eine Re-Edition der Designikone von 1949 heraus und setzt dem Modell aus der Zeit des deutschen Wirtschaftswunders damit ein Denkmal. In der Zeit der Globalisierung und der sich anbahnenden Veränderung durch das Internet ist die Wiederauflage des Klassikers so etwas wie die eiserne Ration an Vertrautheit in einer sich rasant beschleunigenden Welt: ein Kälte- und Ruhepol zugleich. Doch die erneute Auflage des Klassikers ist weit mehr als die nostalgische Erinnerung an eine verloren gegangene Zeit. Ausgestattet mit moderner Technik weckt die Bosch Classic Edition schnell neue Begehrlichkeiten und avanciert wiederum selbst zum Best- und Longseller.

An evolutionary design approach. In 1933, Bosch unveils a surprise in the form of the first Bosch-manufactured refrigerator. It is compact and round. The appliance is launched on the market in 1949, and its typical shape becomes the archetype for refrigerators and goes on to become an iconic design of the Bosch brand. In 2014, Bosch presents a modern reinterpretation of the design classic in the form of the CoolClassic.

Ein evolutionärer Designansatz. 1933 überrascht Bosch mit dem ersten Kühlschrank aus eigener Produktion. Er ist kompakt und er ist rund. 1949 kommt das Gerät auf den Markt, dessen typische Form den Archetyp des Kühlschranks prägte und der zur Designikone der Marke Bosch avancieren sollte. 2014 stellt Bosch mit dem CoolClassic eine moderne Neuinterpretation des Designklassikers vor.

In 2014, Bosch introduced yet another model, the CoolClassic. In contrast to the 1996 edition, which sought the greatest possible loyalty to the original at the level of form, the CoolClassic aimed for a new interpretation and evolution of the refrigerator. While its rounded corners, sturdy metal housing and large Bosch lettering evoke the original model, the new CoolClassic has a charm of its own. Moreover, behind the façade, touchpad controls and a digital display allow for optimal control.

The CoolClassic is exemplary of a design principle that has been applied to many models of leading car brands. The principle pursues two objectives: the right combination of tradition and innovation, of identity and difference; and an evolutionary approach to design that makes the appliances unmistakably identifiable as "a Bosch" while at the same time underscoring their autonomy. Overall, both the Bosch Classic Edition and the CoolClassic are representative of a certain lifestyle that renders them suitable for use as stand-alone appliances in lofts and living areas as well as in offices, agencies and art studios. Their ultimate design message is that comfort and enjoyment already begin before opening the refrigerator door.

In addition, with its "ColorGlass Edition" of reduced, purist and handle-free models featuring coloured glass fronts, the Bosch brand has once again proved to be a design pioneer, particularly in response to the trend of open-concept kitchen and living spaces that call for a more homelike look of appliances.

2014 kommt mit dem CoolClassic dann eine moderne Neuinterpretation auf den Markt. Im Unterschied zur Re-Edition von 1996, die sich formal um eine möglichst große Originaltreue bemüht, geht es beim Re-Design des neuen CoolClassic um eine Überarbeitung und Weiterentwicklung des Kühlschranks. Die abgerundeten Ecken, das robuste Metallgehäuse und der große Bosch-Schriftzug erinnern zwar noch an das Original, der neue CoolClassic versprüht aber seinen eigenen Charme. Hinter den Fronten verbirgt sich moderne Elektronik, die durch berührungsempfindliche Tasten gesteuert wird und digital ablesbar ist.

Hier kommt ein gestalterisches Prinzip zum Ausdruck, das sich auch bei vielen Modellen erfolgreicher Automarken findet: das gelungene Zusammenspiel von Tradition und Innovation, von Identität und Differenz – ein evolutionärer Ansatz in der Gestaltung, der die Hausgeräte eindeutig der Marke Bosch zuordnet und zugleich ihre Eigenständigkeit unterstreicht. Beiden Geräten, der Bosch Classic Edition wie dem CoolClassic ist der Lifestyle-Charakter gemein, den sie als Solitär im Loft und Wohnbereich, aber auch in Büros, Agenturen und Ateliers unterstreichen. Hier beginnt der Genuss eben schon vor dem Öffnen der Kühlschranktür.

Daneben gelingt es der Marke mit ihrer „ColorGlass Edition" – puristisch reduzierten, grifflosen Modellen mit farbigen Glasfronten – gestalterisch nach vorne zu blicken und der Öffnung der Küche zum Wohnraum und damit einhergehenden wohnlicheren Gerätekonzepten als Trendsetter voranzuschreiten.

The Stove and the Oven

As with the refrigerator, the history of the kitchen stove was initially shaped by engineers. Similar to the "cooling machine", stoves were at first called "cooking machines". And as with refrigerator design, the initial aim of stove design was to emulate furniture and to achieve technical feasibility. The electric stove was not marketed until the 1930s, like the refrigerator, and required the concentration of heat within a narrow space accommodating three or four hotplates of different diameters. In retrospect, the evolution of the cast-iron hotplates of the early years to the modern hobs made of ceramic glass is exemplary of a technological optimisation process during which the energy efficiency of the appliances was continuously improved.

In parallel to the technological development, the design of the control elements and the usage of the appliances evolved. After all, the stove and oven continue to be two of the most used appliances in the kitchen – so much so that Robert Sachon seeks inspiration for his work while cooking, possibly explaining his special focus on these two appliances. As a result, he often prefers to work with them on weekends and enjoys "dealing with appliances that allow one to engage directly and creatively with the product." As head of brand design, Sachon is concerned with the question of which design elements lend themselves to being transferred to other products or categories.

The Series 8 built-in appliances, distinguished with a Red Dot: Best of the Best award, are characterised by a simple, sleek design that prioritises the compatibility of different appliances as well as an easy, intuitive operating concept. The defining element is the flush-surface, centralised control ring with which the user can control the wide range of functions with a simple turn. Moreover, trends occurring in other areas are of relevance for the design of home appliances. For example, open-living concepts have repercussions for the design and integration of home appliances. "The current trend is to design concepts that fit well into the new living environments," explains Robert Sachon. "Integration as a whole is an important topic."

The Kitchen Machine

In 1952, Bosch introduced the first food processor on the market. With a name that says it all – the Neuzeit I (Modern Era I) – this product symbolised the beginning of a new era, since it decreased the number of actions and movements required for preparing food. Neuzeit I was able to stir, knead, cut, chop, mash, grate and grind.

In today's busy world, these small home and kitchen devices are indispensable helpers. As with all other equipment categories, here too the main objective is to improve the usefulness and convenience for the user with good design and innovative technology. Robert Sachon summarises the Bosch corporate philosophy as follows: "We simply try to give the appliances the best that there is."

Der Herd und der Ofen

Die Geschichte des Küchenherds ruht wie die Entwicklung des Kühlschranks zunächst in den Händen von Ingenieuren. Analog zur Kältemaschine spricht man zu Beginn der Entwicklung noch von einer Kochmaschine. Und ebenso wie die Form des Kühlschranks orientiert sich auch die Gestaltung des Herdes zunächst an dem Vorbild des Mobiliars und der technischen Machbarkeit. Der Elektroherd setzt sich parallel zum Kühlschrank erst in den 1930er Jahren durch, schließlich musste zunächst die Wärme auf einen engen Raum konzentriert werden, der drei oder vier Herdplatten mit unterschiedlichen Durchmessern Platz bot. Die Entwicklung von den gusseisernen Herdplatten der Anfangsjahre bis zu den modernen Kochfeldern aus Glaskeramik liest sich rückblickend wie ein technischer Optimierungsprozess, bei dem die Energieeffizienz der Geräte laufend verbessert wurde.

Parallel zur technischen Entwicklung hat sich aber auch die Gestaltung der Bedienelemente und der Gebrauch der Geräte verändert. Herd und Backofen sind nach wie vor zwei der am häufigsten genutzten Geräte in der Küche. Und auch Robert Sachon holt sich nicht zuletzt beim Kochen Inspiration für seine Arbeit. Daher gilt beiden Geräten sein besonderes Augenmerk. Gerade an den Wochenenden nimmt er sich Zeit dafür, „weil man es mit Geräten zu tun hat, die es einem unmittelbar erlauben, kreativ mit dem Produkt umzugehen", sagt Sachon. Als Leiter des Markendesigns beschäftigt ihn dabei auch die Frage, „welche gestalterischen Elemente sich eignen, um sie auf andere Produkte oder Kategorien zu übertragen".

Das mit dem Red Dot: Best of the Best ausgezeichnete Einbauprogramm der Serie 8 zeichnet sich durch eine einfache, klare Linienführung aus, die die perfekte Kombinierbarkeit unterschiedlicher Geräte in den Vordergrund stellt, sowie durch ein einfaches, intuitives Bedienkonzept. Prägendes Element ist der flächenbündig integrierte, zentrale Bedienring, mit dem der Nutzer mit einem Dreh die Vielzahl der Gerätefunktionen im Griff hat. Auch Trends aus anderen Bereichen sind für das Hausgeräte-Design von Bedeutung. Offene Wohnkonzepte wirken sich beispielsweise auch auf das Design und die Integration von Hausgeräten aus. „Der Trend geht aktuell zu Gestaltungskonzepten, die sich gut in die neuen Wohnwelten einfügen", erklärt Robert Sachon. „Integration überhaupt ist ein wichtiges Thema."

Die Küchenmaschine

1952 bringt Bosch die erste Küchenmaschine auf den Markt: die „Neuzeit I". Ihr Name ist Programm. Sie steht für den Beginn einer neuen Zeit, da sie den Hausfrauen viele Handgriffe abnimmt und die Zubereitung des Essens erleichtert. Die „Neuzeit I" kann rühren, kneten, hacken, schnitzeln, pürieren, mahlen und reiben.

Heute sind im modernen Alltag die kleinen Haus- und Küchengeräte unverzichtbare Helfer. Wie bei allen anderen Gerätekategorien gilt auch hier die Prämisse, den Nutzen und den Komfort für den Anwender mit guter Gestaltung und innovativer Technik zu verbessern. „Wir versuchen einfach, den Geräten das Beste mitzugeben", bringt Robert Sachon den Anspruch von Bosch auf den Punkt.

A new design language. The Series 8 range of built-in appliances awarded the Red Dot: Best of the Best sets new standards in the design of user interfaces. The defining element is the flush-surface, centralised control ring with which the user can control the wide range of functions with a simple turn.

Eine neue Designsprache. Das mit dem Red Dot: Best of the Best ausgezeichnete Einbauprogramm der Serie 8 setzt neue Maßstäbe in der Gestaltung von User Interfaces. Prägendes Element ist der flächenbündig integrierte, zentrale Bedienring, mit dem der Nutzer mit einem Dreh die Vielzahl der Gerätefunktionen im Griff hat.

Easy to use and featuring high-quality materials and consistent design: the Styline series of small home appliances from Bosch is not only modestly elegant but also uncompromisingly useful.
Hochwertige Materialien, einfache Bedienbarkeit und gestalterisch in einer Linie: Die Styline-Serie der kleinen Hausgeräte von Bosch präsentiert sich nicht nur zurückhaltend elegant, sondern ist auch kompromisslos nützlich.

A good example is the food processor MUM, which received a Red Dot: Best of the Best award as early as 2011. Similar to its predecessor Neuzeit I, its trademark is also versatility and energy-efficient performance. And, with an emphasis on high technological performance, the MUM exhibits the evolutionary design approach unique to Bosch. From one product generation to the next, these appliances affirm and promote the brand identity. Whether large or small home appliances: all are created with the holistic, evolutionary design approach of the Bosch brand, and all follow the uniform design language and uniform design principles derived from the brand values.

The Washing Machine

The first washing machine from Bosch was put into series in 1958, followed in 1960 by the first fully automatic washing machine that allowed clothes to be laundered in a single wash, rinse and spin cycle. In 1972, a new unit was introduced that combined washer and dryer in one. Since then, the washing programme options are being continually refined and expanded. All of these appliances contribute to making the days of hard manual labour a thing of the past. Moreover, the technology and design of the appliances have not only laid the foundation for a better quality of life in the household but also for social change. To the extent that the chore of washing increasingly became a minor matter, women liberated themselves from their role as housewife, and working women were no longer the exception to the rule.

Ein gelungenes Beispiel ist die Küchenmaschine MUM, die bereits 2011 mit dem Red Dot: Best of the Best ausgezeichnet wird. Analog zu ihrem historischen Vorgänger, der „Neuzeit I", ist auch ihr Markenzeichen die Vielseitigkeit und die energieeffiziente Leistung. Im Einklang mit der hohen technischen Leistung spiegelt sich in der Formensprache der Küchenmaschine MUM aber auch der evolutionäre Gestaltungsansatz mustergültig wider. Von Generation zu Generation kommt im Produktdesign auch die Anbindung an die Marke Bosch zum Ausdruck. Ob große oder kleine Hausgeräte – allen wird der ganzheitliche, evolutionäre Designansatz der Marke Bosch zuteil, folgen sie doch sämtlich einer den Markenwerten entspringenden, einheitlichen Designsprache und einheitlichen Gestaltungsprinzipien.

Die Waschmaschine

Die erste Waschmaschine aus dem Hause Bosch geht 1958 in Serie. Es ist noch kein Waschvollautomat. Dieser kommt 1960 auf den Markt und ermöglicht es, die Wäsche erstmals in einem einzigen Durchlauf waschen, spülen und schleudern zu lassen. 1972 werden die Waschmaschine und der Trockner dann eins. Seither wird das Geräteprogramm immer weiter verfeinert und ausgebaut. All diese Geräte sorgen dafür, dass die Zeiten der mühevollen Handarbeit langsam zu Ende gehen. Die Technik und die Gestaltung der Geräte legen nicht nur den Grundstein für mehr Lebensqualität im Haushalt, sondern auch für eine gesellschaftliche Veränderung. Das Thema Waschen wird mehr und mehr zur Nebensache. Die Frauen emanzipieren sich von ihrer Rolle als Hausfrau. Berufstätige Frauen sind keine Ausnahme mehr.

Although women's liberation is usually discussed from the standpoint of a social and political movement, the topic merits further examination from the perspective of home appliance design. Indeed, the changing effect of technology and design extends beyond the material manifestation of the products. On the one hand, design concepts in the home appliance sector are always oriented towards long-term societal trends. Yet on the other hand, the designed products themselves have an impact on society, in particular through their technical function and their user-oriented utility. Here, at the latest, it becomes clear that quality of life is far more than the sum of the products that we own. The importance and impact of design go beyond what is materialised in a product, becoming discernible only in the changing conditions of use and their impact.

The Dishwasher

In 1964, the first dishwasher from Bosch went into serial production. Thanks to technological developments, dishwashers have since become very quiet. So much so that the remaining time of the programme cycles is now explicitly indicated and of late even projected onto the floor so that the user knows when the machine has finished the washing and drying process.

"It's possible to tolerate certain sounds, but it's fair to say that we're probably better off without them," says Robert Sachon. "Essentially, a common priority of all product categories is to ensure that appliances are quiet, so as not to disturb users. At the same time, we have to generate and design the sounds that consumers are supposed to hear which is not always easy." This shows that the team led by Robert Sachon values not only the quality of the product and its exterior design, but also the quality of the experience, from the operation of and interaction with the appliance to the sound design and haptics. Finally, Bosch wants design perfection to be experienceable with all the senses.

The Internet of Things – Design 4.0

Bosch undoubtedly invests heavily in the development and design of new technologies. And Bosch wouldn't be Bosch if it didn't capitalise on emerging technological opportunities to achieve more convenience and ease of use in home appliances.

"Under the slogan the 'Internet of Things', visions have been presented at trade fairs for years," explains Robert Sachon. "With the digitisation and networking of appliances, these visions are now becoming reality and taking on concrete forms," says the chief designer. With the "Home Connect" app, Bosch's digitally connected home appliances can be controlled using a smartphone or tablet PC. Among these is the Series 8 oven, distinguished with a Red Dot: Best of the Best. "'Home Connect' is a good example of how we generate real added value for our customers with new technologies," says Jörg Gieselmann, Executive Vice President Corporate Brand Bosch at BSH Hausgeräte GmbH. He adds: "With the app, setting up the appliance, making or changing basic settings or operating it when away from home becomes child's play. In this way we're creating wholly new possibilities and freedoms in the household."

Während die Emanzipation der Frauen häufig unter dem Gesichtspunkt einer sozialen und politischen Bewegung verhandelt wird, kann es sich durchaus lohnen, das Thema auch aus der Perspektive des Hausgeräte-Designs zu beleuchten. Über die materialisierte Form der Produkte hinaus zeigt sich die verändernde Wirkung von Technik und Design. Auf der einen Seite orientieren sich Gestaltungskonzepte im Hausgerätebereich immer auch an den langfristigen gesellschaftlichen Trends. Auf der anderen Seite nehmen die gestalteten Produkte wiederum selbst Einfluss auf die Gesellschaft, insbesondere über ihre technische Funktion und ihren am Nutzer orientierten Gebrauch. Spätestens hier wird deutlich, dass Lebensqualität weit mehr ist als die Summe der Produkte, die wir besitzen. Die Bedeutung und die Wirkung des Designs gehen über das hinaus, was sich in einem Produkt materialisiert. Sie werden erst in den sich verändernden Bedingungen der Nutzung und des Gebrauchs und deren Wirkung ablesbar.

Der Geschirrspüler

1964 geht der erste Geschirrspüler von Bosch in Serie. Aufgrund der technischen Entwicklung sind die Geschirrspüler inzwischen so leise und geräuscharm geworden, dass die Gestalter über zusätzliche Projektionsflächen die Restlaufzeit der Geräte auf den Boden projizieren, damit der Benutzer weiß, wann das Gerät den Spül- und Trocknungsvorgang beendet hat.

„Man kann mit bestimmten Geräuschen gut leben, man kann ohne sie vielleicht noch etwas besser leben", sagt Robert Sachon. „Es zieht sich im Grunde durch alle Produktkategorien hindurch, dass die Geräte immer angenehm leise arbeiten, damit man eben nicht gestört wird. Und die Geräusche, die der Konsument wahrnehmen soll, versuchen wir bewusst zu gestalten, was nicht immer einfach ist", so Sachon. Hier zeigt sich, dass das Team um Robert Sachon stets versucht, über die Produktqualität hinaus eine Erlebnisqualität zu vermitteln, bei der neben der äußeren Gestaltung auch die Bedienung und Interaktion mit den Geräten, das Sounddesign und die Haptik eine große Rolle spielen. Schließlich will Bosch im Design Perfektion mit allen Sinnen erfahrbar machen.

Das Internet der Dinge – Design 4.0

Bosch gehört ohne Zweifel zu den Unternehmen, die viel in die Entwicklung und Gestaltung neuer Technologien investieren. Und Bosch wäre nicht Bosch, wenn das Unternehmen die sich bietenden technischen Möglichkeiten nicht auch für eine einfachere Handhabung und leichtere Bedienbarkeit der Hausgeräte nutzen würde.

„Unter dem Schlagwort ‚Internet der Dinge' gibt es ja bereits seit Jahren Visionen, die beispielsweise auf Messen vorgestellt werden", erläutert Robert Sachon. „Mit der Digitalisierung und der Vernetzung der Geräte werden diese Visionen heute Realität und nehmen ganz konkrete Formen an", so der Chefdesigner. Mit der „Home Connect"-App können vernetzte Hausgeräte von Bosch mithilfe eines Smartphones oder Tablets gesteuert werden, zum Beispiel der Serie 8 Backofen, der mit dem Red Dot: Best of the Best ausgezeichnet wurde. „Home Connect ist ein gutes Beispiel dafür, wie wir mit den neuen Technologien echte Mehrwerte für unsere Kunden

With the effortless implementation of digitally networked home appliances, Bosch is once again setting a new standard in the design and use of such appliances. And, as in the early twentieth century, the kitchen is today again becoming a new focal point of communication, even if under the changed information and communication environment of the twenty-first century. Of course, these developments also affect the daily work and professional identity of designers. As Robert Sachon explains, "In the past our team was very heavily influenced by industrial designers, while nowadays we are focusing more on the topic of user interface design." Yet this is only to be expected since the perceptions and communication patterns of consumers are evolving. Essentially, it means that technological changes always lead to changes in the design of the future.

generieren", erklärt Jörg Gieselmann, Executive Vice President Corporate Brand Bosch der BSH Hausgeräte GmbH. „Die Verbraucher können mithilfe der App kinderleicht und einfach ihr Gerät in Betrieb nehmen, Grundeinstellungen vornehmen und verändern oder es von unterwegs aus bedienen. So schaffen wir ganz neue Möglichkeiten und Freiheiten im Haushalt."

Mit der mühelosen Benutzung digital vernetzter Hausgeräte setzt Bosch abermals einen neuen Standard in der Gestaltung und im Gebrauch von Hausgeräten. Und wie bereits zu Beginn des 20. Jahrhunderts wird die Küche heute wieder zu einem neuen Mittelpunkt der Kommunikation, wenn auch unter den veränderten Informations- und Kommunikationsbedingungen des 21. Jahrhunderts. Natürlich beeinflussen diese Entwicklungen auch die tägliche Arbeit und das Selbstverständnis der Designer, wie Robert Sachon erläutert: „Wurde unser Team früher sehr stark von Industriedesignern geprägt, so verstärken wir uns heute beim Thema ‚User Interface Design'." Das kann auch nicht anders sein, denn auch die Wahrnehmung und die Kommunikation der Konsumenten entwickeln sich weiter. Insofern ergeben sich grundsätzlich aus den technologischen Veränderungen immer auch Veränderungen für das Design der Zukunft.

The Internet of Things: with the "Home Connect" app, Bosch's digitally connected home appliances can be controlled using a smartphone or tablet PC, true to the Bosch philosophy "Invented for Life".
Das Internet der Dinge: Mit der „Home Connect"-App können vernetzte Hausgeräte von Bosch mithilfe eines Smartphones oder Tablets gesteuert werden, getreu der Bosch-Philosophie „Technik fürs Leben".

Values

When Robert Bosch founded his company in 1886 – a workshop for precision mechanics and electrical engineering in Stuttgart – housework still meant hard work. Hobs were heated with an open fire. Dishes and laundry had to be laboriously cleaned by hand. Food had to be salted or boiled. Housecleaning and the weekly washing day certainly consumed a lot of time and energy.

Robert Bosch recognised this problem and developed technical solutions which others hadn't even imagined. Some 47 years after the founding of the company, in 1933, Robert Bosch moved into the production of home appliances, a division dedicated to making people's everyday lives easier. And still during the Great Depression, he spearheaded a massive overhaul of the Bosch Group, by then globally active in the field of automotive and industrial technology, in order to modernise and diversify its activities.

From the outset it was one of Robert Bosch's principles to produce the best possible quality at all times. As early as 1918, he wrote that "it has always been an unbearable thought that someone might prove, upon examining one of my products, that my performance is inferior in some way. That's why I've always made a point of releasing only products that have passed all quality tests, in other words, that were the best of the best." This maxim essentially guides the corporate philosophy to this day: technology and quality, credibility and trust, along with a sense of responsibility for people's well-being.

And in 1921 he wrote: "I've always acted according to the principle that I would rather lose money than trust. The integrity of my promises, the belief in the value of my products and in my word have always meant more to me than temporary gain." To this day, Robert Bosch and the example he set have a significant impact on the company. The founder is still the reference point that he has always been, and which he will still be tomorrow.

Throughout the years, the design team of Robert Bosch Hausgeräte GmbH has succeeded in finding a modern and appropriate interpretation of the values which its founder embodied and which manifest in the company's brand to this day. It has set pioneering standards in home appliance design and made a significant contribution to improving the quality of life. To acknowledge the design performance of the entire team, which has garnered attention over the years with its high-quality products and consistently outstanding design, the honorary title "Red Dot: Design Team of the Year" for 2015 is bestowed on Robert Sachon and the Bosch Home Appliances Design Team.

Die Werte

Als Robert Bosch im Jahr 1886 sein Unternehmen gründete – eine Werkstatt für Feinmechanik und Elektrotechnik in Stuttgart –, bedeutete Hausarbeit noch Schwerstarbeit. Kochstellen wurden mit offenem Feuer beheizt. Geschirr und Wäsche mussten aufwendig von Hand gereinigt werden. Lebensmittel mussten gepökelt oder eingekocht werden. Der Hausputz und der wöchentliche Waschtag kosteten viel Zeit und Kraft.

Robert Bosch erkannte dieses Problem und entwickelte dafür technische Lösungen, wo andere noch nicht einmal Möglichkeiten erahnten. 47 Jahre nach der Gründung des Unternehmens stieg Robert Bosch im Jahr 1933 in die Produktion von Hausgeräten ein, um mit seinen Produkten den Alltag der Menschen zu erleichtern. Und unter dem Eindruck der Weltwirtschaftskrise verordnete er dem inzwischen weltweit im Bereich der Kraftfahrzeug- und Industrietechnik agierenden Bosch-Konzern einen konsequenten Modernisierungs- und Diversifizierungskurs.

Von Anfang an ist es einer der Grundsätze von Robert Bosch, immer bestmögliche Qualität zu produzieren. „Es war mir immer ein unerträglicher Gedanke, es könne jemand bei der Prüfung eines meiner Erzeugnisse nachweisen, dass ich irgendwie Minderwertiges leiste. Deshalb habe ich stets versucht, nur Arbeit hinauszugeben, die jeder sachlichen Prüfung standhielt, also sozusagen vom Besten das Beste war", schreibt Robert Bosch bereits im Jahr 1918. An dieser Haltung orientiert sich das Unternehmen bis heute: Technik und Qualität, Glaubwürdigkeit und Vertrauen sowie Verantwortungsbewusstsein zum Wohle der Menschen.

„Immer habe ich nach dem Grundsatz gehandelt: Lieber Geld verlieren als Vertrauen. Die Unantastbarkeit meiner Versprechungen, der Glaube an den Wert meiner Ware und an mein Wort standen mir stets höher als ein vorübergehender Gewinn", schreibt Robert Bosch im Jahr 1921. Vieles von dem, was der Firmengründer gedacht und vorgelebt hat, übt bis heute eine große Anziehungskraft aus. Robert Bosch ist der Bezugspunkt des Unternehmens, der er bereits früher war und der er auch morgen noch sein wird.

Dem Designteam der Robert Bosch Hausgeräte GmbH gelingt es seit vielen Jahren, die Werte, die bereits ihr Firmengründer verkörperte und die bis heute in den Markenwerten des Unternehmens zum Ausdruck kommen, auf zeitgemäße Art zu interpretieren. Es hat im Hausgeräte-Design wegbereitende Standards gesetzt und einen wesentlichen Beitrag zu mehr Lebensqualität geleistet. Mit Blick auf die gestalterische Leistung des gesamten Teams, das über Jahre hinweg mit qualitativ hochwertigen Produkten und einem kontinuierlich hohen Gestaltungsniveau auf sich aufmerksam gemacht hat, wird der Ehrentitel „Red Dot: Design Team of the Year" im Jahr 2015 an Robert Sachon und das Bosch Home Appliances Design Team verliehen.

"We simply try to give the
appliances the best that there is."
„Wir versuchen einfach, den
Geräten das Beste mitzugeben."

Robert Sachon, Global Design Director Bosch

Red Dot: Design Team of the Year 2015
Interview: Robert Sachon
Global Design Director
Robert Bosch Hausgeräte GmbH

For generations, the Bosch name has stood for groundbreaking technology and outstanding quality. Home appliances from Bosch have been committed to these standards for over 80 years now. Consumers around the world associate the Bosch brand with efficient functionality, reliable quality and design that has won awards at an international level. The brand has won more than 500 awards in the past 10 years alone. The man behind this design success is Robert Sachon. He started his career in 1999 at Siemens-Electrogeräte GmbH, at the time under the leadership of Gerd Wilsdorf. In 2005 he moved to Robert Bosch Hausgeräte GmbH, where he took over from his predecessor Roland Vetter. Together with a team of roughly 40 employees worldwide, he has shaped the design of the Bosch brand in his role as Global Design Director. Burkhard Jacob met with him for an interview in the Red Dot Design Museum Essen.

Mr Sachon, you have been Global Design Director for the Bosch brand for 10 years now. How would you describe your role as head designer?

The term "head designer" is pretty accurate. It sounds a little like a head chef. Similar to the chef de cuisine, who is in charge of the kitchen crew in fine dining establishments, I lead a team of employees who are responsible for the design of the home appliances.

Do you also get stuck in?

I am one of those designers who not only manage other people but are also happy to get their own hands dirty in order to set out the design direction. We have lots of different product categories, but at the end of the day the point is of course to shape the face of the Bosch brand in a similar way to a signature or a common design language. And that's something I like to stay involved in.

Mr Sachon, you exude a calmness that makes me curious as to what your star sign is?

Aries – a healthy mix of diplomacy and stubbornness.

Der Name Bosch steht seit Generationen für wegweisende Technik und herausragende Qualität. Diesem Anspruch sind die Hausgeräte von Bosch seit über 80 Jahren verpflichtet: Mit der Marke Bosch verbinden Konsumenten weltweit effiziente Funktionalität, verlässliche Qualität und ein international ausgezeichnetes Design. Allein in den letzten 10 Jahren wurden mehr als 500 Auszeichnungen gewonnen. Der Mann hinter diesen Design-Erfolgen ist Robert Sachon. Er begann seine Karriere 1999 bei der Siemens-Electrogeräte GmbH, damals unter Gerd Wilsdorf. 2005 wechselte er zur Robert Bosch Hausgeräte GmbH und beerbte seinen Vorgänger Roland Vetter. Gemeinsam mit einem Team von weltweit rund 40 Mitarbeitern hat er als Global Design Director das Design der Marke Bosch geprägt. Burkhard Jacob traf ihn im Red Dot Design Museum Essen zum Interview.

Herr Sachon, seit 10 Jahren sind Sie nun Global Design Director der Marke Bosch. Wie würden Sie Ihre Tätigkeit als Chefdesigner beschreiben?

Der Begriff „Chefdesigner" trifft es schon ganz gut. Es klingt ein wenig wie Chefkoch. Ähnlich dem Chef de Cuisine, der in der gehobenen Gastronomie die Küchenbrigade leitet, führe ich ein Team von Mitarbeitern, die für das Design der Haushaltsgeräte verantwortlich sind.

Greifen Sie auch selbst zum Kochlöffel?

Ich bin einer der Gestalter, die nicht nur managen, sondern auch selbst zum Stift greifen, um die Gestaltungsrichtung vorzugeben. Wir haben viele unterschiedliche Produktkategorien, aber am Ende des Tages geht es natürlich darum, das Gesicht der Marke Bosch im Sinne einer Art Handschrift, einer gemeinsamen Designsprache zu prägen. Und das möchte ich mir gerne erhalten.

Herr Sachon, Sie strahlen eine Ruhe aus, die die Frage provoziert, welches Sternzeichen Sie sein könnten?

Widder – eine gesunde Mischung aus Diplomatie und Dickköpfigkeit.

What characteristics of an Aries are helpful for the role of designer?

There are two characteristics that help me greatly in my work: a tendency to be a perfectionist and a certain amount of tenacity. Both of these things help me to work on different projects of differing durations and scope in order to bring long-term topics to fruition for the Bosch brand.

Do you have a design role model?

That's a difficult question, because I suppose we ultimately always come back to the heroes of design history. I have to admit I have huge respect for Dieter Rams, even though that's probably something that lots of designers say. His design language resonates with me: its clarity, order and meaningfulness. I am of a similar mindset. Without wanting to overstate their importance, his ten principles of good design are still valid today. I am fascinated by this clear design language, which has become very popular again nowadays in particular, making it all the more fitting for Bosch brand values.

What does the Bosch brand stand for?

Bosch has a long tradition spanning more than 125 years, and has always stood for technical perfection and superior quality. In our design, we take these rational values and make them tangible at an emotional level by means of clear design that showcases high-quality materials and their uncompromising finish in a precise manner down to the smallest detail. This is a holistic approach to design which we apply to all of our products

Welche Eigenschaften des Widders sind denn hilfreich für die Tätigkeit als Designer?

Es gibt zwei Eigenschaften, die mir bei der Tätigkeit sehr entgegenkommen: ein gewisser Hang zum Perfektionismus und eine gewisse Beharrlichkeit. Beides hilft mir, zeitgleich an unterschiedlichen Projekten mit unterschiedlicher Laufzeit und Tragweite zu arbeiten, um langfristige Themen für die Marke Bosch durchzusetzen.

Haben Sie ein Vorbild in Fragen der Gestaltung?

Das ist eine schwierige Frage, weil man vermutlich immer bei den Heroen der Designgeschichte landet. Trotzdem – ich habe einen wahnsinnigen Respekt vor Dieter Rams, auch wenn das wahrscheinlich viele Designer sagen. Aber seine Gestaltungssprache liegt mir schon sehr nahe: die Klarheit, die Ordnung, die Sinnhaftigkeit. Da sehe ich eine ähnliche Geisteshaltung. Seine zehn Gebote des Designs haben ja heute noch ihre Gültigkeit, ohne sie religiös überhöhen zu wollen. Mich fasziniert diese klare Gestaltungssprache, die gerade heute wieder hoch im Kurs steht – und die umso mehr zu den Markenwerten von Bosch passt.

Wofür steht die Marke Bosch?

Bosch hat eine lange Tradition von mehr als 125 Jahren und steht seit jeher für die Themen „Technische Perfektion" und „Überlegene Qualität". Diese rationalen Werte machen wir in unserem Design emotional erfahrbar durch eine klare Gestaltung, welche hochwertige Materialien und deren kompromisslose Verarbeitung präzise bis ins kleinste Detail in Szene setzt.

worldwide. In this regard, it is fair to speak of a uniform design language, or DNA, of the Bosch brand.

So you are a brand manager as well as a designer?

Most definitely. Unlike other companies, where design is part of technical development, design at Bosch benefits from the fact that it is a key part of brand management. Our design team plays an important role and has a clear remit, as it gets involved with the development of product concepts at a very early stage – long before any thoughts of marketing for the products or of an advertising campaign.

Do you base your design decisions on the Bosch brand values?

The aim is always to bring the brand values to life. In the area of home appliances, we see a lot of products that pass on certain design features to the next generation. Consequently there is an underlying evolutionary thought process involved. This goes without saying with a brand like Bosch, which has been conveying the same values for over 125 years.

Basing design language on the brand values is one side of the coin. The other, which you have just described, relates to a product's use and benefit for the consumer.

Absolutely. We pursue a user-centred design approach where consumer monitoring plays a major role. We benefit from the fact that we too are all users of home appliances. We therefore can observe ourselves as well as others. And when observing ourselves, it's important to always be aware of our blind spot.

Who is it that ultimately decides whether or not a product goes into production?

As a designer and as Global Design Director, I don't make lonely decisions, relying instead on my team of specialists. Maybe that is one reason why we were awarded the Design Team of the Year title. But the decision of whether a product goes into production is not made by the design team alone. That is a joint decision made by top management. Our task is to convince all of those involved in the process to take a proposed course of action.

What role does market observation play for design?

The competitive environment for home appliances is very tough. There are over 1,000 home appliances brands worldwide. We know some of those brands very well. After all, we meet our competitors regularly at trade fairs. So it is in our mutual interest to set ourselves apart very clearly from our competitors. Obviously we don't just look at the market from the perspective of the competition, but also always with a view to understanding long-term developments. For example, connectivity is one keyword that shows where the journey is headed. As a consequence, the

Das ist ein ganzheitlicher Gestaltungsansatz, den wir auf all unsere Produkte weltweit übertragen. Insofern kann man durchaus von einer einheitlichen Designsprache, einer DNA der Marke Bosch reden.

Sie sind also nicht nur Designer, sondern auch Markenmanager?

Definitiv. Im Unterschied zu anderen Unternehmen, in denen das Design Teil der technischen Entwicklung ist, profitiert das Design bei uns davon, ein wesentlicher Teil des Markenmanagements zu sein. Unser Designteam hat eine wichtige Rolle und eine klare Aufgabenstellung, da es sich bereits sehr früh mit der Entwicklung von Produktkonzepten befasst; und zwar lange bevor über deren Vermarktung oder eine Werbekampagne nachgedacht wird.

Orientieren Sie Ihre gestalterischen Entscheidungen an den Markenwerten von Bosch?

Es geht immer darum, die Markenwerte erfahrbar zu machen. Im Bereich der Hausgeräte finden wir viele Produkte, die bestimmte gestalterische Merkmale an die nächste Generation weitergeben, denen also ein evolutionärer Gedanke zugrunde liegt. Das kann ja auch nicht anders sein bei einer Marke wie Bosch, die seit mehr als 125 Jahren dieselben Werte vermittelt.

Die Designsprache an den Markenwerten zu orientieren, ist eine Seite der Medaille. Sie beschreiben auch noch eine andere Seite: die Seite des Gebrauchs und des Nutzens für den Konsumenten.

Absolut. Wir verfolgen einen nutzerzentrierten Gestaltungsansatz, bei dem die Beobachtung der Konsumenten eine wichtige Rolle spielt. Dabei kommt uns entgegen, dass wir auch alle selbst Nutzer von Hausgeräten sind. Wir haben also die Ebene der Fremd- und der Selbstbeobachtung. Bei der Selbstbeobachtung muss man aber immer auch seinen blinden Fleck im Visier haben.

Wer entscheidet letztlich, ob ein Produkt in Serie geht?

Als Designer und als Global Design Director treffe ich keine einsamen Entscheidungen, sondern vertraue auf mein Team, das aus Spezialisten besteht. Vielleicht ist das auch ein Grund dafür, warum wir mit der Auszeichnung zum Designteam des Jahres bedacht worden sind. Die Entscheidung, ob ein Produkt in Serie geht, trifft das Designteam aber nicht allein. Das ist eine gemeinsame Entscheidung des Topmanagements. Unsere Aufgabe ist es, alle Prozessbeteiligten davon zu überzeugen, einen vorgeschlagenen Weg zu gehen.

Welche Rolle spielt die Marktbeobachtung für das Design?

Im Hausgerätemarkt finden wir ein sehr starkes Wettbewerbsumfeld vor. Es gibt über 1.000 Hausgerätemarken weltweit. Einige davon kennen wir auch sehr gut. Wir treffen unsere Wettbewerber ja regelmäßig auf Messen. Da liegt eine gute Differenzierung im wechselseitigen Interesse.

competition is never the only source of inspiration. We regularly attend trade fairs to scout out new trends. In doing so, we also learn from other industries such as the automotive industry, where the products and innovation cycles are similar in length to those on the home appliances market. In addition, we observe the developments in interior design, architecture and in consumer electronics. We get an insight into a range of vastly different industries, filtering innovations according to whether they constitute short-term or more long-term trends and how they impact on technical and design development on the home appliances market. Ultimately we want to develop products that are attractive not only on the day they are purchased but for many years afterwards.

What topics will most influence the industry for home appliances in the coming years?

There are some developments that have been influencing the home appliances industry for quite a while. For example, visions in relation to the "Internet of Things" have featured at trade fairs for some years now. With the connectivity of the appliances and the digital transmission of data, these visions are now becoming a reality and are taking shape in a very real way.

To what extent does that also change the work within your design team?

Naturally, developments like these also affect our daily work as designers. In the past our team was very heavily influenced by industrial designers, while nowadays we are focusing more on the topic of user interface design. Although the design of home appliances has always involved the design of functions for use as well as operating elements, digitalisation and connectivity mean that we are also pursuing independent design concepts which in turn result in new forms of operation and user guidance.

Is this also a general indication of how the design of home appliances will develop?

Yes, and that is something which ultimately makes a lot of sense. Coming back to the general developments and influencing factors again, it is fair to say that the change in how we use our living space has a significant role to play. Open-plan living concepts have resulted in kitchens themselves becoming a part of the living area. This of course is also reflected in the concepts behind the appliances. The current trend is to design concepts that fit well into the new living environments. Integration as a whole is an important topic.

To what extent are materials a general topic in the design of home appliances?

The choice and quality of materials are very important to Bosch. These topics run through all categories of appliances and can be found in all product groups. The materials used in home appliances often also have to fulfil technical product characteristics, as they sometimes come into contact with food or are exposed to high temperatures.

Man schaut sich den Markt natürlich nicht nur aus der Perspektive des Wettbewerbs an, sondern immer auch mit Blick auf die langfristigen Entwicklungen. Vernetzung ist beispielsweise ein Stichwort, das zeigt, wo die Reise hingeht. Und insofern ist die Konkurrenz niemals die einzige Quelle der Inspiration. Wir sind regelmäßig als Trendscouts auf Messen. Dabei lernen wir auch von anderen Branchen wie beispielsweise der Automobilindustrie, wo wir es mit ähnlich langlebigen Produkten und Innovationszyklen zu tun haben wie im Hausgerätemarkt. Daneben beobachten wir die Entwicklungen in Interior Design, Architektur und dem Bereich der Consumer Electronics. Wir nehmen Einblick in die unterschiedlichsten Branchen und filtern die Innovationen danach, ob es sich um kurzfristige oder eher langfristige Trends handelt und wie sie die technische und gestalterische Entwicklung im Hausgerätemarkt beeinflussen. Wir wollen ja Produkte entwickeln, die nicht nur im Moment des Kaufs, sondern noch viele Jahre danach attraktiv sind.

Welche Themen werden die Branche der Hausgeräte in den kommenden Jahren besonders beeinflussen?

Es gibt einige Entwicklungen, die auch nicht erst seit gestern Einfluss auf die Hausgeräte-Branche nehmen. Unter dem Schlagwort „Internet der Dinge" gibt es ja bereits seit Jahren Visionen, die auf Messen vorgestellt wurden. Mit der Vernetzung der Geräte und der digitalen Übertragung von Daten werden diese Visionen heute Realität und nehmen ganz konkrete Formen an.

Inwieweit verändert das auch die Arbeit innerhalb Ihres Designteams?

Natürlich beeinflussen solche Entwicklungen auch unsere tägliche Arbeit als Gestalter. Wurde unser Team früher sehr stark von Industriedesignern geprägt, so verstärken wir uns heute beim Thema „User Interface Design". Die Gestaltung von Hausgeräten befasst sich zwar seit jeher mit der Gestaltung von Gebrauchsfunktionen und Bedienelementen, durch die Digitalisierung und Vernetzung verfolgen wir aber auch eigenständige Designkonzepte, die wiederum neue Formen der Bedienung und Benutzerführung mit sich bringen.

Zeigt sich auch hier eine generelle Entwicklung im Design von Hausgeräten?

Ja, was auch letztlich konsequent ist. Wenn wir noch einmal auf die generellen Entwicklungen und Einflussfaktoren zurückkommen, dann spielt die Veränderung der Wohnwelten eine wichtige Rolle. Die Küche hat sich durch offene Wohnkonzepte selbst zu einem Teil des Wohnraums entwickelt. Und das spiegelt sich natürlich auch in den Gerätekonzepten wider. Der Trend geht aktuell zu Gestaltungskonzepten, die sich gut in die neuen Wohnwelten einfügen. Integration überhaupt ist ein wichtiges Thema.

Inwieweit sind Materialien ein generelles Thema im Design von Hausgeräten?

Für Bosch haben Materialien und Materialqualität einen hohen Stellenwert. Sie ziehen sich durch alle Gerätekategorien und finden

As Global Design Director at Bosch, Robert Sachon likes to get stuck in. He is a designer and brand manager in one.

Als Global Design Director Bosch greift Robert Sachon auch selbst zum Zeichenstift. Er ist Designer und Markenmanager in einer Person.

As a result, part of our day-to-day work as designers when dealing with these materials also involves familiarising ourselves with the corresponding technologies for processing the materials.

In order to also express the values of the brand through the quality and processing of the materials?

Most definitely. In some cases, we take processing to the very limits of what is technically feasible, even though many consumers may not be aware of that at first. But even if it is not necessarily the first thing they see, they will notice it when using the products.

Design as a non-verbal means of communication?

Yes, consumer perception is also evolving. For example, this is where influences from the field of consumer electronics and mobile communication come to bear. The materials used in smartphones and tablets as well as their finishing quality change how consumers perceive products. After all, they hold the devices in their hands day after day, and that also makes them more discerning of quality in other product segments.

sich bei allen Produktgruppen. Die verwendeten Materialien müssen im Bereich der Hausgeräte vielfach auch technische Produkteigenschaften erfüllen, da sie teilweise mit Lebensmitteln in Kontakt kommen oder hohen Temperaturen ausgesetzt sind. Insofern gehört es im Umgang mit diesen Materialien auch zur täglichen Arbeit des Gestalters, sich mit entsprechenden Verarbeitungstechnologien auseinanderzusetzen.

Um über die Qualität und die Verarbeitung der Materialien auch die Werte der Marke zum Ausdruck zu bringen?

Definitiv. Dabei gehen wir bei der Verarbeitung teilweise bis an die Grenzen der technischen Machbarkeit, auch wenn es vielen Konsumenten im ersten Moment nicht bewusst sein mag. Aber selbst wenn sie es vielleicht nicht vordergründig wahrnehmen, spüren sie es doch im Umgang mit den Produkten.

Design als nonverbales Mittel der Kommunikation?

Ja, auch die Wahrnehmung der Konsumenten entwickelt sich weiter. Hier kommen beispielsweise die Einflüsse aus dem Bereich der Consumer Electronics und der mobilen Kommunikation zum Tragen. Die verwendeten Materialien im Bereich der Smartphones und Tablets sowie deren Verarbeitungsqualität verändern die Wahrnehmung der Konsumenten. Sie haben die Geräte ja täglich in der Hand, und das macht sie auch sensibler für Qualitäten in anderen Produktsegmenten.

So Bosch is also being forced to be innovative with its home appliances by other industries?

You could see it like that. As a team, we have to ask ourselves the question every day of how we can constantly tweak the quality ethos and the technical perfection of the Bosch brand in order to make the brand visible and tangible again and again. Such efforts include gaps and bending radii that have a major effect on development, production and quality management.

As head designer, are you allowed to have one topic that is particularly close to your heart?

For me that topic is cooking, because it involves appliances that make it possible to be creative with the product in a very immediate way.

And what are the questions in relation to cooking that interest you as a designer?

For example how to combine appliances, and the question of how these appliances relate to each other. What codes and what design elements are suitable to be transferred to other products or categories? That can be very helpful when developing a design DNA for the brand.

Do we even need to be able to cook nowadays?

While our mothers knew exactly how to handle their home appliances, the younger generation is perhaps more likely to use automatic programmes. As a result, today's appliances are geared to meet different user requirements. It was not until sensor technology and new digital display and operating technologies were developed that these options for use and operation became possible. The very question of how the new displays are designed and programmed has become a very exciting field for designers. Our team developed dedicated style guides for images and animations for this purpose that had not existed in that form beforehand.

Don't the technical possibilities automatically lead to a complexity of products that is maybe not even desirable?

The new operating and display technologies have simultaneously given rise to more possibilities when using the appliances. Part of our work is also to prevent the appliances from becoming too complex as a result of more operating possibilities. The whole point is that they should be simple and intuitive to use. Even complex technology must remain manageable. In a technology-based company like Bosch, we as designers also have to act as a control instance vis-à-vis the marketing or technical product development departments, because we keep the user in sight and design an important interface informed by user-focused concepts.

Bosch wird also auch durch andere Branchen zur Innovation im Bereich der Hausgeräte gezwungen?

Wenn man das so sehen will. Als Team müssen wir uns täglich mit der Frage auseinandersetzen, wie wir den Qualitätsgedanken und die technische Perfektion der Marke Bosch stetig nachschärfen können, um diese immer wieder sichtbar oder erfahrbar zu machen. Dazu gehören Spaltmaße und Biegeradien, die sich erheblich auf Entwicklung, Produktion und Qualitätsmanagement auswirken.

Darf man als Chefdesigner auch ein Thema haben, dass einem besonders am Herzen liegt?

Für mich ist es das Thema Kochen, weil man es mit Geräten zu tun hat, die es einem unmittelbar erlauben, kreativ mit dem Produkt umzugehen.

Und welche Fragen interessieren Sie als Designer beim Thema Kochen?

Da geht es beispielsweise um die Kombinierbarkeit von Geräten und die Frage, wie diese Geräte miteinander in Beziehung stehen. Welche Codes und welche gestalterischen Elemente eignen sich, um sie auch auf andere Produkte oder Kategorien zu übertragen? Das kann sehr hilfreich sein, um eine Design-DNA für die Marke zu entwickeln.

Muss man denn heute überhaupt noch kochen können?

Während unsere Mütter noch sehr genau wussten, wie sie mit ihren Hausgeräten umzugehen hatten, greift die jüngere Generation vielleicht eher auf Automatik-Programme zurück. Die Geräte werden also inzwischen unterschiedlichen Nutzeranforderungen gerecht. Diese Möglichkeiten im Gebrauch und in der Bedienung wurden erst durch Sensortechnik und neue digitale Anzeige- und Bedientechnologien eröffnet. Allein die Frage, wie die neuen Anzeigen und Displays gestaltet und bespielt werden, ist ein sehr spannendes Betätigungsfeld für Designer geworden. Unser Team entwickelte dafür eigene Style Guides für Bildwelten und Animationen, die es so vorher nicht gab.

Führen die technischen Möglichkeiten nicht automatisch zu einer Komplexität von Produkten, die vielleicht gar nicht wünschenswert ist?

Mit den neuen Bedien- und Anzeigetechniken wachsen zugleich die Möglichkeiten des Gebrauchs von Geräten. Ein Teil unserer Arbeit besteht auch darin zu verhindern, dass durch mehr Bedienmöglichkeiten die Geräte zu komplex werden. Sie sollen ja gerade einfach und intuitiv zu bedienen sein. Selbst komplexe Technik muss beherrschbar bleiben. In einem technisch geprägten Unternehmen wie Bosch haben wir als Designer also auch die Rolle eines Korrektivs gegenüber dem Marketing oder der technischen Produktentwicklung, weil wir den Benutzer im Blick behalten und durch nutzerorientierte Konzepte eine wichtige Schnittstelle gestalten.

As a designer, are you not always destined to have one foot in the present and one in the future?

We simply try to give the appliances the best that there is. And, depending on the product, we have to look far into the future. To this end, we have developed a dedicated process within the company which is known as "Vision Range". It is roughly comparable with the show cars and concept studies used in the automotive industry. We design an ideal future scenario in order to gear our brand, our products and our design to that scenario from a strategic perspective. This guarantees us a competitive lead, as the content can flow directly into future projects. Maybe that is one of the major advantages of being able to work for one company and with one team on a long-term basis. Because it gives us the freedom to look to the future, quite separately from the specific product.

Thank you for speaking with us, Mr Sachon.

Steht man als Designer nicht permanent mit einem Bein in der Gegenwart und mit dem anderen Bein in der Zukunft?

Wir versuchen einfach, den Geräten das Beste mitzugeben. Und je nach Produkt müssen wir weit vorausschauen. Wir haben dafür in unserem Hause einen eigenen Prozess entwickelt, den wir „Vision Range" nennen. Man kann das in etwa mit den Showcars und Konzeptstudien in der Automobilindustrie vergleichen. Wir entwerfen ein zukünftiges Idealbild, um unsere Marke, unsere Produkte und unser Design strategisch danach auszurichten. Das sichert uns einen Vorsprung, da die Inhalte direkt in künftige Projekte einfließen können. Vielleicht ist es einer der großen Vorzüge, langfristig in einem Unternehmen und mit einem Team arbeiten zu können. Denn es gibt uns die Freiheit, losgelöst vom konkreten Produkt, einen Blick in die Zukunft zu werfen.

Vielen Dank für das Gespräch, Herr Sachon.

The Red Dot: Design Team of the Year 2015 around Robert Sachon, Global Design Director Bosch and Helmut Kaiser, Head of Consumer Products Design Bosch.
Das Red Dot: Design Team of the Year 2015 um Robert Sachon, Global Design Director Bosch und Helmut Kaiser, Head of Consumer Products Design Bosch.

Red Dot: Best of the Best
The best designers of their category
Die besten Designer ihrer Kategorie

The designers of the Red Dot: Best of the Best
Only a few products in the Red Dot Design Award receive the "Red Dot: Best of the Best" accolade. In each category, the jury can assign this award to products of outstanding design quality and innovative achievement. Exploring new paths, these products are all exemplary in their design and oriented towards the future.

The following chapter introduces the people who have received one of these prestigious awards. It features the best designers and design teams of the year 2015 together with their products, revealing in interviews and statements what drives these designers and what design means to them.

Die Designer der Red Dot: Best of the Best
Nur sehr wenige Produkte im Red Dot Design Award erhalten die Auszeichnung „Red Dot: Best of the Best". Die Jury kann mit dieser Auszeichnung in jeder Kategorie Design von außerordentlicher Qualität und Innovationsleistung besonders hervorheben. In jeder Hinsicht vorbildlich gestaltet, beschreiten diese Produkte neue Wege und sind zukunftsweisend.

Das folgende Kapitel stellt die Menschen vor, die diese besondere Auszeichnung erhalten haben. Es zeigt die besten Designer und Designteams des Jahres 2015 zusammen mit ihren Produkten. In Interviews und Statements wird deutlich, was diese Designer bewegt und was ihnen Design bedeutet.

Gino Carollo – Studio 28
Georg Appeltshauser – Designbüro Appeltshauser

"Stay curious!"
„Neugierig bleiben!"

What do you particularly like about your own award-winning product?
To have succeeded in combining form and function in one product.

Is there a certain design approach that you pursue?
To design contemporary products that people can use. Design, function and practical value are of equal importance.

What inspires you?
The tremendous variety of forms, structures and colours displayed by nature. It offers designers an inexhaustible treasure to fall back on.

How do you define quality?
For people who believe in what they are doing, quality means enhanced value. It has a historical and cultural origin and is the result of experience and continuous research.

Was gefällt Ihnen an Ihrem eigenen, ausgezeichneten Produkt besonders gut?
Dass es uns gelungen ist, in einem Produkt Form und Funktion zu vereinen.

Gibt es einen bestimmten Gestaltungsansatz, den Sie verfolgen?
Zeitgemäße Produkte zu entwerfen, die die Leute gebrauchen können. Gestaltung, Funktion und Gebrauchswert haben den gleichen Stellenwert.

Was inspiriert Sie?
Die gewaltige Vielfalt der Formen, Strukturen und Farben, die uns die Natur vor Augen führt. Hier kann jeder Gestalter auf einen unerschöpflichen Schatz zurückgreifen.

Wie definieren Sie Qualität?
Für jemanden, der an das, was er macht, glaubt, bedeutet Qualität einen Mehrwert zu schaffen. Qualität hat ihren Ursprung in der Kultur und Geschichte und ist das Ergebnis von Erfahrungen sowie andauernder Forschung.

reddot award 2015
best of the best

Manufacturer
Draenert Studio GmbH,
Immenstaad, Germany

Fontana
Dining Table of Stone
with Extension Mechanism
Esstisch aus Stein
mit Auszugsmechanismus

See page 72
Siehe Seite 72

Sehwan Bae, Bohyun Nam, Sojin Park, Seunghyun Song
LG Electronics Inc.

"The quality of design lies in the balance between aesthetics and usability and incorporates ideas that arise from a careful consideration of the user's viewpoint."

„Designqualität beruht auf einem ausgewogenen Verhältnis von Ästhetik und Nutzbarkeit und beinhaltet Ideen, die das Ergebnis einer sorgfältigen Erwägung der Nutzerperspektive sind."

What do you particularly like about your own award-winning product?
Firstly, we have made the product more convenient by removing the power cord, which is the most inconvenient part of a vacuum cleaner. We have adopted a new concept that enables users to clean without having to drag the product along behind them. It automatically follows the user. Secondly, we have achieved a luxurious and reliable look thanks to the sturdy image evoked by the metallic materials, while at the same time maintaining the performance level of corded vacuum cleaners.

Is there a product that you have always dreamed about realising someday?
We would like to create products such as robots that are able to free people from housework and offer them greater convenience.

Was gefällt Ihnen an Ihrem eigenen, ausgezeichneten Produkt besonders gut?
Erstens haben wir das Produkt praktischer gemacht, indem wir das Stromkabel entfernt haben. Das ist das am wenigsten zweckmäßige Teil eines Staubsaugers. Wir haben ein neues Konzept angewendet, sodass der Nutzer saubermachen kann, ohne das Produkt hinter sich herziehen zu müssen. Es folgt dem Nutzer automatisch. Zweitens haben wir durch die robuste Ausstrahlung, die Metallic-Materialien hervorrufen, eine luxuriöse und vertrauenerweckende äußere Erscheinung erreicht. Gleichzeitig bietet das Produkt aber dieselbe Leistung wie ein Staubsauger mit Kabel.

Welches Produkt würden Sie gerne einmal realisieren?
Wir würden gerne Produkte realisieren wie beispielsweise Roboter, die Menschen von der Hausarbeit befreien und ihnen einen gesteigerten Nutzerkomfort bieten.

reddot award 2015
best of the best

Manufacturer
LG Electronics Inc., Seoul, South Korea

LG Cordless Canister Vacuum Cleaner CordZero C5 (VK9401LHAN, VK9402LHAN, VK9403LHAN)
LG Kabelloser Bodenstaubsauger CordZero C5 (VK9401LHAN, VK9402LHAN, VK9403LHAN)

See page 96
Siehe Seite 96

Tom Lloyd, Luke Pearson
PearsonLloyd

"Make less, make it in a better way and make it less often."

„Mach weniger, mach es besser und mach es seltener."

Is there a certain design approach that you pursue?
We try to design elegant and simple solutions which respond to natural behaviour. This ambition to create simplicity is in itself a complex process.

Do you have a motto for life?
Good, fast, cheap: Pick any two. This is the designer's holy triangle. In other words: Good service offered cheap won't be fast. Fast, good service won't be cheap. Fast and cheap service won't be good.

What do you see as being the biggest challenges in your industry at present?
Overproduction – in so many fields. If we want to save this planet we need a global shift in objectives.

Gibt es einen bestimmten Gestaltungsansatz, den Sie verfolgen?
Wir versuchen, elegante und einfache Lösungen zu entwerfen, die auf natürliche Verhaltensweisen reagieren. Dieser Ehrgeiz, Einfachheit zu schaffen, ist an sich ein komplexes Unterfangen.

Haben Sie ein Lebensmotto?
Gut, schnell, billig: Wählen Sie zwei davon. Das ist das goldene Dreieck des Designers. Mit anderen Worten: Billiger, guter Service wird nicht schnell sein. Schneller, guter Service wird nicht billig sein. Billiger, schneller Service wird nicht gut sein.

Worin sehen Sie aktuell die größten Herausforderungen in Ihrer Branche?
In der Überproduktion in so vielen Bereichen. Wenn wir diesen Planeten retten wollen, brauchen wir weltweit andere Ziele.

reddot award 2015
best of the best

Manufacturer
Joseph Joseph Ltd, London, Great Britain

Totem
Waste Separation and Recycling Unit
Behälter für Mülltrennung und Recycling

See page 120
Siehe Seite 120

Thomas Starczewski, Michael Fürstenberg, Piero Horn
designship

"Go beyond established patterns of thinking
 in order to create new dimensions of seeing!"
„Bestehende Denkmuster verlassen,
 um neue Dimensionen des Sehens zu schaffen!"

Is there a certain design approach that you pursue?
A focus on the essentials and the creation of a visual field of tension with regard to proportions and materials. Together with a positive, haptic experience and the conscious design of details that should make it fun to use the product.

How do you define design quality?
Quality reveals itself through well-being, through the "feel-good factor" that the use of all manner of products creates in everyday life. Were people the primary consideration in the development process? If lifeblood and devotion played a role during the creative process then the quality will be experienced, felt and become visible.

What do you see as being the biggest challenges in your industry at present?
Preventing the descent of product development into the "gratuitous".

Gibt es einen bestimmten Gestaltungsansatz, den Sie verfolgen?
Die Reduktion auf das Wesentliche und die Schaffung eines visuellen Spannungsfeldes von Proportionen und Materialien, die in Kombination mit einem positiven, haptischen Erlebnis und der bewussten Gestaltung von Details Freude beim Umgang mit dem Produkt bereiten.

Wie definieren Sie Designqualität?
Qualität drückt sich in Wohlbefinden aus, in dem „guten Gefühl" beim Umgang mit Produkten aller Art im täglichen Leben. Stand der Mensch im Mittelpunkt der Entwicklung? Spielten Herzblut und Hingabe bei der Umsetzung eine Rolle, dann wird Qualität erlebt, gefühlt und sichtbar.

Worin sehen Sie aktuell die größten Herausforderungen in Ihrer Branche?
Darin, das Ableiten der Produktentwicklung in die „Beliebigkeit" zu verhindern.

reddot award 2015
best of the best

Manufacturer
WMF AG, Geislingen, Germany

WMF Espresso
Semi-Automatic Coffee Machine
Kaffeehalbautomat

See page 124
Siehe Seite 124

Erwin Meier
Swiss Innovation Products

"Observing how the world changes moment by moment and participating."

„Beobachten, wie sich die Welt jeden Augenblick verändert, und daran teilhaben."

What do you particularly like about your own award-winning product?
It was possible to completely implement the ecological approach, from the capsule right down to mailing it out in an envelope.

What inspires you?
The ever-changing world, altering in every moment.

Is there a product that you have always dreamed about realising someday?
A drinking water filter that people can make themselves using resources available to them free of charge almost anywhere on earth. I am already working on this idea together with a team of people.

What does winning the Red Dot: Best of the Best mean to you?
This award gives me an indescribable energy boost for future projects.

Was gefällt Ihnen an Ihrem eigenen, ausgezeichneten Produkt besonders gut?
Der ökologische Ansatz konnte von der Kapsel bis zum Versand in Briefform perfekt umgesetzt werden.

Was inspiriert Sie?
Die sich in jedem Augenblick verändernde Welt.

Welches Produkt würden Sie gerne einmal realisieren?
Einen selbst herzustellenden Trinkwasserfilter aus Ressourcen, die jedem Menschen an fast jedem Ort der Welt kostenlos zur Verfügung stehen. Ich arbeite bereits mit einem Team an dieser Idee.

Was bedeutet die Auszeichnung mit dem Red Dot: Best of the Best für Sie?
Die Auszeichnung lädt mich mit unbeschreiblicher Energie für meine zukünftigen Projekte.

reddot award 2015
best of the best

Manufacturer
Swiss Innovation Products,
Zürich, Switzerland

mycoffeestar
Reusable Coffee Capsule
for Nespresso®
Wiederbefüllbare Kaffeekapsel
für Nespresso®

See page 130
Siehe Seite 130

Ulrich Goss, Oliver Kraemer, Christoph Ortmann, Robert Sachon
Robert Bosch Hausgeräte GmbH

"We want perfection to be experienced with all the senses."

„Wir wollen Perfektion mit allen Sinnen erfahrbar machen."

What do you particularly like about your own award-winning product?
The new Series 8 built-in appliances epitomise simplicity and quality, both with respect to exterior appearance and to their intuitive user interface. They thus stand for a new generation of future-oriented household appliances under the Bosch brand.

What inspires you?
We create products for everyday life, not for a museum. Therefore, our inspiration is daily life, for example cooking with family and friends.

What do you see as being the biggest challenges in your industry at present?
The household appliance sector is on the threshold of the digital age. Aside from new operating and display technologies, the topic of connectivity presents us with new challenges.

Was gefällt Ihnen an Ihrem eigenen, ausgezeichneten Produkt besonders gut?
Die neuen Einbaugeräte der Serie 8 verkörpern die Themen Einfachheit und Qualität sowohl im Hinblick auf die äußere Gestaltung als auch im Hinblick auf das intuitive User-Interface. Damit stehen sie für eine neue Generation zukünftiger Hausgeräte der Marke Bosch.

Was inspiriert Sie?
Wir entwerfen Produkte für das tägliche Leben, nicht fürs Museum. Daher inspiriert uns der Alltag, z. B. Kochen mit Familie und Freunden.

Worin sehen Sie aktuell die größten Herausforderungen in Ihrer Branche?
Die Hausgeräte-Branche steht immer noch an der Schwelle zum digitalen Zeitalter. Neben neuen Bedien- und Anzeigetechnologien stellt uns das Thema Vernetzung vor neue Herausforderungen.

reddot award 2015
best of the best

Manufacturer
Robert Bosch Hausgeräte GmbH,
Munich, Germany

Series | 8 Built-in Range Black
Built-in Appliance Range
Einbaugerätereihe

See page 148
Siehe Seite 148

Jinwon Kang, Woonkyu Seo, Daesung Lee, Taihun Lim, Sungkyong Han, Junyi Heo
LG Electronics Inc.

"Designers should try to make their sensibility resonate with a lot of people through their sharp insights."

„Designer sollten versuchen, dank ihrer scharfsinnigen Einsichten bei vielen Menschen Widerhall zu finden mit ihrer Sensibilität."

What do you particularly like about your own award-winning product?
We used diamond black stainless steel material for the refrigerator's exterior and also established a new production line to develop a material that distinguishes itself clearly from existing materials and fulfils the required performance criteria. Moreover, this product design offers new features that take account of both aesthetics and usability.

What do you see as being the biggest challenges in your industry at present?
Many people acknowledge the values of past "changes", but worry about unidentified, future "changes". A designer predicts changes that will occur and creates designs that shape this future. Thus, designers help people to embrace future changes.

Was gefällt Ihnen an Ihrem eigenen, ausgezeichneten Produkt besonders gut?
Wir haben für das Äußere des Kühlschranks diamantschwarzen Edelstahl verwendet und auch eine neue Produktionsanlage eingerichtet, um ein Material zu entwickeln, das sich von bestehenden Materialien unterscheidet und die erforderlichen Leistungskriterien erfüllt. Zudem weist dieses Produktdesign Besonderheiten auf, die der Ästhetik wie der Benutzerfreundlichkeit Rechnung tragen.

Worin sehen Sie aktuell die größten Herausforderungen in Ihrer Branche?
Viele Menschen wissen den Wert vergangener „Veränderungen" zu schätzen, sind aber ob der unbekannten, zukünftigen beunruhigt. Ein Designer sagt zukünftige Veränderungen voraus und entwirft Designs, die diese Zukunft prägen. Auf diese Weise helfen Designer Menschen, die Veränderungen der Zukunft zu begrüßen.

reddot award 2015
best of the best

Manufacturer
LG Electronics Inc., Seoul, South Korea

LG Double Door in Door Refrigerator (R-F956EDSB, R-F956VDSB, R-F956VDDN, R-F956VDSR)
LG Doppeltür-Kühlschrank
(R-F956EDSB, R-F956VDSB, R-F956VDDN, R-F956VDSR)

See page 172
Siehe Seite 172

Karl Forsberg, Olle Gyllang
Propeller Design AB

"Bringing together an inspirational mix of users and clients leads
to truly innovative design."

„Das Zusammenbringen von Nutzern und Auftraggebern in einer
inspirierenden Mischung führt zu wirklich innovativem Design."

What do you particularly like about your own award-winning product?
The new Eligo sink solution allows one to optimise the basin size depending on the use scenario. Thanks to its adjustable dividing wall it is suitable for both delicate glasses and large dirty oven pans, while also saving water, energy and washing-up liquid. As the kitchen sink is used many times throughout the day, these improvements have a great impact on the environment and offer a more enjoyable user experience.

Is there a certain design approach that you pursue?
We strive to approach all projects from a holistic and strategic perspective, ensuring to have a full understanding of the entire context. We see ourselves as a catalyst between our clients and their customers, translating the vision of our clients to meet the needs and expectations of the users.

Was gefällt Ihnen an Ihrem eigenen, ausgezeichneten Produkt besonders gut?
Das neue Eligo-Spülenkonzept erlaubt die bestmögliche Nutzung des Spülbereichs je nach Bedarf des Nutzers und ist dank der verstellbaren Trennwand genauso für empfindliche Gläser wie auch für große, verschmutzte Ofenbleche geeignet. Zudem werden sowohl Wasser als auch Energie und Spülmittel gespart. Da eine Küchenspüle mehrmals am Tag genutzt wird, haben unsere Neuerungen eine enorme Auswirkung auf die Umwelt und bieten ein optimiertes Nutzererlebnis.

Gibt es einen bestimmten Gestaltungsansatz, den Sie verfolgen?
Wir sind bestrebt, alle Projekte aus einer ganzheitlichen, strategischen Perspektive anzugehen, um sicherzustellen, dass wir den gesamten Kontext genau erfassen. Wir verstehen uns als Katalysator zwischen unseren Auftraggebern und ihren Kunden. Wir bringen die Vision unserer Auftraggeber dazu, den Anforderungen und Erwartungen der Nutzer zu entsprechen.

red**dot** award 2015
best of the best

Manufacturer
Intra Mölntorp AB, Kolbäck, Sweden

Intra Eligo
Kitchen Sink in Stainless Steel
Küchenspüle aus Edelstahl

See page 180
Siehe Seite 180

Cecilie Manz

"Gut feeling, making sense, necessity, simplification, aesthetics, quality are all parameters in play when designing something new."

„Bauchgefühl, einen Sinn ergeben, Notwendigkeit, Vereinfachung, Ästhetik, Qualität sind alles Parameter, die ins Spiel kommen, wenn man etwas Neues gestaltet."

What do you particularly like about your own award-winning product?
I like the fact that the Easy series is a line of very universal utensils you will find in any period. A simple wooden spoon or spatula on the one hand looks more or less the same everywhere; on the other hand they are all very different in detail, material and craftsmanship.

Is there a certain design approach that you pursue?
If something is not essential to the design it should probably not be there. Simplify just to the limit of banality.

How do you define quality?
Quality is something that withstands both hand- and eye-contact over a long period of time.

Was gefällt Ihnen an Ihrem eigenen, ausgezeichneten Produkt besonders gut?
Mir gefällt, dass die Serie Easy eine Reihe von sehr universalen Utensilien ist, die man in jedem Zeitalter finden kann. Ein einfacher Holzlöffel oder ein Pfannenwender sieht einerseits überall mehr oder weniger gleich aus, andererseits unterscheiden sie sich alle im Detail, im Material und in der Verarbeitung.

Gibt es einen bestimmten Gestaltungs-ansatz, den Sie verfolgen?
Wenn etwas für die Gestaltung nicht unbedingt notwendig ist, sollte es wahrscheinlich nicht da sein. Man sollte bis an die Grenze zur Banalität hin vereinfachen.

Wie definieren Sie Qualität?
Qualität ist etwas, das dem Kontakt von Händen und Augen lange Zeit widerstehen kann.

reddot award 2015
best of the best

Manufacturer
Stelton, Copenhagen, Denmark

Easy
Kitchen Utensils
Küchenutensilien

See page 208
Siehe Seite 208

Komin Yamada
STUDIO KOMIN

"As nature dominates, all products should be designed
to fit into the environment."
„Die Natur ist das Dominierende. Deshalb sollten alle Produkte so konzipiert sein,
dass sie sich in die Umwelt einfügen."

What do you particularly like about your own award-winning product?
The relationship between body and handles and the integral shape which has both handles extending downward from the edge of the body.

What do you see as being the biggest challenges in your industry at present?
To be natural, to naturally fit into the environment.

What does winning the Red Dot: Best of the Best mean to you?
The great opportunity to embody my design motto.

Was gefällt Ihnen an Ihrem eigenen, ausgezeichneten Produkt besonders gut?
Das Verhältnis zwischen Topfkörper und Henkeln sowie die ganzheitliche Form, in der sich beide Henkel vom Rand des Körpers nach unten verlängern.

Worin sehen Sie aktuell die größten Herausforderungen in Ihrer Branche?
Darin, natürlich zu sein und sich auf natürliche Art und Weise in die Umwelt einzugliedern.

Was bedeutet die Auszeichnung mit dem Red Dot: Best of the Best für Sie?
Die großartige Gelegenheit, mein Design-motto verkörpert zu sehen.

reddot award 2015
best of the best

Manufacturer
Sanjo Special Cast. Co., Ltd.,
Sanjo, Niigata, Japan

Unilloy
Enamel-Coated Cast Iron Pot
Emaillierter Topf aus Gusseisen

See page 220
Siehe Seite 220

Ronan & Erwan Bouroullec

"We have a deep interest or even a passion for the small details of life, the way people consider objects, the way they use them etc."

„Wir haben ein echtes Interesse oder sogar eine Leidenschaft für die kleinen Dinge des Lebens, die Art, wie Menschen Objekte einschätzen, die Art, wie sie sie benutzen etc."

Is there a certain design approach that you pursue?
We make a great point of designing pieces that can be integrated into people's interiors, whatever the cultural background may be. We would describe the projects as "not too noisy". But in a sense, we pursue industrial design logic itself, whose target is a huge market. As we do not know exactly in which context each of our objects will end up, we try to take out everything that does not seem absolutely necessary.

What inspires you?
We do not feel inspiration as ever being direct. It is far more complex. However, it is true that observation can be considered as our main external influence. We try accurately to observe people's behaviours in everyday life and to understand common practices and needs.

Gibt es einen bestimmten Gestaltungsansatz, den Sie verfolgen?
Uns ist enorm wichtig, dass wir Produkte entwickeln, die sich leicht in Wohnräume und in jeden kulturellen Hintergrund einfügen. Wir beschreiben unsere Projekte als „nicht besonders laut". Wir folgen aber auch der industriellen Gestaltungslogik, deren Ziel es ist, einen sehr großen Markt zu versorgen. Da wir nicht immer wissen, in welchem Kontext unsere Objekte einmal landen werden, versuchen wir, alles zu beseitigen, was uns nicht unbedingt notwendig erscheint.

Was inspiriert Sie?
Wir nehmen Inspiration nie als unmittelbar gegeben wahr – es ist alles viel komplexer. Es ist allerdings wahr, dass Beobachtungen zu unseren wichtigsten externen Einflussfaktoren zählen. Wir versuchen, das tägliche Verhalten von Leuten genau zu beobachten und alltägliche Bräuche und Anforderungen zu verstehen.

reddot award 2015
best of the best

Manufacturer
Iittala, Helsinki, Finland

Ruutu
Glass Vase
Glasvase

See page 244
Siehe Seite 244

49

Carlo Colombo

"In our times of mass production, it is vital to succeed in creating unique and individual products."

„In unseren Zeiten der Massenproduktion ist es entscheidend, einzigartige und individuelle Produkte zu erschaffen."

What do you particularly like about your own award-winning product?
Luxury means making something come true today that would have been considered impossible until yesterday. I Bordi is the result of an attempt at giving the universal voice of material sense. The result is a wraparound space which is also strikingly simple, a textured setting designed to arouse the senses. Its details are naturally inscribed in the material of its components, making it perfectly suitable for any interpretation, from Carrara marble to the ultramodern Duralight, and also Grey Stone.

Is there a specific approach that is of significance to you and your work?
My philosophy is to put my mark on everything I create, giving it a soul, a character. My designs follow a specific logic of simplicity, elegance, sophistication and attention to detail.

Was gefällt Ihnen an Ihrem eigenen, ausgezeichneten Produkt besonders gut?
Luxus bedeutet, heute etwas wahr werden zu lassen, das bis gestern als unmöglich gegolten hat. I Bordi entstand aus dem Versuch, den universellen Ausdruck der Materie zu versinnbildlichen. Ergebnis ist ein sehr schlichtes, einhüllendes Volumen, ein strukturierter Rahmen, der Emotionen weckt. Die Details sind natürliche Facetten des Materials und passen sich somit jeder Variante an, vom Marmor aus Carrara bis zum hochmodernen Duralight, aber auch Stone Grey.

Gibt es einen bestimmten Ansatz, der für Sie und Ihre Arbeit von Bedeutung ist?
Meine Philosophie besteht darin, alle meine Gestaltungen mit meiner Handschrift zu versehen, ihnen eine Seele und einen Charakter zu verleihen. Meine Gestaltungen folgen einer speziellen Logik der Schlichtheit, Eleganz, Durchdachtheit und Liebe zum Detail.

reddot **award 2015**
best of the best

Manufacturer
Teuco SpA, Montelupone, Italy

I Bordi
Bathtub
Badewanne

See page 252
Siehe Seite 252

Mat Bembridge, Jos van Roosmalen
Royal Philips

"Designing a new product is an opportunity to offer a next level brand experience to customers and, at the same time, bring the full benefit to end-users."

„Die Gestaltung eines neuen Produkts gibt uns die Möglichkeit, Kunden ein neues Markenerlebnis und dem Endverbraucher zugleich den vollen Nutzen zu bieten."

How do you define quality?
Quality lies in the balance between what you want to achieve with a product or service and how it is expressed. It's all about adapting to the context of the user, the right choice of materials and functionality, and fine-tuning proportions and details precisely. Quality is achieved when all the complexities of a concept have been carefully thought through in order to result in a simple product which is easy to use.

What does winning the Red Dot: Best of the Best mean to you?
From the very start of the project, we wanted the product to be noticed as the best design solution possible for functional outdoor lighting. Now we have received the highest level of recognition from within the design industry with the Red Dot: Best of the Best.

Wie definieren Sie Qualität?
Qualität ist der goldene Mittelweg zwischen dem, was man mit einem Produkt oder einer Dienstleistung erreichen will, und wie man das zum Ausdruck bringt. Es geht darum, sich dem Nutzerkontext anzupassen, die richtigen Materialien und Funktionsweisen zu wählen sowie Proportionen und Details genau aufeinander abzustimmen. Qualität erzielt man dann, wenn jegliche Komplexität eines Konzepts sorgsam durchdacht ist und schließlich zu einem schlichten Produkt führt, das einfach zu verwenden ist.

Was bedeutet die Auszeichnung mit dem Red Dot: Best of the Best für Sie?
Gleich zu Beginn des Projekts wollten wir, dass es als beste Designlösung für funktionale Beleuchtung im Außenbereich anerkannt wird. Mit der Auszeichnung „Red Dot: Best of the Best" haben wir die höchste Anerkennung erhalten, die die Designbranche zu vergeben hat.

reddot award 2015
best of the best

Manufacturer
Royal Philips, Eindhoven, Netherlands

LumiStreet
LED Street Light
LED-Straßenleuchte

See page 320
Siehe Seite 320

Ross Lovegrove
Lovegrove Studio

"Trinity between technology, materials science and an intelligent organic form."
„Eine Dreifaltigkeit von Technologie, Werkstoffwissenschaft und einer intelligenten, organischen Form."

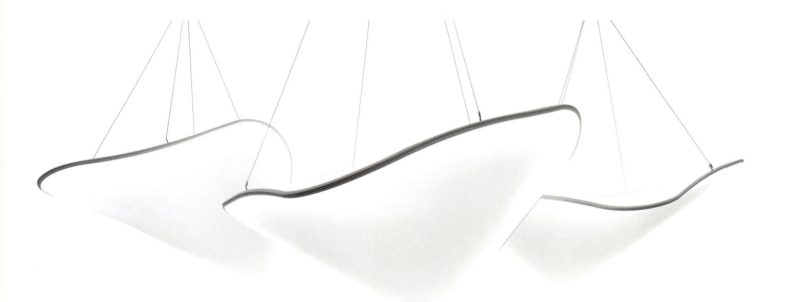

Is there a designer that you particularly admire?
I'm more interested in the architectural and bioengineering debate, as I believe that design, as we know it, should go way beyond the limited borders of furniture design. We sadly lost the ingenious mind of Frei Otto but we now have Achim Menges.

How do you define design quality?
Quality in the conceptual mind, made evident in the physical object.

What do you see as being the biggest challenges in your industry at present?
Global issues but above all the need to create a new economic industrial model based on parametricism and the synergy we are seeing in the coalescence of the digital realm, materials science, energy efficiency and 3D printing.

Gibt es einen Designer, den Sie besonders schätzen?
Ich bin mehr an der architektonischen und biotechnischen Debatte interessiert, da ich meine, dass Design, wie wir es zurzeit verstehen, weit über die Grenzen von Möbeldesign hinausgehen sollte. Leider haben wir die erfinderische Geistesgröße eines Frei Otto verloren, dafür gibt es jetzt Achim Menges.

Wie definieren Sie Designqualität?
Als die in einem physischen Objekt verwirklichte Qualität eines konzeptionellen Gedankens.

Worin sehen Sie aktuell die größten Herausforderungen in Ihrer Branche?
In globalen Fragen, aber vor allem in der Notwendigkeit, ein neues wirtschaftliches Industriemodell zu schaffen, das auf Parametrismus beruht und auch auf der Synergie, die wir zurzeit in der Verschmelzung des digitalen Bereichs mit der Werkstoffwissenschaft, der Energieeffizienz und dem 3D-Druck sehen.

reddot award 2015
best of the best

Manufacturer
Barrisol Normalu SAS, Kembs, France

Barrisol – Lovegrove Manta
Lamp
Leuchte

See page 330
Siehe Seite 330

Klaus Adolph
a·g Licht for Zumtobel Lighting GmbH

"In all our developments we try to create lighting with a reduced appearance that can therefore become an integral part of a building's architecture."

„Mit jeder Entwicklung versuchen wir, Leuchten zu entwerfen, die in ihrem Erscheinungsbild reduziert sind und dadurch zum integralen Bestandteil der Architektur werden können."

What do you particularly like about your own award-winning product?
Working with a team of specialists on developing a product that highlights the possibilities offered by LED technology – intelligence, precision and a high level of efficiency.

Is there a project that you have always dreamed about realising someday?
My own house.

Do you have a motto for life?
Don't take yourself too seriously.

How do you define design quality?
Innovative, but timeless rather than fashionable.

What do you see as being the biggest challenges in your industry at present?
Realising the possibilities offered by technology in a human way.

Was gefällt Ihnen an Ihrem eigenen, ausgezeichneten Produkt besonders gut?
Gemeinsam mit einem Team von Spezialisten ein Produkt zu entwickeln, welches die Möglichkeiten der LED aufzeigt – intelligent, präzise, hocheffizient.

Welches Projekt würden Sie gerne einmal realisieren?
Das eigene Haus.

Haben Sie ein Lebensmotto?
Sich selber nicht so wichtig nehmen.

Wie definieren Sie Designqualität?
Innovativ, jedoch nicht modisch, zeitlos.

Worin sehen Sie aktuell die größten Herausforderungen in Ihrer Branche?
Darin, die Möglichkeiten der Technik menschlich umzusetzen.

reddot award 2015
best of the best

Manufacturer
Zumtobel Lighting GmbH,
Dornbirn, Austria

SEQUENCE
Pendant and Surface-Mounted LED Luminaire
LED-Pendel- und Anbauleuchte

See page 340
Siehe Seite 340

Yu Feng
Deve Build

"Let's part with our cockiness and arrogance,
 and let our heart return to the earth again."

„Wir sollten unsere Großspurigkeit und Arroganz aufgeben und
 unser Herz wieder auf den Boden der Tatsachen zurückkehren lassen."

What do you particularly like about your own award-winning product?
The unique aspect of the Bitter Bamboo Room lies in its criticism and mocking of current so-called mainstream design and also in its undisguised antithesis to the dignified loftiness advocated by the majority of society.

How do you define design quality?
A high-quality, vigorous design must first of all be sincere. Sincerity occurs naturally when the artist faces such a work. All that can result from utilitarianism or a consideration outside of art can only produce a design product, but not a quality piece of work. I think this is a basic standard for judging the value of a piece of work in future.

Was gefällt Ihnen an Ihrem eigenen, ausgezeichneten Produkt besonders gut?
Das Einzigartige am Bitter Bamboo Room ist seine Kritik an und sein Spott über das heutige sogenannte etablierte Design. Außerdem steht es in unverhülltem Gegensatz zu der würdevollen Erhabenheit, die ein Großteil der Gesellschaft befürwortet.

Wie definieren Sie Designqualität?
Ein hochwertiges, dynamisches Designkonzept sollte vor allem ehrlich sein. Ehrlichkeit entsteht ganz von selbst, wenn der Künstler so einem Werk gegenübersteht. Utilitarismus oder eine Betrachtungsweise, die gänzlich außerhalb des Werks liegt, kann nur zu einem Designprodukt führen, das qualitativ nicht hochwertig ist. Ich bin der Meinung, dass dies ein Grundprinzip zur Beurteilung des zukünftigen Wertes einer Arbeit sein sollte.

reddot award 2015
best of the best

Manufacturer
Deve Build, Shenzhen, China

Bitter Bamboo Room
Working Space
Arbeitsraum

See page 386
Siehe Seite 386

Mandy Leeming
Interface European Design Team

"Our products aim to be innovative and desirable to the customer and we set out to enable architects to create healthy spaces that inspire the people working there."

„Unsere Produkte sollen innovativ und für den Kunden begehrenswert sein. Wir wollen Architekten helfen, gesunde Räume zu schaffen, welche die darin arbeitenden Menschen inspirieren."

Is there a certain design approach that you pursue?
We follow environmentally sound practices in everything, from our day-to-day practices to our design and manufacturing processes. Sustainability is at the centre of our development approach.

Is there a designer that you particularly admire?
Working with David Oakey, the worlds' leading designer in biomimetics, has been particularly inspiring. He guides organisations around the globe to demonstrate respect for the future as they develop their own practices.

What trends are currently having a particularly large influence on design?
The natural environment trend is strong within interiors. Designers are aiming to bring the outside in and to create work environments with strong natural elements.

Gibt es einen bestimmten Gestaltungsansatz, den Sie verfolgen?
In allem, was wir tun, nutzen wir umweltschonende Methoden, angefangen bei unseren tagtäglichen Praktiken bis hin zu unseren Gestaltungs- und Herstellungsverfahren. Nachhaltigkeit steht im Zentrum unseres Entwicklungsansatzes.

Gibt es einen Designer, den Sie besonders schätzen?
Die Zusammenarbeit mit David Oakey, dem weltweit führenden Designer in Biomimetik, war besonders inspirierend. Er animiert Organisationen auf der ganzen Welt dazu, beim Aufbau ihrer eigenen Verfahren Respekt für die Zukunft zu zeigen.

Welche Trends beeinflussen das Design zurzeit besonders stark?
Der Trend des natürlichen Lebensraums ist in der Innenarchitektur sehr stark. Designer versuchen, die Außenwelt nach innen zu bringen und ein Arbeitsumfeld mit starken natürlichen Elementen zu schaffen.

reddot award 2015
best of the best

Manufacturer
Interface, Scherpenzeel, Netherlands

Interface Microsfera
Carpet Tiles
Teppichfliesen

See page 398
Siehe Seite 398

Martin A. Dolkowski
Admonter

"Developing solutions for complex problems is always fun!"

„Lösungen für komplexe Aufgabenstellungen zu entwickeln, macht immer Spaß!"

What do you particularly like about your own award-winning product?
A totally linear shape which renders the technology almost invisible.

Is there a designer that you particularly admire?
Nature. Over millions of years it has developed solutions for the widest possible range of different problems – from simple colour combinations to complex statically optimised structures.

How do you define design quality?
Only if aesthetic, practical and economic functionality attain the right equilibrium, can design be called good.

What do you see as being the biggest challenges in your industry at present?
In future design requirements, ecological and economic functionality will merge. This development represents both a challenge and an opportunity.

Was gefällt Ihnen an Ihrem eigenen, ausgezeichneten Produkt besonders gut?
Die absolut geradlinige Formgebung, bei der sich die Technik fast unsichtbar im Hintergrund verbirgt.

Gibt es einen Designer, den Sie besonders schätzen?
Die Natur. Sie hat in Jahrmillionen für unterschiedlichste Aufgabenstellungen Lösungen entwickelt – von simplen Farb-kombinationen bis hin zu komplexen, statisch optimierten Strukturen.

Wie definieren Sie Designqualität?
Nur wenn ästhetische, praktische und ökonomische Funktionalität im richtigen Verhältnis stehen, ist es gutes Design.

Worin sehen Sie aktuell die größten Herausforderungen in Ihrer Branche?
Ökologische und ökonomische Funktio-nalität werden in künftigen Designanfor-derungen immer mehr verschmelzen. Diese Entwicklung ist Herausforderung und Chance gleichermaßen.

reddot award 2015
best of the best

Manufacturer
Admonter, STIA Holzindustrie GmbH, Admont, Austria

Flush-Mount Skirting Board
Wandbündige Sockelleiste

See page 402
Siehe Seite 402

Gilles Rozé, Yves Savinel
Saint-Gobain Glass

"Source of light and softness."

„Quelle von Licht und Leichtheit."

Is there a certain design approach that you pursue?
The approach of the designer must first be honest.

Is there a designer that you particularly admire?
Jean Prouvé, because he was our teacher, and he taught us the rigour and the vanity of selfsatisfaction.

What do you see as being the biggest challenges in your industry at present?
Do not believe that design is enough to make a really good product.

What does winning the Red Dot: Best of the Best mean to you?
It means that we have succeeded in creating a community, thinking and acting with project stakeholders. Design cannot be the product of a lonesome designer but is the result of active collaboration with a company.

Gibt es einen bestimmten Gestaltungsansatz, den Sie verfolgen?
Der Ansatz eines Designers muss vor allem ehrlich sein.

Gibt es einen Designer, den Sie besonders schätzen?
Jean Prouvé, weil er unser Lehrer war und uns die Strenge und die Eitelkeit von Selbstzufriedenheit gelehrt hat.

Worin sehen Sie aktuell die größten Herausforderungen in Ihrer Branche?
Glaube nicht, dass Design ausreicht, um ein wirklich gutes Produkt hervorzubringen.

Was bedeutet die Auszeichnung mit dem Red Dot: Best of the Best für Sie?
Es bedeutet, dass es uns gelungen ist, eine Gemeinschaft aufzubauen, in der wir mit den Interessengruppen des Projekts gemeinsam denken und handeln. Design kann nicht das Produkt eines einzelnen Designers sein, sondern ist das Ergebnis einer aktiven Zusammenarbeit mit einem Auftraggeber.

reddot award 2015
best of the best

Manufacturer
Saint-Gobain Innovative Materials sp. z o.o.
Oddział Glass w Dąbrowie Górniczej, Poland

SGG Master-Soft
Patterned Glass
Ornamentglas

See page 412
Siehe Seite 412

Heiða Gunnarsdóttir Nolsøe, Marie Stampe Berggreen
DropBucket ApS

"User-centred design and simplicity are the key to design quality."
„Orientierung am Nutzer und Einfachheit sind für hochwertiges
Design ausschlaggebend."

What do you particularly like about your own award-winning product?
Our innovative, sustainable rubbish bin is easy to manage and place in large open areas. It makes clean festivals possible while being an appealing product itself.

What inspires you?
Simple ideas with huge potential. During our amazing journey with DropBucket we have learned that a simple piece of recycled cardboard is able to solve the enormous waste management problem during events.

What does winning the Red Dot: Best of the Best mean to you?
That the effort put into our university project is recognised in this way completely overwhelms us, and it is with great honour that we accept the Red Dot: Best of the Best.

Was gefällt Ihnen an Ihrem eigenen, ausgezeichneten Produkt besonders gut?
Unser innovativer, nachhaltiger Abfalleimer ist einfach zu handhaben und findet auf großen Freiflächen gut Platz. Mit ihm ist es möglich, Festivals sauber zu halten und gleichzeitig ist er selbst ansprechend.

Was inspiriert Sie?
Einfache Ideen mit immensem Potenzial. Im Laufe des erstaunlichen Weges, den wir mit DropBucket zurückgelegt haben, haben wir gelernt, dass ein einfaches Stück Recyclingpappe das enorme Abfallmanagementproblem bei Veranstaltungen lösen kann.

Was bedeutet die Auszeichnung mit dem Red Dot: Best of the Best für Sie?
Wir sind vollkommen überwältigt, dass die Bemühungen, die wir in unser Universitätsprojekt investiert haben, auf diese Weise anerkannt wurden. Es ist uns eine große Ehre den Red Dot: Best of the Best zu erhalten.

reddot award 2015
best of the best

Manufacturer
DS Smith Packaging Denmark A/S,
Taastrup, Denmark

DropBucket
Waste Bin
Abfalleimer

See page 442
Siehe Seite 442

Markus Brink, Stefan Brink
Richard Brink GmbH & Co. KG

"We do not sell anything, we wouldn't buy ourselves."

„Wir verkaufen nichts, was wir nicht auch selber kaufen würden."

What inspires you?
Beauty in the smallest and most inconspicuous detail and the will to achieve perfection, particularly in the everyday.

How do you define design quality?
Alongside high-quality materials and their high-quality use, we also see practical function as a critical element of good product design. For us, form also follows function. But in our perception form exists even where others would just leave it at that once there's function.

What do you see as being the biggest challenges in your industry at present?
In proving the design contribution that drainage and guttering can make to architecture. As far as we are concerned, these are not only functional elements of building regulations, but another way of adding well-designed stylish highlights.

Was inspiriert Sie?
Schönheit im kleinsten und unscheinbarsten Detail und der Wille, Perfektion gerade auch im Alltäglichen umzusetzen.

Wie definieren Sie Designqualität?
Neben hochwertigen Materialien und ihrer qualitativen Verarbeitung definieren wir ebenso die praktische Funktion als wesentlichen Bestandteil guten Produktdesigns. So folgt auch bei uns die Form der Funktion. Doch tritt bei uns die Form sogar da auf, wo andere es bei der Funktion belassen würden.

Worin sehen Sie aktuell die größten Herausforderungen in Ihrer Branche?
Darin, den gestalterischen Anspruch von Dränage- und Fassadenrinnen in der Architektur unter Beweis zu stellen. Diese Elemente sind für uns nicht nur zweckmäßige Bestandteile baurechtlicher Verordnungen, sondern eine weitere Möglichkeit, formschöne Akzente zu setzen.

reddot award 2015
best of the best

Manufacturer
Richard Brink GmbH & Co. KG,
Schloß Holte-Stukenbrock, Germany

Gemini
Double-Slit Grates
Doppelschlitzroste

See page 450
Siehe Seite 450

Beds
Cupboards
Seating
Shelving units
Sideboards
Stools
Stoves
Tables

Anrichten
Betten
Hocker
Kaminöfen
Regale
Schränke
Sitzmöbel
Tische

Living rooms and bedrooms
Wohnen und Schlafen

Fontana
Dining Table of Stone with Extension Mechanism
Esstisch aus Stein mit Auszugsmechanismus

Manufacturer
Draenert Studio GmbH,
Immenstaad, Germany

Design
Designbüro Appeltshauser
(Georg Appeltshauser),
Stuttgart, Germany
Studio 28 (Gino Carollo),
Thiene (Vicenza), Italy

Web
www.draenert.de

reddot award 2015
best of the best

The magic of synchronicity

The term synchronicity describes the principle of a simultaneous occurrence of events or procedures, which are subject to a unifying order. An innovative combined lifting and lowering linear rotation mechanism that is part of the Fontana tables makes it possible to extend and retract the table top in a seemingly synchronous movement. The ease with which the heavy extension elements made of natural stone can be moved is breathtaking. They slot into place without a visible change of position. This is doubly impressive, as the apparently magical process is purely mechanical and can easily be achieved by the user. And, what is more, in a self-explanatory way. This operation is easy to carry out and does not require any exertion, as the well thought-out construction of Fontana makes this process very straightforward. The design of the dining table ensures that the overall installation space is scaled-down, resulting in greater leg-room for the people seated at it. In order to make Fontana adaptable to different living areas, there is a choice of 180 different types of stone, various table tops and edge geometries as well as two base types. In its harmonious combination of design vocabulary and functionality, this table embodies the fascination of good design, which can take many different forms.

Die Magie der Synchronität

Der Begriff der Synchronität beschreibt das Prinzip der zeitlichen Übereinstimmung von Ereignissen oder Vorgängen, die dabei einer vereinheitlichenden Ordnung unterliegen. Bei dem Tisch Fontana ermöglicht eine innovative, kombinierte Hub-Senk-Linear-Rotationsmechanik das Ausziehen und Einfahren der Tischplatte in einer derart synchron anmutenden Bewegung. Erstaunlich ist, mit welcher Leichtigkeit dabei die schweren Erweiterungselemente aus Naturstein bewegt werden – ohne sichtbaren Positionswechsel rücken sie an Ort und Stelle. Beeindruckend ist zudem, dass dieser magisch anmutende Vorgang rein mechanisch geschieht und für den Nutzer dennoch einfach und selbsterklärend zu bewerkstelligen ist. Das Betätigen der Funktion erfolgt mit Leichtigkeit und erfordert keinerlei Kraftaufwand. Bei Fontana verbindet sich dieses Erlebnis mit dem Komfort einer durchdachten Konstruktion. Der Bauraum des Esstisches wurde gestalterisch insgesamt verkleinert, weshalb er den an ihm Sitzenden viel komfortable Beinfreiheit bietet. Um Fontana unterschiedlichen Wohnumgebungen anpassen zu können, gibt es eine Auswahl von über 180 Steinsorten, darüber hinaus verschiedene Plattenformen und Kantengeometrien sowie zwei unterschiedliche Sockelvarianten. Im Einklang von Formensprache und Funktionalität verkörpert dieser Esstisch die Faszination guten Designs, das sich immer wieder neu inszenieren lässt.

Statement by the jury

The Fontana dining table combines the aura of a special material – natural stone – with a fascinating and innovative extension mechanism. The user can activate extension and retraction of the massive table top without any effort or exertion. This leads to a seemingly magical synchronous movement of the extension elements. Fontana's perfectly matched design proportions, which offer great leg space, show the high-quality table top to great advantage.

Begründung der Jury

Bei dem Esstisch Fontana verbindet sich die Ausstrahlung des besonderen Materials Naturstein mit dem Erleben einer innovativen Auszugsmechanik. Völlig mühelos und ohne Kraftaufwand kann der Nutzer das Ausziehen und Einfahren der schweren Tischplatte aktivieren, woraufhin sich die Erweiterungselemente auf scheinbar magische Weise synchron bewegen. Die gestalterisch perfekt aufeinander abgestimmten Proportionen dieses viel Beinfreiheit bietenden Esstisches bringen die hochwertige Tischplatte gut zur Geltung.

Designer portrait
See page 28
Siehe Seite 28

Seito
Table
Tisch

Manufacturer
Walter Knoll AG & Co. KG,
Herrenberg, Germany
Design
Wolfgang C. R. Mezger,
Göppingen, Germany
Web
www.walterknoll.de
www.design-mezger.com

The Seito table conveys a playful lightness with the harmonious balance of lines, angles and surfaces. The material selection in particular underlines the dramatic understatement. The table top, made of white quartz, is carried by a base made of elegant nut wood. The filigree table legs are an expression of clear geometry, providing users with generous leg room. Furthermore, the impressively glossy surface of the table is highly pleasant to the touch.

Der Tisch Seito vermittelt aufgrund der Ausgewogenheit von Linien, Winkeln und Flächen eine spielerische Leichtigkeit. Insbesondere die Materialität unterstützt das effektvolle Understatement. So wird die Tischplatte aus weißem Quarzstein von einem Gestell aus edlem Nussbaumholz getragen. Die filigran anmutenden Tischbeine sind Ausdruck einer klaren Geometrie und gewähren den Nutzern eine großzügige Beinfreiheit. Die eindrucksvoll glänzende Tischoberfläche weist zudem eine angenehme Haptik auf.

Statement by the jury
With its particular choice of materials, which is rich in contrast, the design of Seito reaches a formal autonomy that is sure to fascinate.

Begründung der Jury
Die Gestaltung von Seito erreicht insbesondere aufgrund der kontrastreichen Materialwahl eine formale Eigenständigkeit, die begeistert.

KT11
Table
Tisch

Manufacturer
palatti, Trebelovice, Czech Republic
Design
Christian Kröpfl, Vienna, Austria
Web
www.palatti.eu
www.massivholz-design.at
Honourable Mention

The KT11 dining table features a delicate tube structure made of either stainless steel square tubes or painted crude steel. The structure follows the design idea of a truss, allowing generous legroom and the elegant but simple wooden top of the table seems to float above it – reminiscent of acacia tree canopies. The tops come in different sizes, with various edges and timbers including oak, walnut and ash.

Der Esstisch KT11 hat eine filigrane Stabkonstruktion, die optional aus Edelstahlformrohren oder aus lackiertem Rohstahl besteht. Diese Konstruktion folgt der Gestaltungsidee eines Fachwerks und ermöglicht eine große Beinfreiheit. Darüber scheint die elegante und gleichzeitig natürlich anmutende Holzplatte zu schweben – in Anlehnung an Baumkronen von Schirmakazien. Die Tischplatten sind in unterschiedlichen Abmessungen, mit diversen Kanten und wahlweise in Eiche, Nuss oder Esche erhältlich.

Statement by the jury
The branchy base frame of this table achieves a distinct lightness that is quite remarkable.

Begründung der Jury
Das verästelte Untergestell dieses Tisches erreicht eine formale Leichtigkeit, die bemerkenswert ist.

yps
Table
Tisch

Manufacturer
TEAM 7 Natürlich Wohnen GmbH,
Ried im Innkreis, Austria
In-house design
Jacob Strobel
Web
www.team7.at

Resting on flared Y-shaped legs, the yps natural wood table has an elegant and sculptural appeal despite its generous dimensions. As the basic shape for the feet, the "Y" provides a stable base and allows for plenty of legroom. The edges of the tabletop, which are milled with reverse diagonals, endow yps with a modern dynamism. Striking details such as the front band provide evidence of traditional craftsmanship. The natural, hand-selected wooden surface of the stable triple-layer top underlines the manufacturer's high demand on quality.

Statement by the jury
Formally inspired by the letter Y, the self-contained design of the table is characterised by a clear geometry. A consistent realisation.

Auf ausgestellten Y-förmigen Beinen ruhend, wirkt der Naturholztisch yps trotz seiner Größe elegant und skulptural. Das Y als Grundform der Füße sorgt für eine stabile Standfläche und ermöglicht eine große Beinfreiheit. Durch die gegenläufig schräg angefrästen Kanten der Tischplatte erhält yps eine dynamische Wirkung. Markante Details, wie der stirnseitige Anleimer, zeugen von traditioneller Handwerkskunst. Auch das von Hand zusammengestellte Holzbild der formstabilen Dreischichtplatte unterstreicht einen hohen Qualitätsanspruch.

Begründung der Jury
In formaler Anlehnung an den Buchstaben Y prägt eine klare Geometrie den eigenständigen Charakter des Tisches. Eine stringente Umsetzung.

Stable
Desk, Dining Table
Arbeitstisch, Esstisch

Manufacturer
Era Group, Vinkovci, Croatia
Design
Filip Havranek, Zagreb, Croatia
Kristina Lugonja, Zagreb, Croatia
Web
www.era.com.hr
Honourable Mention

Stable is both a desk and a dining table which challenges classic table design: The design elements of legs and frame are connected at the lateral table ridge, resulting in one unit that eliminates the conventional box-type frame structure. In addition, the wood texture of the individual components underlines the table's distinctive character. This design lends the table a high degree of stability and increases its practical use. Furthermore, the table can be disassembled into three parts and reassembled very quickly and intuitively.

Statement by the jury
This desk and dining table convinces with an accurate and consistent realisation of its appealing design concept.

Stable ist ein Arbeits- und Esstisch, der die klassische Tischkonstruktion infrage stellt: Die seitliche Tischkante verbindet die Konstruktionselemente Tischbeine und Rahmen zu einer Einheit, wodurch die klassische Truhenkonstruktion mittels Zargen wegfällt. Zudem unterstreicht die Holzmaserung der einzelnen Bestandteile deren Besonderheit. Diese Bauweise verleiht dem Tisch ein hohes Maß an Stabilität und Zweckmäßigkeit. Zudem lässt sich der Tisch in drei Teile zerlegen und kann schnell und intuitiv aufgebaut werden.

Begründung der Jury
Bei diesem Arbeits- und Esstisch überzeugt die detailgenaue Umsetzung einer ansprechenden Gestaltungsidee.

Wings
Table
Tisch

Manufacturer
MMooD, Malle, Belgium
In-house design
Gust Koyen
Web
www.mmood.be
Honourable Mention

Thanks to a smart mechanism, Wings transforms from a rectangular table into a round one in only a few seconds. It features two particularly smooth-running pull-out wings on both its long sides. The table is entirely made of solid oak, with carefully selected wood from sustainably managed forests. The distinctive design language underlines the environmentally-friendly character of this piece of Belgian made furniture.

Statement by the jury
The high functionality of this pull-out table provides users with comfort and flexibility.

Dank eines raffinierten Mechanismus lässt sich Wings in nur wenigen Sekunden von einem rechteckigen in einen runden Tisch verwandeln. An beiden Längsseiten befinden sich herausziehbare Tischklappen, welche sich besonders leichtgängig hochschieben lassen. Der Tisch ist vollständig aus massivem Eichenholz gefertigt, wobei das ausgewählte Holz aus nachhaltig bewirtschafteten Wäldern stammt. Die prägnante Formensprache unterstreicht den umweltfreundlichen Charakter des in Belgien gefertigten Möbels.

Begründung der Jury
Die hohe Funktionalität dieses ausklappbaren Tisches ermöglicht den Nutzern eine komfortable Flexibilität.

Tama
Occasional Table
Beistelltisch

Manufacturer
Walter Knoll AG & Co. KG,
Herrenberg, Germany
Design
EOOS Design GmbH, Vienna, Austria
Web
www.walterknoll.de
www.eoos.com

In its balance of lightness and heaviness, simplicity and complexity, the Tama occasional table conveys a sculptural appeal. Its horizontal and vertical expanses show flowing lines full of dynamism. The table tops are optionally made of either exquisite marble or rare onyx marble, with textures that are truly eye-catching. The table with its two levels offers plenty of storage space and defines its environment by being a highly distinctive centrepiece.

Statement by the jury
Tama establishes itself at a high level with its sophisticated line design and the immaculate processing of precious materials.

In der Balance von Leichtigkeit und Schwere, Einfachheit und Komplexität vermittelt der Beistelltisch Tama eine skulpturale Anmutung. Seine horizontalen und vertikalen Flächen zeigen eine fließende Linienführung voller Dynamik. Die Tischplatten sind optional aus edlem Marmor oder seltenem Onyx-Marmor gefertigt und erzielen aufgrund ihrer Maserung eine hohe Aufmerksamkeit. Der in zwei Ebenen angeordnete Beistelltisch bietet eine großzügige Abstellfläche und prägt seine räumliche Umgebung als markantes Prunkstück.

Begründung der Jury
Auf hohem Niveau zeichnet sich Tama durch eine raffinierte Linienführung und die makellose Verarbeitung kostbarer Materialien aus.

Halo
Lounge Chair
Lounge-Sessel

Manufacturer
Hypetex™, London, Great Britain
Design
Michael Sodeau Studio
(Michael Sodeau),
London, Great Britain
Web
www.hypetex.com
www.michaelsodeau.com

Halo has been designed to emphasise the aesthetic and technological potential of Hypetex, an innovative coloured carbon fibre. The technically complex, yet refined, chair follows a modernist dictate of "form follows function" with the removal of all unnecessary clutter. This is juxtaposed with the decorative aesthetic of the material, which was developed by engineers from Formula One racing. The contoured shape of the chair interacts harmoniously with light creating vibrant refractions on the surface.

Statement by the jury
The innovative processing of the coloured composite lends the Halo lounge chair a particularly distinctive aesthetic.

Halo wurde entworfen, um das ästhetische und technologische Potenzial von Hypetex, einer innovativen farbigen Kohlefaser, zu betonen. Die Gestaltung des technisch komplexen und dennoch edlen Stuhls folgt dem modernistischen Ansatz „Form folgt Funktion" und verzichtet auf unnötige Details. Dies steht der dekorativen Ästhetik des Materials, das von Ingenieuren im Umfeld der Formel Eins entwickelt wurde, gegenüber. Die klar umrissene Kontur des Stuhls interagiert harmonisch mit den dynamischen Lichtbrechungen auf der Oberfläche.

Begründung der Jury
Die innovative Verarbeitung des farbigen Verbundwerkstoffs verleiht dem Lounge-Sessel Halo eine besonders charakteristische Ästhetik.

OMO Modern Chair and Ottoman 7
OMO Modern Sessel und Ottomane 7

Manufacturer
OMO Modern,
San Leandro, California, USA
In-house design
Jae-Min Kim, Youn Jae Lee
Web
www.omomodern.com

This comfortable memory foam furniture combination of an armchair and an ottoman is manufactured using an innovative method: instead of a heavy wood or steel frame, a solid foam core provides stability. The gentle contours are achieved by filling polyurethane foam into a curved mould. The snugness of the material stimulates the relaxation of the entire body. Both chair and ottoman are upholstered in a fabric that snugly fits around the elastic foam material.

Statement by the jury
This organic armchair together with its ottoman blend an aesthetic appeal with sophisticated seating comfort.

Diese bequeme Sitzmöbelkombination aus Memory-Schaum, bestehend aus Sessel und Ottomane, wird mittels einer innovativen Methode gefertigt: Anstelle von schweren Holz- oder Metallrahmen sorgt ein fester Schaumstoffkern für Stabilität. Die weichen Konturen entstehen dadurch, dass der Polyurethan-Schaum in eine abgerundete Schalung gegossen wird. Die Anschmiegsamkeit des Materials fördert die Entspannung des ganzen Körpers. Sowohl Sessel als auch Ottomane sind mit einem Stoff bezogen, der sich dem elastischen Schaumstoff flexibel anpasst.

Begründung der Jury
Dieser organisch anmutende Sessel mitsamt seinem Ottomanen vereint ein ästhetisches Gesamtbild mit einem gehobenen Sitzkomfort.

Epica
Chair
Stuhl

Manufacturer
Gaber S.r.l.,
Caselle di Altivole (Treviso), Italy
Design
Novus S.r.l. – Marc Sadler,
Milan, Italy
Web
www.gaber.it
www.marcsadler.it

The Epica chair is available either with a fixed or with an adjustable seat. The version with the flexible seat is characterised by the integration of an innovative rotation system that allows the seat to swivel 360 degrees. The rotation of the seat is independent of the backrest and the seat can be turned back to its original position at any time. The chair is made of technopolymer and optionally available in an Eco version that is made of wood composite.

Statement by the jury
Clear lines lend Epica its timeless appeal. The rotatable seat in particular ensures ergonomic and comfortable use.

Der Stuhl Epica ist optional mit einer festen oder verstellbaren Sitzfläche erhältlich. Die flexible Sitzflächen-Version zeichnet sich durch die Integration eines innovativen Rotationssystems aus, wodurch der Sitz um 360 Grad drehbar ist. Die Drehung des Sitzes erfolgt unabhängig von der Rückenlehne, der Sitz lässt sich jederzeit zum Ausgangspunkt zurückstellen. Gefertigt wird der Stuhl entweder aus Polymer oder in seiner Eco-Ausführung aus einem Holzverbund-Werkstoff.

Begründung der Jury
Eine klare Linienführung verleiht Epica seine zeitlose Anmutung. Vor allem seine verstellbare Sitzfläche bietet eine ergonomisch komfortable Nutzung.

Marguerite
Chair
Stuhl

Manufacturer
Joli,
Kuurne, Belgium
Design
Mathias De Ferm,
Merksem, Belgium
Web
www.joli.be
www.mathiasdeferm.be

The design of Marguerite combines an organically shaped bucket seat with four elegant, tapered aluminium legs. The remarkably light chairs are available with or without armrests, and in an extensive colour palette. The chair offers a high degree of seating comfort, and is both suitable for indoor and outdoor use. For additional comfort, it comes with complementary upholstery pads in leather or leatherette in 13 and 8 colours, respectively.

Statement by the jury
Its well-balanced proportions and the diverse optional features characterise Marguerite as a flexible and convertible product solution.

Die Gestaltung von Marguerite kombiniert eine organisch anmutende Sitzschale mit vier eleganten, konischen Aluminium- beinen. Der auffallend leichte Schalenstuhl ist wahlweise mit oder ohne Armlehne sowie in einer breiten Farbpalette erhältlich. Der Stuhl bietet ein hohes Maß an Sitzkomfort und eignet sich gleichermaßen für den Einsatz drinnen und draußen. Um den Komfort zu steigern, bietet sich die Ergänzung mit Sitzschalen- Einlagen aus Leder in 13 Farben oder Torres-Kunstleder in acht Farben an.

Begründung der Jury
Seine ausgewogenen Proportionen und die variantenreichen Ausstattungsmöglichkeiten zeichnen Marguerite als flexibel wandelbare Produktlösung aus.

Fawn
Chair
Stuhl

Manufacturer
Gazzda,
Sarajevo, Bosnia-Herzegovina
In-house design
Mustafa Cohadzic, Salih Teskeredzic
Web
www.gazzda.com

Tactility, lightness and transparency are the central design themes that characterise the Fawn chair, as well as a minimal number of materials and harmonious proportions. Even though the chair is designed for mass production, it conveys the quality of real handcrafted carpentry. Manufactured from solid oak, the chair weighs only 3.7 kg. Its solid design does not need any screws, bolts or other metal parts at all. The chair comes with various seating colours, matching the characteristic wooden framework.

Statement by the jury
The clear lines and trendy colours of the Fawn chair are an expression of its sophisticated design – a consistent creation down to the last detail.

Haptik, Leichtigkeit und Transparenz prägen als zentrale Gestal- tungsthemen den Stuhl Fawn ebenso wie ein minimaler Materi- aleinsatz und harmonische Proportionen. Obwohl er für die Serienproduktion ausgelegt ist, vermittelt er die Qualität echten Tischlerhandwerks. Gefertigt aus massivem Eichenholz, wiegt der Stuhl nur 3,7 kg. Seine stabile Konstruktion kommt ohne die Verwendung von Schrauben, Bolzen oder anderem Metall aus. Fawn ist in verschiedenen Sitzfarben, passend zum charakteristi- schen Holzgestell erhältlich.

Begründung der Jury
Klare Linien und trendige Farben sind beim Stuhl Fawn Ausdruck eines durchdachten Gestaltungskonzepts – eine bis ins Detail stimmige Kreation.

Wing
Chair
Stuhl

Manufacturer
Actiu Berbegal y Formas, S.A.,
Castalla, Spain
Design
Ramos & Bassols, Barcelona, Spain
Web
www.actiu.com
www.ramos-bassols.com

Adapted to more efficient production processes, Wing displays technological evolution in a traditional wooden chair. The robust chair features well-proportioned geometrical shapes and is available in different heights. The curved backrest elegantly wraps around the traditional seat. With its vintage style design, the chair harmoniously blends into a multitude of interior spaces. It is suitable for both indoor and outdoor use, easy to transport and stack.

Statement by the jury
As a successful reinterpretation of a classic wooden chair, Wing offers convincing functionality paired with elegant design.

Wing ist eine an effizientere Produktionsprozesse angepasste Weiterentwicklung eines traditionellen Holzstuhls. Das robuste Sitzmöbel zeigt ausgewogene geometrische Formen und ist in unterschiedlichen Höhen lieferbar. Elegant umschließt die ge- schwungene Rückenlehne den traditionell anmutenden Sitz. Der im Vintage-Stil gestaltete Stuhl fügt sich harmonisch in unter- schiedliche Interieurs ein. Er eignet sich sowohl für den Einsatz drinnen als auch draußen, ist leicht zu transportieren und zu stapeln.

Begründung der Jury
Als gelungene Neuinterpretation eines klassischen Holzstuhls bietet Wing eine überzeugende Funktionalität, gepaart mit einer eleganten Linienführung.

Trea
Chair
Stuhl

Manufacturer
Humanscale,
Piscataway, New Jersey, USA
In-house design
Web
www.humanscale.com

You can also find this product in
Dieses Produkt finden Sie auch in
Working
Page 82
Seite 82

With a reclining mechanism of up to 12 degrees, the backrest of Trea encourages an active sitting posture. The flexible adjustment to the body's natural movement reduces pressure on the intervertebral discs and offers intuitive seating comfort without the need for extra adjustments via knobs and levers. The chair is available as a cantilever as well as with a four-star or four-legged base. Furthermore, it comes with a variety of upholstery and finish options, whilst its modular form allows for easy upgrades.

Statement by the jury
Compact in its design, the Trea multipurpose chair convinces with its formal balance and functional features.

Aufgrund einer Neigungsfunktion von bis zu 12 Grad fördert die Rückenschale von Trea eine aktive Sitzhaltung. Die flexible Anpassung an die natürliche Körperbewegung reduziert den Druck auf die Bandscheiben und bietet intuitiven Sitzkomfort ohne eine zusätzliche Betätigung von Knöpfen und Hebeln. Der Stuhl ist optional mit einem Freischwinger-Gestell, einem vierarmigen Fußkreuz oder einem Vierfußgestell erhältlich. Außerdem stehen eine Vielzahl an Polsterungen und Bezügen zur Auswahl, wobei die modulare Bauweise eine einfache Aufrüstung ermöglicht.

Begründung der Jury
Der kompakt konstruierte Mehrzweckstuhl Trea überzeugt mittels seiner formalen Ausgewogenheit und seiner funktionalen Eigenschaften.

Circle
Folding Chair
Klappstuhl

Manufacturer
Takumi Kohgei Co., Ltd.,
Kamikawa District, Hokkaido, Japan
Design
Konrad Lohöfener,
Aschau im Chiemgau, Germany
Web
www.takumikohgei.com
www.konradlohoefener.de
Honourable Mention

This folding chair was inspired by the minimalist aesthetics of a cable suspension bridge. Its design follows the idea of using the weight of the person sitting on the chair to transform the force into tension. The four legs and the seat cross each other on one axis, and are held in position by only one rope. In the folded position, the rope is used for space-saving hanging on the wall. Circle is made in Japan under the proposition of resource-saving production.

Statement by the jury
The Circle folding chair is characterised by its consistent and well thought-through functionality.

Dieser Klappstuhl wurde von der minimalistischen Ästhetik seilverspannter Hängebrücken inspiriert. Seine Gestaltung folgt der Idee, die einwirkende Gewichtskraft des Sitzenden konstruktiv zu nutzen und in Zugspannung umzuleiten. Die vier Beine und die Sitzfläche kreuzen einander auf einer Achse und werden allein durch ein Seil in ihrer Position gehalten. Im zusammengeklappten Zustand dient das Seil zur platzsparenden Aufhängung an der Wand. Circle wird in Japan unter der Prämisse einer ressourcensparenden Produktion gefertigt.

Begründung der Jury
Der Klappstuhl Circle zeichnet sich durch seine konsequent durchdachte Funktionalität aus.

Indigo Dye Stool
Indigo Dye Hocker

Manufacturer
2in, Caotun Township,
Nantou County, Taiwan
In-house design
Yi Shiang Lin, Chun Wei Chen
Web
www.yslinna.tw
http://yslinna.myweb.hinet.net

The artful dyeing of fabrics inspired the idea of a novel furniture coating with a natural appeal: the indigo dyed stool possesses all fundamental components of a prototypical stool and yet is highly original thanks to its colour gradients. The seat, consisting of two semicircles, is blue, while the four legs feature a colour gradient from blue to white which lightens toward the bottom. Also, the stabilising connecting chains pick up this colour concept.

Statement by the jury
Unconventional colour gradients provide this stool with a visually appealing effect. A stunningly effective design idea.

In Anlehnung an das kunstvolle Einfärben von Textilien entstand die Idee einer neuartigen und natürlich anmutenden Beschichtung von Möbeln: Der in Indigo gefärbte Hocker besitzt alle grundlegenden Komponenten eines prototypischen Hockers und ist dennoch aufgrund seiner Farbverläufe besonders originell. Seine Sitzfläche aus zwei Halbkreisen ist blau, während die vier Hockerbeine einen nach unten heller werdenden Farbverlauf von Blau nach Weiß zeigen. Auch die stabilisierenden Verbindungsketten greifen dieses Farbkonzept auf.

Begründung der Jury
Unkonventionelle Farbverläufe sorgen bei diesem Hocker für einen visuell reizvollen Effekt. Eine verblüffend wirkungsvolle Gestaltungsidee.

Caruzzo
Swivel Chair
Drehsessel

Manufacturer
Leolux, Venlo, Netherlands
Design
Studio Schrofer (Frans Schrofer),
The Hague, Netherlands
Web
www.leolux.com
www.studioschrofer.com

Caruzzo is a reinterpretation of the classic wing chair and was designed according to ergonomic criteria. With its oval curved backrest and the flowing transition to the headrest, the swivel armchair invites users to relax. Details such as the striking embroidered seam on the back of the seat are made by hand. The seating is adjustable by a mechanism. The armchair is available in two seat heights, with a variety of covers and comes with either a lacquered foot or polished aluminium finish.

Caruzzo ist eine Neuinterpretation des klassischen Ohrensessels und wurde nach ergonomischen Kriterien gestaltet. Mit seiner oval geschwungenen Rückenlehne und dem fließenden Übergang zum Kopfteil lädt der Drehsessel zum Entspannen ein. Details wie die auffällige Ziernaht an der Rückseite der Schale sind von Hand gefertigt. Mittels einer Mechanik lässt sich der Sitz einstellen. Der Sessel ist in zwei Sitzhöhen, in diversen Bezügen sowie mit einem lackierten Fuß oder einem aus poliertem Aluminium erhältlich.

Statement by the jury
Caruzzo flexibly adjusts to individual body contours and seating habits, and, furthermore, evokes a sense of cosiness.

Begründung der Jury
Flexibel passt sich Caruzzo den individuellen Körperkonturen und Sitzgewohnheiten an und weckt zudem ein Gefühl der Geborgenheit.

Halia
Armchair
Sessel

Manufacturer
Koleksiyon, Istanbul, Turkey
Design
Studio Kairos, Venice, Italy
Web
www.koleksiyon.com.tr
www.koleksiyoninternational.com
www.studiokairos.net

The design of the Halia armchair is inspired by fairy wings, which reflect in the gently curved lines. The comfortable chair is suitable for both contemporary living and work spaces. It can also be customised for the use in public spaces such as hotels, hospitals and educational institutions. Its legs are available in robust epoxy finish. The armchair is available with either a high or a low backrest and comes in a variety of fabrics and leather seating options.

Die Gestaltung des Sessels Halia ließ sich von Feenflügeln inspirieren, was sich in der geschwungenen Linienführung widerspiegelt. Das komfortable Sitzmöbel eignet sich für zeitgemäße Wohn- und Arbeitsräume. Es lässt sich zudem im Objektbereich auf den Einsatz in Hotels, Krankenhäusern und Bildungseinrichtungen anpassen. Seine Beine sind mit einer robusten Epoxidharzbeschichtung versehen. Der Sessel ist je nach Wunsch mit einer hohen und niedrigen Rückenlehne sowie in verschiedenen Stoff- und Lederausführungen erhältlich.

Statement by the jury
A consistent design language characterises this filigree armchair, which, above all, offers a high level of seating comfort.

Begründung der Jury
Eine stimmige Formensprache charakterisiert diesen filigran anmutenden Sessel, der darüber hinaus einen hohen Sitzkomfort bietet.

Strain

Lounge Chair
Sessel

Manufacturer
prostoria Ltd., Sveti Križ Začretje, Croatia
Design
Simon Morasi Pipercic, Zagreb, Croatia
Web
www.prostoria.eu
www.simonmp.com

Despite a design that communicates lightness, the Strain lounge chair provides comfort otherwise only felt in a large dimension sofa. Thanks to its seat, which is attached by elastic bands, and the lumbar cushion, different sitting positions are possible. The fabric cover of the backrest further enhances comfort. The chair is available in various material combinations – from rough fabrics and leather to transparent mesh fabrics – and easily adapts to any interior or exterior environment.

Trotz seiner Leichtigkeit vermittelnden Konstruktion bietet der Sessel Strain einen Komfort, der ansonsten nur von großformatigen Sitzgarnituren erwartet wird. Dank seiner Sitzfläche, welche mit elastischen Bändern befestigt ist, und einem Lumbalkissen sind individuelle Sitzpositionen möglich. Die Stoffspannung der Rückenlehne verstärkt den Komfort. Der Sessel ist für verschiedene Stoffkombinationen – von groben Textilien über Leder bis zu transparenten Netzstoffen – geeignet, wodurch er sich an jeden Ausstattungsstil anpasst.

Statement by the jury
The modular approach of this lounge chair is appealing not only in its high seating comfort, but also allows for attractive material combinations.

Begründung der Jury
Der modulare Aufbau dieses Sitzmöbels überzeugt nicht nur durch einen hohen Sitzkomfort, sondern ermöglicht zudem reizvolle Materialkombinationen.

Manufacturer
Voglauer Möbelwerk Gschwandtner &
Zwilling GmbH & Co KG,
Abtenau, Austria
Design
Design Ballendat, Simbach am Inn, Germany
Web
www.voglauer.com
www.ballendat.de
Honourable Mention

The V-Solid bench was designed as a distinctive standalone piece of furniture. A straight-forward, protruding cushioned section in a drawn-out Z-shape invites users to sit down in comfort. Arms can casually rest on the wide top. Seemingly floating on air, the pads of the seat rest on a four-legged base, the stabilising frame of which can also be used as a shelf for magazines. Handcrafted wooden joints emphasise the natural solid-wood character of the oak frame.

Die Sitzbank V-Solid wurde als souverän wirkendes Solitärmöbel konzipiert. Ein geradliniger, ausladender Polsterkörper in einer langgezogenen Z-Form lädt zum entspannten Sitzen ein. Lässig können die Arme auf das breite Kopfteil gelegt werden. Das Sitzpolster wird luftig von einem Vierfußgestell getragen, dessen stabilisierender Zargen-Rahmen auch als Ablage für Zeitschriften genutzt werden kann. Handwerkliche Holzverbindungen betonen den natürlichen Massivholzcharakter der Konstruktion aus Eiche Natur.

Statement by the jury
Austere geometric structures in combination with high-contrast materials lend this bench its distinctive aesthetic appeal.

Begründung der Jury
Streng geometrische Strukturen in Verbindung mit kontrastreichen Materialien verleihen dieser Sitzbank ihre markante Ästhetik.

Move

Sectional Sofa
Anbausofa

Manufacturer
Paola Lenti S.r.l., Meda, Italy
Design
Francesco Rota, Milan, Italy
Web
www.paolalenti.it
www.francescorota.com

The Move sectional sofa is suitable for both residential spaces and hospitality areas. Its elements are available in four different seat depths and two heights, allowing for customised combinations. The sofa is complemented by armrests that can be freely positioned on the sofa or serve as back support cushions. The removable cover comes from an exclusive fabric collection while the upholstery is made of stable polyurethane foam.

Das Anbausofa Move eignet sich sowohl für den privaten Wohnbereich als auch für den Objektbereich. Seine Anbauelemente sind in vier unterschiedlichen Sitztiefen und zwei Höhen erhältlich und ermöglichen somit individuelle Kompositionen. Das Sofa lässt sich mit Armlehnen ergänzen, die frei positioniert und zudem als Rückenstützkissen verwendet werden können. Der abnehmbare Polsterbezug stammt aus einer exklusiven Stoffkollektion, während die Polsterung aus formstabilem Polyurethan-Schaum besteht.

Duas
Stove
Kaminofen

Manufacturer
Ganz Baukeramik AG,
Embrach, Switzerland
In-house design
Web
www.ganz.info

With its particularly large surface of angled glass, the Duas wood-burning stove offers a fascinating view of the flames. The innovative firing technology guarantees not only low emission levels and good energy efficiency, but also clearly demonstrates the excellent quality of this stove. The additional thermal storage system is appealing with its combination of quick heat dissipation into and long-lasting heat storage in the room, which ensures a healthy and particularly comfortable indoor climate.

Statement by the jury
This wood-burning stove combines formal independence with high functionality – an energy efficient solution.

Mittels eines großflächigen Glasfensters über Eck ermöglicht Duas eine faszinierende Sicht auf das Kaminfeuer. Eine innovative Feuerungstechnik erreicht nicht nur niedrige Emissionswerte und eine gute Energieeffizienz, sondern stellt zudem die gehobene Qualität des Kaminofens unter Beweis. Dessen zusätzlicher Wärmespeicher überzeugt durch die Kombination von schneller Wärmeabgabe und langanhaltender Speicherwärme. Was für ein gesundes und besonders behagliches Raumklima sorgt.

Begründung der Jury
Auf ansprechende Weise vereint dieser Kaminofen formale Eigenständigkeit mit einer hohen Funktionalität – eine energieeffiziente Produktlösung.

Jøtul F 305
Stove
Kaminofen

Manufacturer
Jøtul AS, Fredrikstad, Norway
Design
Anderssen og Voll, Oslo, Norway
Web
www.jotul.com
www.anderssen-voll.com

Jøtul F 305 is a stylish wood stove with a large glass door that covers 70 percent of the front face, giving an impressive view of the fire. Its horizontal combustion chamber makes filling it with logs easy and guarantees quick heat emission into the room. Its integrated heat convection allows the stove to be safely installed even near combustible materials. A flat top panel is currently under development with the objective of making the stove a multifunctional device.

Statement by the jury
The Jøtul F 305 wood stove owes its high functionality to an easy-access combustion chamber and its efficient thermal properties.

Jøtul F 305 ist ein stilvoller Kaminofen, dessen Front zu 70 Prozent aus einer Glasscheibe besteht und somit einen eindrucksvollen Blick auf das Feuer ermöglicht. Sein horizontal ausgerichteter Korpus erleichtert das Einlegen der Holzscheite und gewährleistet eine schnelle Wärmeabgabe an den Raum. Aufgrund seiner integrierten Wärmeströmung lässt sich der Kaminofen unbedenklich in der Nähe von brennbaren Materialien aufstellen. Die flache Topplatte wird mit der Zielsetzung einer multifunktionalen Nutzung weiterentwickelt.

Begründung der Jury
Seine hohe Funktionalität verdankt der Kaminofen Jøtul F 305 einer leicht zugänglichen Brennkammer und seinen effektiven Wärmeeigenschaften.

Ecofire® Audrey
Pellet Stove
Pelletofen

Manufacturer
Palazzetti Lelio S.p.A.,
Porcia (Pordenone), Italy
Design
MARCARCH Design,
Marco Fumagalli Architetto,
Stresa, Italy
Web
www.palazzetti.it
www.marcarch.it

Audrey is a sealed pellet stove independent in operation from room air. Its design language combines a square with a circle, lending the stove a different appeal when viewed from different angles. The stove features a sealed firing chamber for optimised performance, increased comfort throughout the home and reduced consumption. Audrey heats the air by means of a radial fan which can be turned off when required in order to silently heat the room exclusively by natural convection and radiation.

Statement by the jury
The core of this pellet stove conveys a simple elegance in black and white. Its functionality offers a high ease of use.

Audrey ist ein raumluftunabhängiger Pelletofen. Seine Formensprache vereint das Quadrat mit einem Kreis, wodurch der Ofen je nach Betrachtungsperspektive eine andere Anmutung zeigt. Der Pelletofen ist mit einem raumluftunabhängigen Heizkessel ausgestattet, der die Leistung optimiert, den Wohnkomfort steigert und den Verbrauch senkt. Audrey erwärmt die Luft über ein Radialgebläse, welches bei Bedarf abgeschaltet werden kann, um ausschließlich mittels natürlicher Luftkonvektion und Abstrahlung lautlos zu heizen.

Begründung der Jury
Der Korpus dieses Pelletofens vermittelt eine schlichte Eleganz in Schwarz-Weiß. Seine Funktionalität bietet ein hohes Maß an Bedienungskomfort.

MIA
Pellet Stove
Pelletofen

Manufacturer
Olimpia Splendid S.p.A.,
Cellatica (Brescia), Italy
Design
Sara Ferrari Design (Sara Ferrari),
Gussago (Brescia), Italy
Web
www.olimpiasplendid.com
www.saraferraridesign.com

MIA is a pellet stove that is designed as a customisable piece of furniture. It allows for individual configurations in terms of heat capacity and colour. The body is available in four tones: cream, brick orange, military green and silver tech aluminium. The cover also comes in two different versions. In addition, the stove can be complemented by a matching pellet container, a practical scoop and a 40 x 40 cm or an 80 x 40 cm cabinet.

Statement by the jury
The MIA pellet stove combines sophisticated functionality and a self-contained aesthetic that enhances any living room to a high level.

MIA ist ein Pelletofen, der als personalisierbares Möbelstück konzipiert wurde. Entsprechend besteht die Möglichkeit, zwischen unterschiedlichen Leistungsstärken und Farben individuell auszuwählen. Der Korpus ist in den vier Farbtönen Cremeweiß, Ziegelrot, Militärgrün und Graualuminium erhältlich. Auch seine Abdeckung steht in zwei unterschiedlichen Ausführungen zur Auswahl. Der Ofen lässt sich darüber hinaus mit einem passenden Pelletbehälter, einer praktischen Pelletschaufel und einem Schrank in den Maßen 40 x 40 cm oder 80 x 40 cm bedarfsgerecht ergänzen.

Begründung der Jury
Auf hohem Niveau verbindet der Pelletofen MIA eine ausgereifte Funktionalität mit einer eigenständigen, den Wohnraum bereichernden Ästhetik.

Xilobis-System 24
Modular Bookshelf
Modulares Bücherregal

Manufacturer
Xilobis, Zürich, Switzerland
In-house design
Mario Bissegger,
Muralto/Locarno, Switzerland
Web
www.xilobis.ch

This modular bookshelf was designed
for people with an affinity to design,
who are frequently moving and changing their living space. Without the need
for screws and tools, the individual
shelf units can be removed, altered and
expanded in only a few steps. Precisely
manufactured shelf boards made of non-
toxic laminated birch wood in combination with the back wall form a solid
module. These modules are connected by
a system of mortise and ball-shaped
steel tenon joints. A variety of optional
sliding doors of different materials offers
visual diversity.

Statement by the jury
As a sustainable solution, the modular
Xilobis-System 24 bookshelf offers
a high degree of mobility and flexibility
in design.

Das modulare Bücherregal wurde für
eine designaffine Flexibilitätsgesellschaft
entworfen, die häufig ihren Wohnraum
verändert. Ohne Schrauben und Werkzeuge
können die Regalmodule mit wenigen
Handgriffen aus- und umgebaut werden.
Präzise verarbeitete Regalwände aus
giftfrei verleimtem Birkenschichtholz
ergeben in Kombination mit der Rückwand
ein stabiles Modul. Diese Module werden
dann mittels Kugeln und passgenau
gefräster Mulden miteinander verbunden.
Eine Vielfalt optional einsetzbarer Schiebetüren verschiedener Materialien bietet
visuelle Flexibilität.

Begründung der Jury
Als nachhaltige Produktlösung bietet das
modulare Bücherregal Xilobis-System 24
ein hohes Maß an Gestaltungsfreiheit und
Mobilität.

Wind
Bookcase
Bücherregal

Manufacturer
Rimadesio S.p.A., Giussano, Italy
Design
GB Studio S.r.l.,
Interiors and Industrial Design
(Giuseppe Bavuso),
Seregno, Italy
Web
www.rimadesio.com
www.bavuso-design.com

Wind was developed with the objective of designing a new archetype of a bookcase. The modular aluminium bookcase system features extruded shelves and die-cast uprights. The aesthetics of the system are defined both horizontally and vertically by the filigree aluminium uprights, while the nylon fibre joints ensure a high degree of stability. Closing element of each composition is an aluminium laminated lacquered MDF panel.

Wind entstand unter der Prämisse, einen neuen Archetypus von Bücherregal zu entwerfen. Das modulare Regalsystem aus Aluminium umfasst stranggepresste Fachböden sowie per Druckguss gefertigte Pfosten. Sowohl in der Vertikalen als auch der Horizontalen prägen die filigranen Aluminium-Pfosten die Ästhetik. Verbindungsstücke aus Nylonfaser sorgen an den Pfosten für ein hohes Maß an Stabilität. Jede Komposition schließt mit einer aluminiumlaminierten, lackierten MDF-Platte ab.

Statement by the jury
An innovative combination of materials lends this bookcase an elegant lightness and a high degree of durable stability.

Begründung der Jury
Eine innovative Materialkombination verleiht diesem Bücherregal eine elegante Leichtigkeit und ein hohes Maß an langlebiger Stabilität.

Tumble
Cupboard
Schrank

Manufacturer
StudioVosk, Avelgem, Belgium
In-house design
Web
www.studiovosk.com

This minimalist wall cupboard references the clear geometry of a square box and opens with a surprisingly simple mechanism: upon slightly moving the thin wooden lever on top of the box to the front, the door of the cupboard swings to the side on its own, simply by means of gravity. When the door is closed, the lever automatically returns to its original position. Thanks to this mechanism, the Tumble cupboard does not need any screws or hinges.

Statement by the jury
An innovative design idea, realised in a highly consistent way, makes opening the door of the cupboard comfortably easy.

Der minimalistisch anmutende Hänge-schrank Tumble zitiert die klare Geometrie einer quadratischen Box und lässt sich mit einer überraschend einfachen Mechanik öffnen: Durch das leichte Verschieben des oben sichtbaren Holzsplints nach vorne, schwingt die Schranktür allein aufgrund der Schwerkraft zur Seite. Beim Schließen der Tür verschiebt sich der Splint automatisch wieder in die Ausgangsposition. Aufgrund dieser Mechanik kann auf die Verwendung von Schrauben oder Scharnieren verzichtet werden.

Begründung der Jury
Eine innovative Gestaltungsidee, die stringent umgesetzt wurde, ermöglicht die komfortabel einfache Öffnung der Schranktür.

Stockholm
Sideboard

Manufacturer
Punt, Paterna (Valencia), Spain
Design
Mario Ruiz Design, Barcelona, Spain
Web
www.puntmobles.com
www.marioruiz.es

With its high back and side panels, the Stockholm sideboard showcases an unconventional design language that harmoniously blends into the appearance of its product family. Based on MDF panels, this piece of furniture is available with either natural oak or natural walnut veneer. Both veneers come with different finishes in a wide colour palette, from natural-matt to lacquered black. Doors and drawers of the sideboard close smoothly thanks to an absorption closing system.

Statement by the jury
Clear lines and a minimalist front underline the beauty of the materials used in this sideboard, resulting in a formal and highly attractive design.

Mit seinen hohen Rück- und Seitenwänden zeigt das Sideboard Stockholm eine unkonventionelle Formensprache, die sich harmonisch in das Erscheinungsbild seiner Produktfamilie einfügt. Das aus MDF-Platten gefertigte Möbel ist optional mit Eiche- oder Walnussfurnier verkleidet. Die beiden Holzfurniere sind von naturmatt bis schwarz lackiert in einer breiten Farb-palette erhältlich. Die Türen und Schub-laden des Sideboards lassen sich dank eines Absorptions-Schließsystems leichtgängig schließen.

Begründung der Jury
Bei diesem Sideboard betonen klare Linien und eine minimalistische Front die Schönheit der verwendeten Materialien, ein formal reizvoller Entwurf.

Fawn
Bed
Bett

Manufacturer
Gazzda, Sarajevo, Bosnia-Herzegovina
In-house design
Mustafa Cohadzic, Salih Teskeredzic
Web
www.gazzda.com

The Fawn bed showcases a modern language of form which aims at lightness and mobility. In contrast to heavy solid-wood beds, the 160 x 200 cm solid oak bed frame weighs only 28 kg. Fawn embodies a minimum use of materials and is characterised by a filigree frame with distinctive recesses and an airy distance from the floor. The bed is available in different sizes and its ergonomically shaped headboard is customisable to personal taste.

Das Bett Fawn zeigt eine zeitgemäße Formensprache, welche Leichtigkeit und Beweglichkeit anstrebt. Im Gegensatz zu schweren Betten aus Massivholz wiegt das 160 x 200 cm große Bettgestell aus massiver Eiche nur 28 kg. Fawn verkörpert einen minimalen Materialeinsatz und charakterisiert sich durch sein filigranes Gestell mit akzentuierenden Aussparungen und einem luftigen Bodenabstand. Das Bett ist in verschiedenen Abmessungen erhältlich und sein ergonomisch geformtes Kopfteil wird kundengerecht den jeweiligen Wünschen angepasst.

Statement by the jury
An unconventional design, characterised by delicate structures, lends the Fawn bed an elegant appeal.

Begründung der Jury
Eine unkonventionelle Konstruktion, geprägt von zarten Strukturen, verleiht dem Bett Fawn eine elegante Anmutung.

Cleaning devices   Abfallsysteme
Household appliances   Bügelsysteme
Ironing systems   Haushaltsgeräte
Sewing machines   Nähmaschinen
Tumble dryers   Reinigungsgeräte
Vacuum cleaners   Staubsauger
Washing machines   Trockner
Waste systems   Waschmaschinen

Household
Haushalt

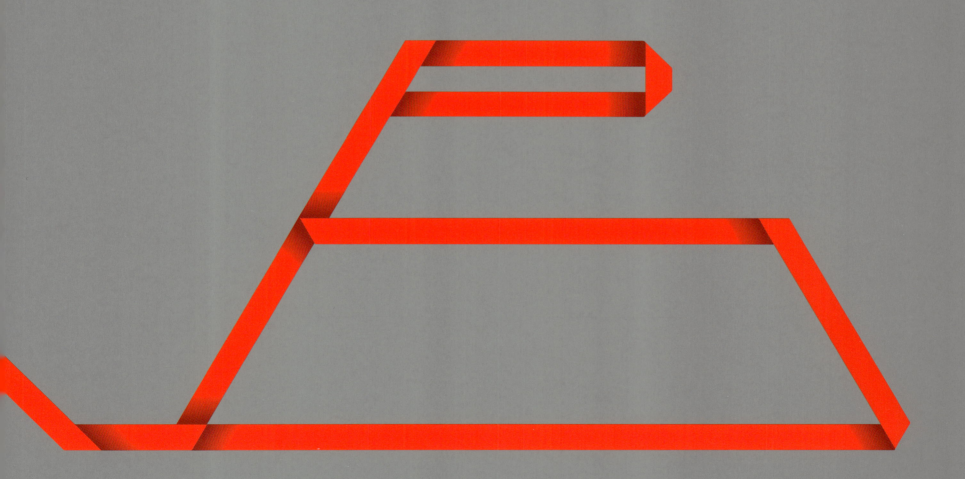

LG Cordless Canister Vacuum Cleaner CordZero C5 (VK9401LHAN, VK9402LHAN, VK9403LHAN)

LG Kabelloser Bodenstaubsauger CordZero C5
(VK9401LHAN, VK9402LHAN, VK9403LHAN)

Manufacturer
LG Electronics Inc.,
Seoul, South Korea

In-house design
Sehwan Bae, Bohyun Nam,
Sojin Park, Seunghyun Song

Web
www.lg.com

reddot award 2015
best of the best

Smart companion

Whereas the very first vacuum cleaners were rather static and bulky devices that demanded a lot of manual labour in handling them, today's models boast high flexibility and manoeuvrability. Against this backdrop, the CordZero C5 impresses with a design concept that reinterprets the principle of mobility of such a device: the vacuum cleaner turns into a sub-serving companion in the living space. Since the CordZero C5 applies complex and sophisticated auto-moving technology, it no longer needs to be dragged or pushed around. Instead the cleaner itself automat-ically follows the user at every step and turn. In add-ition, it features a highly compact design and works wirelessly which prevents it from constantly getting tangled. The vacuum cleaner features an STS blade in the dust canister as well as an automatic dust com-pression function so that the canister can store three times more dust compared to conventional models. It is driven by an integrated smart inverter motor that ensures long durability and high cleaning effective-ness. This is underlined by lending the casing a design that showcases select materials and carefully con-ceived detail. High-quality metallic finishes that are pleasing to the touch on both the wheels and handle provide high comfort for everyday use – vacuum cleaning thus turns into an interactive experience.

Smarter Begleiter

Während die ersten Staubsauger noch statische und überdimensionale Geräte waren, die ein enormes Maß an Körperkraft erforderten, bieten die heutigen Modelle viel Beweglichkeit und Flexibilität. Der CordZero C5 beeindruckt vor diesem Hintergrund mit einem Gestal-tungskonzept, welches das Prinzip der Mobilität eines solchen Gerätes neu interpretiert: Der Staubsauger wird zu einem dienstbaren Begleiter im Wohnraum. Da der CordZero C5 mit einer komplexen und ausgefeilten Bewegungstechnologie ausgestattet ist, muss man ihn nicht mehr selbst ziehen oder schieben, sondern er folgt dem Nutzer automatisch auf Schritt und Tritt. Er ist zudem sehr kompakt gestaltet und arbeitet kabellos, sodass er sich auch nirgendwo auf seinem Weg verhed-dert. Der Bodenstaubsauger verfügt über STS-Flügel in seinem Staubbehälter und eine automatische Staub-verdichtungsfunktion, durch die sein Behälter dreimal mehr Staub aufnehmen kann als herkömmliche Mo-delle. Angetrieben wird er von einem Smart-Inverter-Motor, der dem Nutzer ein hohes Maß an Langlebigkeit und Effektivität beim Saugen bietet. Die Gestaltung des Gehäuses unterstreicht dies durch die Verwendung ausgesuchter Materialien und sorgfältig ausgearbeitete Details. Hochwertige und auch haptisch angenehme Metallic-Beschichtungen auf dem Handgriff und den Rädern sorgen für viel Komfort im Alltag – das Saugen wird zu einem interaktiven Erlebnis.

Statement by the jury
The CordZero C5 delivers users an exciting new ex-perience of vacuum cleaner handling. Equipped with an impressive combination of innovative wireless and auto-moving cleaning technology, this device auto-matically and effortlessly follows the user through the entire house. The carefully crafted details of this overall high-quality designed vacuum cleaner lend it a high degree of durability and aesthetic appeal.

Begründung der Jury
Der CordZero C5 eröffnet dem Nutzer ein spannendes neues Erleben des Gebrauchs eines Bodenstaubsaugers. Ausgestattet mit einer beeindruckenden Kombination aus innovativen Bewegungs- und kabellosen Reini-gungstechnologien, folgt einem dieses Gerät mühelos durch das ganze Haus, ohne dass man es bemerkt. Die sorgfältig ausgeführten Details dieses insgesamt sehr hochwertig gestalteten Staubsaugers verleihen ihm ein hohes Maß an Ästhetik und Langlebigkeit.

Designer portrait
See page 30
Siehe Seite 30

BGS20 Range
Bagless Vacuum Cleaner
Beutelloser Staubsauger

Manufacturer
Robert Bosch Hausgeräte GmbH,
Munich, Germany
In-house design
Daniel Dockner, Helmut Kaiser
Design
Brandis Industrial Design,
Nuremberg, Germany
Web
www.bosch-hausgeraete.de
www.bf-design.de

Central design features of BGS20 Range
are its dynamic lines and striking circular
element. Power is adjusted using an
integrated regulator positioned on top
of the appliance where there is also a
robust handle, which is designed to make
it easy and comfortable to carry. The
dust container is simple to remove and
uncomplicated to empty, clean and then
put back.

Gestalterisches Hauptmerkmal des
BGS20 Range ist neben einer schwungvol-
len Linienführung das markante Kreiselement. Die Einstellung der Leistungsstärke
erfolgt am Regler, der in die Oberseite
des Gerätes integriert ist. Dort ermöglicht
außerdem ein stabiler Griff einen einfachen
und bequemen Transport. Nach dem
Entfernen des zugehörigen Schmutzbe-
hälters kann dieser entleert, gereinigt und
schließlich wieder eingesetzt werden.

Statement by the jury
Its defining characteristic in terms of
outward appearance is the distinctive
circular element of the body, the appliance is remarkable for its great ease of
use.

Begründung der Jury
Das prägnante Kreiselement des Korpus
bestimmt wesentlich das Erscheinungsbild
des BGS20 Range, der durch hohen Bedienkomfort beeindruckt.

BGL80 Range
Vacuum Cleaner
Staubsauger

Manufacturer
Robert Bosch Hausgeräte GmbH,
Munich, Germany
In-house design
Helmut Kaiser, Jörg Schröter
Design
Brandis Industrial Design,
Nuremberg, Germany
Web
www.bosch-hausgeraete.de
www.bf-design.de

The simple BGL80 Range glides across the floor on four soft caster wheels and its slim body, which narrows at the front, follows the user faithfully. During use, the soft protective bumper around the appliance protects any obstacles such as walls and furniture and itself. BGL80's closed shape and optimally designed exhaust areas reduce noise while vacuuming to a minimum.

Der schlicht gehaltene BGL80 Range gleitet auf vier weichen Rollen über den Boden und hat dank eines schmal zulaufenden Korpus einen guten Nachlauf vorzuweisen. Während des Saugens schont ein Rundumschutz mögliche Hindernisse wie Wände oder Möbel und schließlich auch das Gerät selbst. Die geschlossene Form und die optimierten Ausblasbereiche sorgen beim Einsatz des BGL80 Range für einen niedrigen Geräuschpegel.

Statement by the jury
The well-thought-out functionality of this vacuum cleaner makes floor care much easier. Its dynamic use of form is another attractive characteristic.

Begründung der Jury
Mit seiner durchdachten Funktionalität erleichtert der Staubsauger die Bodenpflege. Zudem überzeugt er mit einer dynamischen Formensprache.

LG CordZero C3 (VC7401LHAN)
Cordless Canister Vacuum Cleaner
Kabelloser Bodenstaubsauger

Manufacturer
LG Electronics Inc., Seoul, South Korea
In-house design
Sehwan Bae, Bohyun Nam,
Sojin Park, Seunghyun Song
Web
www.lg.com

The easy handling of this cordless vacuum cleaner makes it a pleasant household tool. With its great suction power and integrated turbo cyclone system, it removes even the finest dust particles. Charging is wireless via an easy to see battery power display, designed to increase safety and usability. The appliance's compact size, its high-gloss finish and metallic-coated handle enhance its elegant appeal.

Statement by the jury
The LG CordZero C3 is appealing thanks to its extremely compact and consistent design, which stands in skilful contrast to the metal control elements.

Der kabellose Bodenstaubsauger wird durch eine einfache Handhabung zum angenehmen Helfer im Haushalt und entfernt mit seiner hohen Saugkraft und integriertem Turbo-Zyklonensystem auch besonders feinen Staub. Die Aufladung erfolgt kabellos, kontrolliert von einer gut erkennbaren Anzeige des Akkus, was die Sicherheit und den Komfort in der Bedienung erhöht. In der Form kompakt, tragen eine hochglänzende Oberfläche und ein Griff mit Metallic-Beschichtung zur Eleganz des Gerätes bei.

Begründung der Jury
Der LG CordZero C3 besticht mit einer äußerst kompakten und homogenen Gestaltung, die durch metallische Bedienelemente gekonnt kontrastiert wird.

UltraFlex
Vacuum Cleaner
Staubsauger

Manufacturer
Electrolux, Stockholm, Sweden
In-house design
Web
www.electrolux.com

140 km/hour is the spinning speed with which UltraFlex separates dust and dirt from air and forces them to the walls of the container. The flaps inside the hose and the insulated motor guarantee a silent bagless cleaning experience. The shape of UltraFlex reflects its powerful cleaning performance, high degree of user-friendliness and a good filtration. Easily manoeuvrable soft wheels guide the appliance around obstacles and corners and the compact design takes up reduced storage space.

Statement by the jury
The UltraFlex bagless vacuum cleaner uses strong design focused on technology to express its great cleaning ability.

Mit einer Geschwindigkeit von 140 km/h trennt UltraFlex Staub und Schmutz von der eingesogenen Luft und drückt ihn an die Wand des Behälters. Die Innenklappen des Schlauches sowie ein isolierter Motor gewährleisten einen nahezu geräuschlosen Betrieb. Die Formgebung des UltraFlex strahlt starke Reinigungsleistung, hohe Benutzerfreundlichkeit, sowie eine gute Filtration aus. Leichtgängige Softräder manövrieren das Gerät gut um Kanten und Ecken, das kompakte Design erlaubt ein platzsparendes Verstauen.

Begründung der Jury
Mit einer kraftvollen, betont technischen Gestaltung kommuniziert der beutellose Staubsauger UltraFlex seine starke Reinigungsleistung.

PowerPro Duo
Vacuum Cleaner
Staubsauger

Manufacturer
Royal Philips, Eindhoven, Netherlands
In-house design
Web
www.philips.com

Thanks to its linear design PowerPro Duo is quick and manoeuvrable when cleaning both hard and soft floors. The handheld unit can be removed, thus making it suitable for use on other surfaces. The TriActive Turbo nozzle provides maximum suction power, thereby boosting the cleaning result both on carpets and hard surfaces. The balanced interplay of technologies lends this battery-driven stick vacuum cleaner an honest feeling of technology and function.

Statement by the jury
The PowerPro Duo stands out not only for its sleek, elegant silhouette but also for its great versatility.

Der PowerPro Duo ist dank seiner schlanken Gestalt schnell und wendig und reinigt harte wie weiche Böden gleichermaßen. Die Handeinheit kann abgenommen werden, sodass er auch auf anderen Oberflächen einsetzbar ist. Die TriActive-Turbodüse unterstützt eine hohe Saugleistung und verbessert so das Reinigungsergebnis auch auf Teppichen oder Hartböden. Das ausgewogene Zusammenspiel seiner Technologien verleiht dem Akku-Stielstaubsauger einen ehrlichen Ausdruck.

Begründung der Jury
Der PowerPro Duo überrascht nicht nur mit seiner schlanken, eleganten Silhouette, sondern auch mit großer Vielseitigkeit.

PowerPro Uno
Vacuum Cleaner
Staubsauger

Manufacturer
Royal Philips, Eindhoven, Netherlands
In-house design
Web
www.philips.com

This sophisticated vacuum cleaner operates using the so-called PowerCyclone technology, which gives it lasting suction power for maximum performance. Thanks to its flexibility, PowerPro Uno is quick and manoeuvrable on hard and soft floors as well as on uneven surfaces. The easy-to-use appliance is bagless and equipped with filters that can be removed after use for washing.

Statement by the jury
This elegant, battery-powered vacuum cleaner is particularly user-friendly thanks to its simple emptying system.

Der mondäne Staubsauger arbeitet basierend auf der sogenannten PowerCyclone-Technologie, mit der eine anhaltende Saugkraft für hohe Leistungsfähigkeit einhergeht. Ob auf hartem oder weichem Untergrund – der PowerPro Uno bewegt sich flink und kommt dank hoher Flexibilität auch auf unebeneren Flächen zurecht. Das handliche Gerät benötigt keine Beutel und besitzt ferner Filter, die nach der Anwendung entfernt und ausgewaschen werden können.

Begründung der Jury
Sein einfaches Entleerungssystem macht diesen eleganten Akku-Staubsauger besonders benutzerfreundlich.

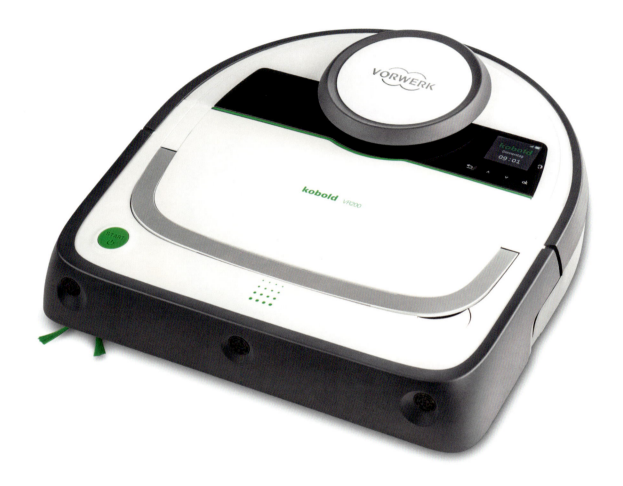

Kobold VR200
Robot Vacuum Cleaner
Saugroboter

Manufacturer
Vorwerk Elektrowerke GmbH & Co. KG,
Wuppertal, Germany
Design
Phoenix Design GmbH & Co. KG,
Stuttgart, Germany
Uwe Kemker, Vorwerk Design,
Wuppertal, Germany
Web
www.vorwerk-kobold.de
www.phoenixdesign.com

The appliance gets its uniform shape from its precise contours, which link the various elements. Colour coding makes operation intuitive. The robot finds its way around with a laser scanner and cleans rooms with rotating brushes. The low-noise, maintenance-free motor ensures that Kobold VR200 is an efficient housekeeping tool.

Seine ausgewogene Formgebung erhält das Reinigungsgerät durch eine prägnante Konturlinie, die die einzelnen Elemente miteinander verbindet. Da diese sich zudem farblich voneinander abheben, erfolgt die Bedienung intuitiv. Per Laserscanner findet der Roboter sich im Raum zurecht und reinigt diesen mit seinen rotierenden Bürsten. Ein wartungsfreier Motor macht den Kobold VR200 zu einem geräuscharmen Helfer im Haushalt.

Statement by the jury
With a harmonious, compact body design this vacuum robot conveys a sense of calm, silent efficiency.

Begründung der Jury
Durch eine harmonische, geschlossene Korpusgestaltung vermittelt dieser Staubsaugerroboter den Eindruck unaufgeregter, lautloser Effizienz.

Scout RX1
Robot Vacuum Cleaner
Saugroboter

Manufacturer
Miele & Cie. KG, Gütersloh, Germany
In-house design
Web
www.miele.de

This discreet vacuum robot glides across the floor using a systematic navigation system and easily reaches even the smallest nooks and crannies. The two brushes attached on either side, as well as a removable additional beater bar, keep rooms clean while consuming very little battery power. If required, Scout RX1 can travel back to the charging station independently and, on completion of his cleaning task, wait there for his next mission.

Der dezent gestaltete Saugroboter gleitet mit systematischer Navigation über den Boden und erreicht problemlos auch kleine Ecken und Winkel. Die zwei seitlich angebrachten Bürsten sowie eine zusätzliche entnehmbare Bürstenwalze halten die Räumlichkeiten bei geringer Akkubelastung zeitsparend sauber. Bei Bedarf fährt der Scout RX1 selbständig in die Ladestation zurück, wo er nach Abschluss der kompletten Reinigung auf den nächsten Einsatz wartet.

Statement by the jury
This vacuum robot skilfully combines complex functionality with an elegant, minimalist body design.

Begründung der Jury
Dieser Saugroboter verbindet eine komplexe Funktionalität gekonnt mit einer minimalistisch-eleganten Gehäusegestaltung.

winbot9s
Window Cleaning Robot
Fensterreinigungsroboter

Manufacturer
Ecovacs Robotics CO., LTD, China
In-house design
Li Xiao Wen, Xie Jing Ya, Liu Jun Shu, Lui Shuai
Web
www.ecovacs.com

A two-part motion system separates the motor and cleaning system of this robot. At the press of a button it begins to clean windows with the help of the pads encircling it and an additional rubber sponge. Within 110 seconds winbot9s is able to thoroughly clean a surface measuring 100 x 60 cm. Its built-in anti-drop sensors ensure it also works on frameless windows.

Mit einem zweiteiligen Bewegungssystem werden Antriebs- und Reinigungssystem des Roboters voneinander getrennt. Auf Knopfdruck beginnt er mittels eines ringsum laufenden Pads und einem zusätzlichen Gummiwischer mit dem Säubern der Fenster. Innerhalb von 110 Sekunden gelingt dem winbot9s so die gründliche Reinigung einer Fläche von 100 x 60 cm, die dank eingebauter Anti-Drop-Sensoren auch bei rahmenlosen Fenstern funktioniert.

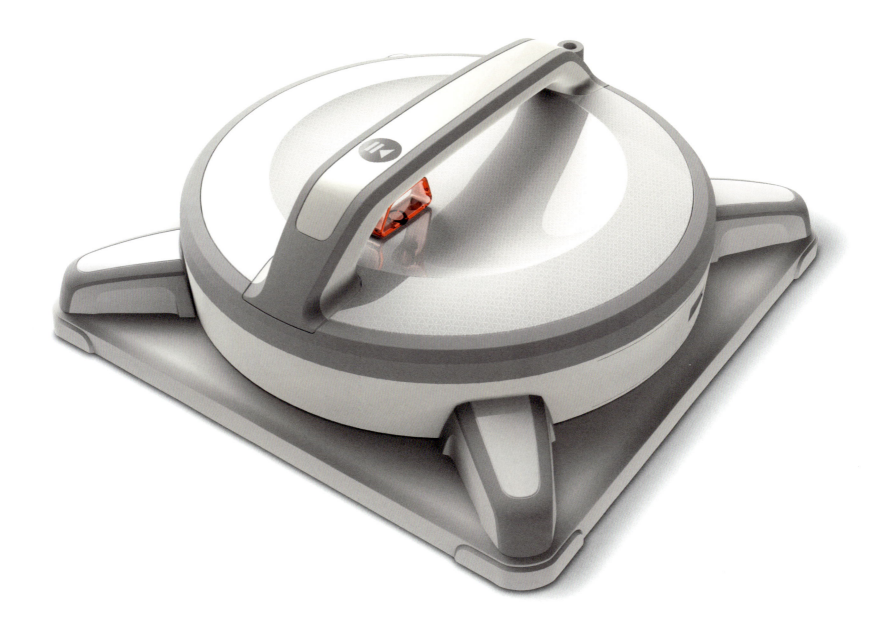

SC 1
Handheld Steam Cleaner
Handdampfreiniger

Manufacturer
Alfred Kärcher GmbH & Co. KG,
Winnenden, Germany
In-house design
Michael Meyer, Christine Blank
Web
www.kaercher.de

The unusually compact shape of the SC 1 enables straightforward handling. This ease of use makes cleaning of even hard-to-reach areas effortless. The steam boiler supplies constant pressure, so that SC 1 is also effective when cleaning floors if used in combination with the matching extension kit. Its great power of up to 3 bar means the steam cleaner can clean every surface. Further accessories make it suitable for additional tasks.

Die außergewöhnlich kompakte Form des SC 1 ermöglicht ein unkompliziertes Arbeiten. Sein leichtes Handling macht selbst das Reinigen von schwer zugänglichen Bereichen zu einer mühelosen Aufgabe. Der Dampfkessel liefert einen konstanten Druck, sodass der SC 1 in Kombination mit einem passenden Verlängerungsset auch zur effektiven Reinigung des Bodens eingesetzt werden kann. Dank einer starken Leistung von bis zu 3 bar säubert der Dampfreiniger jegliche Oberflächen und kann mit weiterem Zubehör ergänzt werden.

Statement by the jury
The ergonomic design of the SC 1 makes using this hand-held steam cleaner very easy and safe.

Begründung der Jury
Die ergonomische Gestaltung des SC 1 macht die Handhabung dieses Handdampfreinigers komfortabel und sicher.

PremiumCare Washer
Washing Machine
Waschmaschine

Manufacturer
Bauknecht, Comerio (Varese), Italy
In-house design
Web
www.bauknecht.de

The highly sophisticated level of technology of this washing machine becomes apparent in the use of side panels that muffle its operating noise. A jet introduces detergent and water directly into the 77-litre drum, which can hold up to 12 kilos of washing. This process avoids excessive use of detergent. The visual effect of the appliance is dominated by the large porthole-formed door in stainless steel and completed by the centrally placed touch user interface in the same colour.

Der Anspruch an die Technologie der Waschmaschine zeigt sich in den sogenannten Side Panels, die das Betriebsgeräusch des Gerätes dämpfen. Bis zu 12 kg Wäsche finden in der 77 Liter fassenden Trommel Platz und werden dort direkt über eine Düse mit Waschmittel und Wasser bespritzt, was eine Überdosierung verhindert. Visuell dominiert das große Bullauge aus Edelstahl, ergänzt durch das gleichfarbige und mittig platzierte Touch-User-Interface.

Statement by the jury
This washing machine combines energy-saving and efficient functionality with a distinctive appearance.

Begründung der Jury
Diese Waschmaschine verbindet eine energiesparende und effiziente Funktionalität mit einem markanten Erscheinungsbild.

WAW28690 Serie | 8
i-DOS
Washing Machine
Waschmaschine

Manufacturer
Robert Bosch Hausgeräte GmbH,
Munich, Germany
In-house design
Florian Metz, Carsten Weber, Robert Sachon
Web
www.bosch-hausgeraete.de

The WAW28690 washing machine's superior energy efficiency ensures perfect laundry results. An easy-to-understand display keeps the operator up to date with electricity and water consumption. Quiet as a whisper in operation it will not disturb users. Its distinctive construction is characterised by the balanced proportions of its individual elements. Metallic applications and an LED-based programme selection round off the overall impression.

Die Waschmaschine WAW28690 hält mit beachtlicher Energieeffizienz die Kleidung sauber. Auf einer übersichtlichen Anzeige behält der Nutzer den Verbrauch von Strom und Wasser stets im Blick und erfährt durch ihren leisen Betrieb keine Störung. Der prägnante Aufbau des Haushaltsgerätes macht sich in einem ausgewogenen Verhältnis der einzelnen Elemente zueinander bemerkbar. Metallische Applikationen und eine durch LED beleuchtete Programmauswahl runden das Bild ab.

C1

Washing Machine
Waschmaschine

Manufacturer
Haier Group, Qingdao, China
In-house design
Kong Zhi, Wang Wentai, Jiang Chunhui,
Jung Yun Tae, Shu Hai
Design
Haier Innovation Design Center,
Qingdao, China
Web
www.haier.com

The use of glass and metal underlines the homogenous design of this drum-type washing machine and contributes to its general appearance. The materials and the colour-coordinated user touchscreen highlight the product's quality. The handle is flush with the door which can be opened without having to bend over. Its finish is reminiscent of diamonds and thus it complements the overall image of a high-quality product.

Die Verwendung von Glas und Metall unterstreicht die homogene Gestalt der Trommelwaschmaschine und trägt zum positiven Gesamteindruck bei. Die Materialien betonen die Qualität des Produktes und werden von einer farblich stimmigen Bedienungsanzeige in ihrer Wirkung unterstützt. Der Griff ist in die Glastür integriert, die in aufrechter Haltung geöffnet werden kann. In ihrem Schliff erinnert sie an einen Diamanten und komplettiert so das Bild eines hochwertigen Gerätes.

Statement by the jury
A prominent design element of this washing machine is the door with a diamond finish which gives this rather functional product an emotional component.

Begründung der Jury
Auffälliges Gestaltungselement dieser Waschmaschine ist der Diamantschliff der Tür, der diesem sachlichen Produkt eine emotionale Komponente hinzufügt.

LG Front Load Washer Titan 2.0 (F14U2FCN8, F14U2TBS2)
Washing Machine
Waschmaschine

Manufacturer
LG Electronics Inc., Seoul, South Korea
In-house design
Jeaseok Seong, Jonghee Han, Youngsoo Ha, Kyungah Lee, On Kim
Web
www.lg.com

Despite a size of 24 inches, this washing machine is of a very discreet appearance as all technical components are hidden by its casing. The controls are very user-friendly and the tinted glass of the floating door allows for a clear view of the inner workings. The centred dial, the display on the right-hand side and the distinctive lighting all help to make for safe use.

Statement by the jury
The large door opening of Titan 2.0, which extends over virtually its entire width, conveys its high load capacity. The washing machine also offers the benefit of intuitive use.

Die komplexe Technik des Gerätes versteckt sich im Gehäuse, sodass die Waschmaschine trotz einer Größe von 24 Zoll von einem unaufdringlichen Erscheinungsbild geprägt ist. In der Bedienung verhält sie sich benutzerfreundlich und erlaubt durch das getönte Glas der frei beweglichen Tür einen Blick ins Innere. Der mittig platzierte Regler, das Display auf der rechten Seite sowie die individuelle Beleuchtung unterstützen eine sichere Handhabung.

Begründung der Jury
Die große, fast die gesamte Breite einnehmende Türöffnung der Titan 2.0 kommuniziert ihr hohes Ladevolumen. Zudem überzeugt sie mit einem intuitiven Bedienkonzept.

Little Swan Beverly
Top Loading Washing Machine
Toplader-Waschmaschine

Manufacturer
Wuxi Little Swan Company Limited, Wuxi, China
In-house design
Web
www.littleswan.com

The interior of the elegantly styled Little Swan Beverly offers ample space to prevent entanglement of different articles of clothing. As the detergent is dispensed automatically and adjusted to varying volumes of washing, this saves the user effort, as well as energy and water. A glass door provides a clear view of the interior of the appliance and is complemented by lighting in a shape reminiscent of a crescent moon.

Statement by the jury
This washing machine is notable for its very user-friendly and sophisticated concept that has been thought through down to the last detail.

Die edel gestaltete Little Swan Beverly bietet ausreichend Platz im Inneren, sodass das Verknoten einzelner Kleidungsstücke vermieden wird. Da das Reinigungsmittel automatisch dosiert und der jeweiligen Wäschemenge angepasst wird, hat der Nutzer einen geringeren Aufwand und spart dabei Strom und Wasser. Eine Tür aus Glas ermöglicht den Blick in das Innenleben der Waschmaschine und wird abgerundet mit einer Beleuchtung, die an einen Halbmond erinnert.

Begründung der Jury
Ein ausgefeiltes, sehr benutzerfreundliches Konzept zeichnet diese Waschmaschine aus.

1528
Washing Machine
Waschmaschine

Manufacturer
Haier Group, Qingdao, China
In-house design
Kong Zhi, Wu Xiaolong, Jiang Chunhui,
Jung Yun Tae, Wu Jian
Design
Haier Innovation Design Center,
Qingdao, China
Web
www.haier.com

The smart ball technology used for the washing of textiles automatically also cleans the interlayer between the inner and outer barrel of the washing machine. This keeps the build-up of dirt inside the body of the machine to a minimum and largely prevents the accumulation of bacteria. The lid made of tempered glass harmonises with the metallic frame and underlines the linear design and use of the product.

Beim Waschen von Textilien werden dank eingesetzter Smart-Ball-Technologie automatisch auch die Innen- und Außenschichten der Waschmaschine gereinigt. So wird die Ansammlung von Schmutz im Gehäuse möglichst gering gehalten und die Entstehung von Bakterien weitestgehend vermieden. Ein Deckel aus gehärtetem Glas harmoniert mit einem Rahmen aus Metall und unterstreicht das geradlinige Konzept in Gestaltung und Anwendung des Produktes.

Statement by the jury
The minimalist design of this functionally well-engineered top loader draws particular attention to the clearly structured control panel.

Begründung der Jury
Die minimalistische Gestaltung dieses funktional ausgereiften Topladers lenkt den Blick besonders auf ein klar strukturiertes Bedienfeld.

PremiumCare Dryer
Dryer
Wäschetrockner

Manufacturer
Bauknecht, Comerio (Varese), Italy
In-house design
Web
www.bauknecht.de

The Premium Care dryer has a volume of 121 litres and can dry up to ten kilos of washing. The motion of the drum is adapted to the programme in use, thus ensuring efficient and gentle drying. Integrated air jets freshen up washing with steam and an additional refreshing ball gives it a pleasant scent. A 15 centimetre raised base makes it easier for the user to load and empty the drum which is sealed with a stainless-steel framed door.

Mit einem Fassungsvermögen von 121 Litern finden bis zu 10 kg Wäsche im PremiumCare Wäschetrockner Platz. Die Bewegung der Trommel wird dem laufenden Programm angepasst und verspricht so eine effiziente und schonende Trocknung. Unter dem Einsatz von Dampf frischen integrierte Luftdüsen die Kleidung auf und verleihen dieser durch einen zusätzlichen Refreshing Ball einen angenehmen Geruch. Ein um 15 cm erhöhter Sockel erleichtert dem Nutzer das Be- und Entladen der Trommel, die mittels einer großen Tür mit Edelstahl-rahmen verschlossen wird.

Statement by the jury
A high-quality, but at the same time very eye-catching body design highlights the large capacity offered by this dryer.

Begründung der Jury
Eine hochwertige und zugleich sehr präg-nante Korpusgestaltung unterstreicht das große Fassungsvermögen dieses Trockners.

LG Front Load Washer/Dryer and Pedestal Compact Washer (WM5000HVA/DLEX5000V/WD100CV)

LG Frontlader-Waschmaschine/Trockner und Podest-Kompaktwaschmaschine (WM5000HVA/DLEX5000V/WD100CV)

Manufacturer
LG Electronics Inc., Seoul, South Korea
In-house design
Jeaseok Seong, Wookjun Chung, Jonghee Han, Kyungah Lee, Yeji Um
Web
www.lg.com

With the successful combination of soft curves and metal, the washing machine and matching dryer embody sturdiness and subtlety at the same time. The deliberately large doors simultaneously highlight the large capacity of the appliance and also make it particularly easy to use. The control system is integrated into the doors and clothes can be loaded and taken out of the machine from above without needing to bend down.

Statement by the jury
This pair of washing machine and dryer in uniform design scores high due to its well-thought-out user concept.

Mit der gelungenen Kombination weicher Kurven und metallischer Materialien verkörpern Waschmaschine und zugehöriger Trockner Robust- und Feinheit zugleich. Die bewusst groß gestalteten Türen verdeutlichen zum einen die hohe Kapazität der Geräte und ermöglichen zum anderen eine besonders komfortable Bedienung. Das Steuerungssystem ist in die Türen integriert und die Wäsche kann ohne Bücken in die Trommel eingefüllt und wieder herausgenommen werden.

Begründung der Jury
Dieses einheitlich gestaltete Ensemble aus Waschmaschine und Trockner überzeugt mit einem durchdachten Bedienkonzept.

LG Top Load Washer/Dryer (WT7700HVA/DLGX7701VE)

LG Toplader-Waschmaschine/Trockner (WT7700HVA/DLGX7701VE)

Manufacturer
LG Electronics Inc. Seoul, South Korea
In-house design
Jeaseok Seong, Hoil Jeon, Nerry Son, Eunyoung Chee, K hyuk Kim
Web
www.lg.com

A rimless glass lid as well as a full-touch user interface round off the simple design of both appliances and thereby create a coherent appearance. The large opening on the washing machine ensures effortless loading and removal of clothing. There is additional storage space in the upper part of the dryer. The doors can be opened from both sides, this makes transferring wash loads quick and easy.

Statement by the jury
The purist design of the front of the washing machine characterises the appearance of both washing machine and dryer, whose functions are well aligned.

Ein randloser Deckel aus Glas sowie eine Full-Touch-Oberfläche runden die schlichte Gestaltung beider Geräte ab und schaffen so ein in sich stimmiges Erscheinungsbild. Die große Öffnung der Waschmaschine sorgt für müheloses Einfüllen und Entnehmen der Kleidung, im oberen Teil des Trockners befindet sich zusätzlicher Stauraum. Da die Türen zu beiden Seiten geöffnet werden können, ist ein schnelles und leichtes Umfüllen der Wäsche möglich.

Begründung der Jury
Eine puristische Frontgestaltung prägt das Erscheinungsbild von Waschmaschine und Trockner, die funktional gut aufeinander abgestimmt sind.

New Zanussi LINDO300
Washing Machine and Tumble Dryer

Waschmaschine und Wäschetrockner

Manufacturer
Electrolux, Stockholm, Sweden
In-house design
Web
www.electrolux.com

The visual presentation of washing machine and dryer reflects their functions: The control area is set into the metal face, while an ergonomic handle and large doors make the opening, loading and emptying of the drums easier. These domestic appliances function as a matching pair whose operation is made clear to the user by the information graphics.

Statement by the jury
A dynamic use of lines, the curved front and a clearly structured control panel characterise the overall appearance of these appliances.

Die optische Aufmachung von Waschmaschine und Trockner spiegelt sich in deren Funktionen wider: Das Bedienungsfeld ist in die Vorderseite aus Metall eingefügt, ein ergonomischer Handgriff und große Türen erleichtern zunächst das Öffnen und schließlich das Be- und Entladen der Trommeln. Die Haushaltsgeräte funktionieren als ein zusammengehöriges Paar, dessen Gebrauch dem Nutzer anhand von Informationsgrafiken erklärt wird.

Begründung der Jury
Eine dynamische Linienführung, die gewölbte Front und ein klar strukturiertes Bedienfeld prägen den hochwertigen Gesamteindruck dieser Geräte.

LG Tromm Slim Styler (S3BER)
Clothes Refresher
Textilerfrischer

Manufacturer
LG Electronics Inc., Seoul, South Korea
In-house design
Jeaseok Seong, Wookjun Chung,
Junghoi Choi, Jaemyung Lim, Byeongkook Kim
Web
www.lg.com

Thanks to a new concept, the clothes refresher not only eliminates odours, but also stains and germs. Pressing the one-touch button results in crease-free clothing that feels like new. The LG Tromm Slim Styler does away with a frame. Its seamless glass appearance results in a design that harmonises with its surroundings. Thanks to a hinge that is integrated in the symmetrical handle, it works as a stand-alone unit, but can, if desired, also be used as a built-in appliance.

Auf Basis des neuen Konzepts werden mit dem Textilerfrischer neben Gerüchen auch Flecken und Keime beseitigt. Die Kleidung hat keine Knitterfalten mehr und fühlt sich nach dem Drücken der One-Touch-Taste wie neu an. Der LG Tromm Slim Styler erhält durch den Verzicht auf einen Rahmen und eine nahtlose Verkleidung aus Glas einen Look, der gut mit anderem Mobiliar harmoniert. Dank eingebautem Scharnier im symmetrischen Griff funktioniert er eigenständig und auf Wunsch auch als Einbaumodell.

Statement by the jury
With its elegant, minimalist appearance, the LG Tromm Slim Styler harmoniously blends in with modern living environments.

Begründung der Jury
Mit seinem elegant-minimalistischen Erscheinungsbild fügt sich der LG Tromm Slim Styler harmonisch in moderne Wohnumgebungen ein.

Redefine Atomist Vapour Iron
Clothes Iron
Bügeleisen

Manufacturer
Morphy Richards,
Rotherham, Great Britain
In-house design
Web
www.morphyrichards.co.uk

The Redefine Atomist Vapour Iron combines technical innovation with an aesthetic revolution. As part of a new collection, its job is to eliminate creases in all textiles with a fine vapour mist while consuming up to 75 per cent less energy and up to 80 per cent less water. A glass soleplate containing a heating element regulates the temperature and distributes the heat evenly. This transparent iron offers users all the technology required in combination with an appealing design.

Statement by the jury
The futuristic design of this steam iron mirrors its innovative functionality and also lives up to environmental expectations.

Das Redefine Atomist Dampfbügeleisen verbindet technische Neuheiten mit einer ästhetischen Wende. Als Teil einer neuen Kollektion entfernt es mit fein vernebeltem Dampf die Falten jeglicher Textilien bei einem um 75 Prozent reduzierten Energie- und um 80 Prozent reduzierten Wasserverbrauch. Eine gläserne Bügelsohle mit Heizelement regelt die Temperatur und verteilt die Wärme gleichmäßig. Das durchsichtige Bügeleisen bietet dem Nutzer alle nötigen Technologien in Kombination mit ansprechendem Design.

Begründung der Jury
Die futuristische Gestaltung dieses Bügeleisens spiegelt seine innovative Funktionalität wider und wird auch ökologischen Aspekten gerecht.

Braun CareStyle 5
Ironing System
Dampfbügelstation

Manufacturer
De'Longhi Braun Household GmbH,
Neu-Isenburg, Germany
In-house design
Web
www.braunhousehold.com

The geometric contours and straight lines of the Braun CareStyle 5 steam station form a complete unit. At the same time, it also fits perfectly with the corresponding iron, which, in contrast, has an ergonomic and flowing shape, thus providing the perfect conditions for precise and comfortable use. The design of the soleplate was inspired by the properties of a snowboard, thus ensuring that it glides easily when ironing.

Statement by the jury
The base and the iron of this steam iron station each have an independent design so that these functional areas are visually clearly distinguished.

Mit seinen geometrischen Konturen und der geraden Linienführung stellt der Aufheizsockel der Braun CareStyle 5 eine eigene Einheit dar und ergänzt zugleich optimal das zugehörige Bügeleisen. Dieses zeigt seine ergonomische Ausstattung in fließenden Formen, die die Voraussetzung für einen präzisen und komfortablen Gebrauch schaffen. Gestalterisch an die Eigenschaften eines Snowboards angelehnt, verspricht die Sohle ein gutes Gleitverhalten beim Bügeln.

Begründung der Jury
Bei dieser Dampfbügelstation weisen Sockel und Bügeleisen jeweils eine eigenständige Gestaltung auf, sodass die Funktionsbereiche visuell klar voneinander getrennt sind.

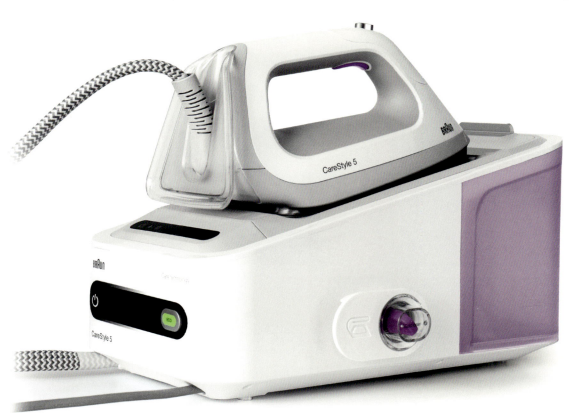

Azur Performer Plus
Steam Iron
Dampfbügeleisen

Manufacturer
Royal Philips, Eindhoven, Netherlands
In-house design
Web
www.philips.com

One of the key design elements of the Azur Performer Plus is its decalcification interface on the back of the appliance. An amber colour code clearly signals the calcification management of this steam iron. Thanks to an optimised soleplate, it glides across the ironing board with great efficiency and removes creases from clothing. Its individual components primarily aim to ensure easy handling and make it pleasant to use for a long period of time.

Statement by the jury
This iron appears exceptionally elegant thanks to its aerodynamic shape. The colour highlighting of the calcification management function is a distinctive design characteristic.

Zu den besonderen Elementen des Azur Performer Plus gehört die Entkalkungs-schn ttstelle auf der Rückseite des Gerätes. Anhand eines bernsteinfarbenen Farbcodes ist das Kalkmanagement des Dampf-bügeleisens deutlich erkennbar. Mit seiner optimierten Sohle gleitet es effizient über das Bügelbrett und entfernt die Falten aus der Kleidung. Die einzelnen Bestandteile zielen in erster Linie auf eine gute Handha-bung und gestalten die Anwendung nach-haltig angenehm.

Begründung der Jury
Dieses Bügeleisen ist mit seiner aero-dynamischen Formgebung ausgesprochen elegant; die farblich betonte Kalk-managementfunktion wird dabei zum markanten Designmerkmal.

Diva
Iron
Bügeleisen

Manufacturer
Royal Philips, Eindhoven, Netherlands
In-house design
Web
www.philips.com

The iron's optimal cable management ensures improved glide properties and thereby ease of use. The illuminated temperature dial and status-light allow for use even in poor light conditions independent of the time of day. The streamlined body of the iron with its modern colours of lilac and white will appeal, above all, to young consumers.

Statement by the jury
The light and graceful appearance of Diva is the result of a felicitous com-bination of precise lines and ergonomic design.

Eine optimierte Kabelführung des Bügel-eisens verbessert das Gleitverhalten und mit ihm die Handhabung des Gerätes. Mit hellem Regler und Statusleuchte ver-sehen kann es auch bei schlechteren Lichtverhältnissen eingesetzt werden und ermöglicht so einen Gebrauch unabhängig von der Tageszeit. All dies mündet schließ-lich im schlanken Korpus, der in modernem Lila und Weiß vor allem ein junges Publi-kum anspricht.

Begründung der Jury
Die leichte, grazile Anmutung von Diva resultiert aus einer gelungenen Kombi-nation von präziser Linienführung und ergonomischer Gestaltung.

OEKAKI50
Sewing Machine
Nähmaschine

Manufacturer
Aisin Seiki Co., Ltd., Kariya, Aichi, Japan
In-house design
Web
www.aisindesign.com

The combination of curved and angular shapes gives this sewing machine, which combines many practical features, a contemporary appearance. The stitch width can be adjusted using a foot pedal whilst the fabric can be freely moved at the same time, thus giving users the necessary creative freedom for their work. Text-free instructions are integrated in the Oekaki50, that makes the machine easy to understand. Its unique three colour variation can give an accent to interior space.

Die Kombination von runden und kantigen Formen macht die Nähmaschine zu einem zeitgemäßen Objekt, das viele praktische Funktionen in sich vereint. So kann etwa die Stichbreite über ein Fußpedal eingestellt und der Stoff gleichzeitig frei bewegt werden, was Hobbyschneidern die nötige kreative Freiheit in der Umsetzung ihrer Arbeit gibt. Dank der, direkt auf dem Gerät angebrachten, textlosen Anleitung ist die Oekaki50 leicht zu verstehen. Die einzigartige Variation dreier Farbtöne kann Innenbereichen einen Akzent verleihen.

Statement by the jury
An unusually colourful design and the use of flowing lines turn this sewing machine into a show-stopper.

Begründung der Jury
Eine ungewöhnlich farbenfrohe Gestaltung und eine fließende Linienführung machen diese Nähmaschine zum Blickfang.

2-in-1 Gift Wrapping Paper Clip
2-in-1-Geschenkpapier-Clip

Manufacturer
Betty Bossi AG, Zürich, Switzerland
In-house design
Thomas Etter
Design
Formfabrik AG (Christoph Jaun),
Zwillikon, Switzerland
Web
www.bettybossi.ch
www.formfabrik.ch

This clever device combines two functions in one. It ensures careful storage of rolled-up gift-wrap that can quickly be removed from the clip when needed and allows for it to be cut cleanly using the clip. This flexible helper is suitable for both thin and thicker paper rolls and makes wrapping presents and other things much easier.

Die kluge Kreation vereint zwei Funktionen in einer: Sie dient dem schonenden Verstauen von zusammengerolltem Geschenkpapier, das bei Bedarf schnell aus dem Clip entfernt und mit diesem sauber geschnitten werden kann. Der flexible Helfer eignet sich sowohl für dünne als auch dickere Papierrollen und erleichtert das Verpacken von Geschenken und anderem maßgeblich.

Vöslauer 8x1l Glas Split-Crate
Returnable Beverage-Crate
Getränke-Mehrwegkiste

Manufacturer
Vöslauer Mineralwasser AG,
Bad Vöslau, Austria
In-house design
Web
www.voeslauer.com

This drinks crate and its light-weight glass bottles are handy and easy to carry, also because of the ergonomic handles which make transportation comfortable. A practical split-mechanism distributes the weight of the contents so that each side weighs no more than seven kilos. The deliberately chosen recycling system highlights the sustainability aspect of this crate and turns this drinks carrier into a product of great user benefit.

Die Getränkekiste inklusive gewichtsoptimierter Glasflaschen ist leicht und einfach zu handhaben. Auch die ergonomisch geformten Griffe begünstigen ein angenehmes Tragen. Ein praktischer Split-Mechanismus verteilt das Gewicht des Inhaltes so, dass keine Seite mehr als 7 kg wiegt. Das bewusst gewählte Mehrwegsystem unterstreicht den Aspekt der Nachhaltigkeit und macht die Mehrwegkiste zu einem Produkt mit hohem Nutzwert.

Statement by the jury
The well-thought-out ergonomic design of this drinks crate turns it into a practical and sustainable means of transporting bottles.

Begründung der Jury
Die durchdachte ergonomische Gestaltung dieser Getränkekiste macht sie zu einem praktischen und nachhaltigen Transporthilfsmittel für Flaschen.

Totem
Waste Separation and Recycling Unit
Behälter für Mülltrennung und Recycling

Manufacturer
Joseph Joseph Ltd,
London, Great Britain

Design
PearsonLloyd,
London, Great Britain

Web
www.josephjoseph.com
www.pearsonlloyd.com

reddot award 2015
best of the best

Multifunctional lightness

Today, pre-sorting waste and recycling at home is a key requirement for effectively recycling all the resources contained in standard domestic waste. However, domestic waste usually comprises a myriad of different materials that are difficult to separate, so people often improvise managing the waste by using a clutter of extra bins and containers. The Totem waste separation unit responds to this problem with a formal and functionally sophisticated concept that offers many separation options in one object. The openness and flexibility of this system is particularly innovative as it allows for individual user configurations to adapt to different people's needs. Featuring a footprint that is comparable to most conventional kitchen bins, Totem delivers a multifunctional solution offering much more capacity for advanced waste separation. The unit comprises a large 36-litre general waste compartment including an odour filter cleverly integrated in the lid, as well as a multi-purpose drawer featuring a removable liner, a divider and carrier bag hooks that lend themselves to easily separating different types of recycling. Also included is a food waste caddy, specially designed to fit into the general waste compartment or the multi-purpose drawer. The wheeled base and rear handle allow Totem to be easily moved and placed where needed. Thanks to its clear design arrangement, this waste bin blends well into almost any kitchen environment, lending the act of separating waste a sense of novel, contemporary ease.

Multifunktionale Leichtigkeit

Um die im Hausmüll enthaltenen Rohstoffe wieder effektiv in Recyclingkreisläufe einbringen zu können, ist eine effiziente Vorsortierung bereits in den Haushalten wichtig. Der Müll ist jedoch durch seine Vielfältigkeit oftmals kaum überschaubar und auch nur schwer zu ordnen, weshalb in den Küchen nicht selten viele unterschiedliche Behälter oder Boxen dafür bereitstehen. Der Abfallbehälter Totem begegnet dieser Problematik mit einem formal wie funktional ausgereiften Konzept, welches viele Möglichkeiten in einem Objekt vereint. Innovativ ist dabei insbesondere die Offenheit und Flexibilität des Systems, denn der Nutzer kann es individuell konfigurieren und damit seinen Bedürfnissen anpassen. Totem ist mit den Grundmaßen eines konventionellen Mülleimers für die Küche gestaltet, der durch eine multifunktionale Aufteilung jedoch viel mehr Platz für eine differenzierte Mülltrennung bietet. Die Einheit besteht aus einer großen 36-Liter-Box für den Restmüll samt klug integriertem Geruchsfilter im Deckel sowie einer Mehrzweckschublade, die durch ein abnehmbares Fach, eine Trennwand und Tragetaschenhaken verschiedenste Arten von Recycling ermöglicht. Ein spezieller Speisereste-Caddy wurde zudem so gestaltet, dass er in das Hauptabfallfach oder die Mehrzweckschublade passt. Durch die fahrbare Unterseite und den hinteren Handgriff kann Totem vom Nutzer leicht bewegt werden. Dank seiner klaren Gestaltung fügt sich dieser Abfallbehälter gut in jedes Küchenumfeld ein und verleiht dabei dem Sortieren von Müll eine neue und sehr zeitgemäße Leichtigkeit.

Statement by the jury

Featuring a multifunctional design, this waste separation unit offers users a myriad of different recycling possibilities. Its innovative, open concept allows users to individually configure all elements of the container and thus adapt the process of separating waste to individual needs and given kitchen space. Totem is easy to clean and with its appeal of clarity and freshness blends in perfectly with most contemporary kitchen environments.

Begründung der Jury

Mit seiner multifunktionalen Gestaltung bietet dieser Behälter zur Mülltrennung dem Nutzer eine enorme Vielzahl von Möglichkeiten. Das innovative offene Konzept ermöglicht es, alle Elemente des Behälters selbst zu konfigurieren und damit das Sortieren des Mülls den eigenen Bedürfnissen und Gegebenheiten anzupassen. Totem ist leicht zu reinigen und fügt sich mit seiner ansprechenden Anmutung von Klarheit und Frische perfekt in die zeitgemäße Küchenumgebung ein.

Designer portrait
See page 32
Siehe Seite 32

Coffee machines	Abzugshauben
Dishwashers	Einbauküchen
Extractor hoods	Kaffeemaschinen
Fitted kitchens	Küchenarmaturen
Kitchen appliances	Küchenausstattung
Kitchen equipment	Küchengeräte
Kitchen fittings	Kücheninseln und -blöcke
Kitchen furnishings	Küchenmöbel
Kitchen isles and blocks	Küchentechnik
Kitchen technology	Kühl- und Gefriergeräte
Microwaves	Mikrowellen
Mixers	Mixer
Ovens and hobs	Öfen und Kochfelder
Refrigerators and freezers	Spülbecken
Sinks	Spülmaschinen
Toasters	Toaster
Water filtration systems	Wasserfiltersysteme

Kitchens
Küche

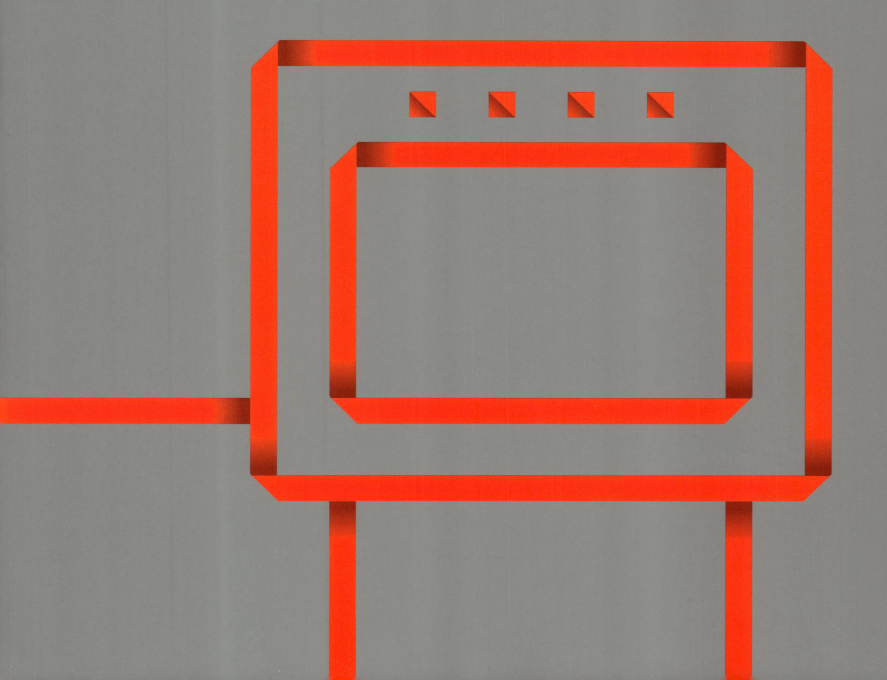

WMF Espresso
Semi-Automatic Coffee Machine
Kaffeehalbautomat

Manufacturer
WMF AG,
Geislingen, Germany

In-house design
Peter Bockwoldt
(Design Management)

Design
designship
(Thomas Starczewski, Piero Horn),
Ulm, Germany

Web
www.wmf.de
www.designship.de

reddot award 2015
best of the best

Formal transition

In traditional coffee houses around the world, such as in Vienna and Rome, coffee is still made by loud machines that first grind the coffee. These places virtually celebrate the preparation and serving of coffee in stylish ambiances reflecting hundreds of years of regional coffee drinking culture. The design language of the WMF semi-automatic coffee machine intends to pick up on this tradition. The formal design reference aims at visualising a new concept of coffee making that delivers a reliable process for perfect coffee and espresso making, with portion by portion freshly ground beans. The familiar appearance of this coffee machine is based on an innovative, seemingly floating cup tray under which a cooled depot for coffee beans is neatly concealed. A further, highly familiar element in terms of form and function is represented by the huge conventional portafilters made of mirrow-polished stainless steel. The successful reinterpretation of the classical bar espresso machine is complemented by a clearly arranged touch display for state-of-the-art user guidance. Details such as the touch display with respective technological intelligence and the intelligent handling concept enable carefree operation for any user of this semi-automatic coffee machine, allowing them to prepare coffee at a premium standard at any time – a familiar form and the association of enjoyment are thus consistently transposed into modern times.

Form im Wandel

In den traditionsreichen Kaffeebars dieser Welt, wie etwa in Wien oder Rom, wird der Kaffee noch heute in kolossalen Maschinen lautstark zubereitet. Im Einklang mit den oft jahrhundertealten regionalen Kaffee-kulturen wird das Zubereiten und stilgerechte Servieren des Kaffees dort regelrecht zelebriert. Der Kaffeehalb-automat von WMF will durch seine Formensprache an diese Art der Zubereitung erinnern. Das Ziel war es, durch die formale Anlehnung ein neues Konzept der Kaffeezubereitung zu visualisieren, bei dem der Nutzer die Möglichkeit hat, zuverlässig perfekt gebrühten Kaf-fee und Espresso aus portionsweise frisch gemahlenen Bohnen zu erhalten. Die vertraute Anmutung dieser Kaffeemaschine beruht einerseits auf der Innovation einer scheinbar schwebend gestalteten Tassenabstell-fläche, unter der ein gekühltes Bohnendepot ange-ordnet ist. Als weiteres, in seiner Form und Funktion überaus vertrautes Element dienen große konventio-nelle Siebträger aus glanzpoliertem Edelstahlguss. Die gelungene Neuinterpretation einer klassischen Bar-Espressomaschine geht einher mit einer zeitgemä-ßen Nutzerführung über ein klar gestaltetes Touch-display. Mittels eines intelligenten Bedienkonzepts kann der Nutzer diese halbautomatische Kaffeemaschine jederzeit fehlerfrei bedienen für einen Kaffee höchster Qualität – eine vertraute Form und die Assoziation von Genuss werden somit schlüssig in die heutige Zeit transportiert.

Statement by the jury

Following a fascinating approach, the design of this semi-automatic coffee machine manages to success-fully merge the appeal of traditional coffee making with innovative, intelligent technologies. Equipped with a touchscreen for easy, carefree operation, this semi-automatic coffee machine offers the highest degree of convenience and enjoyment. Details such as the apparently floating cup tray and a cooled depot for coffee beans both fascinate and emotionalise when in use.

Begründung der Jury

Auf beeindruckende Weise gelingt es der Gestaltung dieser halbautomatischen Kaffeemaschine die Anmu-tung traditioneller Kaffeezubereitung mit innovativen, intelligenten Technologien zu verbinden. Ausgestat-tet mit einem einfach und fehlerfrei zu bedienenden Touchscreen, bietet dieser Kaffeehalbautomat ein Höchstmaß an Komfort und Genuss. Details wie die scheinbar schwebende Tassenabstellfläche und ein gekühltes Bohnenfach begeistern und emotionalisieren im Gebrauch.

Designer portrait
See page 34
Siehe Seite 34

TE60 Range
Fully Automatic Coffee Maker
Kaffeevollautomat

Manufacturer
Siemens-Electrogeräte GmbH,
Munich, Germany
In-house design
Gregor Luippold, Monika Geldhauser
Web
www.siemens-home.de

Characterised outwardly by classical lines and a compact construction, the TE60 fully automatic coffee maker harmonises with office or private kitchens. Construction of the appliance is well-conceived, and operation is user-friendly. An enclosed brewing chamber prevents splashes. The milk foamer is integrated in the brewing unit and allows preparation of various coffee specialities which can be selected by slightly touching the sensor fields. The flat design of the automat also makes it easy to clean.

Äußerlich geprägt von klassischen Linien und einer kompakten Bauweise integriert sich der TE60 Kaffeevollautomat harmonisch in Büro- oder private Küchen. Der Geräteaufbau ist durchdacht, die Bedienung benutzerfreundlich. Ein abgeschlossener Brühraum verhindert Spritzer. Die im Brühkopf integrierte Milchschaumlösung ermöglicht die Zubereitung verschiedener Kaffeespezialitäten, die durch leichtes Berühren der Sensorfelder angewählt werden. Die flächige Gestaltung des Automaten erleichtert zudem die Reinigung.

Statement by the jury
This fully automatic coffee maker makes a professional impression due to clear design and well-conceived construction. Furthermore, it is characterised by uncomplicated operation.

Begründung der Jury
Dieser Kaffeevollautomat vermittelt mit einer klaren Gestaltung und einem durchdachten Aufbau Professionalität und zeichnet sich zudem durch eine unkomplizierte Bedienung aus.

CM 6310
Fully Automatic Coffee Maker
Kaffeevollautomat

Manufacturer
Miele & Cie. KG, Gütersloh, Germany
In-house design
Web
www.miele.de

This fully automatic coffee maker possesses a "One Touch for Two" function, due to which two coffee specialities can be prepared at the same time at the push of a button. Furthermore, in designing the CM 6310, great emphasis was laid on hygiene: when the device is switched off, all milk-carrying components are rinsed automatically. The brew unit can be removed for cleaning; many components can be cleaned in the dishwasher.

Statement by the jury
The CM 6310 is attractive due to its linear, elegant design and an intuitive control panel. Furthermore, it fulfils high demands of hygiene.

Dieser Kaffeevollautomat verfügt über eine sogenannte „One Touch for Two"-Funktion, dank derer zwei Kaffeespezialitäten auf Knopfdruck gleichzeitig zubereitet werden können. Zudem wurde bei der Konzeption des CM 6310 viel Wert auf Hygiene gelegt: Die Spülung aller milchführenden Komponenten erfolgt nach dem Abschalten des Gerätes automatisch. Die Brühgruppe lässt sich zur Reinigung entnehmen. Viele Komponenten können im Geschirrspüler gereinigt werden.

Begründung der Jury
Der CM 6310 gefällt mit einer geradlinigen, eleganten Gestaltung und einem selbsterklärenden Bedienfeld. Außerdem erfüllt er hohe Ansprüche an Hygiene.

C17KS61 Fully Automatic Coffee Maker Design 610
C17KS61 Kaffeevollautomat Design 610

Manufacturer
Constructa-Neff Vertriebs GmbH, Munich, Germany
Design
BSH Hausgeräte GmbH, Neff Design Team, Munich, Germany
Web
www.neff.de

The appearance of this C17KS61 fully automatic built-in coffee maker is purist and characterised by the use of blackened glass and stainless steel. A high-resolution TFT display with ShiftControl offers a high degree of operating convenience. The lighting of buttons and coffee outlet, as well as the concealed yet easily accessible containers for water, milk and coffee beans, allows for convenient operation.

Statement by the jury
With a minimalist designed, black glass front, this fully automatic coffee maker makes an up-market impression, and the attention is drawn to the illuminated operating panel.

Das Erscheinungsbild des Einbau-Kaffeevollautomaten C17KS61 ist puristisch und durch den Einsatz von geschwärztem Glas und Edelstahl geprägt. Ein hochauflösendes TFT-Display mit ShiftControl-Steuerung bietet hohen Bedienkomfort. Die Beleuchtung von Tasten und Kaffeeauslauf sowie verdeckte, aber einfach zugängliche Behälter für Wasser, Milch und Kaffeebohnen erlauben eine komfortable Handhabung.

Begründung der Jury
Mit einer minimalistisch gestalteten, schwarzen Glasfront wirkt dieser Kaffeevollautomat edel, und der Blick wird auf die beleuchtete Funktionsfläche gelenkt.

C15KS61 Fully Automatic Coffee Maker Design 600
C15KS61 Kaffeevollautomat Design 600

Manufacturer
Constructa-Neff Vertriebs GmbH, Munich, Germany
Design
BSH Hausgeräte GmbH, Neff Design Team, Munich, Germany
Web
www.neff.de

Clear lines and high-quality materials such as blackened glass and stainless steel define the elegant aesthetics of the C15KS61 fully automatic coffee maker. Via the colour TFT display and discreetly illuminated operating buttons, every drink speciality for one or two cups can be dispensed with only a press of the button. A height-adjustable coffee outlet with integrated milk frother allows the use of glasses up to 15 cm in height.

Statement by the jury
This fully automatic coffee machine blends well into every kitchen unit due to its elegant impression, compact dimensions and narrow gap clearances.

Eine klare Linienführung und hochwertige Materialien wie schwarzes Glas und Edelstahl bestimmen die elegante Ästhetik des Kaffeevollautomaten C15KS61. Über das farbige TFT-Display und dezent beleuchtete Bedienknöpfe kann jede Getränkespezialität für eine oder zwei Tassen mit nur einem Tastendruck ausgegeben werden. Ein höhenverstellbarer Kaffeeauslauf mit integriertem Milchschäumer ermöglicht die Verwendung von bis zu 15 cm hohen Gläsern.

Begründung der Jury
Dieser Kaffeevollautomat fügt sich durch seine elegante Anmutung, kompakte Maße und schmale Spaltabstände gut in die jeweilige Küchenzeile ein.

Classe 11
Professional Coffee Maker
Gewerbliche Kaffeemaschine

Manufacturer
Rancilio Group S.p.A.,
Villastanza di Parabiago (Milan), Italy
Design
Design Center (Marco Bonetto),
Assago (Milan), Italy
Web
www.ranciliogroup.com
www.bonetto-design.com

The striking design feature of this professional espresso machine is its control panel of tempered glass with multi-touch technology, inspired by smartphone user interfaces. The symbols are clear and user-friendly. The machine can also be operated via a special App and QR codes. The form of the brew groups facilitates work with the filter holder. LEDs are integrated in the covers to display the operating phases of the machine.

Das hervorstechende gestalterische Merkmal dieser professionellen Espresso-maschine ist ihr Bedienfeld aus gehärtetem Glas mit Multi-Touch-Technologie, das von Smartphone-Benutzeroberflächen inspiriert ist. Die Symbole sind klar und benutzerfreundlich. Die Maschine lässt sich zudem mittels einer speziellen App und QR-Codes steuern. Die Form der Brühgruppen erleichtert die Arbeit mit dem Siebträger. In die Abdeckungen sind LEDs integriert, die die Betriebsphasen der Maschine anzeigen.

TELVE
Turkish Coffee Maker
Türkische Kaffeemaschine

Manufacturer
Arçelik A.S., Istanbul, Turkey
In-house design
Arçelik Industrial Design Team
(Nihat Duran)
Web
arcelik.com.tr

The preparation of Turkish coffee, which is served with its grounds (Turkish: telve), is a tradition which goes back hundreds of years. The Telve coffee maker with the gentle curvature of its housing reflects the formal language of the traditional coffee pot. At the same time, its design is emphatically functional, creating a consistently modern appearance. The control buttons, ventilation apertures of perforated metal, and LEDs in the air channel set accents.

Die Zubereitung Türkischen Kaffees, der mit Kaffeesatz (Türkisch: Telve) serviert wird, hat eine jahrhundertelange Tradition. Die Kaffeemaschine Telve zitiert mit der sanften Wölbung des Gehäuses die Formensprache der traditionellen Kaffeekannen. Gleichzeitig hat sie eine betont sachliche Gestaltung, sodass ein durchweg modernes Erscheinungsbild entsteht. Akzente setzen die Bedientasten, Belüftungsöffnungen aus perforiertem Metall und LEDs im Luftkanal.

Statement by the jury
Telve combines a traditional way of coffee making with contemporary technology and conveys these by its design.

Begründung der Jury
Telve kombiniert eine traditionelle Art der Kaffeezubereitung mit zeitgemäßer Technologie und transportiert dies auch durch ihre Gestaltung.

mycoffeestar
Reusable Coffee Capsule for Nespresso®
Wiederbefüllbare Kaffeekapsel für Nespresso®

Manufacturer
Swiss Innovation Products,
Zürich, Switzerland

In-house design
Erwin Meier

Web
www.mycoffeestar.com

reddot award 2015
best of the best

Eco-friendly enjoyment

Nespresso machine coffee has become part of an international lifestyle. It is easy to make and offers consistently good quality flavour. However, the estimated and still rising number of coffee capsules used annually amounts to no less than eight billion, placing an enormous burden on the environment. Against this backdrop, the reusable coffee capsule for Nespresso offers the possibility to enjoy this type of coffee without such grave ecological consequences. The capsule is made of high-quality medical stainless steel that makes it extremely durable and keeps aluminium particles or other unhealthy pollutants from entering the body. Consistently designed and self-explanatory, it is easy to open, fill with coffee and close again. Since the capsule was constructed in three parts that are screwed together, it can be shipped ecologically in a small envelope. An additional, huge benefit of this coffee capsule is that users can fill it with any coffee they like including fair trade coffee, and thus continue using their capsule machine without limitation. The principle of ecologically sound coffee making is thus complemented by a high degree of individuality. The capsule incorporates an engaging design that delivers a poignantly logical and functional solution for a serious ecological issue.

Verantwortungsvoller Genuss

Der Kaffee aus der Nespressomaschine ist Teil eines internationalen Lifestyles. Er ist einfach zuzubereiten und bietet eine gleichbleibend gute Geschmacksqualität. Die geschätzte und immer noch steigende Anzahl der dabei verbrauchten Kaffeekapseln liegt jedoch bei jährlich etwa acht Milliarden, was den Nachteil einer enormen Umweltbelastung mit sich bringt. Vor diesem Hintergrund ermöglicht die wiederbefüllbare Kaffeekapsel für Nespresso, diese Art von Kaffee ohne ökologische Nachteile zu genießen. Die Kapsel ist aus einem sehr hochwertigen medizinischen Edelstahl gefertigt, weshalb sie äußerst langlebig ist und im Körper keinerlei Aluminiumpartikel oder andere Schadstoffe hinterlässt. In sich schlüssig gestaltet und selbsterklärend, ist sie einfach zu öffnen und wieder zu verschließen. Da sie aus drei verschraubbaren Einzelteilen besteht, kann sie zudem ökologisch in einem Briefumschlag verschickt werden. Ein enormer Vorteil dieser Kaffeekapsel ist nicht zuletzt, dass der Nutzer sie mit jeder Sorte Kaffee inklusive Fairtrade-Kaffee befüllen und so seine Kapselmaschine ohne Einschränkung weiterhin verwenden kann. Auf diese Weise verbindet sich das Prinzip einer umweltfreundlichen Art der Kaffeezubereitung mit einem hohen Maß an Individualität. Ein engagiertes Design bietet damit eine bestechend logische und funktionale Lösung für ein ernsthaftes ökologisches Problem.

Statement by the jury

The reusable coffee capsule for Nespresso represents an outstanding product and long-awaited solution, particularly when considering the mountains of garbage caused by this approach toward coffee making. Consisting of three parts that are screwed together, it is functional and can easily be refilled time and again with coffee of any type – users will get the same coffee quality as before. Furthermore, the capsule is made of medical stainless steel that makes it durable and leaves no particle residue.

Begründung der Jury

Die wiederbefüllbare Kaffeekapsel für Nespresso ist vor dem Hintergrund der durch diese Art der Kaffeezubereitung verursachten Müllberge ein herausragendes und lange erwartetes Produkt. Gestaltet aus drei verschraubbaren Einzelteilen, ist sie funktional und kann mit Kaffee jeder Wahl problemlos immer wieder befüllt werden – der Nutzer hat die gleiche Kaffeequalität wie zuvor. Die Kaffeekapsel besteht zudem aus medizinischem Edelstahl, wodurch sie langlebig ist und keinerlei Rückstände hinterlässt.

Designer portrait
See page 36
Siehe Seite 36

Tassimo T32 SUNY Range
Multi Beverage Maker
Multi-Heißgetränkesystem

Manufacturer
Robert Bosch Hausgeräte GmbH,
Munich, Germany
In-house design
Gregor Luippold, Helmut Kaiser
Design
JEFFMILLERinc, New York, USA
Web
www.bosch-hausgeraete.de
www.jeffmillerdesign.com

A compact, slim shape and fresh colours make this multi beverage maker an eye-catcher in k tchens. Preparation of beverages starts immediately without heating-up time as soon as a cup is pressed against the front of the appliance (so-called „SmartStart"). The cup stand is height-adjustable and can be removed completely for large cups and vacuum mugs. The water tank is integrated in the body and can be removed upwards simply by one hand move.

Eine kompakte, schlanke Form und frische Farben machen dieses Multi-Heißgetränke-system zum Blickfang in Küchen. Die Getränkezubereitung startet ohne Aufheizzeit sofort, wenn eine Tasse gegen die Geräte-vorderseite gedrückt wird (sogenannter „SmartStart"). Das Tassenpodest ist höhen-verstellbar und lässt sich für große Tassen und Isolierbecher auch komplett entfernen. Der Wassertank ist in den Korpus integriert und kann mit einem Handgriff einfach nach oben hin entnommen werden.

Statement by the jury
The Tassimo T32 convinces by its space-saving and fresh design as well as by its novel operating method.

Begründung der Jury
Die Tassimo T32 überzeugt sowohl durch eine platzsparende und frische Gestaltung als auch durch die neue Form der Bedienung.

VIVA
Capsule Coffee Machine
Kapsel-Kaffeemaschine

Manufacturer
Delica AG, Birsfelden, Switzerland
Design
2ND WEST GmbH
(Michael Thurnherr, Manuel Gamper),
Rapperswil, Switzerland
Web
www.delica.ch
www.2ndwest.ch

An architecturally strict form in combination with the smooth curves of the structured sides characterises the appearance of the Viva. Control is intuitive via six operating buttons, five of which are programmable to the desired cup size. The cup tray is arranged for various cup sizes. In standby mode and with a warm-up time of only 15 seconds, Viva is energy-saving. The capsules are aluminium-free.

Statement by the jury
Viva manages to combine an aesthetically attractive, purist design concept with environmentally friendly functionality.

Eine architektonisch strenge Formgebung in Kombination mit den weichen Wölbungen der strukturierten Seiten prägt das Erscheinungsbild der Viva. Die Bedienung erfolgt intuitiv über sechs Funktionsknöpfe, von denen fünf auf die gewünschte Tassengröße programmierbar sind. Die Tassenablage ist für Gefäße unterschiedlicher Größe ausgelegt. Mittels Stand-by-Modus und einer Aufwärmzeit von nur 15 Sekunden spart Viva Strom. Die Kapseln sind aluminiumfrei.

Begründung der Jury
Der Viva gelingt es, ein ästhetisch ansprechendes, puristisches Gestaltungskonzept mit einer umweltfreundlichen Funktionalität zu verbinden.

Tchibo Caffissimo Tuttocaffè Crafted by Saeco
Coffee Machine
Kaffeemaschine

Manufacturer
Tchibo GmbH, Hamburg, Germany
Design
Philips Design, Eindhoven, Netherlands
Web
www.tchibo.de
www.philips.com

The Caffissimo Tuttocaffè is a compact coffee machine conceived for Tchibo coffee drinkers. The appliance has a movable cup stand for various cups and glasses and offers a variety of coffee drinks at the touch of a button, thanks to its easily operated user interface. A combined high/low pressure brewing system facilitates preparation of every kind of coffee drink with a compact machine. The appliance is available in black, white and red.

Statement by the jury
This user-friendly capsule machine combines dynamic form with easy operation.

Die Caffissimo Tuttocaffè ist ein kompakter Kaffeeautomat, konzipiert für Tchibo-Kaffee-Trinker. Die Maschine hat einen beweglichen Tassentisch für unterschiedliche Tassen und Gläser und bietet dank ihrer leicht zu bedienenden Benutzeroberfläche verschiedene Kaffeegetränke auf Tastendruck. Ein kombiniertes Hoch-/Niederdruck-Brühsystem ermöglicht die Zubereitung jeder Art von Kaffee mit einer kompakten Maschine. Das Gerät ist in Schwarz, Weiß und Rot erhältlich.

Begründung der Jury
Diese benutzerfreundliche Kapselmaschine verbindet eine dynamische Formgebung mit einer einfachen Bedienung.

Nestlé Oblo
Coffee Maker
Kaffeemaschine

Manufacturer
Nestlé SA, Vevey, Switzerland
Design
Multiple SA (Pierre Struzka),
La Chaux-de-Fonds, Switzerland
Web
www.nestle.ch
www.multiple-design.ch

The Oblo coffee maker places coffee preparation at the centre of its design. The coffee outlet becomes the focal point of the appliance due to the pronounced, circular recess and thus attracts attention. All other elements are discreetly designed and remain unobtrusively in the background. The shelf for the cups is height-adjustable. The water tank is flat so that the machine takes up little space on the worktop.

Statement by the jury
With its concise circular aperture, Oblo sets coffee making in focus – an original design approach which conveys to the product a strong visual identity.

Die Kaffeemaschine Oblo stellt die Kaffeezubereitung in den Mittelpunkt der Gestaltung. Der Kaffeeauslass wird durch die markante kreisrunde Aussparung zum zentralen Punkt der Maschine und zieht die Aufmerksamkeit auf sich. Alle anderen Elemente sind dezent gestaltet und halten sich im Hintergrund. Die Abstellfläche für die Tassen ist höhenverstellbar. Der Wassertank ist flach, sodass die Maschine wenig Platz auf der Arbeitsfläche einnimmt.

Begründung der Jury
Oblo inszeniert mit ihrer prägnanten kreisrunden Öffnung die Kaffeezubereitung – ein origineller Gestaltungsansatz, der dem Produkt eine starke visuelle Identität verleiht.

Redefine
Hot Water Dispenser
Heißwasserspender

Manufacturer
Morphy Richards,
Rotherham, Great Britain
In-house design
Web
www.morphyrichards.co.uk

This appliance combines the versatility of a kettle with the convenience of a water dispenser. The user sets the required volume of water (up to 1.5 litres) and selects the temperature. The appliance then pumps the exact set amount through a filter into the boiling chamber. The heated water is dispensed automatically. The Redefine is operated completely from the front; an LED display and an acoustic system support the intuitive operation.

Dieses Gerät verbindet die Vielseitigkeit eines Wasserkochers mit dem Komfort eines Wasserspenders. Der Benutzer gibt die benötigte Wassermenge (bis zu 1,5 Liter) ein und wählt die Temperatur aus. Das Gerät pumpt exakt die eingegebene Menge durch einen Filter in die Siedekammer. Die Ausgabe des erhitzten Wassers erfolgt automatisch. Die Bedienung erfolgt komplett von vorne, eine LCD-Anzeige und ein Klangsystem unterstützen die intuitive Handhabung.

Statement by the jury
Redefine unites the functions of a kettle and a dispenser and impresses with a design concept where user-friendliness plays the major role.

Begründung der Jury
Redefine vereint die Funktionen von Wasserkocher- und spender und beeindruckt mit einem Gestaltungskonzept, bei dem Benutzerfreundlichkeit die Hauptrolle spielt.

Magic Super
Water Purifier
Wasserreiniger

Manufacturer
Tong Yang Magic Co., Ltd.,
Seoul, South Korea
In-house design
Dongsu Kim, Junyoung Hong
Web
www.magic.co.kr

When designing the "Magic Super"
water purifier, a small glass of clear
water served as model. In order to convey to the user the feeling of drinking
particularly pure water, transparent
materials were used for the encoder and
the tap. Due to dispensing with a water
tank, the water purifier is also particularly hygienic and can dispense cold as
well as hot water. Coloured LEDs indicate
the water temperature in a manner
which the user understands intuitively.

Bei der Gestaltung des „Magic Super"-
Wasserreinigers diente ein kleines, mit
klarem Wasser gefülltes Glas als Vorbild.
Um dem Benutzer das Gefühl zu vermitteln, besonders reines Wasser zu trinken,
wurden für Drehregler und Auslauf transparente Materialien verwendet. Durch
den Verzicht auf einen Wassertank ist der
Wasserreiniger zudem besonders hygienisch und kann sowohl kaltes als auch heißes Wasser spenden. Farbige Leuchtdioden
kommunizieren die Wassertemperatur so,
dass der Nutzer sie intuitiv versteht.

Statement by the jury
The "Magic Super" water purifier
convinces with a high level of intuitiveness, thus making the controls
easy to understand.

Begründung der Jury
Der „Magic Super"-Wasserreiniger
überzeugt mit einer hohen Selbsterklärungsqualität, die eine intuitive
Bedienung ermöglicht.

MES4000 Range
Juicer
Entsafter

Manufacturer
Robert Bosch Hausgeräte GmbH,
Munich, Germany
In-house design
Helmut Kaiser, Tobias Krüger
Design
Brandis Industrial Design,
Nuremberg, Germany
Web
www.bosch-hausgeraete.de
www.brandis-design.de

The defined contours and harmonious curvatures of the die-cast aluminium housing determine the appearance of the MES4000 juicer. This is accentuated by brushed metal applications and a stainless steel juice outlet. The juicer features a powerful, quiet motor and durable ceramic blades which, thanks to the particularly large filler opening, can also chop uncut fruit and vegetables. An innovative electro-polished juice sieve allows for easy and quick cleaning.

Die definierten Konturen und harmonischen Rundungen des Aluminium-Druckguss-Gehäuses bestimmen das Erscheinungsbild des Entsafters MES4000. Akzentuiert wird es von gebürsteten Metallapplikationen und einem Saftauslauf aus Edelstahl. Der Entsafter ist mit einem leistungsstarken, geräuscharmen Motor und langlebigen Keramikmessern ausgestattet, die dank der besonders großen Einfüllöffnung auch nicht vorgeschnittenes Obst oder Gemüse zerteilen können. Ein innovatives elektropoliertes Saftsieb ermöglicht eine einfache und schnelle Reinigung.

Statement by the jury
The MES4000 juicer makes a powerful impression due to its material combination and dynamic lines – an impression justified also by its functionality.

Begründung der Jury
Der Entsafter MES4000 wirkt durch die Materialkombination und eine dynamische Linienführung kraftvoll – ein Eindruck, dem er auch in funktionaler Hinsicht gerecht wird.

HW-SBF15
Juicer
Entsafter

Manufacturer
Hurom Co., Ltd.,
Gimhae City, South Korea
In-house design
Seol Lee, Sung Ha Jung
Web
www.hurom.com

The juicer HW-SBF15 is conceived and designed according to the requirements of professional sellers of fresh juice. The stainless steel surrounding the motor block increases impact resistance and prevents discolouration. Furthermore, it has a high capacity, so that one filling can produce up to four glasses of juice. LEDs in the LCD display panel provide fast and clear information on operating condition and power supply.

Der Entsafter HW-SBF15 ist mit Blick auf die Bedürfnisse von professionellen Frischsaft-Verkäufern konzipiert und gestaltet worden. Der Edelstahl um den Motorblock erhöht die Widerstandsfähigkeit gegen Stöße und Verfärbungen. Zudem hat er ein großes Fassungsvermögen, sodass aus einer Füllung bis zu vier Gläser Saft gewonnen werden können. LED-Leuchten im LCD-Anzeigefeld liefern schnelle und klare Informationen über Betriebszustand und Stromversorgung.

Statement by the jury
This juicer is convincing with its high capacity whilst having a compact design. The choice of material emphasises the demands of professionals.

Begründung der Jury
Dieser Entsafter überzeugt mit großem Fassungsvermögen bei gleichzeitig kompakter Gestaltung. Die Materialwahl unterstreicht den Profi-Anspruch.

Yz15
Slow Juicer
Langsam-Entsafter

Manufacturer
Zhejiang Supor Electrical Appliances Manufacturing Co., Ltd.,
Hangzhou, China
In-house design
Ma Han, Du Cong
Web
www.supor.com

Yz15 is an innovative juicer with which fresh juice is gently pressed out by slow crushing, thus preserving the vitamin, flavour and colour carriers. The juicer can process greater amounts of various fruit or vegetables at the same time, due to its three filling apertures. In addition, fruit ice cream can be made by using frozen fruits, first cutting them up and then feeding them into the device. The housing is gently curved so that vessels for juice and pulp can stand close to the device.

Yz15 ist ein innovativer Entsafter, bei dem frischer Saft schonend durch langsames Zerdrücken herausgepresst wird, sodass Vitamine, Geschmacks- und Farbstoffträger erhalten bleiben. Der Entsafter kann größere Mengen von unterschiedlichem Obst oder Gemüse gleichzeitig verarbeiten, da er drei Einfüllöffnungen hat. Mit dem Entsafter kann außerdem Fruchteis hergestellt werden, indem tiefgefrorene Früchte zunächst etwas zerkleinert und dann in das Gerät gegeben werden. Das Gehäuse ist sanft geschwungen, sodass die Gefäße für Saft und Trester nah am Gerät stehen.

Statement by the jury
With its gentle curves, stylish appearance and well-considered functionality, this slow juicer is an enhancement for the kitchen.

Begründung der Jury
Mit seinen sanften Rundungen, dem modernen Erscheinungsbild und einer durchdachten Funktionalität ist dieser Entsafter eine Bereicherung für die Küche.

BCP600 The Citrus Press
Zitruspresse

Manufacturer
Breville Group, Sydney, Australia
In-house design
Sam Adeloju, Richard Hoare
Web
www.breville.com.au

This innovative citrus press incorporates a reamer designed with undulating fins, to extract juice all the way to the rind, regardless of size. This makes the reamer cone effective for any type of citrus fruit. A pull down handle holds the citrus fruit on top of the cone. The handle requires less grip strength than conventional, hand-held reamers, increases leverage and reduces effort.

Diese innovative Zitruspresse hat einen Presskegel mit wellenförmigen Lamellen, um alle möglichen Arten von Zitrusfrüchten – unabhängig von ihrer Größe – effektiv bis zur Schale auszupressen. Ein Pull-down-Griff fixiert die Früchte auf dem Kegel. Der Griff verstärkt die Hebelwirkung, sodass weniger Kraft benötigt wird als bei normalen Handpressen.

Statement by the jury
Functionally as well as optically, this citrus press is a successful combination of manual and electrical press and thereby radiates freshness.

Begründung der Jury
Funktionell wie visuell ist diese Zitruspresse eine gelungene Kombination aus manueller und elektrischer Presse und strahlt dabei Frische aus.

Braun IdentityCollection
Kitchen Appliances Collection
Küchengeräte-Serie

Manufacturer
De'Longhi Braun Household GmbH,
Neu-Isenburg, Germany
In-house design
Web
www.braunhousehold.com

The Braun IdentityCollection is a kitchen appliance series following a unified design concept, comprising hand blender, steam cooker, stand blender, juicer and a compact food processor. The appliances of the series combine a simple and elegant form language with advanced technology, smart programmes and intuitive operability. The various kitchen appliances are all available in black or white.

Statement by the jury
Due to their compact construction and a consistent, timeless formal language, the appliances blend in with every kitchen without asserting themselves in the foreground.

Die Braun IdentityCollection ist eine einem einheitlichen Gestaltungskonzept folgende Küchengeräte-Serie, die Stabmixer, Dampfgarer, Standmixer, Entsafter und eine Kompakt-Küchenmaschine umfasst. Die Geräte der Serie verbinden eine schlicht-elegante Formensprache mit fortschrittlicher Technologie, intelligenten Programmen und intuitiver Bedienbarkeit. Die verschiedenen Küchengeräte sind jeweils in Schwarz oder Weiß erhältlich.

Begründung der Jury
Mit ihrer kompakten Bauweise und einer durchgängigen, zeitlosen Formensprache fügen sich die Geräte dieser Serie in jede Küche ein, ohne sich in den Vordergrund zu drängen.

Electrolux Masterpiece Collection
Kitchen Appliances Collection
Küchengeräte-Serie

Manufacturer
Electrolux, Stockholm, Sweden
In-house design
Web
www.electrolux.com

Strict geometry characterises the Masterpiece Collection, which comprises a food processor, blender and immersion blender. High-quality materials such as Tritan plastic, stainless steel and die-cast alloy have been used for the appliances to enhance user-friendliness and long life. Use of so-called PowerTilt technology in combination with titanium plated blades increases the performance in blending and leads to improved taste and smoother textures.

Statement by the jury
Powerful, robust and at the same time timelessy designed, the appliances of the Masterpiece Collection score points with long life and efficiency.

Eine strenge Geometrie kennzeichnet die Masterpiece Collection, die sich aus Küchenmaschine, Mixer und Stabmixer zusammensetzt. Hochwertige Materialien wie Tritan-Kunststoff, Edelstahl und Druckgusslegierungen wurden im Hinblick auf eine verbesserte Benutzerfreundlichkeit und Langlebigkeit verwendet. Der Einsatz sogenannter PowerTilt-Technologie in Kombination mit titanbeschichteten Klingen führt zu optimierten Verarbeitungsergebnissen.

Begründung der Jury
Kraftvoll, robust und zugleich zeitlos gestaltet, punkten die Geräte der Masterpiece Collection auch mit Langlebigkeit und Effizienz.

WMF KITCHENminis®
WMF KÜCHENminis®
Breakfast Series
Frühstücksserie

Manufacturer
wmf consumer electric GmbH,
Jettingen-Scheppach, Germany
In-house design
ahackenberg Design Consulting,
Alf Hackenberg, Munich, Germany
Web
www.wmf-ce.de
www.ahackenberg.com

The Kitchenminis are characterised by a reduced and timeless formal language. The series consists of the AromaOne filter coffee maker with 1-cup-function, a 0.8 litre kettle, a one-slice toaster, a two-egg cooker and a 0.8 litre compact blender; it was developed to save space, resources and energy. The contemporary appliance collection is crafted in high-quality Cromargan, fitting nicely in kitchens where space is limited in single and small households.

Statement by the jury
Functionally perfectly suited to single households, this series captivates by a harmonious and timeless overall appearance.

Die Küchenminis sind charakterisiert durch eine reduzierte und zeitlose Formensprache. Die Serie besteht aus der AromaOne-Filterkaffeemaschine mit 1-Tassen-Funktion, einem 0,8-Liter-Wasserkocher, einem 1-Scheiben-Toaster, einem 2-Eier-Kocher und einem 0,8-Liter-Kompaktmixer und wurde entwickelt, um Platz, Ressourcen und Energie zu sparen. Gefertigt aus hochwertigem Cromargan, passt das zeitgemäße Geräteensemble gut in Küchen mit begrenztem Raumangebot in Single- und Kleinhaushalten.

Begründung der Jury
Funktional perfekt auf Single-Haushalte abgestimmt, besticht die Serie durch ein stimmiges und zeitloses Gesamtbild.

TSF01
Toaster

Manufacturer
Smeg S.p.A.,
Guastalla (Reggio Emilia), Italy
Design
deepdesign
(Matteo Leopoldo Bazzicalupo,
Raffaella Mangiarotti),
Milan, Italy
Web
www.smeg.com
www.deepdesign.it

This 2-slice toaster in retro design makes it possible to toast various kinds of bread in different ways. Thanks to its very wide insert openings with automatic centring, it is suitable for every slice size. The toaster has six browning degree settings and can also toast, warm or defrost on one side. Toast slices can be easily removed by means of tongs as accessory; with the grill attachment as accessory, bread rolls etc. can be heated.

Statement by the jury
The TSF01 combines technology and versatility with a consistent, curved 1950's design and thus becomes an eye-catcher in the kitchen.

Dieser 2-Schlitz-Toaster im Retrodesign ermöglicht es, unterschiedliche Brotsorten auf verschiedene Arten zu rösten. Dank sehr breiter Toastschlitze mit automatischer Zentrierung ist er für jede Scheibengröße geeignet. Der Toaster hat sechs Bräunungsstufen und kann auch einseitig toasten, erwärmen oder auftauen. Mit einer Zubehör-Zange können Scheiben leicht entnommen, mit dem Zubehör-Grillaufsatz auch Brötchen etc. erwärmt werden.

Begründung der Jury
Der TSF01 vereint Technologie und Vielseitigkeit mit einem konsequenten, kurvenreichen 1950er-Jahre-Design und wird so zum Blickfang in der Küche.

Braun MultiMix 3
Hand Mixer
Handmixer

Manufacturer
De`Longhi Braun Household GmbH,
Neu-Isenburg, Germany
In-house design
Web
www.braunhousehold.com

For the Braun MultiMix 3, the motor is positioned above the mixing tools for higher efficiency. This type of construction also improves handling, as the mixer is better balanced due to shifting the centre of gravity. The handle of the small and compact device is ergonomically formed; accessories are easy to insert. Due to the enclosed housing concept, the mixer is quiet and easy to clean.

Beim Braun MultiMix 3 ist der Motor für einen höheren Wirkungsgrad oberhalb der Rührstäbe positioniert. Diese Art der Konstruktion verbessert auch die Handhabung, da der Mixer durch die Verlagerung des Schwerpunkts besser ausbalanciert ist. Der Griff des kleinen und kompakten Geräts ist ergonomisch geformt, das Zubehör ist leicht aufsteckbar. Durch das geschlossene Gehäusekonzept ist der Mixer leise und leicht zu reinigen.

Statement by the jury
Optimised ergonomics and reduced form language characterise this mixer and make it a kitchen aid which is not only compact but also practical.

Begründung der Jury
Eine optimierte Ergonomie und eine reduzierte Formensprache kennzeichnen diesen Mixer und machen ihn zum ebenso kompakten wie praktischen Küchenhelfer.

Avance SpeedTouch
Hand Blender
Stabmixer

Manufacturer
Royal Philips, Eindhoven, Netherlands
In-house design
Web
www.philips.com

The Avance SpeedTouch is an ergonomic hand blender which can be held with only fingers and thumbs and is controlled with one hand. The grip and trigger switch provide speed and handling controls. The index finger rests intuitively on the SpeedTouch knob which reacts according to pressure, i.e. the more the pressure, the greater the power. Brushed metal, deep gloss and matt textures define the appearance.

Der Avance SpeedTouch ist ein ergonomischer Stabmixer, der nur mit Fingern und Daumen gehalten werden kann und mit einer Hand gesteuert wird. Griff und Auslöseschalter bieten Geschwindigkeits- und Handhabungskontrolle. Der Zeigefinger ruht intuitiv auf der SpeedTouch-Taste, die auf Druck reagiert, d. h. je stärker der Druck, desto höher die Leistung. Gebürstetes Metall, Tiefenglanz und matte Strukturen bestimmen das Erscheinungsbild.

Statement by the jury
With its slim form, a high-quality surface design and intuitive operation, the Avance SpeedTouch makes an impression of professionalism.

Begründung der Jury
Mit seiner schlanken Form, einer hochwertigen Oberflächengestaltung und einer intuitiven Bedienung hinterlässt der Avance SpeedTouch einen professionellen Eindruck.

MaxxiMUM Range
Food Processor
Küchenmaschine

Manufacturer
Robert Bosch Hausgeräte GmbH,
Munich, Germany
In-house design
Sascha Leng, Tobias Krüger
Web
www.bosch-hausgeraete.de

The sturdy all-metal housing of the MaxxiMUM food processor impresses by its reserved and simultaneously powerful silhouette with chrome applications. The wide base provides a high degree of stability, even when under high stress. Operation is made easy due to automatic functions and a sensor control, whilst an integrated sensor regulates the mix and knead speed of the 1,600 watt motor and stops automatically when, for example, cream or beaten egg white have reached the right consistency.

Das robuste Vollmetallgehäuse der MaxxiMUM Küchenmaschine besticht durch eine dezente und zugleich kraftvolle Silhouette mit Chrom-Applikationen. Die breite Basis sorgt für hohe Standsicherheit auch bei starker Beanspruchung. Verschiedene Automatikfunktionen und eine Sensorsteuerung erleichtern die Bedienung, indem ein integrierter Sensor die Rühr- und Knetgeschwindigkeit des 1.600 Watt starken Motors reguliert und automatisch stoppt, wenn etwa Sahne oder Eischnee die richtige Konsistenz erreicht haben.

Statement by the jury
This food processor scores with its well-conceived functionality in all details as well as its sturdy and simultaneously high quality appearance.

Begründung der Jury
Diese Küchenmaschine punktet mit einer bis ins Detail durchdachten Funktionalität und einem robusten und zugleich hochwertigen Äußeren.

CHEF Sense
Kitchen Machine
Küchenmaschine

Manufacturer
Kenwood Limited, Havant, Great Britain
In-house design
Jamie Weaden, Mark Seidler,
David Lowes, James Corrigan
Web
www.kenwoodworld.com

The Chef Sense is a further development of the Chef kitchen machine. A high level of performance is delivered through an improved motor and gearbox. Externally, the design has been simplified into two main parts, reinforced by the distinctive two-tone colouring. Focus has been given to the key areas of interaction – head lift lever, speed control, bowl insertion and front accessory connection have all been enhanced to improve the product experience.

Die Chef Sense ist eine Weiterentwicklung der Küchenmaschine Chef. Ihre hohe Leistung beruht auf einem verbesserten Motor und Getriebe. Äußerlich wurde das Design auf zwei Hauptteile reduziert und durch die markante Zweifarbigkeit verstärkt. Im Mittelpunkt stehen die Schlüsselbereiche der Interaktion: Hebemechanismus, Geschwindigkeitsregler, Rührschüsselanschluss und der vordere Zubehöranschluss wurden für ein optimiertes Bedienerlebnis verbessert.

Statement by the jury
Soft lines give a modern twist to the otherwise timelessly functional design of the Chef Sense; the overall impression is that of strength and value.

Begründung der Jury
Weiche Linien verleihen der ansonsten zeitlos-funktionalen Gestaltung der Chef Sense einen modernen Twist, der Gesamteindruck vermittelt Stärke und Wertigkeit.

Avance

Kitchen Machine
Küchenmaschine

Manufacturer
Royal Philips, Eindhoven, Netherlands
In-house design
Web
www.philips.com

This compact kitchen machine is versatile thanks to its comprehensive range of accessories. Dough can be prepared with it as well as smoothies, salads or minced meat. The large, metal mixing bowl with handle and splash guard holds four litres. With its seven speeds, a 900 watt motor, pulse function and planetary stirring system, ingredients are thoroughly mixed. Due to the open jointed arm, bowl and device are easily accessible.

Statement by the jury
The Avance kitchen machine combines a compact design with surprisingly great versatility. Machine and accessories meet high demands of functionality.

Diese kompakte Küchenmaschine ist dank umfangreichen Zubehörs vielseitig einsetzbar. Teig kann mit ihr ebenso zubereitet werden wie Smoothies, Salate oder Hackfleisch. Die große Metall-Rührschüssel mit Griff und Spritzschutz fasst vier Liter. Mit sieben Geschwindigkeitsstufen, einem 900-Watt-Motor, Impulsfunktion und Planetenrührsystem werden Zutaten sorgfältig vermischt. Durch den offenen Gelenkarm sind Schüssel und Gerät leicht zugänglich.

Begründung der Jury
Die Avance Küchenmaschine vereint eine kompakte Gestaltung mit überraschend großer Vielseitigkeit. Maschine und Zubehör erfüllen hohe Ansprüche an Funktionalität.

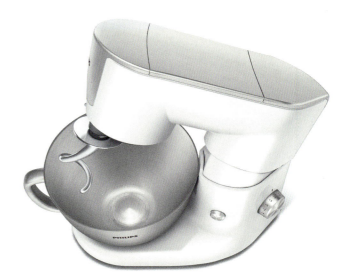

MultiOne

Kitchen Machine
Küchenmaschine

Manufacturer
Kenwood Limited, Havant, Great Britain
In-house design
Jamie Weaden, Nick Jays, Julian Wood, Robert Spencer, Davide Dessi
Web
www.kenwoodworld.com

MultiOne combines the baking emphasis of a mixer with the preparation emphasis of a food processor, giving an equal focus to both types of task to create a hybrid design concept. The large mixing bowl and tools enable whisking, beating, kneading and more. A single outlet on top of the machine allows easy fitting of a range of attachments such as primary food processor, juicer and blender. In this way, the MultiOne is versatile in use and at the same time a space saver. Clean surfaces ensure the product is easily cleaned.

Statement by the jury
The hybrid design concept underlying the MultiOne results in increased functionality and simultaneously reduced space requirement.

Die MultiOne kombiniert die Funktionen eines Rührgeräts mit den Funktionen eines Food-Prozessors für vorbereitende Tätigkeiten. Dank des hybriden Designkonzepts werden beide Funktionsarten gleich gewichtet. Die große Rührschüssel und das passende Zubehör ermöglichen unter anderem das Verquirlen, Aufschlagen oder Kneten von Zutaten. Auf der Oberseite der Maschine können verschiedene Aufsätze wie Multi-Zerkleinerer, Entsafter oder Mixaufsatz angeschlossen werden. So ist die MultiOne vielseitig nutzbar und zugleich platzsparend. Glatte Oberflächen erleichtern zudem die Reinigung.

Begründung der Jury
Das der MultiOne zugrunde liegende hybride Designkonzept mündet in erhöhter Funktionalität bei gleichzeitig reduziertem Platzbedarf.

sono 5
Multislicer
Allesschneider

Manufacturer
ritterwerk GmbH, Gröbenzell, Germany
In-house design
Web
www.ritterwerk.de

A floor panel of high-quality safety glass characterises the elegant appearance of the sono 5. Apart from the floor panel, the multislicer is a metal construction which gives it a high degree of stability. The continuously variable slice thickness settings range up to about 23 mm, the large surface carriage can also be removed. Due to its self-supporting construction, the appliance is easy to clean, and slices can be deposited directly onto the platter. The slicer is driven by an energy-saving 65 watt ecological motor.

Eine Bodenplatte aus hochwertigem Sicherheitsglas prägt die elegante Anmutung der sono 5. Mit Ausnahme der Platte ist der Allesschneider aus Metall gefertigt, was ihm ein hohes Maß an Stabilität verleiht. Die stufenlose Schnittstärkeeinstellung reicht bis ca. 23 mm, der großflächige Schlitten kann abgenommen werden. Durch die freitragende Konstruktion ist das Gerät einfach zu reinigen und die Scheiben können direkt auf einem Teller abgelegt werden. Der Allesschneider wird durch einen energiesparenden 65 Watt Eco-Motor angetrieben.

Statement by the jury
Well-balanced proportions and the use of high-value materials such as a glass floor panel lend an elegant appearance to the sono 5.

Begründung der Jury
Ausgewogene Proportionen und der Einsatz hochwertiger Materialien wie einer Bodenplatte aus Glas verleihen der sono 5 eine elegante Anmutung.

Midea MB-FZ4086 IH
Rice Cooker
Reiskocher

Manufacturer
Midea Group,
Midea Consumer Electric MFG. Co., Ltd.,
Foshan, China
In-house design
Shoucheng Xu, Xunlan Yin
Web
www.midea.com.cn

This rice cooker was specially conceived against the backdrop of an increase in overweight in Asia, in order to boil rice so that it contains less sugar. Functionality of the cooker is taken from the principle of ebb and flood, where sand in the waves is constantly being washed back and forth. Following this principle, rice grains in the cooker are constantly scoured while steaming and boiling, whereby 40 percent of the soluble sugar is washed out.

Statement by the jury
The Midea MB-FZ4086 IH combines innovative functionality with an appealing design so that it is likely to be often in use.

Dieser Reiskocher wurde vor dem Hintergrund einer Zunahme von Übergewicht in Asien konzipiert, um Reis so zu kochen, dass er weniger Zucker enthält. Die Funktionsweise des Kochers ist dem Prinzip von Ebbe und Flut nachempfunden, bei dem der Sand in den Wellen immer wieder hin und her gespült wird. Entsprechend werden die Reiskörner im Kocher beim Dämpfen und Garen ständig durchgespült, wodurch 40 Prozent des löslichen Zuckers herausgeschwemmt werden.

Begründung der Jury
Der Midea MB-FZ4086 IH verbindet eine innovative Funktionsweise mit einer ansprechenden Gestaltung, sodass man ihn gerne häufig nutzt.

IR D5
Food Dehydrator
Dörrgerät

Manufacturer
L'Equip, Seoul, South Korea
In-house design
Jaeyoon Lee
Web
www.lequip.co.kr

This food dehydrator uses infrared radiation in order to dehydrate foods by achieving a similar effect as by dehydration in the sun. Stainless steel is used for the interior of the appliance to maximise penetration by near-infrared radiation. The low-noise food dehydrator contains five levels which are evenly heated. Users can choose between two dehydration methods – sun-drying or air-drying.

Statement by the jury
Strong colour contrasts and gently rounded contours characterise the appearance of the IR D5, which also convinces with sophisticated functionality.

Dieses Dörrgerät arbeitet mit Infrarotstrahlung, um beim Dehydrieren der Lebensmittel einen ähnlichen Effekt zu erzielen, wie beim Trocknen in der Sonne. Im Inneren des Geräts wird Edelstahl eingesetzt, um die Durchdringung mit Nah-Infrarotstrahlung zu maximieren. Der geräuscharme Dörrapparat hat fünf Ebenen, die gleichmäßig erhitzt werden. Nutzer können zwischen zwei Trockenmethoden wählen – sonnengetrocknet oder luftgetrocknet.

Begründung der Jury
Starke Farbkontraste und sanft abgerundete Konturen prägen das Erscheinungsbild des IR D5, der zudem mit einer ausgereiften Funktionalität überzeugt.

Series | 8 Built-in Range Black
Built-in Appliance Range
Einbaugerätereihe

Manufacturer
Robert Bosch Hausgeräte GmbH,
Munich, Germany

In-house design
Robert Sachon,
Ulrich Goss,
Oliver Kraemer,
Christoph Ortmann

Web
www.bosch-hausgeraete.de

reddot award 2015
best of the best

Elegant integration

Contemporary architecture is marked by open kitchens that have often replaced classic space arrangements. The kitchen is defined as the centre of family life, a centre that is open to all sides. This implies a constant challenge for the design of kitchen devices as they have to blend in seamlessly into the existing interior. The built-in oven ranges from the Color Glass series respond to these architectural concepts with an overall purist straightline design paired with innovative user interfaces. All devices are marked by the same clear line design, which allows many possibilities of individual combinations in the kitchen. The high-quality surfaces are held in black to lend the oven ranges an air of premium yet unobtrusive appearance creating an elegant ambiance. The self-explanatory user interface highlights an iconic control ring, which thanks to its sensor technology, the BoschAssist, automatically ensures safe and intuitive operation as well as a high degree of user comfort. State-of-the-art steam and microwave functionality are also integrated. Rounding off the concept, the consistent and highly aesthetic design of these devices is also reflected in their connectability into various combinations that open new possibilities of interaction.

Elegant integriert

In der zeitgenössischen Architektur haben offene Küchenkonzepte vielfach die klassische Raumaufteilung abgelöst. Die Küche wird definiert als ein von allen Seiten zugängliches Zentrum des familiären Lebens. Für das Design der Küchengeräte bedeutet dies eine stetige Herausforderung, da die Geräte sich nahtlos in das Interieur einfügen müssen. Die Einbau-Backöfen aus der Color Glass-Serie begegnen diesen architektonischen Konzepten mit einer puristisch geradlinigen Formensprache und einer innovativen Bedienoberfläche. Alle Geräte verbindet die gleiche klare Linienführung, weshalb sich sehr viele Möglichkeiten der individuellen Kombination in der Küche ergeben. Die hochwertigen Oberflächen in der Farbe Schwarz schaffen dabei ein edel anmutendes und zurückhaltendes Ambiente, welches sich elegant zurücknimmt. Im Mittelpunkt der selbsterklärenden Bedienoberfläche dieser Einbau-Backöfen steht ein zentraler Bedienring. Dieser regelt eine hochentwickelte Sensortechnik, die dank der BoschAssist-Automatik ein hohes Maß an Komfort bietet und den Nutzer sicher und intuitiv agieren lässt. Außerdem ist eine zeitgemäße Dampf- und Mikrowellenfunktionalität integriert. Die stringente und überaus ästhetische Gestaltung der Geräte dieser Serie zeigt sich schließlich auch darin, dass sie untereinander vernetzt sind, wodurch sich neue Möglichkeiten der Interaktion eröffnen.

Statement by the jury

In a fascinating manner, the purist design lends these devices a strong identity. A well-conceived, thoroughly structured interface with a central control ring offers a high level of convenience and an enticing user experience. Designed upon the concept of installation in open kitchens, this series offers many benefits that allow the products to be integrated harmoniously and in many variations in any given environment.

Begründung der Jury

Auf beeindruckende Weise verleiht eine puristische Gestaltung dieser Gerätereihe ihre starke Identität. Ein durchgehend gut strukturiertes Interface mit einem zentral angeordneten Bedienring bietet viel Komfort und ein intensives Nutzererlebnis. Für den Einsatz im Bereich offener Küchenkonzepte weist diese Serie zudem viele Vorteile auf, da sie harmonisch und variabel in jedes Umfeld integriert werden kann.

Designer portrait
See page 38
Siehe Seite 38

Series | 8 Built-in Range Stainless Steel
Built-in Appliance Range
Einbaugerätereihe

Manufacturer
Robert Bosch Hausgeräte GmbH,
Munich, Germany
In-house design
Robert Sachon, Ulrich Goss,
Oliver Kraemer, Christoph Ortmann
Web
www.bosch-hausgeraete.de

The striking feature of this interactive built-in appliance range is the central control ring located in the centre of the high-resolution TFT display. All automatic programmes and functions such as steam cooking or microwave heating can be selected quickly and simply via this control. Operation is intuitive, and cooking is supported by sensor technology. The front design is purist and clearly structured; the combination of black glass and stainless steel elements lends professional aesthetics to the appliances.

Markantes Designmerkmal dieser vernetzten Einbaugerätereihe ist der zentrale Bedienring, der in der Mitte des hochauflösenden TFT-Displays sitzt. Über ihn können sämtliche Automatik-Programme und Funktionen wie Dampfgaren oder Mikrowellenerhitzung schnell und einfach angewählt werden. Die Bedienung ist intuitiv, der Garprozess wird durch Sensortechnik unterstützt. Die Gestaltung der Front ist puristisch und klar gegliedert, die Kombination von schwarzem Glas und Edelstahlelementen verleiht den Geräten eine professionelle Ästhetik.

Statement by the jury
The reduced and at the same time extremely distinctive frontal design of this appliance series indicates quality and technical perfection.

Begründung der Jury
Die reduzierte und zugleich äußerst markante Frontgestaltung dieser Gerätereihe kommuniziert Qualität und technische Perfektion.

Series | 8 Built-in Range White
Built-in Appliance Range
Einbaugerätereihe

Manufacturer
Robert Bosch Hausgeräte GmbH,
Munich, Germany
In-house design
Robert Sachon, Ulrich Goss,
Oliver Kraemer, Christoph Ortmann
Web
www.bosch-hausgeraete.de

This white, high-quality ColorGlass built-in appliance range has been specially conceived for open kitchens and can be integrated harmoniously in the kitchen environment. Continuous lines facilitate the combination of several appliances. The defining feature of the purist series is the central control ring, by means of which the appliances can be controlled with the fingers. Sensor technology and BoschAssist automation simplify preparation of foods. Besides the integrated steam and microwave functionality, the interconnected appliances offer new possibilities of interaction.

Die hochwertige ColorGlass-Einbaugeräte-reihe in Weiß ist speziell für offene Küchen konzipiert und lässt sich harmonisch in das Küchenumfeld integrieren. Eine durchgehende Linienführung ermöglicht die Kombination mehrerer Geräte. Das prägende Element der puristischen Serie ist der zentrale Bedienring, über den sich die Geräte einfach per Finger steuern lassen. Sensortechnik und die BoschAssist-Automatik vereinfachen die Zubereitung von Speisen. Neben integrierter Dampf- und Mikrowellenfunktionalität bieten die vernetzten Geräte auch neue Möglichkeiten der Interaktion.

B56CT64 Built-in Oven Design 600
B56CT64 Einbaubackofen Design 600

Manufacturer
Constructa-Neff Vertriebs GmbH,
Munich, Germany
Design
BSH Hausgeräte GmbH, Neff Design Team,
Munich, Germany
Web
www.neff.de

The B56CT64 oven provides better access to food in the oven due to the so-called Slide&Hide function. "Slide" stands for a rotary knob which stays in the same position in the hand. "Hide" means that the oven door can be easily and completely pushed into a compartment under the oven. Programmes such as the pyrolysis self-cleaning function can be selected simply by touch control via the TFT colour display.

Statement by the jury
This oven impresses due to a clear, purist front design and clever functions such as the retractable door.

Der Backofen B56CT64 ermöglicht mit der sogenannten Slide&Hide-Funktion besseren Zugang zum Gargut. „Slide" steht für einen mitdrehenden Türgriff, der stets in der gleichen Position in der Hand liegt. „Hide" wiederum erlaubt, dass die Backofentür bequem und vollständig in ein Fach unter dem Backofen geschoben werden kann. Programme wie die pyrolytische Selbstreinigungsfunktion lassen sich einfach mittels Touch-Bedienung über das TFT-Farbdisplay anwählen.

Begründung der Jury
Dieser Backofen besticht mit einer klaren, puristischen Frontgestaltung und cleveren Funktionen wie der versenkbaren Türe.

B55VS24 Built-in Oven Design 600
B55VS24 Einbaubackofen Design 600

Manufacturer
Constructa-Neff Vertriebs GmbH,
Munich, Germany
Design
BSH Hausgeräte GmbH, Neff Design Team,
Munich, Germany
Web
www.neff.de

Due to extending the oven by the VarioSteam function, food being cooked is provided with moisture. The water tank is concealed behind the fascia, which lifts by pressing a button, allowing the tank to be removed and refilled. The front design is of high quality and clear; the operating panel with Shift-Control and TFT display is integrated in the fascia which is made of brushed stainless steel.

Statement by the jury
For this oven, a high level of aesthetics combines with a user-friendly design. A particular highlight is the well-conceived positioning of the water tank.

Durch die Erweiterung des Backofens um die Dampffunktion VarioSteam werden Speisen beim Garen mit Feuchtigkeit versorgt. Der Wassertank verbirgt sich hinter der Blende, die sich auf Knopfdruck hebt, sodass der Tank einfach entnommen und befüllt werden kann. Die Frontgestaltung ist hochwertig und klar, das Bedienfeld mit ShiftControl und TFT-Display in die Blende aus gebürstetem Edelstahl integriert.

Begründung der Jury
Eine hochwertige Ästhetik verbindet sich bei diesem Backofen mit einer benutzerfreundlichen Gestaltung. Besonders hervorzuheben ist die durchdachte Positionierung des Wassertanks.

B55CR22 Built-in Oven Design 600
B55CR22 Einbaubackofen Design 600

Manufacturer
Constructa-Neff Vertriebs GmbH,
Munich, Germany
Design
BSH Hausgeräte GmbH, Neff Design Team,
Munich, Germany
Web
www.neff.de

This oven model convinces due to a linear design and a combination of high-quality materials, namely brushed steel and dark glass. The surface is highly resistant and easy to clean. All controls are operated via a generously dimensioned, high-resolution TFT display by intuitive ShiftControl. The rotary knob and the fully retractable oven door (Slide&Hide function) optimise access to the appliance.

Statement by the jury
Its clear design form and the use of high-quality materials in classical colours impart on this oven a timelessly elegant impression.

Dieses Backofenmodell überzeugt durch ein geradliniges Design und eine hochwertige Materialkombination von gebürstetem Edelstahl und dunklem Glas. Die Oberfläche ist unempfindlich und leicht zu reinigen. Die komplette Steuerung erfolgt über ein großzügiges, hochauflösendes TFT-Display per intuitiver ShiftControl-Bedienung. Der mitdrehende Griff und die vollständig versenkbare Backofentür (Slide&Hide-Funktion) optimieren den Zugang zum Gerät.

Begründung der Jury
Seine klare Formensprache und die Verwendung hochwertiger Materialien in klassischen Farben verleihen diesem Backofen eine zeitlos-elegante Anmutung.

B25CR22 Built-in Oven
Design 600
B25CR22 Einbaubackofen
Design 600

Manufacturer
Constructa-Neff Vertriebs GmbH,
Munich, Germany
Design
BSH Hausgeräte GmbH, Neff Design Team,
Munich, Germany
Web
www.neff.de

This built-in oven has a purist frontal design with an elegant combination of brushed stainless steel and blackened glass. A special hot air system (CircoTherm) allows various foods to be cooked together on up to three levels without mixing aromas since the heat is objectively diverted round the foods. A TFT colour display combined with ShiftControl functionality simplifies operation.

Statement by the jury
Behind the elegant appearance of this oven, innovative technologies and a high level of functionality are hidden.

Dieser Einbaubackofen hat eine puristische Frontgestaltung mit einer eleganten Blende aus gebürstetem Edelstahl und geschwärztem Glas. Ein spezielles Heißluftsystem (CircoTherm) erlaubt es, verschiedene Speisen auf bis zu drei Ebenen ohne Aromavermischung zusammen zu garen, indem es die Hitze gezielt um die Speisen herumleitet. Ein TFT-Farbdisplay erleichtert in Kombination mit der ShiftControl-Steuerung die Bedienung.

Begründung der Jury
Hinter dem eleganten Erscheinungsbild dieses Ofens verbergen sich innovative Technologien und eine hohe Funktionalität.

B57VS24 Built-in Oven
Design 610
B57VS24 Einbaubackofen
Design 610

Manufacturer
Constructa-Neff Vertriebs GmbH,
Munich, Germany
Design
BSH Hausgeräte GmbH, Neff Design Team,
Munich, Germany
Web
www.neff.de

Blackened glass and stainless steel in combination with clear lines are the dominating design features of this built-in oven. From a functional viewpoint, it combines the classical oven programmes with a steaming function (VarioSteam), by means of which the food, while cooking, is moisturised. The fascia is easily opened by pressing a button; the newly developed water tank is easy to remove and refill.

Statement by the jury
The elegant impression of this oven results from a minimalist front design and the use of high-quality materials.

Geschwärztes Glas und Edelstahl in Kombination mit einer klaren Linienführung sind die dominierenden Gestaltungsmerkmale dieses Einbaubackofens. In funktionaler Hinsicht kombiniert er die klassischen Backofen-Programme mit einer Bedampfungsfunktion (VarioSteam), durch die die Speisen beim Garen mit Feuchtigkeit versorgt werden. Die Blende lässt sich einfach auf Knopfdruck öffnen, der neu entwickelte Wassertank bequem entnehmen und befüllen.

Begründung der Jury
Die elegante Anmutung dieses Backofens resultiert aus einer minimalistischen Frontgestaltung und dem Einsatz hochwertiger Materialien.

B57CR22
Built-in Oven
Einbaubackofen

Manufacturer
Constructa-Neff Vertriebs GmbH,
Munich, Germany
Design
BSH Hausgeräte GmbH, Neff Design Team,
Munich, Germany
Web
www.neff.de

A clear, purist form language and the use of blackened glass and stainless steel characterise this built-in oven. From an ergonomic viewpoint, the design is also convincing; for instance, when opening and closing the oven door, the doorknob always stays in the same position in the hand. When opened, the oven door disappears under the oven compartment, providing more space in the cooking environment.

Statement by the jury
When designing this oven, great attention was paid to the ergonomics, making operation very convenient.

Eine klare, puristische Formensprache und der Einsatz von geschwärztem Glas und Edelstahl kennzeichnen diesen Einbaubackofen. Die Gestaltung überzeugt auch unter ergonomischen Gesichtspunkten, so liegt der mitdrehende Türgriff beim Öffnen und Schließen der Backofentür stets in gleicher Position in der Hand. Die Backofentür verschwindet beim Öffnen unter dem Backraum und schafft mehr Platz im Kochumfeld.

Begründung der Jury
Bei der Gestaltung dieses Backofens wurde großes Augenmerk auf die Ergonomie gelegt, wodurch die Bedienung sehr komfortabel ist.

B58CT64 Built-in Oven Design 610
B58CT64 Einbaubackofen Design 610

Manufacturer
Constructa-Neff Vertriebs GmbH,
Munich, Germany
Design
BSH Hausgeräte GmbH, Neff Design Team,
Munich, Germany
Web
www.neff.de

The B58CT64 follows a design concept which focuses on a high level of operating convenience. The oven doorknob is designed to rotate in the opposite direction to the door when opening and closing. The oven door is also fully retractable which facilitates not only access to the food inside but also makes cleaning easier. Navigation through the programmes is eased by FullTouch-Control with TFT colour display.

Statement by the jury
This built-in oven is characterised by a high degree of operating convenience and linear, functionally well considered design.

Der B58CT64 folgt einem Gestaltungskonzept, das einen hohen Bedienkomfort in den Mittelpunkt stellt. Der Griff der Backofentür ist so konstruiert, dass er sich während des Öffnens oder Schließens entgegen der Türbewegung mitdreht. Die Backofentür ist zudem voll versenkbar, was den Zugriff auf das Gargut sowie die Reinigung erleichtert. Die Navigation durch die Programme erfolgt mittels FullTouchControl auf dem TFT-Farbdisplay.

Begründung der Jury
Dieser Einbaubackofen zeichnet sich durch einen hohen Bedienkomfort und eine geradlinige, funktional durchdachte Gestaltung aus.

C18QT27 Compact Built-in Oven & N17HH10 Warming Drawer Design 610
C18QT27 Kompakt-Einbaubackofen & N17HH10 Wärmeschublade Design 610

Manufacturer
Constructa-Neff Vertriebs GmbH,
Munich, Germany
Design
BSH Hausgeräte GmbH, Neff Design Team,
Munich, Germany
Web
www.neff.de

Stainless steel strips at the sides together with blackened glass surfaces combine this compact built-in oven and a warming drawer from an optical viewpoint for a harmonious, compact kitchen front. From a functional viewpoint, they can be mounted one above the other without furniture partitions, thanks to the so-called Seamless-Combination. The seamless combination of both appliances makes food preparation more convenient.

Statement by the jury
Due to the stainless steel trims and a harmoniously coordinated design, built-in oven and warming drawer give the appearance of one compact appliance.

Seitlich aufgesetzte Edelstahl-Lisenen sowie schwarze Glasflächen verbinden diesen Kompakt-Einbaubackofen und eine Wärmeschublade in visueller Hinsicht für eine harmonische, kompakte Küchenfront. In funktionaler Hinsicht können sie dank der sogenannten SeamlessCombination ohne Möbelzwischenböden aufeinander gestellt werden. Die nahtlose Kombination der beiden Geräte sorgt für eine komfortable Handhabung bei der Essenszubereitung.

Begründung der Jury
Durch die seitlichen Edelstahlblenden und eine harmonisch aufeinander abgestimmte Gestaltung wirken Einbaubackofen und Wärmeschublade wie ein kompaktes Gerät.

B46FT64, B56VT64 Built-in Ovens Design 600
B46FT64, B56VT64 Einbaubacköfen Design 600

Manufacturer
Constructa-Neff Vertriebs GmbH,
Munich, Germany
Design
BSH Hausgeräte GmbH, Neff Design Team,
Munich, Germany
Web
www.neff.de

This built-in oven combines a classical oven and a steamer in one appliance. By means of the VarioSteam function, food is treated with moisture at three intensity levels. Models with FullSteam function are complete ovens and steamers in one unit and thus provide flexibility when preparing food. All functions operate intuitively by touch via a high-resolution TFT colour display with FullTouchControl.

Statement by the jury
The front of this built-in oven blends harmoniously with a modern kitchen environment due to the use of stainless steel and glass as well as a classical panel design.

Dieser Einbaubackofen vereint klassischen Backofen und Dampfgarer in einem Gerät. Mit der VarioSteam-Funktion werden Speisen in drei Intensitätsstufen mit Feuchtigkeit versorgt. Modelle mit FullSteam-Funktion sind vollwertige Backöfen und Dampfgarer in einem und bieten so Flexibilität bei der Essenszubereitung. Alle Funktionen lassen sich per Fingerberührung intuitiv über ein hochauflösendes TFT-Farbdisplay mit FullTouchControl anwählen.

Begründung der Jury
Die Gerätefront dieses Einbaubackofens ordnet sich durch den Einsatz von Edelstahl und Glas sowie ein klassisches Blendendesign harmonisch in moderne Küchenlandschaften ein.

B58VT68 & C18FT48
Built-in Ovens Design 610
B58VT68 & C18FT48
Einbaubacköfen Design 610

Manufacturer
Constructa-Neff Vertriebs GmbH,
Munich, Germany
Design
BSH Hausgeräte GmbH, Neff Design Team,
Munich, Germany
Web
www.neff.de

The special feature of this built-in solution is that even appliances of different heights merge, in terms of design, into one unit due to the so-called Seamless-Combination. The ovens integrate harmoniously into the kitchen front by utilising the available space. The different kitchen appliances such as oven and steamer with various features can be individually combined.

Statement by the jury
The compatibility of these ovens, even in cases of limited available space, is convincing, and they present a harmonious overall appearance.

Das Besondere an dieser Einbaulösung ist, dass auch Geräte von unterschiedlicher Höhe durch die sogenannte SeamlessCombination in gestalterischer Hinsicht eine Einheit bilden. Die Öfen integrieren sich unter Ausnutzung des vorhandenen Raums harmonisch in die Küchenfront. Die verschiedenen Kochgeräte, darunter Backöfen und Dampfgarer mit unterschiedlichen Ausstattungen, lassen sich individuell zusammenstellen.

Begründung der Jury
Diese Öfen überzeugen durch ihre Kombinierbarkeit auch bei begrenztem Platzangebot und zeigen dabei ein stimmiges Gesamtbild.

B47FS22, C17FS42
Built-In Ovens Design 610
B47FS22, C17FS42
Einbaubacköfen Design 610

Manufacturer
Constructa-Neff Vertriebs GmbH,
Munich, Germany
Design
BSH Hausgeräte GmbH, Neff Design Team,
Munich, Germany
Web
www.neff.de

These built-in ovens can be combined in a vertical configuration seamlessly, without the necessity of furniture partitions. Thanks to continuous strips of stainless steel at the sides, even appliances of different heights merge optically into one unit. Among the compatible appliances of this series are also microwaves, ovens and steamers with various equipment items such as the FullSteam function or the VarioSteam steam support.

Statement by the jury
This built-in solution facilitates various appliance combinations in a small space while assuring a homogenous appearance.

Diese Einbauöfen lassen sich vertikal nahtlos miteinander kombinieren, ohne dass Möbelzwischenböden zur Separierung notwendig sind. Dank durchgängiger Edelstahlblenden an den Seiten bilden auch Geräte von unterschiedlicher Höhe eine visuelle Einheit. Zu den kombinierbaren Geräten dieser Linie gehören Mikrowellen, Backöfen und Dampfgarer mit unterschiedlicher Ausstattung wie der Volldampffunktion FullSteam oder der Dampfunterstützung VarioSteam.

Begründung der Jury
Diese Einbaulösung ermöglicht unterschiedliche Gerätekombinationen auf engem Raum unter Wahrung eines homogenen Erscheinungsbildes.

B48FT68 Built-in Oven &
N17HH20 Warming Drawer
Design 610
B48FT68 Einbaubackofen &
N17HH20 Wärmeschublade
Design 610

Manufacturer
Constructa-Neff Vertriebs GmbH,
Munich, Germany
Design
BSH Hausgeräte GmbH, Neff Design Team,
Munich, Germany
Web
www.neff.de

This combination of built-in oven and warming drawer shows a harmonious design, facilitated by the built-in solution "SeamlessCombination". It allows even appliances of different heights to appear as one unit so that they merge seamlessly into the kitchen front and make optimal use of the available space. In the warming drawer, food can be defrosted, kept warm or gently cooked.

Statement by the jury
This built-in appliance combination makes a formally harmonious and functionally mature appearance and thus contributes to a peaceful impression of the kitchen front.

Diese Kombination von Einbaubackofen und Wärmeschublade zeigt eine harmonische Gestaltung, die durch die Einbaulösung „SeamlessCombination" ermöglicht wird. Sie erlaubt es, auch Geräte von unterschiedlicher Höhe visuell zu einer Einheit zu verbinden, sodass sie sich nahtlos in die Küchenfront einpassen und vorhandener Raum optimal ausgenutzt wird. In der Wärmeschublade können Speisen angetaut, warmgehalten oder sanft gegart werden.

Begründung der Jury
Diese Einbaugerätekombination präsentiert sich formal harmonisch und funktional ausgereift und trägt so zu einem ruhigen Erscheinungsbild der Küchenfront bei.

B27CR22
Built-in Oven
Einbaubackofen

Manufacturer
Constructa-Neff Vertriebs GmbH,
Munich, Germany
Design
BSH Hausgeräte GmbH, Neff Design Team,
Munich, Germany
Web
www.neff.de

The purist appearance of the B27CR22 built-in oven is created by the homogenous front of blackened glass which is interrupted only by side trims of stainless steel and the stainless steel handle. The high-resolution display with convenient ShiftControl operation is intuitive. Furthermore, due to a special air circulation technology (CircoTherm), different foods can be cooked, baked and roasted together on three levels.

Statement by the jury
The overall appearance of the oven is characterised by its minimalist black glass front. It convinces additionally due to its intuitive user interface.

Das puristische Erscheinungsbild des Einbaubackofens B27CR22 entsteht durch die homogene Front aus geschwärztem Glas, die nur durch seitliche Edelstahlblenden und den Griff aus Edelstahl aufgebrochen wird. Das hochauflösende Display mit komfortabler ShiftControl-Bedienung ist selbsterklärend. Darüber hinaus können dank einer speziellen Luftzirkulationstechnik (CircoTherm) verschiedene Speisen auf drei Ebenen zusammen garen, backen und braten.

Begründung der Jury
Das Gesamtbild des Backofens wird durch seine minimalistische schwarze Glasfront geprägt. Zudem überzeugt er durch eine selbsterklärende Bedienoberfläche.

B48FT68, B58VT68, B48FT78 Built-in Ovens Design 610
B48FT68, B58VT68, B48FT78
Einbaubacköfen Design 610

Manufacturer
Constructa-Neff Vertriebs GmbH,
Munich, Germany
Design
BSH Hausgeräte GmbH, Neff Design Team,
Munich, Germany
Web
www.neff.de

The basic design features of this built-in oven are an extensive front surface design, clear lines and reduction to essentials. Touch operation of the high-resolution TFT colour display is intuitive; programmes such as "VarioSteam", which supplies moisture to the food while it is cooking, or the steam cooking function "FullSteam" are easily selectable.

Statement by the jury
These built-in ovens with their extensive, reduced design blend discreetly in modern living environments and impress with a complex functionality.

Grundlegende Designmerkmale dieser Einbaubacköfen sind eine großflächige Frontgestaltung, klare Linienführung und Reduktion auf das Wesentliche. Die Bedienung per Fingerberührung auf dem hochauflösenden TFT-Farbdisplay ist intuitiv, Programme wie „VarioSteam", das die Speisen während des Erhitzens mit Feuchtigkeit versorgt, oder die Dampfgarfunktion „FullSteam" sind einfach anwählbar.

Begründung der Jury
Diese Einbaubacköfen fügen sich mit ihrer großflächigen, reduzierten Gestaltung dezent in moderne Wohnwelten ein und beeindrucken mit einer komplexen Funktionalität.

C18MT27
Compact Built-in Oven
Kompakt-Einbaubackofen

Manufacturer
Constructa-Neff Vertriebs GmbH,
Munich, Germany
Design
BSH Hausgeräte GmbH, Neff Design Team,
Munich, Germany
Web
www.neff.de

The appearance of this compact built-in oven is characterised by an elegant, reduced design and the combination of blackened glass and stainless steel elements. The control panel with TFT display and touch control is integrated discreetly in the black glass panel. In spite of its compact size, the oven offers steam support by the VarioSteam function. The water tank is located behind the panel and is easy to remove.

Statement by the jury
The harmonious overall appearance of this compact oven is based on a technically as well as aesthetically convincing design.

Das Erscheinungsbild dieses Kompakt-Einbaubackofens ist geprägt durch ein elegantes, reduziertes Design und die Kombination von geschwärztem Glas und Edelstahlelementen. Das Bedienfeld mit TFT-Display und Touch-Steuerung ist dezent in die Blende aus schwarzem Glas integriert. Trotz seiner kompakten Größe bietet der Ofen mit der VarioSteam-Funktion eine Dampfunterstützung. Der Wassertank ist hinter der Blende platziert und lässt sich leicht entnehmen.

Begründung der Jury
Das stimmige Gesamtbild dieses Kompakt-Einbaubackofens basiert auf einer technisch wie ästhetisch überzeugenden Gestaltung.

CS4574M
Built-in Combi Steam Oven
Einbau-Kombidampfgarer

Manufacturer
ATAG Nederland BV, Duiven, Netherlands
In-house design
Design
Van Berlo, Eindhoven, Netherlands
Web
www.atag.nl
www.vanberlo.nl

Black glass with discrete metallic paint and a sculptural handle characterise the front of this oven of the Magna series. It is framed by precision-milled profiles which match seamlessly with the furniture. The characteristic line on the control panel continues in the touchscreen, and adjacent menu keys allow intuitive control. The oven volume is sufficient for simultaneous preparation of several foods with the steam cooking function or a combination of steam and hot air.

Statement by the jury
The front design of this combi steam oven delights with a continuously maintained, distinctive design, which includes also the graphic design of the user interface.

Schwarzes Glas mit dezenter Metalliclackierung und ein skulpturaler Griff prägen die Front dieses Ofens aus der Magna-Serie. Eingefasst wird sie von präzisionsgefrästen Profilen, die nahtlos ans Mobiliar anschließen. Die prägnante Linie des Bedienpanels setzt sich im Touchscreen fort, angrenzende Menütasten erlauben eine intuitive Bedienung. Das Volumen des Ofens ermöglicht die gleichzeitige Zubereitung mehrerer Speisen mit der Dampfgarfunktion oder einer Kombination von Dampf und Heißluft.

Begründung der Jury
Die Frontgestaltung dieses Kombidampfgarers begeistert mit einer durchgehend klaren, markanten Gestaltung, die auch die Grafik der Benutzeroberfläche mit einschließt.

DGM 6800
Steam Oven with Microwave
Dampfgarer mit Mikrowelle

Manufacturer
Miele & Cie. KG, Gütersloh, Germany
In-house design
Web
www.miele.de

The DGM 6800 combines a steam cooker and microwave in one built-in appliance. The stainless steel cooking interior is illuminated by LEDs and is generously dimensioned with 40 litres capacity. The water tank is located flush with the front frame; to remove it, only a light pressure on the back of the handle is needed. The appliance is equipped with MultiSteam technology. Automatic programmes and steamer baskets specially designed to fit the oven contribute to user convenience.

Statement by the jury
A well-conceived design makes it possible for the DGM 6800 to integrate two cooking appliances innovatively in one built-in module.

Der DGM 6800 kombiniert Dampfgarer und Mikrowelle in einem Einbaugerät. Der mit LEDs ausgeleuchtete Edelstahl-Garraum ist mit 40 Litern Fassungsvermögen großzügig bemessen. Der Wassertank sitzt flächenbündig im Frontrahmen, zur Entnahme genügt ein kurzes Drücken auf den Griffrücken. Das Gerät ist zur Dampferzeugung mit MultiSteam-Technologie ausgestattet. Automatikprogramme und speziell auf den Garraum zugeschnittene Garbehälter machen die Benutzung komfortabel.

Begründung der Jury
Eine durchdachte Gestaltung ermöglicht bei dem DGM 6800 die innovative Kombination zweier Gargeräte in einem Einbaumodul.

C17WR01
Built-in Microwave
(with Grill) Design 610
C17WR01 Einbau-Mikrowelle
(mit Grill) Design 610

Manufacturer
Constructa-Neff Vertriebs GmbH,
Munich, Germany
Design
BSH Hausgeräte GmbH, Neff Design Team,
Munich, Germany
Web
www.neff.de

These built-in microwaves with or without grill are characterised by a purist form language and a material mix of blackened glass and stainless steel. The appliances blend in well with the kitchen ambience due to their compact appearance and narrow gap clearance. The cooking chamber is of stainless steel, and operation is by touch-buttons which, in combination with a high-definition TFT colour and text display, facilitate quick navigation.

Statement by the jury
Functionality, contemporary equipment details and a purist appearance characterise these built-in microwaves.

Diese Einbau-Mikrowellen mit oder ohne Grill sind durch eine puristische Formensprache und einen Materialmix aus geschwärztem Glas und Edelstahl geprägt. Die Geräte fügen sich dank ihrer kompakten Erscheinung und schmaler Spaltabstände gut in die Küchenumgebung ein. Der Garraum ist aus Edelstahl gefertigt, die Bedienung erfolgt über Touchtasten, die in Verbindung mit einem hochauflösenden TFT-Farb- und Textdisplay ein schnelles Navigieren ermöglichen.

Begründung der Jury
Funktionalität, zeitgemäße Ausstattungsdetails und ein puristisches Erscheinungsbild zeichnen diese Einbau-Mikrowellen aus.

WD1674M
Built-in Warming Drawer
Einbau-Wärmeschublade

Manufacturer
ATAG Nederland BV, Duiven, Netherlands
In-house design
Web
www.atag.nl

In this warmer drawer from the Magna series, cups and plates can be warmed, and food can be kept warm or cooked gently at low temperature. Standing 15 cm high, it fits under appliances with a height of 45 cm so that together they stand flush with built-in appliances 60 cm in height placed next to them. The front consists of black glass with metallic paint specially developed for the series.

Statement by the jury
With its minimalist glass front, which is interrupted only by the horizontal line typical for the series, the WD1674M is perfectly combinable with other appliances of the series.

In dieser Wärmeschublade aus der Magna-Serie können Tassen oder Teller erwärmt und Speisen warm gehalten oder bei niedriger Temperatur schonend gegart werden. Mit einer Höhe von 15 cm passt sie unter 45 cm hohe Geräte, sodass beide zusammen bündig mit daneben platzierten 60 cm hohen Einbaugeräten abschließen. Die Front besteht aus einem eigens für die Serie entwickelten schwarzen Glas mit Metalliclackierung.

Begründung der Jury
Mit ihrer minimalistischen Glasfront, die nur durch die serientypische horizontale Linie unterbrochen wird, ist die WD1674M perfekt mit anderen Geräten der Serie kombinierbar.

Breville BMO700 Quick Touch™ Crisp
Microwave
Mikrowelle

Manufacturer
Breville Group, Sydney, Australia
In-house design
Pierce Barnard , John Lee
Web
www.breville.com.au

Pies or pizzas often become soggy when warmed up in the microwave. An intelligent "Smart Cook & Grill" function has been developed to alternate between microwave and grill energy which has been designed to work with a heat absorbing crisper pan to brown the bottom of the food and crisp the top. The pan has adjustable legs to allow the food to get closer to grill or folded down for storage purposes.

Pasteten oder Pizzen werden bei der Erhitzung in der Mikrowelle oft weich. Eine eigens entwickelte, sogenannte „Smart Cook & Grill"-Funktion wechselt zwischen Mikrowellen- und Grillfunktion hin und her. Eine spezielle Knusperpfanne absorbiert die Hitze, bräunt den Boden mit und sorgt dafür, dass das Essen kross wird. Die Pfanne hat verstellbare Beine, mit deren Hilfe Speisen näher unter dem Grill platziert werden können. Zur Lagerung können die Beine platzsparend eingeklappt werden.

Statement by the jury
Aesthetically attractive in design, this microwave convinces with intelligent functionality and very convenient operation.

Begründung der Jury
Ästhetisch ansprechend gestaltet überzeugt diese Mikrowelle mit einer intelligenten Funktionalität und hohem Bedienkomfort.

JZY/T/R-JACB
Gas Hob
Gaskochstelle

Manufacturer
Ningbo Fotile Kitchen Ware Co., Ltd.,
Ningbo, China
In-house design
Shixing Wang, Qing Yao
Design
R&D Design Co., Ltd.
(Zongheng Wei, Yun Zhang),
Hangzhou, China
Web
www.fotile.com
www.rddesign.cc

An elegant coffee-coloured glass surface, impressive metallic pan rests and cylindrical control knobs characterise the clear aesthetics of this gas hob. The innovative gas burners can be regulated for an exact degree of heat supply. Below the pan rests, flat metal drip-trays are located for catching fat and liquids, making it easy to clean.

Eine elegante kaffeebraune Glasoberfläche, markante metallische Topfträger und zylindrische Bedienknöpfe prägen die klare Ästhetik dieser Gaskochstelle. Die innovativen Gasbrenner lassen sich für eine genaue Dosierung der Wärmeleistung exakt regeln. Unterhalb der Topfträger sind flache Metallschalen zum Auffangen von Fett und Flüssigkeiten eingelassen, die die Reinigung erleichtern.

Statement by the jury
From a functional viewpoint well-conceived, this hob convinces also due to its clear geometric surface design, which lends a harmonious effect.

Begründung der Jury
In funktionaler Hinsicht durchdacht, überzeugt diese Gaskochstelle auch durch eine klar geometrische, harmonisch wirkende Flächengestaltung.

N.EX.T.
Gas Hob
Gaskochfeld

Manufacturer
Smalvic S.p.A., Sarcedo (Vicenza), Italy
Design
brogliatotraverso Design Studio
(Loris Gasparini), Vicenza, Italy
Web
www.smalvic.it
www.brogliatotraverso.com

Three gas burners upraising from the smooth, black vitroceramic surface and the cast iron, ring-shaped pan holders characterise the distinctive appearance of this one metre wide hob. The high efficiency of the three burners (2 x 5 kW, 1 x 4.3 kW) allows for professional use. It is operated by a touch control panel. The hob is easy to clean due to the smooth surface and the upraised burners.

Statement by the jury
Due to the upraising burners with well visible flame, and its easy-to-clean surface, cooking with N.EX.T. is a pleasant experience.

Drei aus der glatten, schwarzen Glaskeramikfläche herausragende Gasbrenner und die gusseisernen, ringförmigen Topfträger prägen das markante Erscheinungsbild dieses ein Meter breiten Gaskochfeldes. Die hohe Leistungsfähigkeit der drei Brenner (2 x 5 kW, 1 x 4,3 kW) erlaubt eine professionelle Nutzung, die Bedienung erfolgt über ein Touch-Control-Panel. Das Kochfeld ist dank der glatten Oberfläche und der über dem Feld thronenden Brenner leicht zu reinigen.

Begründung der Jury
Durch die etwas höher stehenden Brenner mit gut sichtbarer Flamme und das leicht zu reinigende Kochfeld wird das Kochen auf dem N.EX.T. zum Erlebnis.

IGT9472MBA
Built-in Induction Gas Hob
Einbau-Induktions-
Gaskochstelle

Manufacturer
ATAG Nederland BV,
Duiven, Netherlands
In-house design
Web
www.atag.nl

The IGT9472MBA combines gas and induction cooking zones on a matt black glass ceramic surface. The inner flame of the wok burner is located exactly in the centre in order to heat the wok optimally. A digital control element makes it possible to switch to the outer flame as well as to set the power and programmes. The control of the three induction zones only operates when a pot is identified on one of the zones. It then lights up and waits on standby for an input.

Statement by the jury
The matt black surface characterises the purist appearance of this versatile hob. It convinces furthermore by a well-considered operating concept.

Der IGT9472MBA kombiniert Gas- und Induktionskochstellen auf einer mattschwarzen Glaskeramikfläche. Die innere Flamme des Wok-Brenners ist genau in der Mitte platziert, um den Wok optimal zu erhitzen. Ein digitales Bedienelement ermöglicht das Umschalten auf die Außenflamme sowie das Einstellen der Leistungsstufen und Programme. Die Bedienung der drei Induktionszonen reagiert erst, wenn ein Topf auf einer Zone erkannt wird. Dann leuchtet sie auf und wartet im Stand-by auf eine Eingabe.

Begründung der Jury
Die mattschwarze Oberfläche prägt das puristische Erscheinungsbild dieser vielseitigen Kochstelle. Zudem überzeugt sie mit einem durchdachten Bedienkonzept.

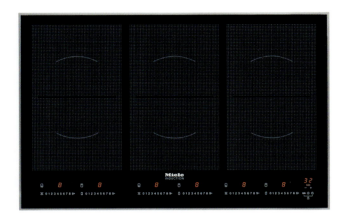

KM 6366-1
Induction Hob
Induktionskochfeld

Manufacturer
Miele & Cie. KG, Gütersloh, Germany
In-house design
Web
www.miele.de

KM 6366-1 is an induction hob which is independent of an oven. It has a stainless steel frame and a discrete raster print, allowing it to integrate well in modern kitchens. Six cooking zones with three so-called PowerFlex cooking surfaces heat even large pots and pans quickly; the individual cooking zones can be controlled by direct selection. Via the so-called "Con@ctivity" function the Miele extraction hood automatically reacts to the settings of the hob.

Statement by the jury
Minimalist design and the use of innovative technologies characterise this induction hob.

KM 6366-1 ist ein herdunabhängiges Induktionskochfeld, das sich mit seinem Edelstahlrahmen und einer dezenten Rasterbedruckung gut in moderne Küchen integriert. Sechs Kochzonen mit drei sogenannten PowerFlex-Kochbereichen erhitzen auch großes Kochgeschirr schnell, die einzelnen Kochzonen lassen sich über eine Direktwahl steuern. Über die sogenannte „Con@ctivity"-Funktion kommuniziert das Kochfeld mit der entsprechenden Miele Dunstabzugshaube und regelt automatisch deren Leistung.

Begründung der Jury
Eine minimalistische Gestaltung und der Einsatz innovativer Technologien zeichnen dieses Induktionskochfeld aus.

ERA-B313E
Island Electric Hob
Elektroherd für Kochinseln

Manufacturer
Tong Yang Magic Co., Ltd.,
Seoul, South Korea
In-house design
Dong Su Kim, Jong Yoon Yu
Web
www.magic.co.kr

When designing this elegant electric hob, which is suitable for island kitchens as well as built-in installation, great attention has been given to safety and energy-saving. In standby mode, the ERA-B313E consumes only 1 watt; if it is not used, it switches off automatically. Furthermore, the appliance uses sensor technology as protection against overheating and reacts to foreign bodies in the operating panel. The thickness of the glass frame has been minimised, contributing to a high level of aesthetics.

Statement by the jury
With a surface design reduced to the essentials, this hob combines energy efficiency with safety in use.

Bei der Konzeption dieser eleganten Elektrokochstelle, die sich sowohl für Kochinseln als auch für den Einbau eignet, wurde großes Augenmerk auf Sicherheit und Energieeinsparung gelegt. Im Stand-by-Betrieb verbraucht der ERA-B313E lediglich 1 Watt, wird er nicht gebraucht, schaltet er sich automatisch ab. Außerdem nutzt das Gerät Sensortechnologie als Überhitzungsschutz und reagiert auf Fremdkörper im Bedienfeld. Die Glasrahmendicke wurde minimiert, was zu einer hochwertigen Ästhetik beiträgt.

Begründung der Jury
Eine auf das Wesentliche reduzierte Oberflächengestaltung verbindet sich bei dieser Kochstelle mit Energieeffizienz und Sicherheit bei der Benutzung.

SOLIS Combi-Grill 3 in 1
Table Grill
Tischgrill

Manufacturer
Solis of Switzerland AG,
Glattbrugg (Zürich), Switzerland
In-house design
Fabian Zimmerli, Martin Hurter,
Thomas Nauer
Web
www.solis.ch

Solis Combi-Grill 3 in 1 is a multifunctional, stainless steel grill with which raclette, barbeque and fondue can be prepared easily on the dining table. With its two separate, continuously adjustable heating systems, the grill assures that only those zones which are actually used heat up. In a cold zone under the grill hot pans can cool down safely. The grill and base plate are removable for cleaning.

Solis Combi-Grill 3 in 1 ist ein Multifunktionsgrill aus Edelstahl, mit dem Raclette, Grillgut oder Fondue bequem auf dem Esstisch zubereitet werden können. Mit seinen zwei separaten, stufenlos regulierbaren Heizsystemen stellt der Grill sicher, dass sich nur die Zonen aufheizen, die auch genutzt werden. In einer Kaltzone unter dem Grill können heiße Pfännchen sicher auskühlen. Für eine einfache Reinigung sind Grill- und Bodenplatte abnehmbar.

Statement by the jury
Thanks to its functionally well-conceived design and its high-quality stainless steel construction, one can gladly make this grill the focal point of the table.

Begründung der Jury
Dank seiner funktional durchdachten Gestaltung und seiner hochwertigen Edelstahlausführung macht man diesen Grill gerne zum Mittelpunkt des Tischs.

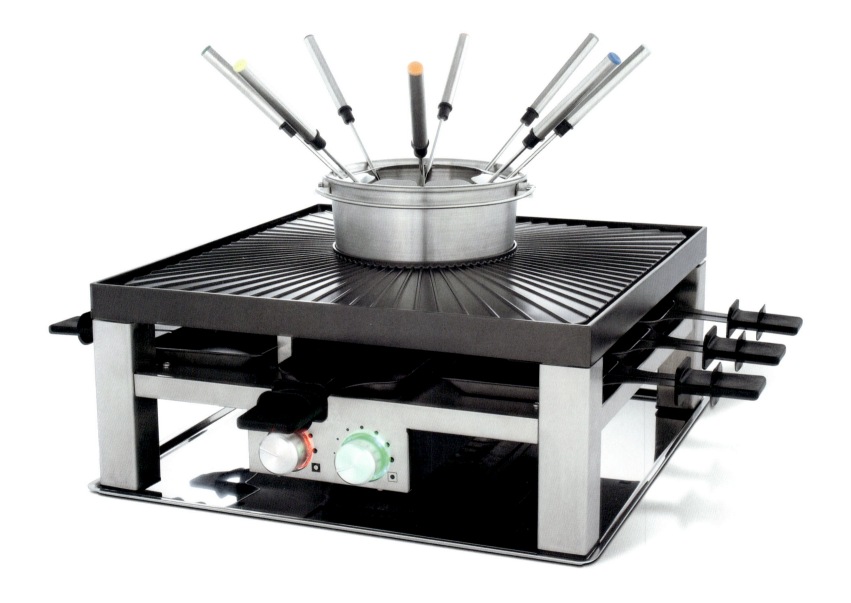

Manufacturer
Ningbo Fotile Kitchen Ware Co., Ltd.,
Ningbo, China
In-house design
Yilong Dong, Wang Yanhui
Design
Ideo (Keith Tan, Tony Wong),
Shanghai, China
R&D Design Co., Ltd. (Run Wang, Yun Zhang),
Hangzhou, China
Web
www.fotile.com
www.ideo.com
www.rddesign.cc

With this elegant range extractor hood, oil and vapours above the hob are suctioned along the wing plates into cavities integrated at the sides. This is effected by means of powerful extraction technology (3D Perimeter Extraction Technology), which covers a large area. The innovatively designed extraction vent in conjunction with microporous filters reduces noise development to a maximal 48 decibels at the highest setting. The control panel design is reduced in order to ease operation of the complex technology.

Bei dieser eleganten Dunstabzugshaube werden die Öle und Dämpfe oberhalb des Kochfeldes entlang der Flügelplatten in seitlich integrierte Hohlräume gesogen. Dahinter steckt eine leistungsfähige Extraktionstechnologie (3D Perimeter Extraction Technology), die eine große Fläche abzudecken vermag. Die innovativ gestaltete Abzugsöffnung reduziert in Verbindung mit mikroporösen Filtern die Geräuschentwicklung auf maximal 48 Dezibel bei höchster Stufe. Das Kontrollfeld ist reduziert gehalten, um die komplexe Technik leicht bedienbar zu machen.

Statement by the jury
The elegant aesthetics of this hood are characterised by an innovative extraction solution and high quality engineering of the surfaces without visible transitions.

Begründung der Jury
Eine innovative Abzugslösung und eine hochwertige Verarbeitung der Oberflächen ohne sichtbare Übergänge prägen die elegante Ästhetik dieser Haube.

WS9074M
Range Hood
Dunstabzugshaube

Manufacturer
ATAG Nederland BV, Duiven, Netherlands
In-house design
Design
Van Berlo, Eindhoven, Netherlands
Web
www.atag.nl
www.vanberlo.nl

Due to high airflow speeds at the outer edging, the WS9074M offers effective vapour extraction at low noise level in efficiency class A+. LED strips illuminate the hob well. The glass cover on the bottom is opened by use of telescopic springs; the stainless steel grease filters and carbon filter can be cleaned in the dishwasher. The black glass with discrete metallic paint used for the front and the cover harmonises with modern kitchens.

Statement by the jury
Strict, yet simultaneously elegant lines emphasise the convincing functionality of this extractor hood.

Durch hohe Strömungsgeschwindigkeiten an den Außenkanten bietet die WS9074M eine effektive und geräuscharme Dunstextraktion in der Energieeffizienzklasse A+. LED-Leisten leuchten das Kochfeld gut aus. Die Glasabdeckung an der Unterseite lässt sich teleskopfederunterstützt öffnen, Fett- und Karbonfilter aus Edelstahl werden im Geschirrspüler gereinigt. Das für die Front und Abdeckung verwendete schwarze Glas mit dezenter Metalliclackierung harmoniert mit modernen Küchen.

Begründung der Jury
Eine strenge und gleichzeitig elegante Linienführung unterstreicht die überzeugende Funktionalität dieser Dunstabzugshaube.

D79MT64N1 Extractor Hood Design 610
D79MT64N1
Dunstabzugshaube Design 610

Manufacturer
Constructa-Neff Vertriebs GmbH, Munich, Germany
Design
BSH Hausgeräte GmbH, Neff Design Team, Munich, Germany
Web
www.neff.de

This stainless steel and dark glass extractor hood offers convenient operation, thanks to an LED display with touch control. When switched off, the user interface remains in the background; when activated, it becomes available and is intuitively operable. The low-noise high-power fan can also be adjusted via the "Silence" button. Integrated dimmer lights (Softlights) allow stageless setting of the light intensity.

Statement by the jury
Clear lines and high quality materials contribute to the elegant appearance of this hood, while the dimmable lighting creates a cosy ambiance.

Diese Dunstabzugshaube aus Edelstahl und dunklem Glas bietet hohen Bedienkomfort dank eines LCD-Displays mit Touch-Steuerung. Im ausgeschalteten Zustand hält sich die Benutzeroberfläche dezent im Hintergrund, aktiviert ist sie präsent und intuitiv bedienbar. Das geräuscharme Hochleistungsgebläse kann über die Silence-Taste zusätzlich justiert werden. Integrierte, dimmbare Leuchten (Softlights) erlauben eine stufenlose Einstellung der Lichtintensität.

Begründung der Jury
Eine klare Linienführung und hochwertige Materialien tragen zum eleganten Erscheinungsbild dieser Haube bei, während die dimmbaren Leuchten ein behagliches Ambiente schaffen.

I99CM67N0
Range Hood
Dunstabzugshaube

Manufacturer
Constructa-Neff Vertriebs GmbH, Munich, Germany
Design
BSH Hausgeräte GmbH, Neff Design Team, Munich, Germany
Web
www.neff.de

This extractor hood is integrated in the ceiling and thus merges seamlessly with the overall ambiance of the kitchen landscape. Thanks to edge extraction, the filters disappear behind a high-quality LED glass panel which illuminates the hob and is easy to clean. The hood has three power settings as well as an intensive setting which can be selected conveniently by remote control.

Statement by the jury
Objective minimalism in the form language characterises this hood for ceiling mounting so that it blends discreetly into any kitchen environment.

Diese Dunstabzugshaube wird in die Raumdecke integriert und fügt sich so nahtlos in das Gesamtambiente der Küchenlandschaft ein. Dank der Randabsaugung verschwinden die Filter hinter einem hochwertigen LED-Glas-Panel, das das Kochfeld ausleuchtet und leicht zu reinigen ist. Die Haube hat drei Leistungsstufen sowie eine Intensivstufe, die bequem über eine Fernbedienung angewählt werden können.

Begründung der Jury
Ein konsequenter Minimalismus in der Formgebung kennzeichnet diese Haube für den Deckeneinbau, sodass sie sich dezent in jede Küchenumgebung einfügt.

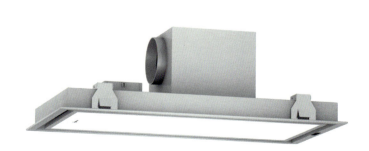

berbel Glassline
Headroom Hood
berbel Kopffreihaube Glassline

Manufacturer
berbel Ablufttechnik GmbH,
Rheine, Germany
Design
Studio Ambrozus (Stefan Ambrozus),
Cologne, Germany
Web
www.berbel.de
www.studioambrozus.de

Prominent highlight of this headroom hood is the gently curving front glass in which the control panel is seamlessly integrated. It becomes fully visible at the first touch. The output air slot is housed in a stainless steel strip; special technology prevents by means of secondary ventilation condensation forming at the front. The colour temperature of the LED lighting can be regulated; the integrated rear wall effect lighting sets visual accents.

Statement by the jury
Balanced proportions and the elegant arch of the front glass enforce the minimalist overall look of this headroom hood which also generates atmospheric lighting ambience.

Besonderes Merkmal dieser Kopffreihaube ist ihr sanft geschwungenes Frontglas, in welches das Bedienfeld nahtlos integriert ist. Es wird erst bei Berührung vollständig sichtbar. Der Abluftschlitz ist in ein Edelstahlband eingefasst, eine spezielle Technologie verhindert mittels Sekundärbelüftung eine Kondensatbildung an der Front. Die LED-Beleuchtung lässt sich in der Farbtemperatur regulieren, die eingebaute Rückwand-Effektbeleuchtung setzt Akzente.

Begründung der Jury
Ausgewogene Proportionen und die elegante Wölbung des Frontglases prägen das minimalistische Gesamtbild dieser Kopffreihaube, die zudem ein stimmungsvolles Lichtambiente erzeugt.

Gorenje+ Hood
Range Hood
Dunstabzugshaube

Manufacturer
Gorenje d.d., Velenje, Slovenia
Design
GORENJE DESIGN STUDIO, d.o.o.
(Lidija Pritrznik, Uros Bajt, Ursa Kovacic),
Velenje, Slovenia
Web
www.gorenje.com
www.gorenjedesignstudio.com

This elegant vertical hood allows extraction power and light intensity to be set stagelessly, simply by sliding the fingertips across the user interface. In automatic mode, a sensor regulates the extraction power. A perimetric suction system extracts the air not only through the central section of the hood but directs it also to the edges. In this way, suction performance is enhanced, and the energy consumption, as well as the noise emission, is reduced.

Statement by the jury
A slide/touch operating panel and sensors make operation of this range hood convenient; furthermore, the hood convinces with a timeless design.

Abzugsleistung und Leuchtintensität lassen sich bei dieser eleganten Kopffreihaube stufenlos einstellen, indem einfach mit der Fingerspitze über das Bedienfeld gestrichen wird. Im Automatikbetrieb übernimmt ein Sensor die Regulierung der Abzugsleistung. Ein perimetrisches Saugsystem saugt die Luft nicht nur durch den zentralen Abschnitt der Haube ein, sondern leitet sie auch zu den Rändern. So wird die Saugleistung verbessert und Stromverbrauch sowie Geräuschentwicklung reduziert.

Begründung der Jury
Ein Slide-Touch-Bedienfeld und Sensoren machen die Bedienung dieser Dunstabzugshaube komfortabel, zudem überzeugt sie mit einer zeitlosen Gestaltung.

Series | 6 Inclined Range Color

Range Hood
Dunstabzugshaube

Manufacturer
Robert Bosch Hausgeräte GmbH,
Munich, Germany
In-house design
Oliver Kraemer, Robert Sachon
Web
www.bosch-hausgeraete.de

The "Series | 6" glass, inclined range hoods in the colours black and white as well as stainless steel are components of the Series 8 inbuilt appliance range in a design which matches the ovens. Their appearance is defined by a clear and functional form language. The wide front glass calms the entire impression and simultaneously promotes easy cleaning due to its seamless construction with touch control. The metal application at the rear glass reflects the holistic design approach of Series 8.

Die „Series | 6"-Glasschrägessen in den Farben Schwarz und Weiß sowie Edelstahl gehören zur Serie-8-Einbaugerätereihe und sind im Design auf die Backöfen abgestimmt. Ihr Erscheinungsbild ist durch eine klare und sachliche Formensprache definiert. Das breite Frontglas beruhigt die gesamte Anmutung und unterstützt durch seinen fugenlosen Aufbau mit Touch-Bedienung und Randabsaugung zugleich eine einfache Reinigung. Die Metallapplikation am hinteren Glas greift den ganzheitlichen Gestaltungsansatz der Serie 8 auf.

Statement by the jury
With their greatly reduced design, the hoods of this series merge harmoniously in every kitchen ambience.

Begründung der Jury
Mit ihrer sehr reduzierten Gestaltung fügen sich die Essen dieser Reihe harmonisch in jede Küchenlandschaft ein.

berbel Skyline Edge Ceiling Lift Hood

berbel Deckenlifthaube
Skyline Edge

Manufacturer
berbel Ablufttechnik GmbH,
Rheine, Germany
Design
Studio Ambrozus (Stefan Ambrozus),
Cologne, Germany
Web
www.berbel.de
www.studioambrozus.de

Ceiling light and hood in one unit, the Skyline Edge moves at the touch of a button to the desired working height or stops automatically at an individually set distance from the hob. The integrated JetStream technology can be coupled with an EcoSwitch function, to combine the advantages of circulation and extraction operations. The hood can also be installed in a ceiling recess, so that when not in use, only an elegant ceiling light remains visible. The dazzle-free LED hob illumination with adjustable light colours provides pleasant atmospheric lighting.

Statement by the jury
This surprisingly versatile hood seems to float in the room – an impression which is further emphasised by the plain, circumferential light band.

Deckenleuchte und Haube in einem, fährt die Skyline Edge auf Knopfdruck auf die gewünschte Arbeitshöhe oder stoppt automatisch in einem individuell einstellbaren Abstand zum Kochfeld. Die integrierte JetStream-Technologie kann mit einer EcoSwitch-Funktion kombiniert werden, um die Vorteile von Umluft- und Abluftbetrieb zu verbinden. Die Haube lässt sich auch in Deckenvertiefungen einbauen, sodass im ungenutzten Zustand nur eine elegante Deckenleuchte sichtbar bleibt. Die blendfreie LED-Kochfeldbeleuchtung mit regulierbarer Lichtfarbe sorgt für ein stimmungsvolles Ambiente.

Begründung der Jury
Diese überraschend vielseitige Haube wirkt, als würde sie im Raum schweben – ein Eindruck der durch das schlichte, umlaufende Lichtband noch betont wird.

SLIDE-DOWN
Range Hood
Dunstabzugshaube

Manufacturer
Silverline Built-in Appliances,
Bayrampasa/Istanbul, Turkey
In-house design
Beyza Dogan
Web
www.silverline.com

The concave, black glass panel is visually and functionally the focal point of this range hood. When the hood is switched on, the concave glass panel slides down to guide cooking fume from the hob to the filter with maximum performance and acts as a backsplash to protect the kitchen wall behind the hob. When the hood is switched off, the concave glass panel rises again and covers the filter discreetly.

Statement by the jury
This hood impresses with a charming, minimalist design and provides a high level of functionality.

Der gewölbte, schwarze Glasschirm steht bei dieser Dunstabzugshaube visuell und funktional im Mittelpunkt. Wenn die Haube angeschaltet wird, gleitet der Schirm automatisch nach unten und leitet die Kochdünste vom Herd in den Hochleistungsfilter. Gleichzeitig dient er als Spritzschutz an der Wand hinter dem Herd. Wenn die Haube ausgeschaltet wird, fährt der konkave Glasschirm wieder hoch und verdeckt den Filter diskret.

Begründung der Jury
Diese Haube besticht mit einer anmutigen, minimalistischen Gestaltung und bringt ein hohes Maß an Funktionalität mit sich.

Maris Plus FMPL 906 BK B
Range Hood
Dunstabzugshaube

Manufacturer
Franke Kitchen Systems,
Bad Säckingen, Germany
In-house design
Web
www.franke.com

Symmetrical proportions and a centrally located glass panel with a continuous, slim stainless steel strip characterise the high-performance range hood Maris Plus and emphasise its horizontal alignment. The A+ hood is controlled via a touch control display with four power stages plus an intensive stage. The control elements are invisible when switched off and are activated simply by touch or by remote control.

Statement by the jury
The Maris Plus hood combines elegant aesthetics, characterised by the glass panel with stainless steel strip, with energy-efficient technology.

Symmetrische Proportionen und ein mittig platziertes Glaspaneel mit durchlaufendem, schmalem Edelstahl-Band kennzeichnen die leistungsstarke Kopffreihaube Maris Plus und betonen ihre horizontale Ausrichtung. Gesteuert wird die A+-Haube mit vier Leistungsstufen plus Intensivstufe über ein Touch-Control-Display. Unsichtbar im ausgeschalteten Zustand, lässt es sich einfach durch Berührung mit dem Finger oder per Fernbedienung aktivieren.

Begründung der Jury
Die Kopffreihaube Maris Plus verbindet eine elegante, durch das Glaspaneel mit Edelstahlband geprägte Ästhetik mit energieeffizienter Technik.

Édith
Range Hood
Dunstabzugshaube

Manufacturer
Elica, Fabriano (Ancona), Italy
In-house design
Fabrizio Crisà
Web
www.elica.com

Soft, round forms dominate the appearance of the Édith range hood. The integrated perimeter extraction system with high-performance motor provides good air quality and also leaves sufficient room in the housing for a large LED to illuminate the hob. Accentuating details such as the narrow metal frame emphasise the hood's form. Édith can be mounted on a wall or can be suspended.

Statement by the jury
Édith is a successful mix of lamp and compact extractor hood. With its formal balance, it provides the island unit with an atmosphere of well-being.

Weiche, runde Formen dominieren das Erscheinungsbild der Inselhaube Édith. Das integrierte Randabsaugungssystem mit Hochleistungsmotor sorgt für eine hohe Luftqualität und lässt im Gehäuse zudem viel Platz für eine große LED-Leuchtfläche zur Ausleuchtung des Kochfelds. Akzentuierende Details wie die schmale Metallumrandung betonen die Form der Haube. Édith kann als Wandhaube oder frei hängend installiert werden.

Begründung der Jury
Édith ist eine gelungene Mischung aus Leuchte und kompakter Dunstabzugshaube. Mit ihrer formalen Ausgewogenheit verleiht sie der Kochinsel Wohlfühlatmosphäre.

Amica IN.
Kitchen Appliance Range
Küchengeräte-Serie

Manufacturer
Amica S.A., Wronki, Poland
Design
CODE architecture & design,
Katowice, Poland
Web
www.amica.pl

Amica IN. is an intelligent kitchen
appliance series which comprises ovens,
induction and gas hobs, dish washers
and refrigerators. The appliances were
developed with a view to easing the
routine house chores and are equipped
with innovative technology – among
these are cooking programmes, a slow-
cooking function, a fast heat-up mode
and energy-saving functions. Modern
materials, matt surfaces and gentle forms
in combination with a sophisticated
ambient light concept characterise the
elegant and harmoneous appearance of
the appliance series.

Statement by the jury
The appliances of this series present
themselves in a timeless-linear design and
are state of the art.

Amica IN. ist eine intelligente Küchenge-
rätereihe, die Backöfen, Induktions- und
Gaskochfelder, Dunstabzugshauben, Spülmaschinen und Kühlschränke umfasst. Die
Geräte wurden mit dem Anspruch entwickelt, die tägliche Hausarbeit zu erleichtern,
und sind mit innovativer Technik ausgestattet – darunter Kochprogramme, eine
Slow-Cooking-Funktion, ein Schnellaufheiz-
Modus oder Energiesparfunktionen. Moderne Materialien, matte Oberflächen und
sanfte Formen prägen in Verbindung mit
einem ausgereiften Ambiente-Lichtkonzept
das elegante und harmonische Erscheinungsbild der Gerätereihe.

Begründung der Jury
Die Geräte dieser Reihe präsentieren sich in
einem zeitlos-geradlinigen Design und sind
auf dem neuesten Stand der Technik.

LG Double Door in Door Refrigerator (R–F956EDSB, R–F956VDSB, R–F956VDDN, R–F956VDSR)

LG Doppeltür-Kühlschrank (R-F956EDSB, R-F956VDSB, R-F956VDDN, R-F956VDSR)

Manufacturer
LG Electronics Inc.,
Seoul, South Korea

In-house design
Jinwon Kang, Woonkyu Seo,
Daesung Lee, Taihun Lim,
Sungkyong Han, Junyi Heo

Web
www.lg.com

reddot award 2015
best of the best

Everything at a glance

After orderly shopping is done, putting the groceries into the refrigerator mostly follows an individual sense of order. However, haphazardly, some may end up in the back of the fridge compartment, other at the front. In order to allow ordering of groceries for daily use in a more meaningful manner, the compartments of the LG double door in door refrigerator have been thoroughly redesigned. The unit features an innovative and sophisticated space system (Magic Space) that divides the left- and right-hand doors into different functions to better suit both the needs and the eating and drinking habits of a family. Integrated into the left is the "secret space", while to the right is the "family space" for use by all family members. The "secret space" is designed exclusively for the home-maker of the house, providing space to keep food ingredients that are used on a regular basis, whereas the "family space" is used for the storage of beverages and favourite snacks. This allows direct access depending on what is needed without having to search the fridge. This arrangement system is facilitated by integrating separate doors in the doors so that both left- and right-hand spaces can be accessed when the main doors are open. This allows all foodstuffs in the fridge to be accessed from all sides. Based on a unique approach to compartment arrangement and a highly consistent layout, the daily use of a refrigerator has been redefined.

Alles im Blick

Nach dem Einkauf bleibt das Einordnen der Lebensmittel in den Kühlschrank meist dem individuellen Ordnungsempfinden überlassen. Man kann sie entweder nach vorne oder nach hinten sortieren. Um die Dinge des täglichen Gebrauchs wesentlich sinnvoller zu organisieren, wurde die innere Aufteilung des LG Doppeltür-Kühlschranks grundlegend überdacht. Innovativ ist insbesondere sein ausgeklügeltes Raumsystem (Magic Space), welches der linken und rechten Seite unterschiedliche Funktionen zuweist, um so die Bedürfnisse und das Ess- und Trinkverhalten einer Familie besser abzubilden. Auf der linken Seite wurde das „Geheimfach" integriert, während sich rechts der Kühlschrankplatz für die gesamte Familie befindet. Das „Geheimfach" ist definiert als Bereich für den „Haushaltsmanager" und dient der Aufbewahrung häufig genutzter Lebensmittel. Der Familienbereich hingegen kann für Getränke und Lieblingssnacks verwendet werden. Auf diese Weise besteht ein direkter Zugriff je nach Bedarf, ohne dass lange gesucht werden muss. Um dieses Raumsystem zu ermöglichen, wurden bei diesem Kühlschrank separate Türen hinzugefügt, damit sowohl der linke als auch der rechte Bereich geöffnet werden kann, wenn die Haupttüren geöffnet sind. So lassen sich Lebensmittel von allen Seiten entnehmen. Auf der Basis einer anders interpretierten Innenaufteilung und eines sehr schlüssigen Layouts wurde der tägliche Gebrauch eines Kühlschranks dadurch neu definiert.

Statement by the jury

The LG double door in door refrigerator fascinates with an innovative compartment layout that perfectly reflects and responds to a wide variety of modern lifestyle scenarios. With its clear lines and well-balanced proportions, it embodies a harmonious design that offers highly functional details. The well-conceived idea of a "secret space" for foodstuffs of daily use is both time- and energy-saving.

Begründung der Jury

Der LG Doppeltür-Kühlschrank begeistert mit einer innovativen Art der Innenaufteilung, wodurch er unterschiedliche Szenarien des aktuellen Lifestyles perfekt abbildet. Mit klaren Linien und ausgewogenen Proportionen ist er harmonisch gestaltet und bietet hochfunktionale Details. Die gut umgesetzte Idee eines „Geheimfachs" für häufig genutzte Lebensmittel spart Zeit und Energie.

Designer portrait
See page 40
Siehe Seite 40

LG Curved Glass Design Refrigerator (R-U956VBLB, R-U956VDLB, R-U956VDTS, R-U956EBLB, R-956VBAW)
Refrigerator
Kühlschrank

Manufacturer
LG Electronics Inc., Seoul, South Korea
In-house design
Jinwon Kang, Woonkyu Seo, Daesung Lee,
Jihyun Lee, Hyein Yang, Jinhee Park,
Soonho Jung
Web
www.lg.com

This four-door refrigerator-freezer combination attracts attention due to its front of rounded, reinforced glass. For the so-called curved glass design, a method was used by means of which the glass edges have not been framed with injection moulded material. Fine patterns and prints have been applied to the glass, whereby a metallic lustre is created. The handles and the edges of the appliance are also rounded and blend harmoniously into the front design.

Statement by the jury
The front of this refrigerator-freezer combination delights with elegant aesthetics, in which the approach of the curved glass design has been thoroughly defined.

Diese viertürige Kühl-Gefrier-Kombination macht durch eine Front aus abgerundetem, verstärktem Glas auf sich aufmerksam. Für das sogenannte Curved-Glass-Design wurde eine Methode verwendet, dank derer die Glaskanten nicht mit Spritzguss-material eingefasst werden. Das Glas hat feine Muster und Prägungen, wodurch ein metallischer Glanz entsteht. Die Griffe und Kanten der Geräte sind ebenfalls ab-gerundet und fügen sich harmonisch in die Frontgestaltung ein.

Begründung der Jury
Die Front dieser Kühl-Gefrier-Kombination begeistert mit einer eleganten Ästhetik, bei der der Ansatz des Curved-Glass-Designs konsequent durchdekliniert wurde.

LG French Door Refrigerator (LMXS30776S)
LG Viertüriger Doppeltüren-Kühlschrank (LMXS30776S)

Manufacturer
LG Electronics Inc., Seoul, South Korea
In-house design
Jinwon Kang, Wocnkyu Seo, Daesung Lee,
Jaeyoung Kim, Seungjin Yoon, Anthony Ogg
Web
www.lg.com

The upper two-winged door of this four-door refrigerator can be opened wide. The door-in-door design of the appliance divides the refrigerator door into several storage positions. Drinks and snacks which are more often needed can simply be stored on the outside and are readily at hand. A refrigerator drawer half-way up facilitates four-stage temperature regulation. The freezer provides storage space at three different sized levels.

Statement by the jury
With its intelligent partitioning and ample space design, this refrigerator is optimally conceived for the require-ments of families.

Dieser viertürige Kühlschrank hat im oberen Bereich geteilte Türen, die sich weit öffnen. Das Tür-in-Tür-Design des Geräts unterteilt die Kühlschranktür in mehrere Lagen. Getränke und Snacks, die häufiger benötigt werden, können ganz einfach außen gelagert werden und sind schnell griffbereit. Die Kühlschublade auf halber Höhe ermöglicht eine vierstufige Tempe-raturregelung. Der Gefrierschrank bietet Lagerraum auf drei unterschiedlich großen Ebenen.

Begründung der Jury
Mit seiner intelligenten Aufteilung und dem großzügigen Raumangebot ist dieser Kühlschrank optimal auf die Bedürfnisse von Familien zugeschnitten.

Ref-SmartWindow
Refrigerator
Kühlschrank

Manufacturer
Haier Group, Qingdao, China
Design
Haier Innovation Design Center
(Chang Yonghun, Yin Jingxuan,
Huang Zeping, Cheng Yongli,
Li Xia, Chi Shasha, Li Ruizhao),
Qingdao, China
Web
www.haier.com

The design of the Ref-SmartWindow is based on the idea that a refrigerator can also be a status symbol. Its appearance is accordingly up-market, whereby integrated transparent areas at the front are the main feature. When the user approaches this Smart Window, it automatically lights up and presents a view of the food which is stored within easy reach behind the first door. At the same time, the refrigerator thus gives an impression of technical sophistication and clear interior design which is characterised by transparent areas and metal details.

Die Gestaltung des Ref-SmartWindow beruht auf der Idee, dass ein Kühlschrank auch Statussymbol sein kann. Entsprechend hochwertig ist sein Erscheinungsbild, wobei die in die Front integrierte transparente Fläche im Vordergrund steht. Nähert sich der Nutzer diesem Smart Window, leuchtet es automatisch auf und gewährt einen Blick auf die Lebensmittel, die griffbereit hinter der ersten Türe lagern. Gleichzeitig vermittelt der Kühlschrank damit einen Eindruck seiner technischen Ausgereiftheit und klaren Innenraumgestaltung, die von transparenten Flächen und Metalldetails geprägt ist.

Statement by the jury
This refrigerator interacts to a certain degree with the user, leading to a novel user experience.

Begründung der Jury
Dieser Kühlschrank interagiert über das Smart Window gewissermaßen mit dem Nutzer, was zu einem neuen Gebrauchserlebnis führt.

Siemens SbS Integrated-M
Refrigerator
Kühlschrank

Manufacturer
BSH Home Appliances (China) Co., Ltd.,
Nanjing, Jiangsu, China
In-house design
Christoph Becke, Yao Xingen, Max Eicher,
Christine Hartwein, Wang Yueru
Web
www.bsh-group.com

In this side-by-side appliance, operating panel and handle form one unit so that a minimalist and simultaneously independent appearance is created. Behind the homogeneous metal front, which is interrupted only by the dark electronics/ handle combination, a bright and well-structured interior is hidden. The drawers rest on high quality mechanisms. All elements are reinforced at handles and high-wear areas by sturdy metal applications.

Bei diesem Side-by-Side-Gerät bilden Bedienfeld und Griff eine Einheit, sodass ein minimalistisches und zugleich eigenständiges Erscheinungsbild entsteht. Hinter der homogenen Metallfront, die nur durch die dunkle Elektronik-Griff-Kombination unterbrochen wird, verbirgt sich ein heller und gut strukturierter Innenraum. Die Schubladen sind auf hochwertigen Auszügen gelagert. Alle Elemente sind an Griffen und belasteten Bereichen mit robusten Materialapplikationen verstärkt.

Statement by the jury
Elegant minimalism and the combination of handle and electronics are the core features of this standalone appliance.

Begründung der Jury
Ein eleganter Minimalismus und die Kombination von Griff und Elektronik sind die Hauptmerkmale dieses Solitärgeräts.

Neo
Fridge-Freezer
Kühl-Gefrier-Kombination

Manufacturer
Vestel Beyaz Esya San. ve Tic. A.S.,
Manisa, Turkey
In-house design
Vestel White Goods Industrial Design Team
(Elif Ezgi Oguz)
Web
www.vestel.com.tr

A distinctive design element of the Neo is a continuous handle integrated at the front, with a chrome trim and touch button controls in the coated stainless steel handle area. They emphasise the clear frontal design with a combination of soft curves and straight lines. The user-friendly display allows control of the temperature and the refrigerator mode. The fridge-freezer also provides ample space with a height of 180 cm and a width of 70 cm.

Statement by the jury
The purist front of this fridge-freezer provides a very homogenous effect due to its integrated handle and gives Neo a high degree of formal quality.

Markantes Designelement des Neo ist ein durchgängiger, in die Front integrierter Griff mit Chromleiste und Sensortasten im beschichteten Edelstahlbereich. Sie betonen die klare Frontgestaltung mit einer Kombination aus sanften Kurven und geraden Linien. Das nutzerfreundliche Display ermöglicht die Steuerung von Temperatur und Kühlschrankmodus. Die Kühl-Gefrier-Kombination bietet zudem mit einer Höhe von 180 cm und einer Breite von 70 cm viel Platz.

Begründung der Jury
Die puristische Front dieser Kühl-Gefrier-Kombination wirkt durch den integrierten Griff sehr homogen und verleiht Neo eine hohe formale Qualität.

Combi Pro Fresh Plus Range – AEG
Free-Standing Refrigerator
Freistehender Kühlschrank

Manufacturer
Electrolux, Stockholm, Sweden
In-house design
Web
www.electrolux.com

The free-standing combi refrigerators of this series combine an elegant appearance with environmentally friendly functionality at the energy efficiency rating A+++. The outer door incorporates large metal handles and a touch-panel. An innovative LED lighting solution assures uniform lighting in the interior. Thanks to the Mulitflow ventilation systems, each level is provided evenly with cold air.

Statement by the jury
The environmentally friendly appliances of the Combi Pro Fresh Plus Range make a modern, purist impression. The interior is convincing with plentiful space and good illumination.

Die freistehenden Kombi-Kühlschränke dieser Serie vereinen ein elegantes Erscheinungsbild mit einer umweltfreundlichen Funktionalität in der Energieeffizienzklasse A+++. Die Außentür ist mit großen Handgriffen aus Metall und einem Touch-Bedienfeld versehen. Im Innenraum sorgt eine innovative LED-Lichtlösung für eine einheitliche Beleuchtung. Dank der Mulitflow-Lüftungssysteme wird jede Ebene gleichermaßen mit kühler Luft versorgt.

Begründung der Jury
Die umweltfreundlichen Geräte der Combi Pro Fresh Plus Range haben eine moderne, puristische Anmutung. Das Innere überzeugt mit viel Platz und einer guten Ausleuchtung.

Cooling appliance series K 30.000
Kühlgeräteserie K 30.000

Manufacturer
Miele & Cie. KG, Gütersloh, Germany
In-house design
Web
www.miele.de

An innovative lighting solution characterises the interior of this built-in cooling appliance series: the invisible light sources can be positioned flexibly so that the interior is illuminated well and without dazzle. Due to the touch control incorporated in all design features of Miele inbuilt kitchen appliances, a strong homogeneity of design is achieved in them. Thanks to the "PerfectFresh Pro" functionality, foods can be stored with care and for a considerable time.

Statement by the jury
The K 30.000 cooling appliance series convinces with a novel lighting concept, high-quality materials and clear lines.

Eine innovative Lichtlösung prägt das Innere dieser Einbau-Kühlgeräteserie: Die nicht sichtbaren Lichtquellen lassen sich flexibel positionieren, sodass der Innenraum gut und blendfrei ausgeleuchtet wird. Durch die Touch-Bedienung, die alle Gestaltungsmerkmale der Miele Küchen-einbaugeräte aufweist, wird eine starke Designhomogenität zu diesen erzielt. Dank der „PerfectFresh Pro"-Funktionalität lassen sich Lebensmittel besonders schonend und lange lagern.

Begründung der Jury
Die Kühlgeräteserie K 30.000 überzeugt mit einem neuen Lichtkonzept, hochwertigen Materialien und einer klaren Linienführung.

IKBP 3550
Built-in Refrigerator
Einbaukühlschrank

Manufacturer
Liebherr-Hausgeräte GmbH,
Ochsenhausen, Germany
In-house design
Design
Prodesign Brüssing GmbH & Co. KG,
Neu-Ulm, Germany
Web
www.liebherr.com
www.prodesign-ulm.de

When this refrigerator was designed, special attention was given to the combination of energy efficiency and convenience of operation. When the door is opened, the LED columns integrated in the side walls provide pleasant light which gradually becomes brighter. The BioFresh compartments are illuminated, fully extendable and fitted with self-retraction and a soft-stop closing mechanism. Stainless steel profiles of the shelves and inserts as well as the GlassLine interior door highlight its elegant appearance.

Statement by the jury
The IKBP 3550 combines well-considered equipment details and a harmonious lighting concept with high-quality design.

Bei der Konzeption dieses Kühlschranks wurde Wert auf die Kombination von Energieeffizienz und Bedienkomfort gelegt. Beim Öffnen der Tür spenden in die Seitenwände integrierte, sanft heller werdende LED-Säulen angenehmes Licht. Die BioFresh-Safes sind ausgeleuchtet, voll ausziehbar und mit Selbsteinzug und Schließdämpfung ausgestattet. Edelstahlprofile an Ablageflächen und Einsätzen sowie die GlassLine-Innentür prägen sein elegantes Aussehen.

Begründung der Jury
Der IKBP 3550 kombiniert durchdachte Ausstattungsdetails und ein stimmiges Lichtkonzept mit einer hochwertigen Gestaltung.

Metacarpal
Wine Cabinet
Weinschrank

Manufacturer
Sensis Innovation Ltd., Hong Kong
In-house design
Web
www.sensisworld.com

The visual and haptic contrast between the cool glass front and the warm, extendable wooden shelves for storing wine bottles inside defines the appearance of the wine cabinet. The layered glass door protects the bottles from UV radiation; two temperature zones with humidity indicator, controlled via a touchscreen panel, facilitate correct storage of up to 154 bottles of white and red wines.

Statement by the jury
Contrasting materials, a high level of operating convenience and sophisticated functionality characterise this generously dimensioned wine cabinet.

Der visuelle und haptische Kontrast zwischen der kühlen Glasfront und warmen, ausziehbaren Holzböden zur Lagerung der Weinflaschen im Inneren bestimmen das Erscheinungsbild des Weinschranks. Die beschichtete Glastür schützt die Flaschen vor UV-Einstrahlung, zwei Temperaturzonen mit Feuchtigkeitsmessung, steuerbar über ein Touchscreen-Bedienfeld, ermöglichen die fachgerechte Lagerung von bis zu 154 Flaschen weißen und roten Weins.

Begründung der Jury
Kontrastierende Materialien, ein hoher Bedienkomfort und eine ausgereifte Funktionalität prägen diesen großzügigen Weinschrank.

UWT 1682 Vinidor
Under-Worktop Multi-Temperature Wine Cabinet
Unterbau-Weintemperierschrank

Manufacturer
Liebherr-Hausgeräte GmbH,
Ochsenhausen, Germany
In-house design
Web
www.liebherr.com

The UWT 1682 has two wine safes for which the temperature can be set separately to the exact degree. When tapped lightly, the glass door opens 7 cm and can thereafter be opened wide. If this is not done, the door closes automatically after three seconds. The energy-efficient wine cabinet provides an even illumination and room for 34 bottles. The wooden shelves can be widely extended for good overview and access.

Statement by the jury
This wine cabinet gains points with two compartments which can be individually temperature controlled and a sophisticated door solution.

Der UWT 1682 bietet zwei Weinsafes, für die die Temperatur separat gradgenau eingestellt werden kann. Die Glastür öffnet sich durch Antippen 7 cm weit und kann dann bequem vollständig geöffnet werden. Geschieht dies nicht, schließt sie nach drei Sekunden selbsttätig. Der energieeffiziente Weinschrank bietet eine gleichmäßige Ausleuchtung und Platz für 34 Flaschen. Die Borde aus Holz lassen sich für einen guten Überblick und Zugriff weit ausziehen.

Begründung der Jury
Dieser Weinschrank punktet mit zwei unabhängig voneinander temperierbaren Fächern und einer ausgereiften Türlösung.

Intra Eligo
Kitchen Sink in Stainless Steel
Küchenspüle aus Edelstahl

Manufacturer
Intra Mölntorp AB,
Kolbäck, Sweden

In-house design
Bjørn Høiby, Mattias Robson

Design
Propeller Design AB
(Karl Forsberg, Olle Gyllang),
Stockholm, Sweden

Web
www.intra-teka.com
www.propeller.se

reddot award 2015
best of the best

Interactive efficiency

Especially when serving a party with many different courses, used pots alongside other kitchen utensils tend to pile up in the kitchen sink. Often the available space is simply not sufficient and chaos starts to reign. Such a scenario was the starting point for the design of the Intra Eligo kitchen sink. The innovative kitchen sink concept follows a fundamentally new approach aimed at making work in the kitchen easier. The sink features several smart functions that can be activated and operated using a piezo switch through the actual stainless steel surface. These functions include opening the water for the dishwasher, which also reduces the risk of water damage, or the "scraper" for lifting sink bottom residue and easy garbage removal. Another highly functional solution for daily kitchen work is presented by the flexible wall, which allows users to divide the sink into two separate bowls. Other sophisticated details include an easy-to-clean colander, a cutlery basket and a hinged chopping board with full- or half-size use. The Intra Eligo kitchen sink features soft and sharp radiuses, promises a pleasant tactile user experience and showcases a thoroughly clean design. With a language of form that skilfully merges soft and hard surfaces with user-friendly layout and functionality, this kitchen sink incorporates a real life-proven interpretation of these desirable elements.

Interaktive Effizienz

Insbesondere wenn mehrere Gänge serviert wurden, stapeln sich nach dem Kochen die Töpfe und andere Kochutensilien im Spülbecken. Der vorhandene Platz reicht dann nicht aus, schnell wird es unübersichtlich. Ein solches Szenario war der Ausgangspunkt für die Gestaltung der Küchenspüle Intra Eligo. Im Mittelpunkt des innovativen Konzepts für ein Spülbecken steht eine grundsätzlich neue Herangehensweise, die die Arbeit in der Küche erleichtern soll. Unterschiedliche intelligente Funktionen lassen sich bei diesem Spülbecken durch die Stahloberfläche hindurch per Touchbedienung aktivieren und steuern: Das Geschirrspülerventil beispielsweise, was zusätzlich gegen Wasserschäden schützt, oder der „Scraper", der dem Abziehen des Beckenbodens dient, wodurch sich Speiserückstände leichter entfernen lassen. Eine für die Arbeit in der Küche sehr funktionale Lösung ist zudem eine flexible Abtrennung, die es erlaubt, die Aufteilung des Beckens zu verändern. Weitere durchdachte Details sind ein einfach zu reinigendes Restebecken, ein Besteckkorb sowie ein ebenfalls teilbares Schneidbrett. Die Küchenspüle Intra Eligo ist mit weichen und engen Radien gestaltet und bietet dem Nutzer eine angenehme Haptik und insgesamt sehr klare Anmutung. Mit einer weiche und harte Oberflächen gekonnt vereinenden Formensprache und seiner nutzerfreundlichen Ausstattung erfährt das Spülbecken hier eine situationsnahe Interpretation.

Statement by the jury

The concept of the Intra Eligo kitchen sink offers users enchanting novel functionality based on smart technologies. Several features are operated via a touch panel, and the sink bowl can divide into two, rearranging the available space for a more effective work process. Designed with soft radiuses and clear lines, this concept also integrates a wide range of practical solutions that bring more ease and convenience to everyday work in the kitchen.

Begründung der Jury

Das Konzept der Küchenspüle Intra Eligo bietet dem Nutzer eine bestechend neue Funktionalität auf der Basis intelligenter Technologien. Verschiedene Funktionen werden über ein Touchpanel gesteuert, und diese Spüle lässt sich teilen, woraufhin sich die Platzverhältnisse für ein effektives Arbeiten entsprechend ändern. Gestaltet mit weichen Radien und klaren Linien, integriert dieses Konzept darüber hinaus eine Vielzahl weiterer Problemlösungen, die dem Alltag in der Küche mehr Komfort und Leichtigkeit verleihen.

Designer portrait
See page 42
Siehe Seite 42

Siro 90
Sink
Spüle

Manufacturer
systemceram GmbH & Co. KG,
Siershahn, Germany
Design
Dieter Pechmann Industriedesign,
Essen, Germany
Web
www.systemceram.de

The Siro 90 ceramic sink is characterised by sloping surfaces and consistency of form. The tilt of the large, smooth draining surface defines the appearance of the sink. The front and rear rims also slope gently to the basin and draining surface. The tap hole bank located at the side is sloped and holds four possible bores. The sink is thus reversible and suitable for the installation in front of a window.

Statement by the jury
Siro 90 impresses by its clear geometric form and the functional tap hole bank located at the side, allowing flexible installation of the sink unit.

Die Keramikspüle Siro 90 ist geprägt durch schräge Flächen und formale Konsequenz. Die Neigung der großen, glatten Abtropffläche bestimmt ihr Erscheinungsbild. Vorderer und hinterer Spülenrand bilden ebenfalls eine Schräge, die sich sanft zu Becken und Abtropffläche neigt. Auch die seitlich platzierte Hahnlochbank mit vier möglichen Bohrungen ist abgeschrägt. Damit ist die Spüle reversibel einsetzbar und eignet sich auch für den Einbau vor einem Fenster.

Begründung der Jury
Siro 90 besticht durch ihre klare geometrische Formgebung und die funktionale, seitlich liegende Hahnlochbank, durch die die Spüle flexibel einbaubar ist.

BLANCO JARON XL 6 S-IF
Sink
Spüle

Manufacturer
BLANCO GmbH + Co KG,
Oberderdingen, Germany
In-house design
Brigitte Ziemann
Web
www.blanco-germany.com/de

Characteristic for this minimalist, stainless steel sink is the symmetric arrangement of two equally sized squares for basin and drainer as well as the elegantly centrally placed mixer tap ledge. The frame is flat-fitting with gently rounded, parallel contoured corners and is matched to the balanced geometry of the basin. Thanks to the clear form language and high-quality impression of the material, the sink merges harmoniously into modern kitchens.

Statement by the jury
Sensuous minimalism and strictly geometric lines characterise the appearance of the Blanco Jaron XL 6 S-IF elegant stainless steel sink.

Charakteristisch für diese minimalistische Edelstahlspüle ist die symmetrische Anordnung von zwei gleich großen Quadraten für Becken und Tropffläche sowie die elegant mittig positionierte Armaturenbank. Der flache Einbaurand bildet mit sanft gerundeten, konturparallelen Ecken den Rahmen und ist auf die ausgewogene Beckengeometrie abgestimmt. Dank der klaren Formensprache und der edlen Materialanmutung fügt sich die Spüle harmonisch in moderne Küchen ein.

Begründung der Jury
Sinnlicher Minimalismus und eine streng geometrische Linienführung prägen das Erscheinungsbild der eleganten Edelstahlspüle Blanco Jaron XL 6 S-IF.

Sink
Spüle

Manufacturer
Foster S.p.A.,
Brescello (Reggio Emilia), Italy
In-house design
Web
www.fosterspa.com

Characteristic design elements of this stainless steel sink are its curved side edges which form a slight wave and interrupt the minimalist impression of the arc-welded sink. The draining board has a gently rounded form which allows the water to drain off more easily. The generously dimensioned, quadratic sink with its diamond-ground bottom and a brushed surface is easy to clean. A continuous overflow edge and the large tap area offer space for accessories.

Charakteristisches Gestaltungselement dieses Edelstahl-Spülbeckens sind seine geschwungenen Seitenkanten, die eine leichte Welle bilden und die minimalistische Anmutung des bogengeschweißten Beckens aufbrechen. Die Abtropffläche hat eine langgezogene Rundung, die das Wasser leichter abfließen lässt. Das großzügige quadratische Spülbecken mit seinem Diamantschliff-Boden und einer gebürsteten Oberfläche ist leicht zu reinigen. Eine rundum laufende Überlaufkante und der große Armaturenbereich bieten Platz für Zubehör.

Statement by the jury
The wave element of the side edges, the distinctive draining board and the diamond ground of the basin impart an exclusive impression to this stainless steel sink.

Begründung der Jury
Das Wellenelement der Seitenkanten, die markante Abtropffläche und der Diamantschliff des Beckens verleihen diesem Edelstahlspülbecken eine exklusive Anmutung.

Franke Box BXX 210-50 TL
Sink
Spülbecken

Manufacturer
Franke Kitchen Systems,
Bad Säckingen, Germany
In-house design
Web
www.franke.com

The elegant stainless steel sinks of the Franke Box series are characterised by high-quality workmanship and a clear design. The extra deep bowls are provided either under-mounted or with tap unit in the "Slim-Top" and "Flush" versions. The linear design also incorporates the integral waste water outlet and the concealed overflow and is further emphasised by the tight radius of only 12 mm. With their depth of 20 cm, the sinks provide a generous volume.

Statement by the jury
These stainless steel sinks make a high-value and timeless impression and form an elegant unit with the worktop.

Die eleganten Edelstahlspülen der Serie Franke Box zeichnen sich durch hochwertige Verarbeitung und klares Design aus. Die extra tiefen Becken gibt es als Unterbaubecken oder mit Armaturenbank in den Ausführungen „SlimTop" und „Flächenbündig". Die geradlinige Gestaltung bezieht auch den Integralablauf und den verdeckten Überlauf mit ein und wird durch den engen Radius von 12 mm noch betont. Mit 20 cm Tiefe bieten die Becken viel Volumen.

Begründung der Jury
Diese Edelstahlspülen wirken hochwertig und zeitlos und bilden eine elegante Einheit mit der Arbeitsplatte.

Franke Crystal CLV 214
Sink
Spülbecken

Manufacturer
Franke Kitchen Systems,
Bad Säckingen, Germany
Design
marioferrarini designprojects
(Mario Ferrarini), San Fermo d.B. (Como), Italy
Web
www.franke.com
www.marioferrarini.com

The clear, minimalistic form of the Crystal sink and its distinctive, coloured glass inserts form a prominent mix of material and colour. Two insert panels, each 6 mm thick and made of shatterproof glass, cover the drain outlet and overflow. They give the bowl a clear geometry and can be removed for cleaning. The generously dimensioned basin with stringent zero radii and gently declined drip well is encased in a flat stainless steel frame.

Statement by the jury
The Crystal sink convinces with an exciting interaction of stainless steel and glass as well as stringent geometric lines.

Die klare, minimalistische Formgebung von Crystal und ihre markanten, kolorierten Glaseinsätze bilden einen auffälligen Material- und Farbmix. Zwei je 6 mm starke Einsätze aus bruchfestem Glas verdecken Ab- und Überlauf. Sie verleihen dem Becken eine klare Geometrie und können zum Reinigen einfach herausgenommen werden. Das geräumige Becken mit stringenten Null-Radien und leicht abfallender Tropfmulde ist in einen flachen Edelstahl-Rahmen gefasst.

Begründung der Jury
Die Spüle Crystal überzeugt durch ein spannungsreiches Zusammenspiel aus Edelstahl und Glas sowie eine streng geometrische Linienführung.

Reginox Niagara
Sink
Spülbecken

Manufacturer
Reginox B.V., Rijssen, Netherlands
In-house design
Wim ter Steege
Web
www.reginox.com

Niagara is a stainless steel single sink which can be fitted in three different ways: underbuilt, integrated or semi-integrated. The sink has two different bottom levels for water saving. Thanks to its rounded corners, it is also easy to clean. Available with a number of smart, innovative gadgets such as a bottom grid that is also a convenient hot plate, plus a colander that can be used as a small storage box or as a waste bin, it is very versatile.

Statement by the jury
This stainless steel sink gains points due to its well-conceived functionality and a stringent design concept, which also applies to the accessories.

Niagara ist eine Edelstahl-Einbeckenspüle die als Unterbau, flächenbündig oder Flacheinbau eingesetzt werden kann. Die Spüle hat zwei verschiedene Bodenniveaus für eine wassersparende Nutzung. Dank abgerundeter Kanten ist sie zudem einfach zu reinigen. Durch eine Reihe von sinnvollen Ergänzungen für den täglichen Gebrauch, wie ein Abtropfgitter, das für ein einheitliches Bodenniveau sorgt, sowie ein Restebecken mit einer Abdeckung, das auch als Abfallbehälter verwendbar ist, ist sie vielseitig nutzbar.

Begründung der Jury
Diese Edelstahlspüle punktet mit durchdachter Funktionalität und einem stringenten Gestaltungskonzept, was sich auch auf das Zubehör erstreckt.

Prolific
Sink and Accessories
Spüle und Zubehör

Manufacturer
Kohler Co., Kohler, Wisconsin, USA
In-house design
Web
www.kohler.com

Prolific is a sink which can serve simultaneously as a functional worktop; it was designed with ergonomics in mind. The stainless steel sink basin has shelves at three different levels so that the bamboo cutting board, the two multi-purpose racks, the colander and the wash bin can be flexibly positioned for extended functionality. The drain is cone-shaped and thus makes cleaning easier; a silencer (SilentShield) reduces noise.

Prolific ist eine Spüle, die gleichzeitig als funktionelle Küchenarbeitsfläche dienen kann und unter ergonomischen Gesichtspunkten entworfen wurde. Die Edelstahlspüle hat Auflageflächen auf drei verschiedenen Ebenen, sodass das Bambus-Schneidebrett, die zwei Mehrzweckroste, das Abtropfsieb und die Schüssel für eine erweiterte Funktionalität flexibel platziert werden können. Der Abfluss ist kegelförmig gestaltet und erleichtert damit die Reinigung, eine Schalldämpfung (SilentShield) reduziert die Geräuschentwicklung.

Statement by the jury
With its accessories, which can be positioned on three levels, this linear stainless steel sink becomes a versatile workstation.

Begründung der Jury
Mit ihrem Zubehör, das sich auf drei Ebenen platzieren lässt, wird diese geradlinige Edelstahlspüle zu einem vielseitig nutzbaren Arbeitsplatz.

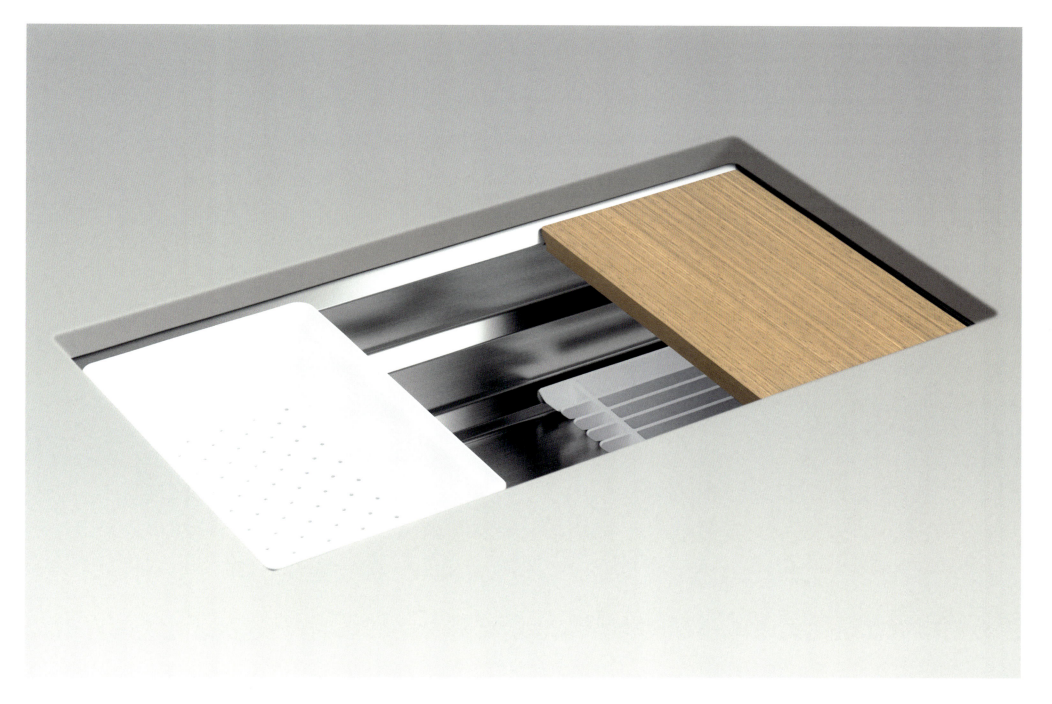

SF-WCH120
Boiling Water Tap (Mono Type)
Wasserhahn für kochendes Wasser

Manufacturer
Lixil Corporation, Tokyo, Japan
In-house design
Kosuke Yasuma, Yoshihiko Ando
Web
www.global.lixil.co.jp

The SF-WCH120 is a strictly linear designed boiling water tap which was conceived for office kitchen units. A double layer construction with an insulating air pocket ensures that the outer surface of the water tap does not get hot, in spite of its metallic impression. The operating control is located at the user-facing end of the tap in front of the outflow spout so that the user does not need to reach past the hot water to turn it on and off.

Der SF-WCH120 ist ein streng linear gestalteter Wasserhahn für kochendes Wasser, der für Büroküchen konzipiert wurde. Eine Doppelwandkonstruktion mit einer isolierenden Luftschicht sorgt dafür, dass das Äußere des Wasserhahns trotz seiner metallenen Anmutung nicht heiß wird. Das Bedienelement ist an dem nutzerzugewandten Ende des Hahns vor der Ausflussöffnung platziert, sodass der Nutzer nicht an dem heißen Wasser vorbeigreifen muss, um es an- oder abzustellen.

Statement by the jury
Handling boiling hot water involves particular dangers, which this water tap minimises thanks to a well-conceived, reduced design.

Begründung der Jury
Die Handhabung von kochend heißem Wasser birgt besondere Gefahren, die bei diesem Wasserhahn dank einer durchdachten, reduzierten Gestaltung minimiert werden.

NAVISH
Filter Kitchen Tap
Küchenarmatur mit Wasserfilter

Manufacturer
Lixil Corporation, Tokyo, Japan
In-house design
Yuichiro Komatsu
Web
www.global.lixil.co.jp

This gooseneck-shaped kitchen tap with water filter is designed primarily with hygiene issues in mind. Thanks to the integrated sensor, contactless operation keeps this tap clean and operating levers could be omitted. The purist form is characterised by a gently arched pipe which is easy to clean. A light in the sensor indicates when the filter cartridge has to be replaced or fresh water has to be refilled.

Diese an einen Schwanenhals erinnernde Küchenarmatur mit Wasserfilter ist vor allem unter hygienischen Gesichtspunkten gestaltet. Dank eines integrierten Sensors erfolgt die Bedienung berührungslos; die Armatur bleibt sauber und auf Hebel konnte ganz verzichtet werden. Die puristische Formgebung ist durch ein sanft gebogenes, schlankes Rohr geprägt, das leicht zu reinigen ist. Ein Lichtsignal im Sensor zeigt an, wann die Filterkartusche gewechselt oder frisches Wasser nachgefüllt werden muss.

Statement by the jury
Nothing distracts from the elegant as well as purist appearance of the Navish. The formal clarity is reflected functionally in its hygienic operating concept.

Begründung der Jury
Bei Navish lenkt nichts vom ebenso eleganten wie puristischen Erscheinungsbild ab. Diese formale Klarheit spiegelt sich funktional in ihrem hygienischen Bedienkonzept wider.

Brizo® Artesso™
Kitchen Tap
Küchenarmatur

Manufacturer
Brizo, Indianapolis, USA
In-house design
Seth Fritz
Web
www.brizo.com

The design of the Brizo Artesso kitchen tap series is inspired by the metal works of the early 20th century and combines hand crafted elements with modern aesthetics. The series comprises a single-lever mixer with or without shower head, a two-handle bridge-design tap and a soap dispenser. The single-lever mixer with shower head is available with SmartTouch technology, by means of which the water flow can be activated by touching the tap anywhere on the spout, hub or handle.

Statement by the jury
The taps of the Artesso series become a highlight in modern kitchens due to their industrial and at the same time high-quality appearance.

Die Gestaltung der Brizo-Artesso-Küchenarmaturenserie ist inspiriert von den Metallarbeiten des frühen 20. Jahrhunderts und verbindet handgeschmiedete Elemente mit einer modernen Ästhetik. Die Serie umfasst einen Einhebelmischer mit oder ohne Brause, eine Zweigriff-Brückenarmatur und einen Seifenspender. Der Einhebelmischer mit Brause ist mit einer SmartTouch-Technologie verfügbar, bei der der Wasserfluss aktiviert werden kann, indem man die Armatur irgendwo am Ausfluss, am Mittelstück oder am Hebel berührt.

Begründung der Jury
Die Armaturen der Artesso-Serie setzen durch ihre industrielle und zugleich edle Anmutung Akzente in modernen Küchen.

Fern
Kitchen Tap
Küchenarmatur

Manufacturer
Damixa ApS, Odense, Denmark
Design
Schmelling industriel design ApS
(Steffen Schmelling), Hørsholm, Denmark
Web
www.damixa.com
www.schmelling.com

Fern is a flexible kitchen tap which reduces water consumption by 40 to 50 per cent. When set to "Save", the water consumption is reduced from 10 or 12 litres to about 6 litres, enough for washing. An aerator mixes air into the issuing water so that the difference between the two settings is hardly perceptible. The outlet can be located in front of, next to or behind the tap body.

Statement by the jury
This well-conceived tap facilitates a reduction in water consumption and offers great flexibility of installation.

Fern ist eine flexible Küchenarmatur, die den Wasserverbrauch um 40 bis 50 Prozent reduziert. In der Einstellung "Save" verringert sich der Wasserbrauch von 10 bis 12 Litern auf ca. sechs Liter, was zum Waschen ausreicht. Ein Perlator reichert das austretende Wasser mit Luft an, sodass kein Unterschied zwischen den beiden Einstellungen fühlbar ist. Der Auslauf kann vor, neben oder hinter dem Armaturenkörper positioniert werden.

Begründung der Jury
Diese durchdachte Armatur erleichtert eine Reduzierung des Wasserverbrauchs und bietet große Flexibilität bei der Montage.

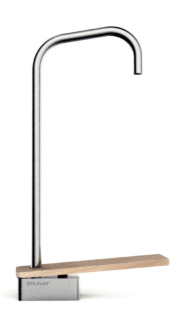

Hand-free Cup Filler
Filter Kitchen Tap
Küchenarmatur mit
Wasserfilter

Manufacturer
Bravat China GmbH, Guangzhou, China
In-house design
Yu Zheng Tie
Web
www.bravat.com

This purist water dispensing tap functions on the principle of a seesaw and makes use of the lever principle to fill glasses or cups. The weight of the vessel activates the water flow, removing it stops the water flow automatically. By this means, the user always has one hand free, since he no longer needs to operate a lever, thus making the tap particularly hygienic. The tap is made of high-quality lead-free brass and stainless steel.

Statement by the jury
This tap with water filter is based on an original operating concept which has a high degree of intuitiveness.

Diese puristische Wasserspender-Armatur funktioniert nach dem Prinzip einer Wippe und macht sich die Hebelwirkung zunutze, um Gefäße zu befüllen. Das Gewicht des Gefäßes aktiviert den Wasserfluss; wird es entfernt, stoppt das Wasser automatisch. Auf diese Weise hat der Nutzer stets eine Hand frei, denn er muss keine Hebel bewegen, was diese Armatur besonders hygienisch macht. Die Armatur wird aus hochwertigem bleifreien Messing und Edelstahl gefertigt.

Begründung der Jury
Dieser Armatur mit Wasserfilter liegt ein originelles Bedienkonzept zugrunde, das eine hohe Selbsterklärungsqualität hat.

Spirit
Kitchen Tap
Küchenarmatur

Manufacturer
AM.PM Produktionsgesellschaft S&T mbH,
Berlin, Germany
Design
Design3 GmbH, Hamburg, Germany
Web
www.ampm-germany.com
www.design3.de

The form language of the Spirit Collection moves between organic softness and geometric precision. The single-lever mixer with its compact and clear form facilitates convenient handling. The lever's softness provides a particularly good feel for precise water regulation. An integrated water and energy saving function assures economic consumption.

Statement by the jury
The taps of the Spirit Collection impress with convenient, precise operation and a distinctive purist design.

Die Formensprache der Spirit Collection bewegt sich zwischen organischer Weichheit und geometrischer Strenge. Die Einhebel-Spültischarmatur mit ihrer kompakten und klaren Form ermöglicht eine komfortable Handhabung. Die Weichheit des Hebels sorgt für eine besonders gute Haptik zur präzisen Wasserregulierung. Eine integrierte Wasser- und Energiesparfunktion gewährleistet einen ökonomischen Verbrauch.

Begründung der Jury
Die Armaturen der Spirit Collection beeindrucken mit einer komfortablen, präzisen Bedienung und einer markant puristischen Gestaltung.

Accurate Water
Kitchen Tap
Küchenarmatur

Manufacturer
Xiamen Solex High-Tech Industries Co., Ltd.,
Xiamen, China
In-house design
Chen Zhida, Lin Jingbin, Wang Min,
Cheng Xumin, Xiao Chao
Web
www.solex.com.cn

This kitchen tap makes it possible to preset the exact amount of water required. By means of a regulator located at the outlet, the desired volume of water can be set; setting "C" will provide a constant water flow. The tap is fitted with a novel cartridge (valve) and the flow velocity, as well as the temperature, is regulated with one hand via a single lever mixer.

Statement by the jury
With the possibility of setting the desired volume of water before opening the tap, this minimalist designed faucet provides real added value.

Diese Küchenarmatur ermöglicht es, die benötigte Wassermenge exakt vorzugeben. Mittels eines am Auslauf angebrachten Reglers kann das gewünschte Volumen eingestellt werden, die Einstellung „C" hingegen sorgt für einen dauerhaften Wasserfluss. Der Wasserhahn ist mit einer modernen Kartusche (Ventil) versehen, und über den Einhandmischer lassen sich sowohl Durchflussgeschwindigkeit als auch Temperatur mit einem Handgriff regulieren.

Begründung der Jury
Mit der Möglichkeit, die gewünschte Wassermenge bereits vor Öffnen des Hahns festzulegen, bietet diese minimalistisch gestaltete Armatur einen echten Mehrwert.

Electrolux Quicksource Tap
Kitchen Tap
Küchenarmatur

Manufacturer
Electrolux, Stockho m, Sweden
In-house design
Design
Gessi S.p.A., Seravalle Sesia (Vercelli), Italy
Web
www.electrolux.com
www.gessi.com

The Quicksource tap provides boiling water straight from the faucet. For safety, the single-lever mixer for normal operation is located on one side of the tap while the handle for the boiling water function is located on the other side and requires a two-stage activation. Since the tap is fitted with changeable design rings, it can be made to match individual kitchen styles.

Statement by the jury
Quicksource scores points with a safe operating concept and the possibility of personalising this otherwise purist tap by an option as simple as it is charming.

Beim Quicksource-Wasserhahn kommt kochend heißes Wasser direkt aus der Armatur. Für eine sichere Bedienung befindet sich der Einhebelmischer für die normale Funktion auf der einen Seite der Armatur, während der Griff für die Kochend-Wasser-Funktion auf der anderen Seite platziert ist und eine zweistufige Aktivierung erfordert. Mit austauschbaren Designringen kann die Armatur an den individuellen Küchenstil angepasst werden.

Begründung der Jury
Quicksource punktet mit einem sicheren Bedienkonzept und einer ebenso einfachen wie charmanten Individualisierungsmöglichkeit dieser ansonsten puristischen Armatur.

Hansgrohe Talis Select S
Product Family Kitchen Mixer
Hansgrohe Talis Select S
Produktfamilie Küchenmischer

Manufacturer
Hansgrohe SE, Schiltach, Germany
Design
Phoenix Design GmbH & Co. KG, Stuttgart, Germany
Web
www.hansgrohe.com
www.phoenixdesign.com

This kitchen mixer series is characterised by a conical body, giving it an elegant appearance. The tap is available with an up to 50 cm pull-out sprinkler or with swivel outlet. In both versions, the water flow is activated via the handle and the temperature and flow rate are set. Due to a newly developed shut-off valve, the water flow can then be opened or closed via the setting knob which is located above the spout.

Statement by the jury
With their conical basic form and precise transitions, these mixers make an elegant impression; furthermore, they captivate with a coherent novel operating concept.

Diese Küchenmischer-Reihe zeichnet sich durch einen konischen Körper aus, der ihr ein elegantes Aussehen verleiht. Die Armaturen gibt es mit einer bis zu 50 cm weit ausziehbaren Brause oder mit Schwenkauslauf. Bei beiden öffnet der Nutzer den Wasserfluss am Griff und stellt Temperatur und Wassermenge ein. Dank eines neu entwickelten Absperrventils kann der Wasserfluss anschließend über den am Auslauf platzierten Select-Knopf geöffnet oder geschlossen werden.

Begründung der Jury
Mit ihrer konischen Grundform und präzisen Übergängen wirken diese Armaturen elegant; zudem bestechen sie mit einem schlüssigen neuen Bedienkonzept.

BLANCO CARENA-S / JURENA-S
Kitchen Taps
Küchenarmaturen

Manufacturer
BLANCO GmbH + Co KG,
Oberderdingen, Germany
In-house design
Brigitte Ziemann
Web
www.blanco-germany.com/de

The Blanco Carena tap with its rounded arch and the linear Blanco Jurena tap are based on a common design concept. The contrast between the raised spout and compact body of the taps creates an exciting highlight. The angled lever facilitates ergonomically pleasant operation. Both models are equipped with a concealed integrated spray hose.

Statement by the jury
Both taps are convincing with their clear lines and convey elegant professionalism. A nicely solved detail is the concealed spray hose.

Die schwungvoll gebogene Armatur Blanco Carena und die geradlinige Armatur Blanco Jurena basieren auf einem gemeinsamen Designkonzept. Der Kontrast zwischen erhabenem Auslaufrohr und kompaktem Armaturenkörper setzt jeweils einen spannungsvollen Akzent. Die schräg angesetzten Hebel ermöglichen eine ergonomisch angenehme Bedienung. Beide Modelle sind mit einer versteckt integrierten Schlauchbrause ausgestattet.

Begründung der Jury
Beide Armaturen überzeugen mit einer klaren Linienführung und vermitteln elegante Professionalität. Ein schön gelöstes Detail ist die versteckte Brause.

BLANCO FELISA
Kitchen Tap
Küchenarmatur

Manufacturer
BLANCO GmbH + Co KG,
Oberderdingen, Germany
In-house design
Brigitte Ziemann
Web
www.blanco-germany.com/de

Slender silhouettes and harmonious, flat transitions between cylindrical and rectangular forms characterise the purist aesthetic of this kitchen mixer tap family while precise edges and light reflections provide the taps with elegance and dynamics. The gently rising, rectangular outlet and the parallel aligned control lever are tapered towards the front. The water outlet is emphasised by a flush integrated Caché perlator.

Statement by the jury
Precise lines define the well-balanced and simultaneously very practical appearance of this tap family.

Schlanke Silhouetten und harmonische, flächige Übergänge zwischen zylindrischen und rechteckigen Formen prägen die puristische Ästhetik dieser Küchenarmaturen-Familie, während präzise Kanten und Lichtreflexe den Armaturen Eleganz und Dynamik verleihen. Der leicht ansteigende, rechteckige Auslauf und der parallel ausgerichtete Bedienhebel verjüngen sich nach vorne hin. Der Wasseraustritt wird durch einen bündig integrierten Caché-Perlator betont.

Begründung der Jury
Eine präzise Linienführung bestimmt das ausgewogene und zugleich sehr sachliche Erscheinungsbild dieser Armaturenfamilie.

S517T80, S527T80 Dishwashers
S517T80, S527T80 Geschirrspüler

Manufacturer
Constructa-Neff Vertriebs GmbH,
Munich, Germany
Design
BSH Hausgeräte GmbH, Neff Design Team,
Munich, Germany
Web
www.neff.de

Thanks to integrated TouchControl and high-definition TFT colour display, these dishwashers offer convenient operation. With the push of a button, information on programme selection and special functions is provided. In the interior, there is a VarioFlexPro basket system with hinged elements, retractable cutlery drawer and three-stage height adjustment for the top basket, allowing flexible loading. The adjustable elements are easily recognisable by their coloured markings.

Diese Spülmaschinen bieten dank integrierter Touch-Bedienung und hochauflösendem TFT-Farbdisplay eine komfortable Bedienung. Per Tastendruck lassen sich Informationen zu Programmauswahl und Sonderfunktionen abrufen. Im Inneren erlaubt das VarioFlexPro-Korbsystem mit klappbaren Elementen, absenkbarer Besteckschublade und dreistufiger Höhenverstellung des Oberkorbs eine flexible Beladung. Die justierbaren Elemente sind durch eine farbige Markierung erkennbar.

Statement by the jury
The dishwashers of this series score with a variable interior design and easy operability.

Begründung der Jury
Die Geschirrspüler dieser Serie punkten mit einer variablen Innenraumgestaltung und einer einfachen Bedienung.

SMI 88TS01E
Integrated Dishwasher
Integrierter Geschirrspüler

Manufacturer
Robert Bosch Hausgeräte GmbH,
Munich, Germany
In-house design
Robert Sachon, Thomas Ott
Web
www.bosch-hausgeraete.de

This dishwasher has, apart from the ergonomic touch panel control integrated in the upper door edge, two displays which not only show the time but also provide information and explanatory texts as user guide. The front display is harmoniously set in the minimalist designed stainless steel cover. Thanks to its clear graphics, it is easily read even from a greater distance. A flexible basket system, sensor-optimised programmes and interior lighting make this dishwasher complete.

Dieser Spüler hat neben der ergonomisch in den oberen Türrand integrierten Touch-Bedienfläche zwei Displays, die sowohl die Zeit als auch Informationen und Hinweistexte zur Nutzerführung anzeigen. Das vordere Display ist stimmig in die minimalistisch gestaltete Edelstahlblende eingelassen. Dank seiner klaren Grafik lässt es sich auch aus größerer Entfernung einfach ablesen. Ein flexibles Korbsystem, sensoroptimierte Programme und eine Innenbeleuchtung komplettieren diesen Spüler.

Statement by the jury
Convenient controls, flexibility of loading and contemporary technology combine here to produce a very user-friendly dishwasher.

Begründung der Jury
Eine komfortable Bedienung, Flexibilität bei der Beladung und zeitgemäße Technologien vereinen sich hier zu einem sehr benutzerfreundlichen Geschirrspüler.

SBE 88TD02E
Fully Integrated Dishwasher
Vollintegrierter Geschirrspüler

Manufacturer
Robert Bosch Hausgeräte GmbH,
Munich, Germany
In-house design
Robert Sachon, Thomas Ott
Web
www.bosch-hausgeraete.de

This fully integrated dishwasher features a new control and display concept. The TFT colour display not only shows the time but also provides a good deal of information and explanatory texts to be used as operating aid and user guide. Furthermore, sensor controlled programmes optimise water consumption during the dishwashing cycle. Partitioning in the interior is variable, thanks to flexible baskets; interior lighting further simplifies loading and unloading.

Dieser vollintegrierte Geschirrspüler bietet ein neues Bedien- und Anzeigekonzept. Das TFT-Farbdisplay liefert neben der Zeitanzeige viele Informationen und Hinweistexte als Bedienhilfe und zur Nutzerführung. Darüber hinaus optimieren sensorgesteuerte Programme den Wasserverbrauch während des Spülvorgangs. Die Aufteilung im Innenraum ist dank flexibler Körbe variabel, eine Innenbeleuchtung vereinfacht das Be- und Entladen zusätzlich.

Statement by the jury
This dishwasher can be discreetly integrated in every kitchen unit and convinces with intelligent functions and a user-friendly control and display concept.

Begründung der Jury
Dezent in jede Küchenzeile integrierbar, überzeugt dieser Geschirrspüler mit intelligenten Funktionen und einem benutzerfreundlichen Bedien- und Anzeigekonzept.

Green&Clean
Professional Dishwasher
Haubenspülmaschine

Manufacturer
Electrolux Professional, Pordenone, Italy
In-house design
Electrolux Group Design, Professional Sector
(Michele Cadamuro, Davide Benvenuti)
Web
www.professional.electrolux.com

The Green&Clean is a high-performance hood-type dishwasher for restaurants. The innovative angular display offers good visibility and protection against accidental shocks. The LED bar on the top of the hood reports on the status of the machine even to the chef working at distance increasing the efficiency of the workflow in the kitchen. The innovative automatic opening and closing eliminates the need for handles and improves the quality of the working environment for the operators.

Die Green&Clean ist eine leistungsstarke Haubenspülmaschine für Restaurants. Dank einer innovativen, abgeschrägten Eckausführung ist das Bedienfeld gut einsehbar und zugleich vor Stößen geschützt. Ein LED-Display im oberen Bereich der Haube zeigt jederzeit den Status des Reinigungszyklus an und ist auch aus einiger Distanz gut ablesbar, was zu effizienteren Arbeitsabläufen in der Küche führt. Das Öffnen und Schließen der Haube erfolgt automatisch. Dadurch konnte nicht nur auf Griffe verzichtet werden, sondern auch die Qualität des Arbeitsumfelds für den Benutzer verbessert werden.

Statement by the jury
A concise appearance emphasises the professional demand of this dishwasher. A highlight is the well-conceived positioning of the control panel.

Begründung der Jury
Ein prägnantes Erscheinungsbild unterstreicht den professionellen Anspruch dieser Spülmaschine. Besonders hervorzuheben ist die durchdachte Positionierung des Bedienfeldes.

Poggenpohl P'7350
Design by Porsche Design Studio
Kitchen Furniture
Küchenmöbel

Manufacturer
Poggenpohl Möbelwerke GmbH,
Herford, Germany
In-house design
Design
Porsche Design GmbH, Zell am See, Austria
Web
www.poggenpohl.com
www.porsche-design.at

The P'7350 departs from the horizontal lines of classical kitchens. This effect is created by an industrial mitring of the front and body. The front merges with the body in the mitre and forms a filigree, vertical line which is emphasised by an aluminium profile in stainless steel optic. Solid, vertical design trims of brushed aluminium, so-called blades, are fitted to interrupt the front surfaces and as additional emphasis on the verticals.

Die P'7350 bricht die horizontale Linienführung der klassischen Küche auf. Dieser Effekt entsteht durch eine industrielle Gehrungsbearbeitung von Front und Korpus. Die Front läuft in der Gehrung mit dem Korpus zusammen und bildet eine filigrane, vertikale Linie, die durch ein Aluminiumprofil in Edelstahloptik hervorgehoben wird. Massive, vertikale Designblenden aus gebürstetem Aluminium, sogenannte Blades, werden eingesetzt, um die Frontflächen zu unterbrechen und die Vertikale zusätzlich zu betonen.

Statement by the jury
With the contrasting interplay of solid blades and filigree vertical lines, the P'7350 succeeds in breaking from the traditional lines in kitchens.

Begründung der Jury
Mit dem kontrastreichen Zusammenspiel von massiven Blades und filigranen vertikalen Linien gelingt mit der P'7350 eine Durchbrechung der traditionellen Linienführung in Küchen.

zeyko Flybridge
Kitchen Island
Kücheninsel

Manufacturer
Zeyko Möbelwerk GmbH & Co. KG,
Mönchweiler, Germany
In-house design
Boris Pokupec, Tobias Hollerbach
Web
www.zeyko.de

The zeyko Flybridge statics system allows whole sections of cabinets to float seemingly in the air. In the version with handle-less zeyko Metal-X^2 fronts and a solid worktop of stainless steel, the impression of a closed metal kitchen block is created, whereas the cantilever and the bottom illumination give an effect of floating lightness. The contrasting design of the kitchen island is completed by walnut elements which connect overhead cabinets, counter and block.

Statement by the jury
zeyko Flybridge captivates not only with an interesting statics solution, but also with a skilful interplay of opposites.

Das Statiksystem zeyko Flybridge ermöglicht es, ganze Partien von Unterschränken scheinbar schweben zu lassen. In der Ausführung mit grifflosen zeyko-Metal-X^2-Fronten und einer massiven Edelstahlarbeitsplatte entsteht der Eindruck eines geschlossenen, metallischen Küchenblocks, während die Auskragung und Unterleuchtung schwebende Leichtigkeit vermitteln. Ergänzt wird die kontrastreiche Gestaltung der Kücheninsel durch Nussbaumelemente, die Hochschränke, Theke und Block verbinden.

Begründung der Jury
zeyko Flybridge besticht nicht nur mit einer interessanten Statiklösung, sondern auch mit einem gekonnten Spiel mit Gegensätzen.

Functional Channel in the Worktop
Funktionskanal in der Arbeitsplatte

Manufacturer
Warendorfer Küchen GmbH,
Warendorf, Germany
Design
dietzproduktgestaltung
(Jörg Dietz), Bielefeld, Germany
Web
www.warendorf.com
www.dietzproduktgestaltung.de

These functional worktops facilitate integration of a variety of functions discreetly at the right position of the working surface. Components of the system are integrated flush with the working surface and are made of the same material as the worktop. Elements such as knife block, utensil compartment, herb garden or dish warmer all have the same dimensions. A stainless steel channel under the worktop encircles the elements and can be easily wiped clean.

Statement by the jury
With elements which are harmoniously integrated in the worktop, the strictness of the working surface is interrupted, and a lively, coherent overall appearance is created.

Diese Funktionsarbeitsplatten ermöglichen es, eine Vielzahl an Funktionen dezent an der richtigen Stelle in die Arbeitsfläche zu integrieren. Die Systembestandteile werden flächenbündig eingelassen und sind aus demselben Material wie die Arbeitsplatte gefertigt. Elemente wie Messerblock, Utensilienfach, Kräutergarten oder Rechaud haben die gleichen Abmessungen. Ein Edelstahlkanal unter der Arbeitsfläche umfasst die Elemente und lässt sich leicht auswischen.

Begründung der Jury
Mit harmonisch in die Arbeitsplatte integrierten Elementen wird die Strenge der Arbeitsfläche aufgebrochen und es entsteht ein lebendiges, stimmiges Gesamtbild.

Shelving Element
Konturregal

Manufacturer
Warendorfer Küchen GmbH,
Warendorf, Germany
Design
dietzproduktgestaltung (Jörg Dietz),
Bielefeld, Germany
Web
www.warendorf.com
www.dietzproduktgestaltung.de

These open shelving elements in light design, used as end shelving in the cabinet area, take the interior of the kitchen to the outside and facilitate opening of the kitchen into the living and dining area. The shelves are flexible to use. In the rear panel, system rails are located which can accommodate, apart from shelves of glass or oak, also drawers or storage boxes and utensil holders.

Diese offenen, leicht gestalteten Konturregale bringen als Endregale im Schrankbereich die Innenausstattung der Küche nach außen und ermöglichen eine Öffnung der Küche in den Wohn- und Essbereich. Die Regale sind flexibel in der Nutzung. In der Rückwand befinden sich Systemschienen, in die neben Regalböden aus Glas oder Eiche auch Schubkasten- oder Vorratsdosenboxen und Utensilienbehälter eingehängt werden können.

Statement by the jury
The shelving elements set accents in modern, open kitchens by providing a pleasant contrast to the closed kitchen front.

Begründung der Jury
Die Konturregale setzen Akzente in modernen, offenen Küchen, indem sie einen schönen Kontrast zu den geschlossenen Küchenfronten bilden.

Worktop Extender
Arbeitsflächen-Erweiterung

Manufacturer
Nobia AB, Stockholm, Sweden
In-house design
Web
www.nobia.com

The Worktop Extender is a floor cupboard designed for smaller kitchens, offering one metre of additional stable worktop space without losing any storage space. The floor cupboard is simply pulled out, revealing a secondary worktop. Thanks to the included space for legroom, it is even possible to sit down to work or eat.

Der Worktop-Extender ist ein für kleinere Küchen konzipierter Unterschrank, der bei Bedarf über einen Meter zusätzliche stabile Arbeitsfläche bietet, ohne dabei zusätzlichen Stauraum zu beanspruchen. Der Unterschrank wird dafür einfach herausgezogen und eine Arbeitsfläche kommt zum Vorschein. Dank eines offen gestalteten Fußbereichs kann man sich zum Arbeiten oder Essen auch hinsetzen.

Statement by the jury
This worktop extender is a clever, practical design solution for small kitchens with limited worktop space.

Begründung der Jury
Diese Arbeitsflächen-Erweiterung ist eine clevere, praktische Gestaltungslösung für kleine Küchen mit begrenzter Arbeitsfläche.

MultiMatic Aluminium
Interior Accessories System
Innenausstattungssystem

Manufacturer
SieMatic Möbelwerke GmbH & Co. KG,
Löhne, Germany
Design
speziell®, Offenbach, Germany
Web
www.siematic.de
www.speziell.net

The basis of this interior accessories
system for kitchen cabinets consists of
aluminium trays and frames which are
hung on the inside of doors into the
corresponding SieMatic multifunction
tracks without visible mounting brackets.
The system can be extended with
wooden inserts and elements such as
roll, towel or shelf holder, receptacle
for spice jars or small items as well as
elements for storing glasses and bottles,
thus providing a means for individual
space usage.

Statement by the jury
MultiMatic Aluminium with its linear
elements creates order in cabinets and
at the same time offers scope for many
individual designs.

Basis dieses Innenausstattungssystems
für Küchenschränke sind Aluminiumschalen
und -rahmen, die auf der Innenseite
von Türen ohne sichtbare Befestigungs-
elemente in die zugehörige SieMatic-Multi-
funktionsschiene eingehängt werden. Das
System lässt sich durch Holzeinsätze und
Elemente wie Rollen-, Tuch- oder Brett-
halter, Gewürzdosen- oder Kleinteileschale
sowie Elemente zur Glas- und Flaschenauf-
bewahrung ergänzen und ermögl cht eine
individuelle Raumausnutzung.

Begründung der Jury
MultiMatic Aluminium schafft mit geradlinigen Elementen Ordnung in Schränken
und bietet dabei viel individueller
Gestaltungsspielraum.

AVENTOS HK-XS
Lift System
Klappensystem

Manufacturer
Julius Blum GmbH, Höchst, Austria
In-house design
Web
www.blum.com

Aventos HK-XS is a compact fitting which can be combined with various materials and front thicknesses of stay lifts in high and top cabinets. By means of the mechanical Tip-on opening support, furniture items even without handles can be easily and conveniently opened. A stageless stop and so-called Blumotion dampening provide a smooth movement sequence and allow soft and quiet closing.

Statement by the jury
Due to its slim-line construction, this flap lift system offers a great margin for design options; its sophisticated functionality provides a high degree of operating ease.

Aventos HK-XS ist ein kompakter Beschlag, der mit unterschiedlichen Materialien und Frontstärken von Hochklappen in Hoch- und Oberschränken kombinierbar ist. Mit der mechanischen Öffnungsunterstützung Tip-on lassen sich auch grifflose Möbel leicht und komfortabel öffnen. Ein stufenloser Stopp sowie die sogenannte Blumotion-Dämpfung sorgen für einen reibungslosen Bewegungsablauf und erlauben sanftes und leises Schließen.

Begründung der Jury
Durch seine schlanke Bauweise bietet dieses Klappensystem viel Gestaltungsspielraum, seine ausgereifte Funktionsweise sorgt für hohen Bedienkomfort.

TIP-ON BLUMOTION für LEGRABOX
Motion Technology
Bewegungstechnologie

Manufacturer
Julius Blum GmbH, Höchst, Austria
In-house design
Web
www.blum.com

Tip-on Blumotion is a motion technology which facilitates convenient opening of handle-less fronts. It involves a mechanical solution which combines two functions: easy opening with one touch (Tip-on) and dampened, quiet closing (Blumotion). In spite of its minimum front gap of only 2.5 mm, doors and drawers can be reliably opened. The gap alignment can be adjusted accurately due to four-dimensional setting options.

Statement by the jury
Tip-on Blumotion technology increases options for the furniture industry in a logical manner and is characterised by its precise functionality.

Tip-on Blumotion ist eine Bewegungstechnologie, die ein komfortables Öffnen griffloser Fronten ermöglicht. Es ist eine mechanische Lösung, die zwei Funktionen kombiniert: leichtes Öffnen durch Antippen (Tip-on) und gedämpftes, leises Schließen (Blumotion). Trotz eines Mindestfrontspalts von nur 2,5 Millimetern lassen sich Türen und Schubladen zuverlässig öffnen. Das Fugenbild kann dank vierdimensionaler Einstellmöglichkeiten exakt justiert werden.

Begründung der Jury
Die Bewegungstechnologie Tip-on Blumotion erweitert auf sinnvolle Weise die Möglichkeiten der Möbelindustrie und zeichnet sich durch ihre präzise Funktionsweise aus.

Free-Series
Flap Fittings
Möbelbeschläge

Manufacturer
Häfele GmbH & Co KG, Nagold, Germany
Design
A1 Productdesign Reindl + Partner GmbH
(Leon Seydel, Lucian Reindl),
Cologne, Germany
Web
www.haefele.de
www.a1-productdesign.com

The flap fitting family Free is a new, independent family of brackets which is modular in design and flexible in use. The Free series, from the viewpoint of design, is characterised by compact construction and, with Free flap, Free fold, Free up und Free swing, covers all conventional types of openings with flaps for wall cabinets. The fittings offer scope for design, can be used with various materials, are easily mounted, adjustable and, thanks to a multi-positioning stop function, they are convenient to operate.

Bei der Klappenbeschlagfamilie Free handelt es sich um eine neue, eigenständige Beschlagfamilie, die als Baukasten konzipiert und flexibel einsetzbar ist. Die Free-Serie ist in gestalterischer Hinsicht geprägt durch eine kompakte Bauweise und deckt mit Free flap, Free fold, Free up und Free swing alle gängigen Öffnungsarten für Klappen am Oberschrank ab. Die Beschläge bieten Gestaltungsfreiraum, sind mit verschiedenen Materialien verwendbar, leicht zu montieren, zu justieren und dank Multipositionsstopp-Funktion komfortabel zu bedienen.

Statement by the jury
The Free flap fittings convince due to their functional, space saving design and convenient method of mounting as well as their ease of movement.

Begründung der Jury
Die Free-Klappenbeschläge überzeugen durch ihre funktionale, platzsparende Gestaltung, eine angenehme Handhabung bei der Montage und ihre Leichtgängigkeit.

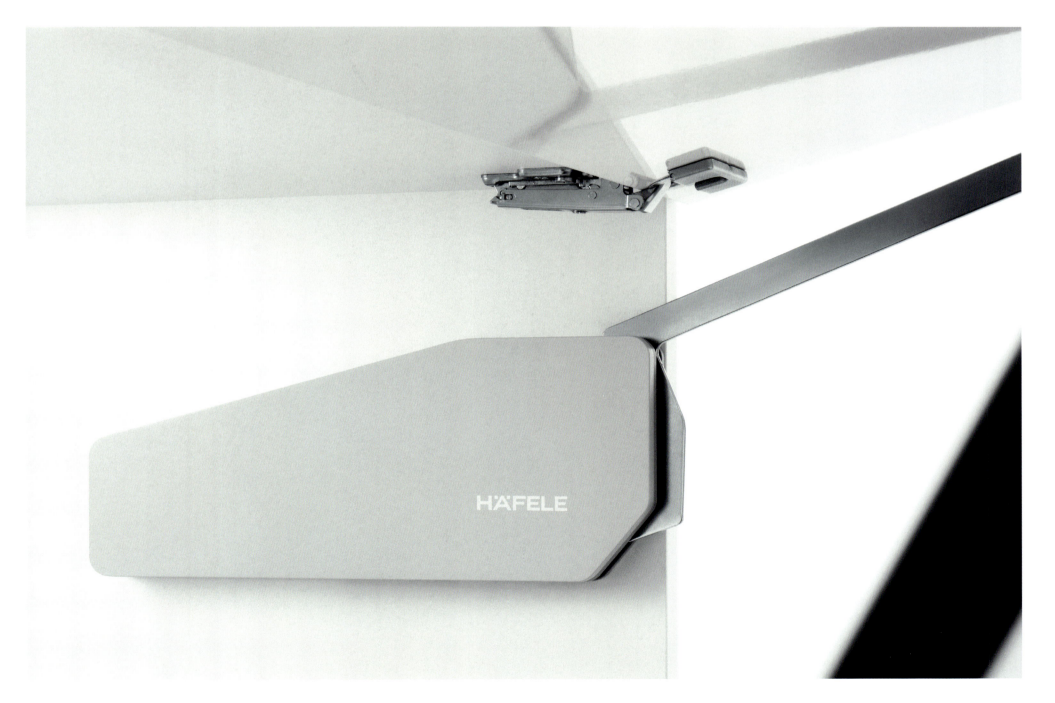

EVOline BackFlip
Electrification Module
Elektrifizierungsmodul

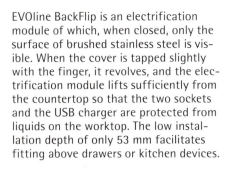

Manufacturer
Schulte Elektrotechnik GmbH & Co. KG,
Lüdenscheid, Germany
Design
Siegfried Schulte
Web
www.evoline.com

EVOline BackFlip is an electrification module of which, when closed, only the surface of brushed stainless steel is visible. When the cover is tapped slightly with the finger, it revolves, and the electrification module lifts sufficiently from the countertop so that the two sockets and the USB charger are protected from liquids on the worktop. The low installation depth of only 53 mm facilitates fitting above drawers or kitchen devices.

Statement by the jury
Discreetly integrated in the worktop when not in use, the purist electrification module offers safety and three connection options in its open position.

EVOline BackFlip ist ein Elektrifizierungsmodul, von dem in geschlossenem Zustand nur die Oberfläche aus gebürstetem Edelstahl sichtbar ist. Auf leichten Fingerdruck hin dreht sich der Deckel zurück, und das Modul erhebt sich hoch genug aus der Arbeitsfläche, damit die zwei Steckdosen und der USB-Charger auf der Arbeitsplatte geschützt sind. Die geringe Einbautiefe von nur 53 mm ermöglicht auch eine Montage über Schubladen oder Küchengeräten.

Begründung der Jury
Dezent in die Arbeitsfläche integriert, wenn es nicht genutzt wird, bietet das puristische Elektrifizierungsmodul ausgeklappt Sicherheit und drei Anschlussmöglichkeiten.

EVOline V-Port
Electrification Module
Elektrifizierungsmodul

Manufacturer
Schulte Elektrotechnik GmbH & Co. KG,
Lüdenscheid, Germany
Design
Siegfried Schulte
Web
www.evoline.com

The EVOline V-Port electrification module is specially conceived for vertical installation in corners as well as in and under overhead kitchen cabinets. When closed, the V-Port is an elegant aluminium housing in stainless steel optic. Upon slight pressure, the electrification module glides smoothly down and outwards, and a variably useable multiple socket with USB charging port becomes visible.

Statement by the jury
The V-Port provides the opportunity to make sensible use of dead corners under overhead cabinets and scores with its clear, high-quality design.

Das Elektrifizierungsmodul EVOline V-Port ist speziell für die senkrechte Anbringung in Ecken sowie in und unter Küchenoberschränken konzipiert. Geschlossen ist der V-Port ein elegantes Aluminiumgehäuse in Edelstahloptik. Auf leichten Druck gleitet das Elektrifizierungsmodul sanft nach unten heraus und eine variabel bestückbare Mehrfachsteckdose mit USB-Ladeanschluss wird sichtbar.

Begründung der Jury
Der V-Port bietet eine Möglichkeit, tote Ecken unter Oberschränken sinnvoll zu nutzen, und punktet mit einer klaren, hochwertigen Gestaltung.

Sound Unit
Integrated Kitchen Sound System
Integriertes Küchen-Soundsystem

Manufacturer
Nobia AB, Stockholm, Sweden
In-house design
Web
www.nobia.com

The development of the Sound Unit resulted from the challenge of finding a suitable place in the kitchen for a music system. Sound Unit is an active Bluetooth speaker, conceived as minimalist built-in module. At only 10 cm wide, it is designed to efficiently use up that left over bit of space, whilst still delivering an excellent sound experience. The interface has been reduced to a simple but beautifully machined aluminium touchpad.

Die Entwicklung der Sound Unit beruht auf der Herausforderung, in der Küche einen geeigneten Platz für eine Musikanlage zu finden. Sound Unit ist ein aktiver Bluetooth-Lautsprecher, der als minimalistisches Einbaumodul konzipiert ist. Mit einer Breite von nur 10 cm füllt die Sound Unit ungenutzte Lücken effizient und bietet gleichzeitig hohe Klangqualität. Die Benutzeroberfläche ist auf ein einfaches, hochwertig gefertigtes Aluminium-Touchpad reduziert.

Statement by the jury
The Sound Unit provides high sound quality and, due to its narrow body and minimalist design, fits seamlessly into kitchen units.

Begründung der Jury
Die Sound Unit bietet hohe Klangqualität und fügt sich mit ihrem schmalen Korpus und einer minimalistischen Gestaltung nahtlos in Küchenzeilen ein.

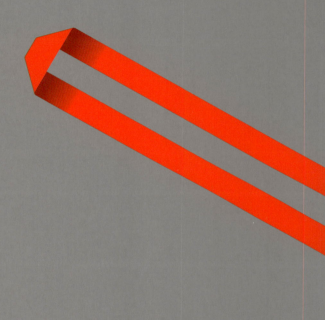

Tableware and cooking utensils
Tableware und Kochutensilien

Easy
Kitchen Utensils
Küchenutensilien

Manufacturer
Stelton,
Copenhagen, Denmark

Design
Cecilie Manz,
Copenhagen, Denmark

Web
www.stelton.com
www.ceciliemanz.com

reddot award 2015
best of the best

Pure functionality

Everyday kitchen utensils such as spatulas and cooking spoons are used way more often than one may think. However, it is often the case that metal spatulas, in particular, leave ugly scratches in pots and pans, whereas traditional silicone models are not particularly long-lasting and become deformed and unsightly. The Easy kitchen utensils collection lends this product category a new aesthetic both in terms of the utensil's form and functionality. They are made of specially treated white oak that ensures robustness and a long service life. The material rests pleasantly in the hand and spares pots and pans from scratches when stirring or turning food. When given proper care, the utensils even develop a beautiful patina over time. The clear design of these kitchen utensils is complemented by elegantly rounded handles with a pleasing surface that rests comfortably and ergonomically in the hand. The individual pieces of the collection are well matched, allowing for good control while cooking. The wide spatulas for instance make light work of turning pancakes, while the narrow spatulas offer a good grip and control when making scrambled eggs or omelets. The Easy collection replaces classical kitchen utensils by a successful purist design – enhancing everyday cooking with a sense of lightness and visual appeal based on the pure beauty of the material.

Pure Funktionalität

Während des Kochens benötigt man Küchenutensilien wie Pfannenwender oder auch Kochlöffel weitaus häufiger als gedacht wird. Oft ist es dabei jedoch der Fall, dass etwa Pfannenwender aus Metall unschöne Kratzer in Töpfen und Pfannen hinterlassen, während die Modelle aus Kunststoff nicht unbedingt langlebig sind und mit der Zeit unansehnlich werden. Die Küchenutensilien Easy verleihen dieser Produktgattung eine neue Ästhetik, sowohl hinsichtlich ihrer Form wie auch ihrer Funktionalität. Gefertigt sind sie aus einem speziell behandelten und deshalb robusten weißen Eichenholz. Dieses Material ist angenehm in seiner Haptik und hinterlässt keinerlei Kratzer beim Rühren oder Wenden. Vielmehr entwickelt es bei richtiger Pflege im Laufe der Jahre eine schöne Patina. Ausgeführt sind diese klar gestalteten Küchenutensilien mit einer stilvoll abgerundeten Grifffläche. Wie Handschmeichler liegen sie daher ergonomisch in der Hand des Nutzers. In ihrem Einsatz ist diese Kollektion von Kochutensilien zudem gut aufeinander abgestimmt. So können etwa die großzügig gestalteten Spatel für das Wenden von Pfannkuchen eingesetzt werden, während die etwas kleineren sich für die Zubereitung von Rührei eignen. Durch ihre gelungene puristische Gestaltung ersetzt die Kollektion Easy klassische Kochutensilien – der Arbeit in der Küche verleihen sie Leichtigkeit und zeigen dabei die Schönheit des Materials Holz auf.

Statement by the jury

With its purist design, the Easy collection enhances almost any kitchen interior. The kitchen utensils showcase a successful reduction to the essentials. They are made of high-quality white oak, making them highly robust, durable and age with grace to become ever more beautiful over time. Offering users many functional benefits for everyday cooking, they rest well in the hand with their rounded handles and leave no scratches in pot and pans.

Begründung der Jury

Mit ihrer puristischen Gestaltung bereichert die Kollektion Easy das Küchenambiente. Diese Kochutensilien zeigen eine gelungene Reduktion auf das Wesentliche. Gefertigt aus hochwertigem weißem Eichenholz, sind sie äußerst robust und langlebig, wobei sie mit der Zeit immer schöner werden. Dem Nutzer bieten sie viele funktionale Vorteile, da sie keinerlei Kratzer in Pfannen und Töpfen hinterlassen und mit ihrer abgerundeten Stielfläche während des Kochens gut in der Hand liegen.

Designer portrait
See page 44
Siehe Seite 44

Cook It
Kitchen Tools
Küchenhelfer

Manufacturer
Norbert Woll GmbH, Saarbrücken, Germany
Design
Friemel Design, Darmstadt, Germany
Web
www.woll-cookware.com
www.friemeldesign.de

Cook It is a series of kitchen tools that are heat resistant up to 260 degrees centigrade and dishwasher proof. Their silicone surrounded easy grip handles are reinforced with glass fibre, which make the handling of the respective tool more comfortable. The Jumbo spatula is equipped with a cutting function. In addition, the flexible soup ladle adjusts to different surfaces. The design in black and red achieves a high-contrast effect.

Statement by the jury
The design concept with its concise colours harmoniously blends into the brand appearance. Furthermore, the materials allow for a high ease of use.

Cook It ist eine Serie von Küchenhelfern, die bis 260 Grad Celsius hitzebeständig und spülmaschinenfest sind. Ihre von Silikon eingefassten easy-grip-Griffe sind glasfaserverstärkt und erleichtern die Handhabung der einzelnen Utensilien. Der Jumbo-Pfannenwender ist zusätzlich mit einer Schneidefunktion ausgestattet. Darüber hinaus passt sich die Schöpfkelle der jeweiligen Oberfläche flexibel an. Mit den Farben Schwarz und Rot erzielt die Gestaltung die kontrastreiche Wirkung.

Begründung der Jury
Das farblich prägnante Gestaltungskonzept fügt sich harmonisch in die Markenwelt ein. Zudem ermöglichen die Materialien einen hohen Bedienkomfort.

Tong Tools
Fork/Spoon Set, Fork/Turner Set
Gabel/Löffel-Set, Gabel/Wender-Set

Manufacturer
Kuhn Rikon AG, Rikon, Switzerland
In-house design
Philipp Beyeler
Web
www.kuhnrikon.com

The concept of Tong Tools offers multiple kitchen tools in one set. It offers a myriad of options depending on whether users need the set for whisking, flipping or serving the food. Tong Tools can either be used as two separate tools or they can be connected to form a pair of kitchen tongs. Employing soft materials, Tong Tools is also suitable for safe use with non-stick cookware. The set is heat resistant up to 220 degrees centigrade and dishwasher safe.

Statement by the jury
The functional characteristics of the flexibly usable Tong Tools are underscored by their high-contrast colours and distinctive line design.

Die Konzeption von Tong Tools bietet mehrere Küchenhelfer in einem Set. Seine Kombinationsmöglichkeiten sind vielfältig und abhängig davon, ob die Nutzer das Set zum Wenden, zum Quirlen oder zum Servieren von Speisen verwenden. Tong Tools lassen sich sowohl in ihre beiden Einzelteile als auch zusammengesteckt als Küchenzange nutzen. Ihr weiches Material lässt sich selbst bei beschichtetem Kochgeschirr problemlos einsetzen. Es ist bis 220 Grad Celsius hitzebeständig und spülmaschinentauglich.

Begründung der Jury
Kontrastreiche Farben und eine prägnante Linienführung betonen die funktionalen Eigenschaften der flexibel nutzbaren Tong Tools.

Curvo All-Purpose Tongs
Curvo Mehrzweck-Küchenzange

Manufacturer
Nambé LLC, Santa Fe, New Mexico, USA
Design
Cozzolino Studio (Steve Cozzolino), New York, USA
Web
www.nambe.com
www.cozzolinostudio.com

The Curvo All-Purpose Tongs merge stylish design language with sophisticated functionality. They can be used in many different ways, from serving pasta to turning mussels when stir-frying. The gently scooped tongs feature a slightly jagged edge to aid gripping. Thanks to their oval contours, the tongs rest pleasantly in the hand, and are polished to a mirror finish inside and out.

Statement by the jury
These stylishly scooped tongs appeal with both easy handling and a design language that focuses on detail.

Die Curvo Mehrzweck-Küchenzange verbindet eine stilvolle Formensprache mit einer ausgereiften Funktionalität. Sie ist variabel nutzbar und eignet sich sowohl für das Servieren von Pasta als auch für das Wenden von Miesmuscheln in der Pfanne. Die sanft geschwungene Zange ist am Ende mit einer filigran gezackten Greifleiste ausgestattet. Dank seiner ovalen Kontur liegt das innen und außen hochglänzende Küchenutensil angenehm in der Hand.

Begründung der Jury
Diese stilvoll geschwungene Küchenzange überzeugt sowohl aufgrund ihrer einfachen Handhabung als auch ihrer detailgenauen Formensprache.

Kitchen by Thomas
Kitchen Tools and Accessoiries
Küchenhelfer und Accessoires

Manufacturer
Rosenthal GmbH, Thomas, Selb, Germany
Design
Johanna Kleinert, Munich, Germany
Office for Product Design, Hong Kong
Web
www.thomas-porzellan.de
www.johannakleinert.de
www.officeforproductdesign.com

Kitchen by Thomas is a collection of accessories comprising approximately 40 pieces that can be combined with each porcelain series of the manufacturer. It includes for example a decanter with matching glasses and aroma-sealed food containers, a lemon press as well as two-piece bowls. The entire collection features a timeless design vocabulary with attractive highlights in the colours grey and green. The material mix of porcelain, glass, metal, wood and silicon is easy to clean and robust.

Kitchen by Thomas umfasst rund 40 miteinander und zu allen Porzellanserien des Herstellers kombinierbare Kollektionsteile. Dazu gehören unter anderem neben einer Karaffe mit passenden Gläsern auch aromadichte Aufbewahrungsboxen, eine Zitronenpresse sowie zweiteilige Schüsseln. Die gesamte Kollektion zeigt eine zeitlose Formensprache, welche mit den Farben Grau und Grün ansprechende Akzente setzt. Der Materialmix aus Porzellan, Glas, Metall, Holz und Silikon ist sowohl pflegeleicht als auch robust.

Statement by the jury
The Kitchen by Thomas collection owes its self-contained appearance to a trend-conscious design concept. The combination of materials is also eye-catching.

Begründung der Jury
Einem trendbewussten Gestaltungskonzept verdankt die Kollektion Kitchen by Thomas ihre eigenständige Anmutung, auch die Materialkombination fällt auf.

Orca 11"
5-in-1 Chef's Knife
5-in-1 Küchenmesser

Manufacturer
Songson Industrial Arts Co Ltd (Tony Chen),
Tainan, Taiwan
Design
Re-Wish Ltd. (Chao-Shun Liang),
London, Great Britain
Web
www.songson.com.tw
www.re-wish.co.uk

The Orca 11" chef's knife combines five functions in one cutting tool and can additionally be used as a mezzaluna knife. The unconventionally curved contour of the blade is reminiscent of the whale with the same name. An ergonomic handle makes cutting and chopping easier. The balance between the handle and the blade protects the wrist when the full length of the blade is used. The German stainless steel blade is highly durable and its seamless design provides bacteria control.

Das Orca 11" Küchenmesser kombiniert fünf Funktionen in einem Schneidwerkzeug, es ist unter anderem als Wiegemesser einsetzbar. Seine unkonventionell geschwungene Kontur erinnert an die des gleichnamigen Schwertwals. Ein ergonomischer Handgriff erleichtert das Schneiden. Sobald die Klinge in ihrer vollen Länge genutzt wird, schont die Gewichtsbalance zwischen Griff und Klinge das Handgelenk. Die aus deutschem Edelstahl gefertigte Klinge ist lange haltbar und hält aufgrund ihrer nahtlosen Gestaltung eine mögliche Bakterienverbreitung unter Kontrolle.

212

JIU
Chef's Knife
Kochmesser

Manufacturer
Kuhn Rikon AG, Rikon, Switzerland
In-house design
Valeria Hiltenbrand
Web
www.kuhnrikon.com

The versatile JIU chef's knife is made in Switzerland and named after the Chinese word for "long-lasting", as its very sharp stainless steel blade guarantees a long-lasting performance. It comes with a precisely fitting sheath for safe storage. The distinctively etched blade features a dynamic mountain panorama. The non-slip handle, made of high-grade material, rests well in the hand, allowing for comfortable handling even for long periods of time.

Statement by the jury
The JIU knife convinces with its self-sufficient and elegant design language, which is an expression of a high standard of quality.

Das vielseitig einsetzbare Kochmesser JIU wird in der Schweiz gefertigt und ist nach dem chinesischen Begriff für „lang anhaltend" benannt. Seine sehr scharfe Klinge aus rostfreiem Edelstahl garantiert eine konstante Schneidleistung. Zur sicheren Aufbewahrung dient der passgenaue Klingenschutz. Die markant geätzte Klingenfläche zeigt ein dynamisch wirkendes Bergpanorama. Der abrutschsichere Griff aus edlem Material liegt gut in der Hand und erlaubt ein angenehmes Arbeiten auch über längere Zeit.

Begründung der Jury
Das JIU Kochmesser überzeugt dank seiner eigenständigen und eleganten Formensprache, welche einem hohen Qualitätsanspruch Ausdruck verleiht.

Arctic
Knife Series
Messerserie

Manufacturer
OBH Nordica Group AB, Sundbyberg, Sweden
In-house design
Toni Luukkonen Fredäng
Design
Oxyma Innovation AB (Lars Håkansson), Solna, Sweden
Web
www.obhnordica.com
www.oxyma.se

The design of the Arctic knife series responds to the fact that many chefs grip the knife on the blade and not entirely on the handle. Consequently, all knives feature a rounded top edge and an ergonomic handle. The smooth, polished handle allows hanging the knife for storage and features a recessed area for the thumb. Each blade was designed for its specific purpose whilst the design language harmoniously blends into the overall appearance of the series.

Statement by the jury
The geometrical lines of the handles lend this knife series a dynamic and distinctive appearance.

Die Gestaltung der Messerserie Arctic berücksichtigt, dass viele Köche ein Messer nicht nur am Griff, sondern auch oben an der Klinge anfassen. Entsprechend verfügen alle Messer über eine abgerundete obere Kante und einen ergonomischen Griff. Der glatt polierte Griff ermöglicht die hängende Lagerung und besitzt eine innenliegende Daumenkerbe. Jede Klinge wurde für ihren spezifischen Zweck entwickelt, wobei sich die jeweilige Formensprache harmonisch in das Gesamtbild der Serie einfügt.

Begründung der Jury
Die geometrische Linienführung der Griffe verleiht dieser Messerserie eine dynamische und zugleich unverwechselbare Anmutung.

Krinkle Knife
Buntschneidemesser

Manufacturer
Kuhn Rikon AG, Rikon, Switzerland
In-house design
Carmela Weder-Niederhauser
Web
www.kuhnrikon.com

The Krinkle Knife makes crinkle cutting potatoes, cucumbers, carrots and other vegetables a breeze. The knife is equipped with a very sharp non-stick blade made of Japanese stainless steel. Designed according to ergonomic criteria, its round handle is suitable for both right and left-handed use. The complementary protective sheath allows for safe storage after use. The colours of the knife and the sheath are matched with each other harmoniously.

Statement by the jury
The functionally sophisticated Krinkle Knife catches the eye with its appealing colour and distinctive lines.

Mittels dieses Buntschneidemessers gelingt das Wellenschneiden von Kartoffeln, Gurken, Karotten und anderem Gemüse auf einfache Weise. Das Messer ist mit einer sehr scharfen, antihaftbeschichteten Klinge aus japanischem Edelstahl ausgestattet. Sein nach ergonomischen Kriterien gestalteter Rundgriff eignet sich sowohl für Rechts- als auch für Linkshänder. Der passende Klingenschutz ermöglicht nach dem Gebrauch eine sichere Aufbewahrung. Messer und Klingenschutz sind farblich aufeinander abgestimmt.

Begründung der Jury
Dieses funktional ausgereifte Buntschneidemesser erreicht aufgrund seiner ansprechenden Farbe und Linienführung eine hohe Aufmerksamkeit.

Twist Grater
2-in-1 Grater
2-in-1 Reibe

Manufacturer
Joseph Joseph Ltd, London, Great Britain
In-house design
Web
www.josephjoseph.com

Twist Grater has been designed to adapt to different grating tasks and foods with ease. It features a two-style stainless-steel blade set that ensures versatile application. Its handle has two different grating positions which offer enhanced user-friendliness. Locking the handle in the straight position makes it possible to grate food directly over plates and dishes. Twisting and locking the handle at 90 degrees allows the grater to rest diagonally on the worktop providing more support when grating.

Statement by the jury
This grater is marked by functionally sophisticated design details such as the flexible handle that ensures convenience in use.

Twist Grater wurde entwickelt, um das Zerkleinern unterschiedlicher Lebensmittel zu erleichtern. Zum einen ist der zwei-geteilte Schneideinsatz aus Edelstahl variabel nutzbar. Darüber hinaus lässt sich der Griff in zwei unterschiedlichen Positionen arretieren und bietet somit mehr Komfort: In der geraden Griffposition eignet sich die Reibe zur Anwendung direkt über Tellern oder Schüsseln. Wird der Griff hingegen um 90 Grad gekippt, liegt die Reibe diagonal zur Arbeitsplatte auf und kann leichter festgehalten werden.

Begründung der Jury
Funktional durchdachte Gestaltungsdetails wie ein flexibler Griff zeichnen diese Reibe aus und ermöglichen eine komfortable Handhabung.

Michel BRAS Mandoline
Michel BRAS Gemüsehobel

Manufacturer
Kai Corporation, Tokyo, Japan
Design
Michel Bras, Laguiole, France
Igarashi Design Studio
(Hisae Igarashi), Tokyo, Japan
Web
www.kai-group.com
www.braskai.net

The Michel BRAS Mandoline includes five different blades made of VG-10 steel. It sets up diagonally and delivers particularly precise cutting of slices, crinkles, wafers as well as Julienne cuts. The freely adjustable blade allows flexible slice thickness of 1-10 mm and Julienne cuts of 3, 7 or 10 mm. A special food cart enables safe and easy handling while a convenient case safely stores and protects the blades after use.

Statement by the jury
As a functionally well thought out product solution, the Michel BRAS Mandoline convinces with its manifold usage possibilities.

Der Michel BRAS Gemüsehobel umfasst fünf verschiedene Klingen aus VG-10-Stahl. Das diagonal aufstellbare Küchenutensil ermöglicht besonders präzise Scheiben-, Wellen-, Waffel- sowie Julienne-Schnitte. Eine stufenlose Einstellung des Klingenabstands ermöglicht eine flexible Scheibenstärke von 1-10 mm sowie Julienne-Schnitte von 3, 7 oder 10 mm. Ein spezieller Schnittguthalter sorgt für sicheres und leichtes Arbeiten, während eine praktische Aufbewahrungsbox die Klingen nach dem Gebrauch sicher verstaut.

Begründung der Jury
Als funktional durchdachte Produktlösung überzeugt der Michel BRAS Gemüsehobel dank seiner variablen Nutzungsmöglichkeiten.

Magnetic Cutting Board
Magnetisches Schneidbrett

Manufacturer
International Trading Co., Ltd., Ningbo, China
Design
Ningbo Galachy Creative Product Design Co., Ltd.,
Ningbo, China
Web
www.galachy.cn
Honourable Mention

This bamboo cutting board is magnetically connected to a plastic storage box. The storage box creates an attractive colour contrast and conveniently stores chopped food or collects food waste. When cutting, bowl and cutting board form a custom-fit unit. The connection smoothly disconnects for emptying the bowl. Both sides of the board are suitable for cutting while the bottom of the box serves as an extension of the cutting surface.

Dieses Schneidbrett aus Bambus wird per Magnet mit einer Kunststoffschale verbunden. Die farblich abgesetzte Auffangschale ermöglicht eine komfortable Aufbewahrung von bereits zerkleinerten Lebensmitteln oder auch von Schnittabfall. Während des Arbeitens bilden Schneidbrett und Schale eine passgenaue Einheit. Zum Entleeren der Schale lässt sich die Verbindung leicht lösen. Das Schneidbrett kann beidseitig verwendet werden, wobei dann die Rückseite der Schale die Schneidfläche verbreitert.

KMN Home
Rolling Pin
Nudelholz

Manufacturer
KMN Home LLC,
Traverse City, Michigan, USA
In-house design
Web
www.kmnhome.com

The rolling pin is made of aircraft grade, anodised aluminium. With its tapered shape, the pin rests well in the hand. Its hollow core makes sure that the weight is just right; it is neither too heavy nor too light. The anodised surface, which comes in red, blue, slate and black, features laser engraved graduated measurements in both inches and millimetres. This allows rolling out dough to just the right size. In addition, it stays cool after being chilled to keep butter from melting into the dough.

Statement by the jury
Functional material characteristics combine with an appealing design to support the comfortable use of this rolling pin.

Das Nudelholz ist aus eloxiertem Aluminium gefertigt in dem Grad, wie es auch im Flugzeugbau verwendet wird. Mit seiner konischen Kontur liegt das Nudelholz gut in der Hand. Für ein ausgewogenes Gewicht sorgt ein Hohlkern im Inneren. Auf der eloxierten Oberfläche in Rot, Blau, Schiefergrau oder Schwarz sind Maßeinheiten in Zoll und Millimeter per Laser eingraviert. Somit lässt sich leicht die gewünschte Teiggröße ausrollen. Zudem hält das gekühlte Nudelholz lange seine Temperatur, sodass die Butter im Teig nicht schmilzt.

Begründung der Jury
Funktionale Materialeigenschaften in Verbindung mit einer ansprechenden Gestaltung unterstützen bei diesem Nudelholz einen komfortablen Gebrauch.

Delícia
Cookie Stamp
Keks-Stempel

Manufacturer
Tescoma s.r.o., Zlín, Czech Republic
In-house design
David Veleba
Web
www.tescoma.com

This cookie stamp with integrated cutter allows the cutting of round cookies out of dough and the imprinting of their surfaces with decorative motifs. It comes with a selection of six different animal motifs, which are stored inside the stamp. After use, the stamp can be inserted into the cutter for space-saving storage. Christmas, Easter and other timeless motifs are also available. All motifs remain clearly recognisable and undistorted even after the baking process.

Statement by the jury
This compact product allows for simple and comfortable handling. The design of the various stamp motifs is emotionally appealing.

Mithilfe dieses Keks-Stempels mit integriertem Ausstecher lässt sich Teig kreisrund ausstechen und seine Oberfläche dekorativ prägen. Es stehen Aufsätze mit sechs unterschiedlichen Tiermotiven zur Auswahl, die im Stempel aufbewahrt werden. Nach Gebrauch lässt sich der Stempel platzsparend in den Ausstecher stecken. Neben den Tiermotiven sind auch Weihnachts-, Oster- und weitere zeitlose Motive verfügbar. Alle Motive bleiben auch nach dem Backen klar erkennbar und unverändert.

Begründung der Jury
Diese kompakte Produktlösung ermöglicht eine einfache und komfortable Handhabung. Die Motivgestaltung des Keks-Stempels ist emotional ansprechend.

Caesar

Kitchen Scales
Küchenwaage

Manufacturer
Zhongshan Camry Electronic Company Limited,
Zhongshan, China
Design
Junwei Li, Zhongshan, China
Web
www.camry.com.cn

Caesar are kitchen scales with an innovative design language. Via a rotating hinge, the arms of these scales fold out up to 90 degrees to form a cross. The user can put a bowl on it to start weighing. The matte zinc alloy scales are particularly strong and equipped with heat resistant silicon mats that serve as coasters on the top. The digital display at the front is easy to read. After use, the scales are quickly folded for space-saving storage.

Caesar ist eine Küchenwaage mit einer innovativen Formensprache. Mittels eines Drehscharniers lassen sich vor dem Wiegen die Aufstellarme um 90 Grad zu einem Kreuz ausklappen, um darauf eine Schüssel abzustellen. Die mit einer matten Zink-Legierung veredelte Küchenwaage ist stabil und mit hitzebeständigen Silikonmatten auf den Stellflächen ausgestattet. Die vorne positionierte Digitalanzeige ist einfach abzulesen. Nach dem Gebrauch kann die Waage schnell wieder zusammengeklappt und platzsparend verstaut werden.

Statement by the jury
An unconventional and convincingly realised design concept is the basis of these space-saving and user-friendly kitchen scales.

Begründung der Jury
Eine unkonventionelle und überzeugend ausgeführte Gestaltungsidee liegt dieser platzsparend und ‹omfortabel zu handhabenden Küchenwaage zugrunde.

Twixit Seal & Pour
Clip
Verschlussklammer

Manufacturer
Lindén International AB, Värnamo, Sweden
Design
Splitvision Design AB, Stockholm, Sweden
Web
www.lindenint.se
www.splitvision.com

Twixit Seal & Pour is an innovative bag clip that not only seals food bags airtight but also provides a practical pouring spout. Its transparent plastic lid allows users to look inside the bag even when it is closed. For easy and controlled pouring of foods such as muesli, flour or sugar, users simply need to flip the lid open. A dual-link hinge, which moves the spout away when opened, was developed to facilitate the application of the clamp.

Statement by the jury
This innovative design idea combines two functions in one clip. Convincing in its application, Twixit Seal & Pour offers a high ease-of-use.

Twixit Seal & Pour ist ein innovativer Clip, der eine Lebensmitteltüte nicht nur luftdicht verschließt, sondern zudem eine praktische Schütte bietet. Deren transparenter Kunststoffdeckel erlaubt auch im verschlossenen Zustand einen Einblick in die Tüte. Um Lebensmittel wie beispielsweise Müsli, Mehl oder Zucker kontrolliert zu entnehmen, wird der Deckel einfach aufgeklappt. Um die Anwendung der Klammer zu erleichtern, wurde eigens ein Doppelgelenkscharnier entwickelt, welches die Schütte beim Öffnen verschiebt.

Begründung der Jury
Diese innovative Gestaltungsidee vereint zwei Funktionen in einem Clip. Überzeugend in seiner Anwendung bietet Twixit Seal & Pour einen hohen Bedienkomfort.

Uno Vino
Wine Aerator
Weinbelüfter

Manufacturer
Tescoma s.r.o., Zlín, Czech Republic
In-house design
Ladislav Skoda
Web
www.tescoma.com
Honourable Mention

Thanks to its exchangeable valves, the Uno Vino can oxygenate various types of wines in an appropriate way while the wine is being poured into the glass: the red valve, which has nine openings, is perfect for red wines, while white wines are oxygenated using the green valve, which has three openings. The anthracite valve, which has no openings, serves as a spout for rosé wine. The aerator as well as the valves are kept in a convenient stand. Users simply need to attach the appropriate valve before serving the wine.

Statement by the jury
The Uno Vino aerator convinces with its three exchangeable valves, answering to the special requirements of different types of wine.

Mittels seiner unterschiedlichen Austauschventile belüftet Uno Vino verschiedene Weinsorten während des Einschenkens auf adäquate Weise: Das rote Ventil mit neun Luftlöchern eignet sich für Rotweine, während Weißweine mit dem grünen Ventil, das drei Löcher aufweist, belüftet werden. Das anthrazitfarbene Ventil ohne Löcher dient als Trichter für Rosé. Der Weinbelüfter sowie seine drei Ventile werden in einem praktischen Behältnis aufbewahrt. Das passende Ventil wird vor dem Ausschenken einfach aufgesteckt.

Begründung der Jury
Der Weinbelüfter Uno Vino überzeugt aufgrund einer bedarfsgerechten Ausführung mit drei funktionalen Austauschventilen.

Ridge Coffee Pod Carousel
Coffee Pod Organizer
Kaffeekapsel-Organizer

Manufacturer
Nambé LLC, Santa Fe, New Mexico, USA
Design
Alvaro Uribe, New York, USA
Web
www.nambe.com
www.alvarouribedesign.com

The Ridge Coffee Pod Carousel is a stylish coffee pod organizer for up to 25 coffee pods. With a minimal countertop footprint, it delivers users the option to choose between different types of coffee. Its sculptural design conveys a sense of dynamic flow and lends it a pleasing tactility and feel. The harmoniously curved ridges create an enticing interplay of dynamic light reflections. The Coffee Pod Carousel comprises five levels that have been brought together with innovative laser welding.

Statement by the jury
Employing high-quality materials, the Ridge Coffee Pod Carousel has managed to emerge as a consistent implementation of innovative design.

Ridge Coffee Pod Carousel ist ein stilvolles Edelstahl-Behältnis für 25 Kaffeekapseln. Es erleichtert auf minimaler Stellfläche die individuelle Auswahl zwischen unterschiedlichen Kaffeesorten. Seine skulptural anmutende Formensprache vermittelt eine schwungvolle Dynamik und bietet zudem haptische Reize. Dank der harmonisch geschwungenen Ausbuchtungen entsteht ein reizvolles Spiel von Licht und Schatten. Der Kaffeekapsel-Organizer besteht aus fünf Ebenen, die mittels innovativer Lasertechnik zusammengeschweißt sind.

Begründung der Jury
Unter Verwendung hochwertiger Materialien gelingt beim Ridge Coffee Pod Carousel die stringente Umsetzung eines innovativen Gestaltungskonzeptes.

JOHNNY CATCH Magnet
Wall-Mounted Bottle Opener
Wandmontierter Flaschenöffner

Manufacturer
höfats GmbH, Kraftisried, Germany
In-house design
Christian Wassermann, Thomas Kaiser
Web
www.hoefats.com

Johnny Catch is a bottle opener of minimalist appearance that mounts to the wall. The opener needs neither screws nor dowels as it comes equipped with high-quality, double-faced scotch tape. The tape is highly durable and even adheres to uneven surfaces. The distinctive Johnny Catch bottle opener design also features a surprising function: after opening a bottle, the opener easily "catches" the cap thanks to its excellent magnetic properties.

Statement by the jury
This style-conscious bottle opener is based on an innovative and consistently integrated design idea. It delivers a convincing, convenient user experience.

Johnny Catch ist ein minimalistisch anmutender Flaschenöffner, der an der Wand befestigt wird. Seine Montage kommt ohne Schrauben und Dübel aus, da er mit einem hochwertigen, doppelseitigen Klebeband ausgestattet ist. Dieses Klebeband hält dauerhaft und selbst auf unebenen Flächen. Der markant gestaltete Flaschenöffner bietet eine überraschende Zusatzfunktion: Nach dem Öffnen einer Flasche „fängt" Johnny Catch den Kronkorken dank eines sehr starken Magneten einfach auf.

Begründung der Jury
Dieser stilvolle Flaschenöffner basiert auf einer innovativen Gestaltungsidee, die stringent umgesetzt wurde. Er bietet einen überzeugenden Bedienkomfort.

Ganbei
Bottle Opener
Flaschenöffner

Manufacturer
Chi-Hsing Metal Co. Ltd.,
Changhua City, Taiwan
Design
Office for Product Design,
Hong Kong
Web
www.no30-inc.com
www.officeforproductdesign.com

The bottle opener Ganbei, which literally means "Cheers!" in both Chinese and Japanese, features the contours of a flat pebble. Its material mix, comprising stainless steel and a rough zinc alloy, lends this kitchen utensil a matte, stone-like appearance. The haptically pleasing shape of Ganbei makes it easy to grasp and feels reassuring in the hand. Thanks to its innovative 360-degree design, it allows opening bottles from any angle or direction without the need for particular alignment.

Statement by the jury
The haptically and visually enticing Ganbei bottle opener follows a unique design approach that also appeals emotionally to the user.

Der Flaschenöffner namens Ganbei, was im Chinesischen und Japanischen so viel wie „Prost" bedeutet, zeigt die Kontur eines flachen Kieselsteins. Sein Materialmix aus Edelstahl und einer rauen Zink-Legierung verleiht dem Küchenutensil ein mattes, steinartiges Erscheinungsbild. Haptisch angenehm ermöglicht Ganbei einen intuitiven Gebrauch und fühlt sich dabei sicher in der Hand an. Dank seiner 360-Grad-Gestaltung können Flaschen aus jedem Winkel, ohne dass auf den richtigen Ansatz geachtet werden muss, geöffnet werden.

Begründung der Jury
Der haptisch und visuell reizvolle Flaschenöffner Ganbei folgt einer originellen Gestaltungsidee, welche die Nutzer auch emotional anspricht.

Heineken Vector Merchandise Range
Bottle Holder
Flaschenhalter

Manufacturer
Miles Promocean,
's-Hertogenbosch, Netherlands
Design
GRO design, Eindhoven, Netherlands
Heineken International
(Caroline van Hoff),
Amsterdam, Netherlands
Web
www.milespromocean.eu
www.grodesign.com
www.heineken.com

The design concept of this bottle holder was developed in collaboration with talented young designers from around the world. Accentuating the contours of a beer bottle, it allows users to serve beer in style. The distinctive grid design picks up the lines of the brand logo, contrasting with the harmoniously curved contour that encloses the bottle. The distinctively shaped handle also conveys a high ease-of-use. Appealing colour highlights are created by a red star on glossy stainless steel.

Statement by the jury
This distinctively shaped bottle holder reaches a high degree of recognition. Its characteristic lines capture the zeitgeist.

Das Gestaltungskonzept für diesen Flaschenhalter entstand in Zusammenarbeit mit Nachwuchs-Designern aus der ganzen Welt. Den Konturen einer Bierflasche schmeichelnd ermöglicht er ein stilvolles Ausschenken. Ein prägnantes Gitternetz greift die Linienführung des Markenlogos auf. Im Kontrast dazu umschließt eine harmonisch geschwungene Kontur die Bierflasche. Der markant gestaltete Griff vermittelt zudem einen hohen Bedienkomfort. Reizvolle Farbakzente setzt der rote Stern auf glänzendem Edelstahl.

Begründung der Jury
Einen hohen Wiedererkennungseffekt erreicht dieser markant anmutende Flaschenhalter. Seine charakteristische Linienführung trifft den Zeitgeist.

Unilloy
Enamel-Coated Cast Iron Pot
Emaillierter Topf aus Gusseisen

Manufacturer
Sanjo Special Cast. Co., Ltd.,
Sanjo, Niigata, Japan

Design
Komin Yamada,
Kodaira, Tokyo, Japan

Web
www.unilloy.com

reddot award 2015
best of the best

New lightness

Since cast iron pots feature high heat retention and thermal conductivity, both professional chefs and private households have relied on them for centuries. However, being made of iron, these tradition-steeped pots are often heavy in weight and difficult to handle. Especially when filled with food they hardly allow being moved around. Unilloy is a cast iron pot that has been developed to keep the good characteristics of cast iron and merge them with new qualities. The pot is surprisingly light and highly functional. With a thickness of only 2 mm, its wall is half that of conventional cast iron pots, yet without the risk of cutting one's hand on the pot's rim or surface. The pot is made of a material that was originally used in the production of industrial machinery parts and further developed to open up an entirely new area of application. The innovative material properties of Unilloy are matched by a language of form that is both elegant and well-balanced. The pot showcases a design of flowing lines and a handle of almost delicate appearance. Different in appearance to that of a cast iron product at first glance, the pot can be lifted easily, thanks to its light weight, and without leaving any kind of marks on the hands or fingers. The combination of innovatively utilised materials with a sensually appealing design reinterprets the design of cast iron pots – cooking enjoyment with highly elegant, contemporary cookware.

Neue Leichtigkeit

Da gusseiserne Töpfe die Wärme gut leiten und auch lange halten, verwenden Profiköche und Privathaushalte sie schon seit Jahrhunderten in ihren Küchen. Diese traditionsreichen Töpfe sind allerdings wegen ihres Eisenanteils oftmals sehr schwer und unhandlich. Insbesondere mit Inhalt kann der Koch sie deshalb kaum von der Stelle bewegen. Mit Unilloy entstand ein gusseiserner Topf, der die positiven Eigenschaften von Gusseisen mit neuen Qualitäten verbindet. Dieser Topf ist überraschend leicht und sehr funktional. Mit nur 2 mm Dicke ist seine Wand etwa halb so stark wie die anderer gusseiserner Töpfe, dies jedoch ohne jede Gefahr, sich an seinem Rand in die Hand zu schneiden. Das dafür verwendete, ursprünglich für Industriekomponenten eingesetzte Material wurde entsprechend weiterentwickelt, um damit nun einen neuen Anwendungsbereich zu erschließen. Die innovativen Materialeigenschaften des Unilloy gehen zugleich einher mit einer eleganten und ausgewogenen Formensprache. Er besitzt fließende Linien und ist mit einem filigran anmutenden Griff gestaltet. Für den Betrachter ist dies bei einem gusseisernen Topf zunächst irritierend, wegen seines geringen Gewichts kann er jedoch problemlos angehoben werden, ohne Spuren an den Händen zu hinterlassen. In der Verbindung eines innovativ eingesetzten Materials mit einer sinnlich ansprechenden Gestaltung erfährt der gusseiserne Topf damit eine neue Interpretation – hin zu einem zeitgemäßen und überaus eleganten Kochgeschirr.

Statement by the jury

Unilloy is a cast iron pot with a surprisingly light weight, yet retains the same excellent heat-retentive characteristics as conventional models. Well-balanced in its proportions and featuring a filigree handle design, this pot boasts an appearance that appeals to the senses. The pot has emerged as a product that utilises a material that is innovative in cookware design to set new standards in this field.

Begründung der Jury

Unilloy ist ein gusseiserner Topf mit einem verblüffend geringen Gewicht, der dennoch die gleichen exzellenten wärmeleitenden Eigenschaften besitzt wie die üblichen Modelle. Sehr ausgewogen in seinen Proportionen und mit einer filigranen Griffgestaltung, hat dieser Topf eine die Sinne ansprechende Anmutung. Unter Einsatz eines für Kochgeschirr innovativen Materials entstand so ein Topf, der neue Maßstäbe in diesem Bereich setzt.

Designer portrait
See page 46
Siehe Seite 46

STONELINE® Future
Cookware
Kochgeschirr

Manufacturer
WARIMEX GmbH, Neuried, Germany
In-house design
Web
www.warimex.de
Honourable Mention

An inner coating of clearly visible stone particles characterises the Stoneline Future cookware set. The convenient lid holder at the rim allows condensed liquids to flow back into the cookware. The rim of the lids, which are made of glass, features a fine and coarse sieve function so that meat juices, vegetable broth or pasta water can be easily poured off after cooking without the need to remove the lid. In addition, two spouts on each piece of the set help users to cleanly pour off liquids.

Eine Innenbeschichtung mit deutlich erkennbaren Steinpartikeln kennzeichnet das mehrteilige Kochgeschirr-Set Stoneline Future. Die praktische Deckel-halterung am Topf- oder Pfannenrand sorgt dafür, dass das Kondenswasser zurück in das Kochgeschirr fließt. Im Rand der Glasdeckel ist eine Fein- und Grob-Siebfunktion integriert, sodass Bratenfond, Gemüsesud oder Nudelwasser komfor-tabel einfach abgeschüttet werden kann, ohne dafür den Deckel abnehmen zu müssen. Zudem erleichtern jeweils zwei Ausgüsse an den Töpfen und den Pfannen das saubere Abseihen von Flüssigkeiten.

Ultima

Cookware
Kochgeschirr

Manufacturer
Tescoma s.r.o., Zlín, Czech Republic
In-house design
Frantisek Fiala
Web
www.tescoma.com

The Ultima cookware series features an innovative cover system. Different positions of the lid allow the closing of the cookware to suit individual needs: hermetically sealed for quickly warming up food, or in a slightly opened position to release steam and prevent boiling-over. For straining water, the lid is securely fixed in a third position. After opening the pot, it can be placed onto the handle in an upright position, which allows condensed water to run back into the pot.

Das Kochgeschirr Ultima bietet ein innovatives Schließsystem. Mittels unterschiedlicher Deckelpositionen lässt sich der Topf bedarfsgerecht verschließen: optional hermetisch dicht zum schnellen Erhitzen von Speisen und des Weiteren in einer leicht geöffneten Position, die Dampf entweichen lässt und ein Überkochen verhindert. Zum Abseihen von Wasser wird er in einer dritten Position sicher fixiert. Nach dem Öffnen des Kochtopfs kann der Deckel hochkant in den Griff abgelegt werden, wodurch das Kondenswasser in den Topf zurückläuft.

Statement by the jury
A functionally well thought-through design concept characterises this cookware series. The flexible handling of the lid enhances usability considerably.

Begründung der Jury
Ein funktional durchdachtes Gestaltungskonzept zeichnet dieses Kochgeschirr aus. Die flexible Handhabung des Deckels erhöht den Bedienkomfort erheblich.

T Chef Series Pure Cookware

Cookware
Kochgeschirr

Manufacturer
Tupperware Belgium N.V.,
Aalst, Belgium
Design
Tupperware Worldwide Product Development Team
Web
www.tupperwarebrands.com

T Chef Series Pure Cookware is a high quality cookware range that cooks with a minimum amount of water. With the objective of providing users with a health-conscious way of cooking, a valve system with three different positions was developed: the open position allows steam to escape without splashing, the closed position keeps the steam inside and supports cooking at low temperatures, and the whistle position gives an audible signal when the water is boiling. The lid can be hooked onto any place of the rim in an upright position. In addition, measurement marks can be found on the inside of the cookware.

T Chef Series Pure Cookware ist eine hochwertige Kochgeschirr-Serie, die mit einem Minimum an Wasser gart. Mit der Zielsetzung, ein gesundheitsbewusstes Kochen zu ermöglichen, wurde ein Ventil-system mit drei verschiedenen Einstellungen entwickelt: Beim geöffneten Ventil kann der Dampf spritzfrei entweichen. Beim geschlossenen Ventil verbleibt der Dampf im Topf und begünstigt ein Garen bei niedriger Temperatur. Die Signalton-Einstellung meldet akustisch, sobald das Wasser kocht. Der Deckel kann an jeder Stelle des Topfrandes aufrecht eingehängt werden, zudem finden sich im Topf Maß-angaben.

Statement by the jury
This cookware range offers a high degree of functionality. In addition, the distinctive form language emphasises a high quality standard.

Begründung der Jury
Ein hohes Maß an Funktionalität bietet diese Kochgeschirr-Serie. Darüber hinaus betont die eigenständige Formensprache einen hohen Qualitätsanspruch.

Micro-Pressure Stockpot
Mikro-Schnellkochtopf

Manufacturer
Guangdong Master Group, Yunfu, China
Design
Xivo Design (Chao Jing, Linghan Li),
Shenzhen, China
Web
www.mastergroup.com.cn
www.xivodesign.com

This micro-pressure stockpot cooks food more quickly and uses less energy than traditional pressure cookers. It is intuitive in use while the arch-shaped design of the lid's handle allows users to lock the pot in a smooth and safe way. Produced from 0.7 mm thick stainless steel, the cooker can be used on any type of stove. The multilayered aluminium alloy bottom of the pot has a high heat conductivity. Thanks to the heat-resistant handles, it is easy to carry the pot.

Mit diesem Mikro-Schnellkochtopf können Speisen im Vergleich zu herkömmlichen Töpfen schneller und energiesparender gegart werden. Seine Handhabung ist intuitiv, wobei der Topf aufgrund des bogenförmigen Griffes des Deckels leichtgängig und sicher zu verschließen ist. Gefertigt aus 0,7 mm dickem Edelstahl kann der Topf auf verschiedenen Herdarten benutzt werden. Der Topfboden aus einer mehrschichtigen Aluminiumlegierung besitzt eine hohe Wärmeleitfähigkeit. Die hitzebeständigen Griffe erleichtern den Transport.

Statement by the jury
This micro-pressure stockpot owes its high ease-of-use to an innovative locking mechanism while the arch above the lid and the handles constitute a formal unity.

Begründung der Jury
Seinen hohen Bedienkomfort verdankt der Mikro-Schnellkochtopf einem innovativen Verschluss, wobei der Bogen über dem Deckel und die beiden Griffe formal eine Einheit bilden.

Chinese Wok
Chinesische Wokpfanne

Manufacturer
BK Cookware BV, Delft, Netherlands
In-house design
Nadia Bartels-Wijstma
Web
www.bkcookware.com

This wok, with a diameter of 36 cm, is supplied with sturdy heat-resistant handles. Its relief base ensures that oil or fat is well distributed to give the best possible cooking results. The wok is crafted entirely from 3-ply, an innovative three-layer structure incorporating a layer of aluminium sandwiched between two layers of high quality stainless steel, which optimises heat conduction. The wok is suitable for use on all heat sources. Separately available accessories include an insert for steaming as well as a grid rack.

Statement by the jury
The wok owes its high degree of user comfort to a design concept that makes full use of the specific benefits of the materials.

Diese mit hitzebeständigen Griffen ausgestattete Wokpfanne hat einen großzügigen Durchmesser von 36 cm. Ihr Reliefboden sorgt für eine gleichmäßige Fettverteilung und entsprechend gute Bratergebnisse. Die Wokpfanne ist komplett aus 3-Ply hergestellt, einer innovativen 3-Lagen-Konstruktion, bestehend aus zwei Schichten rostfreiem Edelstahl und einer Zwischenschicht aus Aluminium, was die Wärmeleitung optimiert. Der Wok eignet sich für jeden Herd. Weiteres Zubehör, wie ein Einsatz zum Dünsten und ein Ablagegitter, ist separat erhältlich.

Begründung der Jury
Ihr hohes Maß an Bedienkomfort verdankt die Wokpfanne einem Gestaltungskonzept, welches die spezifischen Vorteile des Materials stringent nutzt.

Essenso Bombé
Grill Pan
Grillpfanne

Manufacturer
Essenso Housewares, Istanbul, Turkey
In-house design
Efe Erinç Erdogdu, Ali Cagatay Afsar
Web
www.essenso.com.tr

The Essenso Bombé cast iron grill pan sets itself apart from conventional grill pans through the convex shape of its bottom. This characteristic shape makes the wok an energy-saving kitchen tool, especially when it is used on a gas stove. It is another benefit of the convex base that all the fat is collected at the rim of the bottom so that it can easily be drained. This prevents the fat from burning and hence compromising the taste of the food. The manufacturer also advises to fill some water into the edge of the pan to prevent the collected fat from burning there.

Statement by the jury
The innovative idea of a convex base was convincingly implemented into this grill pan. The contrasting colours give it another appealing touch.

Die Essenso Bombé Grillpfanne unterscheidet sich dank ihres konvexen Bodens von anderen Pfannen. Der nach oben gewölbte Boden ist bei der Verwendung auf einem Gasherd besonders energiesparend. Ein weiterer Vorteil der Wölbung ist, dass sich das Fett am Pfannenrand sammelt und sich leicht ausgießen lässt. Somit wird verhindert, dass das Fett verbrennt und den Geschmack des Bratguts beeinträchtigt. Der Hersteller empfiehlt zudem, etwas Wasser in den Pfannenrand zu füllen, damit das Fett nicht einbrennen kann.

Begründung der Jury
Die innovative Gestaltungsidee eines konvexen Bodens wurde bei dieser Grillpfanne überzeugend umgesetzt. Ansprechend sind zudem die Kontrastfarben.

Just Cook
Frying Pan
Pfanne

Manufacturer
Norbert Woll GmbH, Saarbrücken, Germany
In-house design
Web
www.woll-cookware.com

The Just Cook frying pan follows a concise design language. The dishwasher proof frying pan features a drip free rim for pouring sauces cleanly while quick and even heat distribution allows for more effective cooking. The strikingly coloured handle is ovenproof up to 180 degrees centigrade. The 5 mm thick base is suitable for all kinds of cooktops and features an abrasion-proof, three layered non-stick coating, which is PFOA free.

Statement by the jury
Just Cook is a product that is both visually appealing and convenient. The red handle highlights the dynamic look of the frying pan.

Die Pfanne Just Cook folgt einer prägnanten Formensprache. Die spülmaschinenfeste Pfanne ermöglicht mittels ihres Schüttrands ein sauberes Ausgießen von Soßen. Eine schnelle und gleichmäßige Hitzeverteilung steigert darüber hinaus die Kocheffektivität. Der farblich auffallende Stiel ist bis 180 Grad Celsius backofentauglich. Der fünf Millimeter dicke Boden eignet sich für alle Herdarten und ist mit einer abriebfesten, dreilagig verstärkten Antihaft-Versiegelung beschichtet. Diese Versiegelung ist PFOA-frei.

Begründung der Jury
Just Cook ist eine gleichsam visuell ansprechende wie komfortable Produktlösung. Einen besonders dynamischen Akzent setzt der rote Pfannenstiel.

Wave
Food Storage Container
Lebensmittelbehälter

Manufacturer
Lock & Lock Co., Ltd., Seoul, South Korea
In-house design
Web
www.locknlock.com
Honourable Mention

Wave is a food storage container with an integrated steam hole. The vividly coloured microwave-proof lid adds charm to the kitchen and emphasises the functionality of the design. The heat-resistant glass container with its wavy relief pattern can be used to cook food in the oven and store it in completely hygienic conditions. Opening the steam hole releases hot steam and ensures that the food stays juicy. Wave is available in different sizes, shapes and colour combinations.

Wave ist ein Lebensmittelbehälter mit integriertem Dampfloch. Der in knalligen Farben gestaltete, mikrowellengeeignete Silikondeckel erweckt Sympathie und betont seine Funktionalität. Das hitze-beständige Glasgefäß mit Wellenrelief ermöglicht sowohl eine Zubereitung von Speisen im Ofen als auch deren hygie-nische Aufbewahrung. Durch das Öffnen des Dampflochs kann heißer Dampf entweichen und das Essen bleibt saftig. Wave ist in unterschiedlichen Größen, Formen und Farbkombinationen erhältlich.

Statement by the jury
The emotionally appealing design effectively underlines the benefits of these food storage containers.

Begründung der Jury
Die emotional ansprechende Gestaltung dieser Lebensmittelbehälter bringt deren vorteilhafte Handhabung wirkungsvoll zur Geltung.

Smart Savers
Storage Container
Vorratsbehälter

Manufacturer
Tupperware India Pvt Limited,
Dehradun, India
Design
Tupperware Worldwide Product Development Team
Web
www.tupperwarebrands.com

The stackable Smart Savers are designed to store a large amount of food in a limited space. The transparent material makes it easy to identify contents while the tight sealing lids keep the stored food fresh longer. The contour of the storage containers is made through the transition of a softer radius at the base to a tighter radius at the top, which allows for easy portioning of the food. The non-slip surface ensures comfortable handling.

Die stapelbaren Smart Savers sind so konzipiert, dass sie große Mengen an Lebensmitteln auf einem möglichst geringen Raum verstauen. Ihr transparentes Material erlaubt einen schnellen Überblick, während die zuverlässig abdichtenden Deckel die gelagerten Lebensmittel länger frisch halten. Die Kontur der Vorratsbehälter geht von einem weichen Radius am Boden in einen engeren Radius an der Oberseite über, was die Dosierung der Lebensmittel vereinfacht. Die griffige Oberfläche begünstigt eine komfortable Handhabung.

Statement by the jury
A consistent design down to the last detail provides the Smart Savers with convincing features. A highly versatile product series.

Begründung der Jury
Eine bis ins Detail schlüssige Gestaltung stattet die Smart Savers mit überzeugenden Eigenschaften aus. Eine vielseitig nutzbare Produktserie.

BreadSmart II
BrotMax 2
Bread Box
Brotkasten

Manufacturer
Tupperware France S.A.,
Joué-les-Tours, France
Design
Tupperware Worldwide Product Development Team
Web
www.tupperwarebrands.com

The key element in the creation of the BreadSmart II was the concept of a cover that doubles as a serving tray and allows the bread box to be accessed from the top. The product is extended by additional features such as the option to store the cutting board either inside or on top of the container. An integrated filter system regulates the humidity level inside. The slightly curved surfaces and the refined hole pattern on the rim lend the box its distinctive appearance.

Das Hauptaugenmerk bei der Gestaltung des BrotMax 2 lag in der Konzeption eines Deckels, der gleichzeitig als Serviertablett fungiert und zudem das Öffnen des Brotkastens von oben ermöglicht. Der Produktaufbau wurde mit zusätzlichen Funktionen erweitert, wie der Option, das Schneidbrett sowohl in als auch auf dem Behälter aufzubewahren. Ein integriertes Filtersystem reguliert die Feuchtigkeit im Inneren des Behälters. Charakteristisch wirken die leicht gebogene Strukturlinie der Seitenwände sowie das Lochmuster im Rahmen.

BISTRO Bread Box
BISTRO Brotkasten

Manufacturer
Bodum AG, Triengen, Switzerland
Design
Pi-Design AG, Triengen, Switzerland
Web
www.bodum.com

With its large capacity, the Bistro bread box will accommodate even oversized loaves of bread. The box keeps bread and rolls fresh for a long time. The lid is made of durable bamboo. It serves to catch crumbs and doubles as a serving tray and cutting board. Made of BPA-free plastic, the bread box is robust and easy to clean. Manufactured in Europe, its brand-distinctive design vocabulary harmoniously blends into even the most unique kitchen interiors.

Statement by the jury
This bread box showcases a robust and premium-quality appearance. The contrast of the materials underlines the visually appealing bamboo structure.

Mit seinen großzügigen Ausmaßen bietet der Brotkasten Bistro viel Platz, selbst für übergroße Brotlaibe. Die dort aufbewahrten Brote und Brötchen bleiben lang anhaltend frisch. Der Deckel aus nachhaltigem Bambus dient einerseits als Krümelsammler und andererseits als Tablett und Schneidebrett. Der Brotkasten aus BPA-freiem Kunststoff ist stabil und leicht zu reinigen. Er wird in Europa gefertigt. Seine markentypische Formensprache fügt sich harmonisch in unterschiedliche Küchenwelten ein.

Begründung der Jury
Eine robuste und hochwertige Anmutung vermittelt dieser Brotkasten. Der Kontrast der Materialien betont die reizvolle Bambusstruktur.

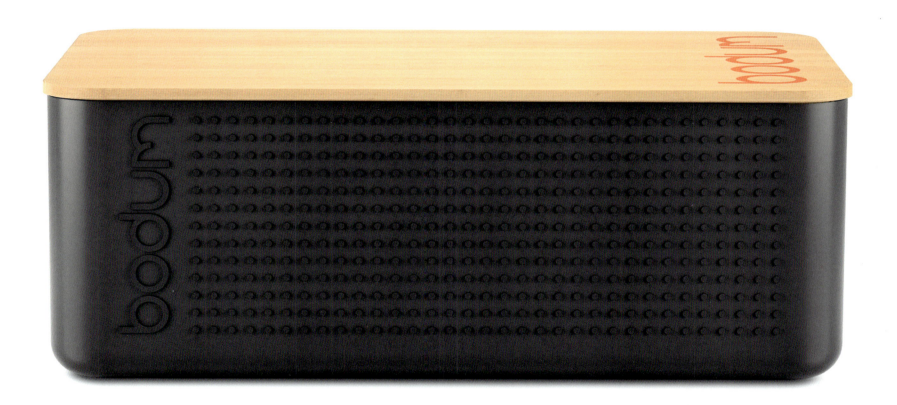

Orquestra
Tableware
Geschirrserie

Manufacturer
Vista Alegre Atlantis, SA, Ilhavo, Portugal
In-house design
Web
www.myvistaalegre.com

The design concept of the Orquestra tableware series was inspired by the complexity of the harmonies and sounds of a symphony orchestra. The design language creates its own complexity with a varied combination of geometric patterns. Each line, in its different direction and density, reflects a different soundscape. Highlights are set by the discrete colours on the rim of the tableware. The visualisation of musical rhythms and harmonies also has an emotional appeal on the beholder.

Statement by the jury
Orquestra combines aesthetic lines with elegant contours and harmonious colours. Overall, a highly consistent work.

Das Gestaltungskonzept der Geschirrserie Orquestra wurde von der Klangvielfalt eines Symphonieorchesters inspiriert. Die Formensprache kreiert mittels einer variantenreichen Kombination von geometrischen Mustern eine eigenständige Komplexität. Jede einzelne Linie spiegelt in ihrer jeweiligen Stärke und Ausrichtung unterschiedliche Klangwelten wider. Dezente Farben am Rand der einzelnen Geschirrteile setzen Akzente. Die Visualisierung von musikalischen Rhythmen und Harmonien spricht die Betrachter auch emotional an.

Begründung der Jury
Eine ästhetische Linienführung verbindet sich bei Orquestra mit eleganten Konturen und harmonischen Farben, eine insgesamt stimmige Kreation.

Ultimate BBQ
Platters
Servierplatten

Manufacturer
Villeroy & Boch AG, Mettlach, Germany
In-house design
Web
www.villeroy-boch.de

Ultimate BBQ includes selected products around the topic BBQ. Among other articles, the collection includes platters in two sizes which are designed with ridges to allow meat juice to drain away from rare or medium grilled steaks. Skewers can easily be hooked in the notches. The large platter offers a separate area for sauces or vegetable garnishes which makes sure that the ingredients do not mix.

Statement by the jury
These platters convey an overall appearance that is both formally and functionally sophisticated. The practical details are convincing.

Ultimate BBQ umfasst ausgewählte Produkte rund um das Thema Grillen. Dazu gehören unter anderem Servierplatten in zwei Größen. Durch die von innen nach außen abfallenden Abtropfrillen bleiben englisch oder medium gebratene Steaks saftig ohne durchzuweichen, wenn beim Schneiden Fleischsaft austritt. Die integrierten Spieß-Halter verhindern das Ineinanderrutschen. Die große Servierplatte bietet darüber hinaus einen separaten Bereich, in dem Saucen oder Gemüsebeilagen vom Fleisch getrennt serviert werden können.

Begründung der Jury
Ein gleichsam formal wie auch funktional durchdachtes Gesamtbild vermitteln diese Servierplatten. Es überzeugen die praktischen Details.

Seagull
Tableware
Geschirrserie

Manufacturer
Guangdong Songfa Ceramics Co., Ltd., Chaozhou, China
In-house design
Web
www.songfa.com

The Seagull tableware is fired at a temperature of 1,340 degrees centigrade and distinguishes itself with its eye-catching, pure white surfaces, which are pleasing to the touch. The design was inspired by a seagull soaring with the waves of the ocean to conceive this tableware creation with its appealing curved lines. The flowing contours merge harmoniously into the two ergonomically well-thought out handles. The bowls, which are different in size, are stackable and thus store away protected from breaking.

Statement by the jury
Inspired by flying seagulls, this product has emerged as a successful creation of outstandingly aesthetic high-quality tableware.

Die Geschirrserie Seagull wird bei 1.340 Grad Celsius gebrannt und zeichnet sich durch seine auffallend reinweiße, haptisch angenehme Oberflächenbeschaffenheit aus. Aus dem Meer aufsteigende Möwen inspirierten zu einer Gestaltung mit einer anmutig geschwungenen Linienführung. Die fließenden Konturen gehen harmonisch in die beiden ergonomisch durchdachten Griffe über. Die unterschiedlich großen Schüsseln lassen sich platzsparend und bruchsicher stapeln.

Begründung der Jury
In Anlehnung an fliegende Möwen gelingt hier die überzeugende Kreation eines bemerkenswert ästhetischen und qualitativ hochwertigen Tafelgeschirrs.

Lufthansa Premium Economy Class
Inflight Tableware
Flugzeuggeschirr

Manufacturer
Schönwald, Schönwald, Germany
Design
Spiriant (Daniel Knies, Volker Klag, Jochen Bittermann),
Neu-Isenburg, Germany
Web
www.schoenwald.com
www.spiriant.com

The clear and modern design of this in-flight porcelain tableware is part of a Premium Economy Class concept and was developed for the new long-haul service of Lufthansa. The items' sizes are modular designed so that the tableware fits different tray sizes and service variations. The conical contour of the bowls, plates and cups conveys a discreet elegance and high quality. All pieces of the tableware match and are easy to clean in industrial dishwashers.

Dieses klar und zeitgemäß gestaltete Flugzeuggeschirr aus Porzellan ist Teil eines Premium Economy Class Konzepts und wurde für den neuen Langstrecken Service der Lufthansa entworfen. Die Produktgröße ist modular gestaltet, sodass das Geschirr zu unterschiedlichen Tablettgrößen und Servicevarianten passt. Die kegelförmige Kontur der Schalen, Teller und Tassen vermittelt eine dezente Eleganz und Hochwertigkeit. Sämtliche Geschirrteile sind aufeinander abgestimmt und einfach in der Industriespülmaschine zu säubern.

Statement by the jury
The sophisticated form language of this inflight tableware conveys a timeless elegance and a superior quality standard. A harmonious overall appearance.

Begründung der Jury
Eine zeitlose Eleganz sowie einen gehobenen Qualitätsanspruch vermittelt die ausgereifte Formensprache dieses Flugzeuggeschirrs. Ein harmonisches Gesamtbild.

Sphere
Crystal Glass Series
Kristallglas-Serie

Manufacturer
Nachtmann GmbH,
Neustadt a.d. Waldnaab, Germany
Design
Roman Kvita,
Kobylá nad Vidnavkou, Czech Republic
Web
www.nachtmann.com
www.romankvita.ccm

Sphere was developed as part of the NextGen design projects, in collaboration with the Academy of Arts, Architecture & Design in Prague. Thanks to its spherical convex form language, the crystal glass series plays with the transparency of the material and creates fascinating light refractions. Comprising drinking glasses, plates, bowls, vases and a votive, the series is highly versatile. Hi-tech glass techniques are used to create high quality products at a competitive price point.

Statement by the jury
This crystal glass series conveys impressive aesthetics. Its organically appealing contours generate both visual and tactile stimuli.

Im Rahmen der NextGen-Projektarbeit mit der Prager Academy of Arts, Architecture & Design entstand Sphere. Mittels ihrer sphärisch konvexen Formensprache produziert die Kristallglas-Serie faszinierende Lichtbrechungen und spielt mit der Transparenz des Materials. Bestehend aus Trinkgläsern, Tellern, Schalen, Vasen und einem Votivlicht, bietet Sphere Produkte für viele Anlässe. Durch moderne Fertigungstechniken können die hochwertigen Kristallglas-Produkte preisgünstig am Markt angeboten werden.

Begründung der Jury
Diese Kristallglas-Serie vermittelt eine beeindruckende Ästhetik. Ihre organisch anmutenden Konturen erzeugen zugleich visuelle und haptische Reize.

Pebble Bagasse
Disposable Tableware
Einweggeschirr

Manufacturer
Duni AB, Malmö, Sweden
In-house design
Simi Gauba
Web
www.duni.com

This elegant disposable tableware is inspired by the organic shapes of pebbles. The plates and bowls are made from bagasse, a by-product of sugar cane fibre, which is left after the juice has been extracted from the sugar cane. The tableware is shaped in a high heat and pressure process. After use, the plates and bowls are 100 per cent biodegradable and compostable, and return to the soil in only eight weeks.

Statement by the jury
The Pebble Bagasse disposable tableware convinces as a particularly eco-friendly and visually appealing product.

Inspiriert von den organischen Formen von Kieselsteinen, entstand dieses elegante Einweggeschirr. Die Teller und Schalen sind aus Bagasse gefertigt. Als ein Nebenprodukt der Zuckerrohrfaser bleibt Bagasse übrig, nachdem der Saft aus dem Zuckerrohr gewonnen wurde. Das Geschirr wird per Druckverfahren bei hoher Hitze in seine Form gepresst. Die Teller und Schalen sind nach ihrem Gebrauch vollständig biologisch abbaubar und kompostierbar. Auf dem Kompost baut sich das Produkt innerhalb von nur acht Wochen ab.

Begründung der Jury
Das Einweggeschirr Pebble Bagasse überzeugt als besonders umweltfreundliche und zugleich visuell ansprechende Produktlösung.

Integrale
Cutlery
Besteck

Manufacturer
Amefa Stahlwaren GmbH,
Solingen, Germany
Design
Isabel Heubl Research & Design
(Isabel Heubl), Bad Aibling, Germany
Web
www.amefa-gastro.de
www.dancing-colours.de
Honourable Mention

Integrale is a cutlery set that was especially designed for people with gripping difficulties and limited range of wrist motion. The palm of the hand can embrace the ball shaped handle in different ways and can thus operate in the most comfortable position. The ball adapts to the favoured grip position and to the individual needs of the user. In this way, the cutlery creates a strong basis for independent eating. In addition, its aesthetic appearance motivates users to use this cutlery set.

Statement by the jury
Modelled after classic cutlery, Integrale is a successful advancement of ergonomically meaningful contours.

Integrale ist ein Besteck, welches speziell für Menschen mit Bewegungseinschränkungen der Handmotorik entwickelt wurde. Die Handfläche kann die Kugelform auf unterschiedliche Weise umschließen und somit in der jeweils geforderten Position agieren. Die Kugel passt sich der bevorzugten Halteposition und den Bedürfnissen des Nutzers individuell an. Somit schafft das Besteck die Voraussetzungen für ein eigenständiges Einnehmen einer Mahlzeit. Zudem motiviert die ästhetische Linienführung zur Nutzung des Bestecks.

Begründung der Jury
In Anlehnung an klassische Bestecke gelingt bei Integrale eine innovative Weiterentwicklung ergonomisch sinnvoller Konturen.

Evoque
Cutlery
Besteck

Manufacturer
WMF AG, Geislingen, Germany
Design
Daniel Eltner,
Baiersdorf-Igelsdorf, Germany
Web
www.wmf.de
www.destudio.de
Honourable Mention

The design of the Evoque cutlery reflects the human desire for natural originality. The simple, classic lines are a reflection of reduced complexity and inherent product quality that the user can experience directly. The carefully rounded edges and cross-sections further enhance the overall impression of the haptically pleasing shapes. Careful attention has also been paid to the well-balanced weight distribution. Each piece of the cutlery is matt-finished by hand and special surface hardening methods are employed to protect it against wear.

Statement by the jury
Thanks to an elaborate matt-finish and its harmoniously balanced contours, this cutlery conveys an emotionally appealing haptic quality.

Die Gestaltung des Bestecks Evoque reflektiert das Bedürfnis nach Natürlichkeit. Reduzierte Komplexität sowie eine unmittelbar wahrnehmbare Wertigkeit spiegeln sich in der schlichten Linienführung wider. Die sorgfältige Abrundung aller Kanten und Querschnitte unterstützt den Gesamteindruck der haptisch ansprechenden Konturen. Zudem sind Gewicht und Balance sorgfältig austariert. Das Besteck wird einzeln von Hand mattiert und mit einer speziellen Oberflächenhärtung dauerhaft vor Abnutzung geschützt.

Begründung der Jury
Dank seiner aufwendig mattierten Oberfläche und seinen harmonischen Konturen vermittelt dieses Besteck eine emotional ansprechende Haptik.

Zens-BMW Zisha Travel Tea Set
Zens-BMW Zisha Reise-Teegeschirr

Manufacturer
Guangzhou Zens Houseware Co., Ltd.,
Guangzhou, China
In-house design
Qi Ming Liu, Qian Wu Xu
Web
www.zens.asia

The design of this travel tea set was developed in interdisciplinary cooperation between the porcelain manufacturer and the car manufacturer BMW. Based on the idea of a modern mobile lifestyle, this tea set allows users to enjoy the Far Eastern tradition of drinking tea with others wherever they are. Four teacups, two small teapots, a serving tray and further ingredients come in a compact case. Custom-fit moulding within the suitcase guarantees safe transport.

Das Gestaltungskonzept für dieses Reise-Teegeschirr entstand in der interdisziplinären Zusammenarbeit des Porzellanherstellers mit dem Automobilkonzern BMW. Basierend auf dem Trend einer mobilen Lebensführung, ermöglicht dieses Set seinen Benutzern, die fernöstliche Tradition des gemeinsamen Teetrinkens auch unterwegs zu pflegen. Entsprechend finden vier Teetassen, zwei Kännchen, ein Serviertablett und weitere Zutaten in dem handlichen Koffer Platz. Passgenaue Einschübe ermöglichen einen sicheren Transport.

Statement by the jury
The functionally and formally sophisticated design of this travel tea set leads to a high ease-of-use and an overall attractive aesthetic.

Begründung der Jury
Beim Reise-Teegeschirr führt eine funktional und formal durchdachte Gestaltung zu einem hohen Bedienkomfort und einem ästhetischen Gesamtbild.

Bamboo Fibre Tableware
Bambusfaser-Geschirr

Manufacturer
International Trading Co., Ltd.,
Ningbo, China
Design
Ningbo Galachy
Creative Product Design Co., Ltd.,
Ningbo, China
Web
www.galachy.cn

This innovative tableware suit is manu-factured from bamboo fibre composites. Bamboo fibre particles are pressed under high temperature and high pressure to form vessels, cups and plates. Featuring lines with a natural appearance, the tableware references the characteristic-ally curved rings of bamboo. The table-ware is dishwasher safe and versatile in use. Free of petroleum components, the material is biologically degradable, hence not harming the environment after use.

Statement by the jury
Thanks to its natural components, this bamboo fibre tableware suit convinces as an ecologically friendly product solu-tion. Its asymmetrical form language has a highly original appeal.

Dieses Geschirr wird auf innovative Weise aus Bambusfaser-Verbundwerkstoffen gefertigt. Dabei werden zerteilte Bambus-fasern unter dem Einfluss hoher Tempera-turen und starkem Druck zu Gefäßen, Tassen und Tellern gepresst. Die natürlich anmutende Linienführung des Geschirrs zitiert die charakteristisch gewölbten Ringe eines Bambushalms. Das robuste Geschirr ist spülmaschinenfest und vielseitig nutz-bar. Frei von Erdölbestandteilen ist das Material nach seiner Verwendung unbe-denklich biologisch abbaubar.

Begründung der Jury
Dank seiner natürlichen Bestandteile über-zeugt dieses Geschirr aus Bambusfasern als nachhaltige Produktlösung. Die asym-metrische Formensprache wirkt originell.

Double Wall Mug
Doppelwand-Becher

Manufacturer
Guangdong Songfa Ceramics Co., Ltd.,
Chaozhou, China
In-house design
Web
www.songfa.com

This elegant double wall mug is strikingly thin thanks to its innovative material. Even though the so-called eggshell por-celain is less than one millimetre thick, the porcelain does not conduct any heat to the outside. However, differing from the porcelain around the area where the user holds the cup, the upper part conducts heat, allowing users to test the water temperature. Thus, the cup suc-cessfully prevents scalding both when drinking and touching.

Statement by the jury
The innovative use of traditional Chinese porcelain manufacturing techniques lends this double wall mug a high level of user convenience.

Dieser elegante Doppelwand-Becher ist aufgrund seines innovativen Materials auf-fallend dünnwandig. Obwohl die Stärke dieses sogenannten Eierschalen-Porzellans weniger als einen Millimeter beträgt, leitet das Porzellan keinerlei Hitze nach außen. Anders als in Griffhöhe des Bechers ist das Porzellan des oberen Becherrands wärmeleitend, damit die Benutzer dort die Wassertemperatur testen können. Somit werden sowohl beim Anfassen des Bechers als auch beim Trinken Verbrühungen er-folgreich verhindert.

Begründung der Jury
Der innovative Einsatz einer traditionellen chinesischen Porzellan-Fertigung erreicht bei diesem Doppelwand-Becher einen hohen Bedienkomfort.

Self-Cooling Cool-ID Tumbler
Self-Cooling Tumbler
Selbstkühlendes Bierglas

Manufacturer
Magisso Oy, Helsinki, Finland
Design
Simon Stevens Design Studio,
London, Great Britain
Web
www.magisso.com
www.simonstevensdesigns.com

This self-cooling tumbler keeps bever-ages cool for long even in very high temperatures. For the cooling effect to start, the tumbler simply needs to be soaked in water for a minute. The un-glazed and porous outer surface of the tumbler absorbs the water and the evaporation process cools it for hours. The tumbler is also dishwasher safe and, thanks to its shape, suitable for dif-ferent beers. Additionally, its outer sur-face can be marked by using chalk.

Statement by the jury
This tumbler stands out with its sur-prisingly long-lasting cooling effect. Furthermore, it is also convincing in terms of its formal design in contrasting colours.

In diesem selbstkühlenden Bierglas bleiben Getränke auch bei heißen Temperaturen lange kalt. Um die Kühlwirkung zu erzielen, muss der Becher nur eine Minute lang unter Wasser gehalten werden. Seine nicht glasierte Außenfläche nimmt dabei Wasser auf, welches dann langsam verdunstet und so den Becher über Stunden kühlt. Dank seiner Form ist das spülmaschinen-feste Glas für verschiedene Biersorten geeignet, die Außenfläche lässt sich mit Kreide beschriften.

Begründung der Jury
Ein überraschend lang anhaltender Kühl-effekt zeichnet das Bierglas aus. Darüber hinaus überzeugt die formale Gestaltung in Kontrastfarben.

Brewhouse
Tumbler
Becher

Manufacturer
Sahm GmbH & Co. KG,
Höhr-Grenzhausen, Germany
In-house design
Sylvia Weber
Web
www.sahm.de

The Brewhouse tumbler quotes classic beer glass design and adapts its characteristics to the demands of international urban bar culture. The organically appealing shape offers a high level of practicality. The design of the contours was conceived to allow the mugs to be stacked on top of each other. The glass wall and crystallised base lend the Brewhouse tumbler sound robustness, making it popular for tastings in the up-and-coming craft-beer scene.

Statement by the jury
A functionally as well as formally consistent design concept turns the Brewhouse tumbler into a convenient and appealing product solution.

Der Becher Brewhouse zitiert ein klassisches Bierglas-Design und passt seine Eigenschaften den Anforderungen einer internationalen, urbanen Bar-Szene an. Seine organisch anmutende Formensprache bietet ein hohes Maß an Funktionalität: Die Konturen-Ausgestaltung ist so beschaffen, dass der Becher unbedenklich stapelbar ist. Glaswandung und Eisboden verleihen Brewhouse eine Robustheit, was ihn gerade in der jungen Craft-Beer-Szene für die Verkostung von Bierspezialitäten beliebt macht.

Begründung der Jury
Ein funktional und formal schlüssiges Gestaltungskonzept macht den Becher Brewhouse zu einer komfortablen und ansprechenden Produktlösung.

Craft Beer Glasses
Craft-Beer-Gläser

Manufacturer
Kristallglasfabrik Spiegelau GmbH,
Neustadt a.d. Waldnaab, Germany
In-house design
Web
www.spiegelau.com

The Craft Beer Glasses were developed together with an expert tasting panel of master brewers and industrial professionals. Through this special approach these glasses bring out the best of the complex aroma profiles of the craft beers India Pale Ale, American Wheat Beer and Stout. The functional designed glass shapes correspond with each beer style and successfully deliver the aromas to the nose and palate while presenting the best possible texture and flavour of the beer.

Statement by the jury
The Craft Beer Glasses series blend formal and functional features into an appealing unity.

Die Craft-Beer-Gläser wurden gemeinsam mit einem Expertenteam aus Braumeistern und Spezialisten der Bierbranche entwickelt. Durch diese besondere Vorgehensweise sind die aromenspezifischen Gläser bestens auf die komplexen Geschmacksprofile der handwerklich gebrauten Biersorten India Pale Ale, American Wheat Beer und Stout abgestimmt. Aufgrund der funktionalen, speziell für die jeweilige Biersorte gestalteten Glasformen gelingt es, die Aromen gleichmäßig an Nase und Zunge weiterzugeben sowie Geschmack und Textur der Biere bestmöglich darzustellen.

Begründung der Jury
Bei dieser Serie von Craft-Beer-Gläsern verbinden sich formale und funktionale Eigenschaften zu einer ansprechenden Einheit.

Pepsi Axl Glass Bottle
Pepsi Axl-Glasflasche

Manufacturer
PepsiCo, Purchase, New York, USA
In-house design
PepsiCo Design & Innovation Center, New York, USA
Design
Tether, Seattle, USA
Web
www.pepsico.com
www.tetherinc.com

This single-serve glass bottle was launched in the course of the brand's first design update since 1996. It is being used for the entire trademark portfolio including Pepsi, Diet Pepsi, Pepsi MAX and Pepsi True. The etched, gripable bottom of the bottle allows for a visible and tactile brand experience and expresses the brand's character. The bold swirl and elevated profile of the glass bottle reflects the brand's youthful spirit and appeals to consumers in an emotional way.

Statement by the jury
The innovative design concept, which has been consistently and successfully realised, underlines the distinctive and formally autonomous character of this glass bottle.

Im Rahmen des ersten Design-Updates seit 1996 wurde diese Portions-Glasflasche auf den Markt gebracht. Sie wird für das gesamte Markenportfolio eingesetzt, darunter Pepsi, Diet Pepsi, Pepsi MAX und Pepsi True. Der gewellte, gut greifbare untere Teil der Flasche ermöglicht sowohl eine sichtbare wie auch greifbare Markenerfahrung und drückt den Markencharakter aus. Die markant gedrehte Silhouette der Glasflasche vermittelt den jugendlichen Esprit der Marke und spricht die Konsumenten emotional an.

Begründung der Jury
Eine innovative Gestaltungsidee, die stringent umgesetzt wurde, unterstützt den markanten und formal eigenständigen Charakter dieser Glasflasche.

Pepsi Drinking Glass
Pepsi Trinkglas

Manufacturer
PepsiCo, Purchase, New York, USA
In-house design
PepsiCo Design & Innovation Center, New York, USA
Web
www.pepsico.com

Following the launch of the packaging redesign in 2015, the Pepsi Drinking Glass features an innovative design language. The inspiration for the conical upper branding area and the lower grip area derives from the contour of the Pepsi Axl glass bottle. The bold swirl of the Axl glass' profile reflects the brand's youthful spirit, allowing users to have tactile interaction with the glass. The drinking glass is being leveraged globally currently coming in sizes ranging from 270 ml to 570 ml.

Statement by the jury
The impressive design language of this drinking glass conveys a unique, brand-specific dynamic.

Passend zum Relaunch des Verpackungsdesigns zeigt seit 2015 auch das Pepsi Trinkglas eine innovative Linienführung. Die Gestaltung des konischen oberen Markenbereichs wie auch des unteren Haltebereichs orientiert sich an der Kontur der Pepsi Axl-Glasflaschen. Die markant gedrehte Silhouette des Axl-Glases spiegelt den jugendlichen Esprit der Marke wider und ermöglicht den Konsumenten eine taktile Interaktion mit dem Trinkgefäß. Das Glas wird derzeit zur weltweiten Einführung in unterschiedlichen Größen von 270 ml bis 570 ml angeboten.

Begründung der Jury
Die eindrucksvolle Formensprache vermittelt bei diesem Trinkglas eine markenspezifische Dynamik, die unverwechselbar ist.

55°
Cup
Becher

Manufacturer
Beijing 55 Technology Co., Ltd.,
Beijing, China
Design
LKK Design Beijing Co., Ltd.,
Beijing, China
Web
www.55du.cc
www.lkkdesign.com

This innovative cup cools boiling water down to 55 degrees centigrade within only one minute. Through this, it considerably reduces the time the water needs to reach a temperature at which it is drinkable. In addition, the cup can be closed with the lid and shaken in order to dissolve honey or tea powder quickly. The harmonious round lines in combination with the bright colours lend the vessel a friendly overall appearance.

Statement by the jury
This cup blends a convincing functionality and an attractive form language into a consistent product solution.

Dieser innovative Becher kühlt kochendes Wasser innerhalb einer Minute auf 55 Grad Celsius ab. Somit wird die Zeitspanne, die Wasser normalerweise braucht, um eine trinkbare Temperatur zu erreichen, deutlich verkürzt. Zudem lässt sich der Becher mit einem Deckel verschließen und schütteln, sodass sich Honig oder Teepulver schnell im Gefäß auflösen können. Die harmonisch abgerundete Linienführung in Verbindung mit hellen Farben schafft ein sympathisch wirkendes Gesamtbild.

Begründung der Jury
Eine überzeugende Funktionalität und eine ansprechende Formensprache vereinen sich bei diesem Becher zu einer stimmigen Produktlösung.

Waterever Smart Cup
Drinking Cup
Trinkgefäß

Manufacturer
Shenzhen IPINTO Technology Co., Ltd.,
Shenzhen, China
Design
Shenzhen ARTOP Design Co., Ltd.,
Shenzhen, China
Web
www.ipinto.com
www.artopcn.com

Waterever Smart Cup is a portable drinking cup that measures the water temperature and the amount contained. In addition, it allows users to monitor their behaviour when drinking water and to share and document it on an online platform. Inspired by a waterfall, a concealed light strip simulates the effect of flowing water. To the right of it, there is a flush LED display showing the current temperature. The lid opens at a gentle touch and is leak-proof when closed.

Statement by the jury
Innovative features characterise the Waterever Smart Cup. Its interactive functions enhance the ease of use in a remarkable way.

Waterever Smart Cup ist ein tragbares Trinkgefäß, welches die Wassertemperatur und -menge misst. Zudem ermöglicht es dem Benutzer, sein Trinkverhalten zu kontrollieren und dies auf einer Online-Plattform mitzuteilen und zu dokumentieren. Inspiriert von einem Wasserfall simuliert eine Lichtleiste den Effekt von fließendem Wasser. Rechts davon zeigt ein flächenbündiges LED-Display die aktuelle Temperatur an. Der Deckel öffnet sich, sobald man darauf drückt und ist in seiner geschlossenen Position auslaufsicher.

Begründung der Jury
Eine innovative Ausstattung zeichnet den Waterever Smart Cup aus. Seine interaktiven Funktionen erweitern den Bedienkomfort auf bemerkenswerte Weise.

Manufacturer
Pacific Market International, Shanghai, China
In-house design
Cacica Tang
Web
www.migo.com

The Enjoy Sport water bottle series combines superior functionality with an appearance that places the material variety at the centre of attention. The combination of glass, Tritan plastics and stainless steel aims to suit the needs of different users. The lid is designed for comfortable and hygienic drinking as well as easy disassembly for cleaning. The durable carrying strap is flexible and adapts to individual needs.

Die Wasserflaschen-Serie Enjoy Sport verbindet eine gehobene Funktionalität mit einer Ästhetik, welche die Materialvielfalt in den Mittelpunkt der Betrachtung stellt. Die Kombination aus Glas, dem Kunststoff Tritan und Edelstahl möchte den Bedürfnissen unterschiedlicher Nutzer gerecht werden. Der Deckel wurde für ein bequemes und hygienisches Trinken sowie ein leichtes Zerlegen zwecks Reinigung konzipiert. Das robuste Trageband ist flexibel und passt sich dem jeweiligen Bedarf an.

Statement by the jury
The eye-catching colour concept as well as the elegant language of form are an expression of the high ease of use of this water bottle series.

Begründung der Jury
Ein aufmerksamkeitsstarkes Farbkonzept sowie eine elegante Formensprache verleihen dem Bedienkomfort dieser Wasserflaschen-Serie Ausdruck.

Contigo® AUTOCLOSE™ Shake & Go™
Tumbler
Trinkbecher

Manufacturer
Newell Rubbermaid, Atlanta, USA
In-house design
Web
www.newellrubbermaid.com

The Shake & Go tumbler provides an easy way to prepare and enjoy drinks. It is suitable for water enhancers, powdered beverages, and iced coffees and teas. The patent pending Autoclose technology reliably seals the tumbler, allowing users to effectively mix in flavour enhancers and powders by shaking the tumbler. A spring-loaded mechanism snaps closed when the straw is removed.

Statement by the jury
The Shake & Go tumbler blends a convincing functionality with a slim profile and attractive colours.

Im Shake & Go Trinkbecher lassen sich Getränke komfortabel zubereiten und genießen. Er eignet sich für aromatisiertes Wasser, Getränkepulver sowie Eiskaffee und -tee. Der zum Patent angemeldete Autoclose-Verschluss hält den Trinkbecher zuverlässig dicht, sodass sich wasserlösliche Pulver und Aromen durch Schütteln effektiv untermischen lassen. Sobald der Trinkhalm entfernt wird, verschließt ein federunterstützter Mechanismus die Öffnung.

Begründung der Jury
Eine überzeugende Funktionalität vereint sich beim Shake & Go Trinkbecher mit einer schlanken Gefäßkontur und ansprechenden Farben.

Contigo® AUTOSEAL®
Pitcher
Krug

Manufacturer
Newell Rubbermaid, Atlanta, USA
In-house design
Web
www.newellrubbermaid.com

This pitcher features an infuser stick and an ice core as well as an innovative sealing system. Its patented Autoseal guarantees freshness; the pitcher has a reliable spill- and leak-proof design. Users simply press a button to pour and then release the button to reseal the pitcher. The infuser stick makes it easy to add flavours from fruits, herbs and vegetables while the ice core chills any beverage without watering down its flavour. The BPA-free pitcher is shatter-resistant and top rack dishwasher-proof.

Statement by the jury
Following a sophisticated design concept, the innovative sealing system of this pitcher provides a high ease-of-use.

Dieser Krug ist mit einem Aroma- und einem Kühleinsatz sowie einem innovativen Verschlusssystem ausgestattet. Dessen patentierte Autoseal gewährleistet Frische, wobei der Krug zuverlässig auslaufsicher ist. Auf Knopfdruck kann das Getränk ausgegossen werden, wird der Knopf losgelassen, ist der Krug wieder versiegelt. Der Aroma-Einsatz für Früchte, Kräuter und Gemüse reichert die Getränke geschmacklich an. Der Kühleinsatz kühlt das Getränk, ohne es dabei zu verwässern. Der BPA-freie Krug ist bruchsicher und geeignet fürs obere Spülmaschinenfach.

Begründung der Jury
Einer ausgereiften Gestaltungsidee folgend bietet dieser Krug dank seiner innovativen Verschlussautomatik einen hohen Bedienkomfort.

BRITA Fill&Serve
Water Filter Carafe
Wasserfilter-Karaffe

Manufacturer
Brita GmbH, Taunusstein, Germany
Design
yellow design, Cologne, Germany
Web
www.brita.net
www.yellowdesign.com

The Fill&Serve water filter carafe delivers a stylish approach to filtering and serving tap water. The activated carbon filter reduces substances that otherwise contaminate the smell and taste of tap water. The carafe is designed in such a way that, when water is being filtered, the inner funnel disappears through the effect of light refraction. As a sustainable alternative to water bottles, Fill&Serve can be cleaned in the dishwasher and thus used over and over again.

Statement by the jury
The high functionality of this water filter carafe is underlined by an elegantly appealing design vocabulary – a fascinating design creation.

Die Fill&Serve Wasserfilter-Karaffe ermöglicht ein stilvolles Filtern und Servieren von Leitungswasser. Der Aktivkohlefilter reduziert die Stoffe, welche den Geruch und Geschmack von Leitungswasser beeinträchtigen können. Die Wasserfilter-Karaffe ist so konzipiert, dass nach dem Wassereinfüllen der innere Trichter durch den Effekt der Lichtbrechung nicht mehr zu sehen ist. Als nachhaltige Alternative zu Flaschenwasser kann die Karaffe einfach in der Spülmaschine gereinigt und langfristig eingesetzt werden.

Begründung der Jury
Eine ansprechend elegante Formensprache unterstreicht die hohe Funktionalität dieser Wasserfilter-Karaffe – ein faszinierender Entwurf.

Tea-Jane
Tea Maker
Teezubereiter

Manufacturer
blomus GmbH, Sundern, Germany
Design
Floez Industrie-Design GmbH
(Oliver Wahl), Essen, Germany
Web
ww.blomus.com
www.floez.de
Honourable Mention

Tea-Jane allows users to make tea in a traditional way: first, the teapot is filled with the desired amount of water; then, the tea leaves are brewed with a little water in the upper part of the pot. After the steep time, the richly coloured brew flows into the hot water below to release its full flavour. The tea leaves remain in the strainer so the tea no longer steeps. The tea can be served directly at the table without having to remove the strainer.

Statement by the jury
Thanks to its functionally sophisticated design, Tea-Jane turns tea making into a visually appealing experience.

Tea-Jane ermöglicht eine traditionelle Tee-zubereitung: Zunächst befüllt man die Kanne mit der gewünschten Wassermenge, danach werden die Teeblätter im oberen Teil der Kanne mit wenig Wasser aufge-brüht. Nach dem Ziehen fließt der farben-prächtige Teesud in das heiße Wasser und vermischt sich dort zu einem vollaroma-tischen Tee. Die benutzten Teeblätter ver-bleiben im Filter, sodass das Getränk nicht nachziehen kann. Ohne den Filter entfernen zu müssen, kann der Tee direkt am Tisch serviert werden.

Begründung der Jury
Dank ihrer funktional durchdachten Gestal-tung macht Tea-Jane die Teezubereitung zu einer reizvollen visuellen Erfahrung.

Ossidiana
Espresso Coffee Maker
Espressomaschine

Manufacturer
Alessi S.p.A., Crusinallo, Verbania, Italy
Design
Mario Trimarchi, Milan, Italy
Web
www.alessi.it
www.mariotrimarchi.eu

The design concept of Ossidiana follows the objective of combining sculptural aesthetics with the functionality of an espresso coffee maker. The asymmetric contours look as if carved while indi-vidually accentuated surfaces quote the appearance of a sculptured stone. Nevertheless, Ossidiana conveys the emotional familiarity of an everyday object. The handle made of thermo-plastic resin rests ergonomically well in the hand, providing a secure grip when handling the pot.

Statement by the jury
With its artful craftsmanship, this espresso coffee maker reaches a high level of quality. An original design.

Das Gestaltungskonzept von Ossidiana folgt der Zielsetzung, eine skulpturale Ästhetik mit der Funktionalität einer Espressomaschine zu verbinden. Ihre unsymmetrische Kontur wirkt wie gemei-ßelt, einzeln akzentuierte Flächen zitieren dabei die Anmutung eines bearbeiteten Steins. Ossidiana vermittelt dennoch den emotionalen Gehalt eines vertrauten Alltagsgegenstands. Der Griff aus thermo-plastischem Harz schmiegt sich der Hand ergonomisch sinnvoll an und bietet eine hohe Stabilität bei der Handhabung der Kanne.

Begründung der Jury
Die Espressomaschine erreicht aufgrund der handwerklich kunstvollen Fertigung eine eigenständige Qualität. Ein origineller Entwurf.

Ruutu
Glass Vase
Glasvase

Manufacturer
Iittala,
Helsinki, Finland

Design
Ronan & Erwan Bouroullec,
Paris, France

Web
www.iittala.com
www.bouroullec.com

reddot award 2015
best of the best

Precious geometry

Glass vases from the studio of French artist René Lalique once captivated the fascination of their contemporaries. They were highly precious, hand-made creations featuring organically playful forms typical of Art Deco style. Ruutu, the new Iittala vase created by Ronan and Erwan Bouroullec, signifies pieces of glass art of similarly high craftsmanship quality, yet fully epitomises today's zeitgeist. The vases with their flowing lines are mouthblown by Finnish craftsmen, turning each single vase into a master-piece of artistic appearance. Another central aspect of Ruutu's distinctive aesthetic lies in the shape of a rhombus marking their design. The vases unfold their delicate character both when arranged individually on a table and when grouped together into geometric arrangements, putting no boundaries on user creativ-ity. It is the sheer limitless possibilities of combination that capture the fascination of Ruutu. The collection comprises ten distinctive, unique vases of different sizes that are available in seven shining colours. The Ruutu vases thus invite users to express their creativ-ity in ever new installation. The vases can be arranged freely and come to life with their colour play in any setting. When placed side by side they create a spe-cial effect: the contrasting properties of the glass, its stiffness and durability on the one hand, and its delicate fragility on the other, merge into harmonious unity.

Kostbare Geometrie

Die Glasvasen aus den Ateliers des französischen Künstlers René Lalique begeisterten einst ihre Zeitge-nossen. Waren es doch überaus kostbare, handgefer-tigte Kreationen mit den organisch verspielten Formen des Jugendstils. Ruutu, die neue Iittala-Vase der Desi-gner Ronan und Erwan Bouroullec, steht für kunstvolle Objekte von ähnlich hoher handwerklicher Qualität, die dennoch ganz dem heutigen Zeitgeist entsprechen. Gefertigt werden die fließend anmutenden Vasen von finnischen Glasbläsern in Handarbeit, wodurch jede einzelne Vase auch ein filigranes Einzelstück ist. Ein zentraler Aspekt der besonderen Ästhetik von Ruutu liegt außerdem in der Form des Rhombus, der ihrer Ge-staltung zugrunde liegt. Die Vasen wirken dadurch ein-zeln aufgestellt auf dem Tisch und lassen sich ebenso in Gruppen zu vielen geometrischen Figuren anordnen, wobei der Kreativität keinerlei Grenzen gesetzt sind. Gerade in diesen schier unendlichen Möglichkeiten ihrer Kombination liegt die Faszination von Ruutu. Die Kollektion besteht aus insgesamt zehn Vasen in unter-schiedlichen Größen, die zudem in sieben leuchtenden Farben erhältlich sind. Die Ruutu-Vasen fordern ihre Besitzer damit immer wieder zu neuen Kreationen heraus. Sie lassen sich völlig frei arrangieren, wobei sie jedes Mal durch ihr Farbenspiel begeistern. Nebenein-ander positioniert ergeben sie ein besonderes Bild: Die Gegensätzlichkeiten des Glases, seine Härte und Stabi-lität einerseits sowie filigrane, zerbrechliche Materiali-tät andererseits, verschmelzen zu einer Einheit.

Statement by the jury

Based on the simple shape of a rhombus, the Ruutu glass vase collection amalgamates a filigree design concept with the tradition of glass blowing crafts-manship. The fascinating collection allows for a myriad of individual and ever changing arrangements. Coming in different heights, sizes and colours, they seamlessly arrange into appealing compositions. Made by hand to high standards of craftsmanship, each sin-gle piece turns into a true eye-catcher in any interior.

Begründung der Jury

Basierend auf der einfachen Grundform des Rhombus vereint die Vasenkollektion Ruutu ein feinsinniges Gestaltungskonzept mit der Tradition des Glasbläser-handwerks. Diese faszinierende Kollektion ermöglicht vielfältige neue Konstellationen des Zusammenstel-lens. Durch unterschiedliche Höhen, Größen und auch Farben entsteht jedes Mal ein neues Bild. Die auf einem hohen kunsthandwerklichen Niveau gefertigten Einzel-stücke werden damit zu einem Blickfang im Raum.

Designer portrait
See page 48
Siehe Seite 48

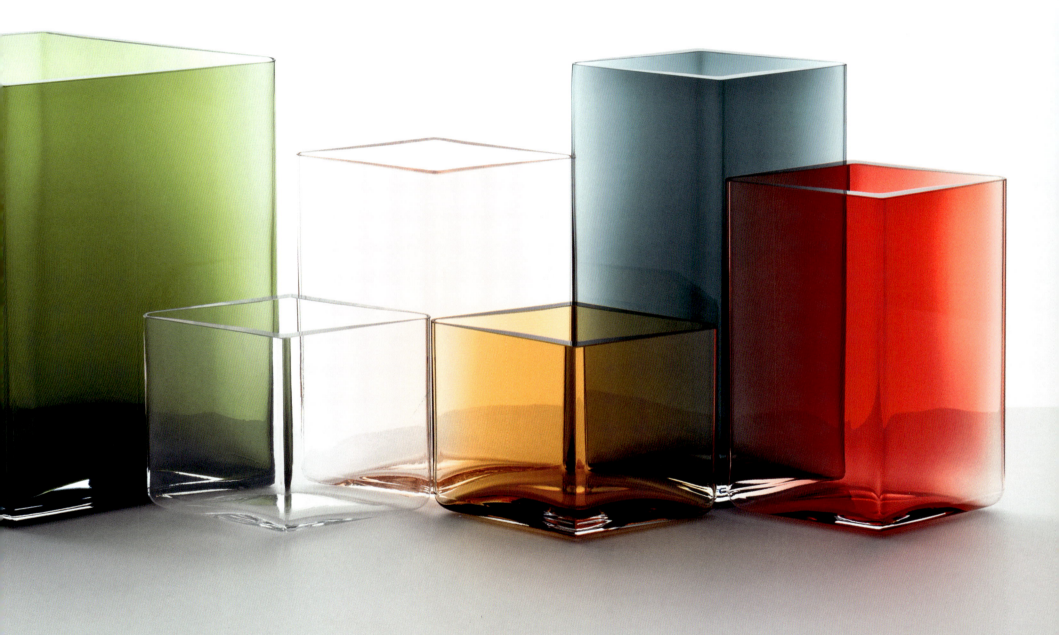

Shallows –
Flower Blooms in the Puddle
Vase

Manufacturer
Critiba, Fukuoka, Japan
In-house design
Kazunaga Sakashita
Web
www.critiba.com

The Shallows crystal vase showcases the surface tension of water. With its cylindrical hole in the centre, it is especially suitable for arranging a single flower or twig. The space between the hole and the outer rim of the vase features a machine-cut recess, so that the vase can be filled with water over the rim. The overflowing water creates the illusion of a water column, within which only the contour of the cylinder is visible.

Statement by the jury
Inspired by nature, the design of Shallows creates a visual effect that is both fascinating and emotionally appealing.

Die Kristallvase Shallows inszeniert die Oberflächenspannung von Wasser. Mit ihrem schmalen, zylinderförmigen Loch in der Mitte eignet sie sich für das dekorative Arrangement einer einzelnen Blüte oder eines Zweiges. Der Raum zwischen dem zylinderförmigen Loch und dem äußeren Vasenrand ist mit einer maschinell gefertigten Vertiefung versehen, sodass das Wasser überrandvoll aufgefüllt werden kann. Das überlaufende Wasser lässt dann das Trugbild einer Wassersäule entstehen, in der nur noch die Kontur des Zylinders zu sehen ist.

Begründung der Jury
Die von der Natur inspirierte Gestaltung von Shallows ermöglicht einen visuellen Effekt, der zugleich faszinierend und emotional ansprechend wirkt.

Diamonds
Glass Series
Glas-Serie

Manufacturer
Zwiesel Kristallglas AG, Zwiesel, Germany
In-house design
Bernadett King
Web
www.zwiesel-kristallglas.com
Honourable Mention

The Diamonds series showcases a clear, reduced form language, providing plenty of options for decorating, serving as either vases or lanterns. The mouth-blown glass series comes in harmonious shades of green, blue and smoky. Each of those variations is available in either a transparent version or with a satin finish. The wide, inclined opening of these home accessories creates a dynamic tension. Thanks to their expressive aesthetics, the hand-made vases and bowls convey elegance, as either individual pieces or a group.

Statement by the jury
The design implementation of the Diamonds glass series is convincing with a distinctive diversity of forms and colours.

Die Serie Diamonds zeigt eine klare, reduzierte Formensprache und bietet als Vase oder Windlicht einen großen Deko-Spielraum. Die mundgeblasene Glas-Serie wird n harmonischen Grün-, Blau- und Rauchtönen gefertigt. Jede Farbausführung ist jeweils transparent oder satiniert erhältlich. Die weite, schräge Öffnung der Wohnaccessoires erzeugt eine dynamische Spannung. Dank ihrer ausdrucksstarken Ästhetik vermitteln die handgefertigten Vaser und Schalen Eleganz, sowohl als Einzelstück als auch in der Gruppe.

Begründung der Jury
Die gestalterische Umsetzung der Glas-Serie Diamonds beeindruckt mittels einer eigenständigen Vielfalt in Form und Farbe.

Stockholm Aquatic
Vases, Bowls
Vasen, Schalen

Manufacturer
Stelton, Copenhagen, Denmark
Design
Bernadotte & Kylberg, Stockholm, Sweden
Web
www.stelton.com
www.bernadottekylberg.se

Stockholm Aquatic comprises three vases and four bowls of different sizes. The design line showcases an innovative material mix of enamel and aluminium. Its organically shaped contours are combined with expressive patterns and colours. These elegant aesthetics reflect the beauty of the Stockholm Archipelago and the Baltic Sea coast. Each piece is finished by hand, giving the item its very own individual expression.

Statement by the jury
A maritime colour palette and an effective material combination characterise these vases and bowls, resulting in a distinctively unique overall appearance.

Stockholm Aquatic umfasst drei Vasen und vier Schalen in verschiedenen Größen. Die Gestaltungslinie präsentiert sich in einem innovativen Materialmix aus Emaille und Aluminium. Ihre organisch anmuten-den Konturen werden mit ausdrucksstarken Mustern und Farben kombiniert. Diese elegante Ästhetik spiegelt die Schönheit der Stockholmer Schären sowie der Ost-seeküste wider. Alle miteinander kombinier-baren Stücke werden nach der Fertigung von Hand perfektioniert und erhalten so ihren individuellen Ausdruck.

Begründung der Jury
Eine maritime Farbwelt und eine effekt-volle Materialkombination zeichnen diese Vasen und Schalen aus. Es entsteht ein eigenständiges Gesamtbild.

Plissé
Porcelain Vessel
Porzellangefäß

Manufacturer
Vista Alegre Atlantis, SA, Ilhavo, Portugal
In-house design
Web
www.myvistaalegre.com

The creation of Plissé was inspired by the idea of taking familiar crystal glass engravings and translating them onto a porcelain vessel. The innovative manu-facturing process was developed as part of a collaboration between two com-panies. Leaving the porcelain factory smooth and glazed, the pieces are then handed to highly skilled master crystal engravers who cut each groove manu-ally. Employing grinding and cutting techniques with diamond-edged wheels, the porcelain surfaces are sculpted to obtain a texture that is highly defined and detailed.

Statement by the jury
An innovative engraving technique highlights the flowing contours of Plissé to an outstanding aesthetic effect, achieving an overall artistic expression.

Inspiriert von der Idee, einen für Kristall-glas typischen Schliff auch bei Porzellange-fäßen einzusetzen, entstand Plissé. Das innovative Herstellungsverfahren wurde im Rahmen einer Kooperation zweier Firmen möglich: Nach der Fertigung in der Porzel-lanfabrik kerben hochspezialisierte Graveu-re die charakteristische Struktur in das bereits glasierte Porzellan. Aufgrund ver-schiedener Schleifprozesse mit Diamant-schleifrädern erhält die Porzellanoberfläche ihre detailliert definierte Beschaffenheit.

Begründung der Jury
Ausgesprochen ästhetisch unterstreicht eine innovative Gravur die fließenden Konturen von Plissé und erreicht ein ins-gesamt kunstvolles Gesamtbild.

Evo
Planter
Pflanzgefäß

Manufacturer
Fiskars Garden Oy Ab, Helsinki, Finland
In-house design
Fiskars R&D Team
Web
www.fiskarsgroup.com

Evo is a versatile living accessory that can be used as a planter or storage solution anywhere in the house. It can, for instance, be used as a conventional cachepot for indoor plants and kitchen herbs, while the waterproof saucer protects the surface it is placed on against moisture. Thanks to an appealing colour combination, Evo harmoniously blends in with different ambiances. The textile material is washable, UV proof, breathable and highly tear resistant.

Statement by the jury
Versatile in use, Evo follows an original design idea. Its materials create both visual and haptic pleasures.

Evo ist ein vielseitiges Wohnaccessoire, das als Pflanzgefäß oder Aufbewahrungslösung im ganzen Haus genutzt werden kann. Es lässt sich beispielsweise wie ein herkömmlicher Übertopf für Zimmerpflanzen und Küchenkräuter verwenden, wobei ein wasserdichter Untersetzer den jeweiligen Stellplatz vor Feuchtigkeit schützt. Mittels seiner ansprechenden Farbkombinationen fügt sich Evo harmonisch ins jeweilige Ambiente ein. Das textile Material ist waschbar, UV-stabil, atmungsaktiv und von hoher Reißfestigkeit.

Begründung der Jury
Flexibel nutzbar folgt Evo einer originellen Gestaltungsidee. Seine Materialität schafft sowohl visuelle als auch haptische Reize.

Plantui Smart Garden
Planter
Pflanzgefäß

Manufacturer
Plantui Ltd., Turku, Finland
In-house design
Janne Loiske
Web
www.plantui.com

Plantui Smart Garden is a gardening device to grow greens inside the house all year round. The innovative planter does not need any soil and can be used to grow herbs as well as salad greens. Featuring an intelligent LED light system and an automatic watering pump, the device takes care of the plants nearly completely by itself. Plantui Smart Garden offers space for six plants and its low-voltage 12 V battery consumes less than 60 kWh per year.

Statement by the jury
The innovative idea behind Plantui Smart Garden has been realised in a formally appealing design, paired with a high ease of use.

Plantui Smart Garden dient der ganzjährigen Aufzucht von Küchenpflanzen in der Wohnung. Das innovative Pflanzgefäß kommt ohne Erde aus und eignet sich sowohl für Kräuter als auch Blattsalate. Ausgestattet mit einem intelligenten LED-Beleuchtungssystem sowie einer automatischen Bewässerungspumpe sorgt das Gefäß weitestgehend autark für die Pflanzen. Plantui Smart Garden bietet Platz für sechs Pflanzen, sein 12-V-Niederspannungsgerät verbraucht bei ständigem Betrieb weniger als 60 kWh im Jahr.

Begründung der Jury
Die innovative Gestaltungsidee von Plantui Smart Garden wurde hier in einen formal ansprechenden Entwurf, der mit einem hohen Bedienkomfort einhergeht, umgesetzt.

Luna
Gel Fire Pit
Gel-Feuerstelle

Manufacturer
blomus GmbH, Sundern, Germany
Design
Floez Industrie-Design GmbH
(Oliver Wahl), Essen, Germany
Web
ww.blomus.com
www.floez.de
Honourable Mention

The Luna gel fire pit creates the appropriate atmosphere for evening parties
on a terrace. With its well-balanced
proportions, it can be positioned as an
aesthetic eye-catcher within sight of
a dinner party. In addition, the elegant
black sculpture is a treat for the eyes
also during daylight. When the fire is lit
in the dark, the contour takes a back-
seat to the dancing of the flames, which
catch the full attention of the beholder.
A 500 ml gel tin allows the fire to burn
for approximately three hours.

Statement by the jury
The Luna gel fire pit is characterised by
an effective functionality paired with
formal simplicity.

Die Gel-Feuerstelle Luna schafft die
passende Stimmung für abendliche Feste
auf der Terrasse. Mit ihren ausgewogenen
Proportionen lässt sie sich als ästhetischer
Blickfang beispielsweise in Sichtweite einer
Tischgesellschaft positionieren. Tagsüber
schmeichelt die elegante schwarze Skulptur
dem Auge. Sobald im Dunkeln das Feuer
brennt, tritt die Kontur in den Hintergrund
und überlässt dem Feuerspiel die Aufmerksamkeit des Betrachters. Eine 500 ml
fassende Gel-Dose lässt das Feuer circa
drei Stunden lang brennen.

Begründung der Jury
Eine effektvolle Funktionalität, gepaart
mit einer formalen Stringenz, zeichnet die
Gel-Feuerstelle Luna aus.

Bathrooms and spas
Bad und Wellness

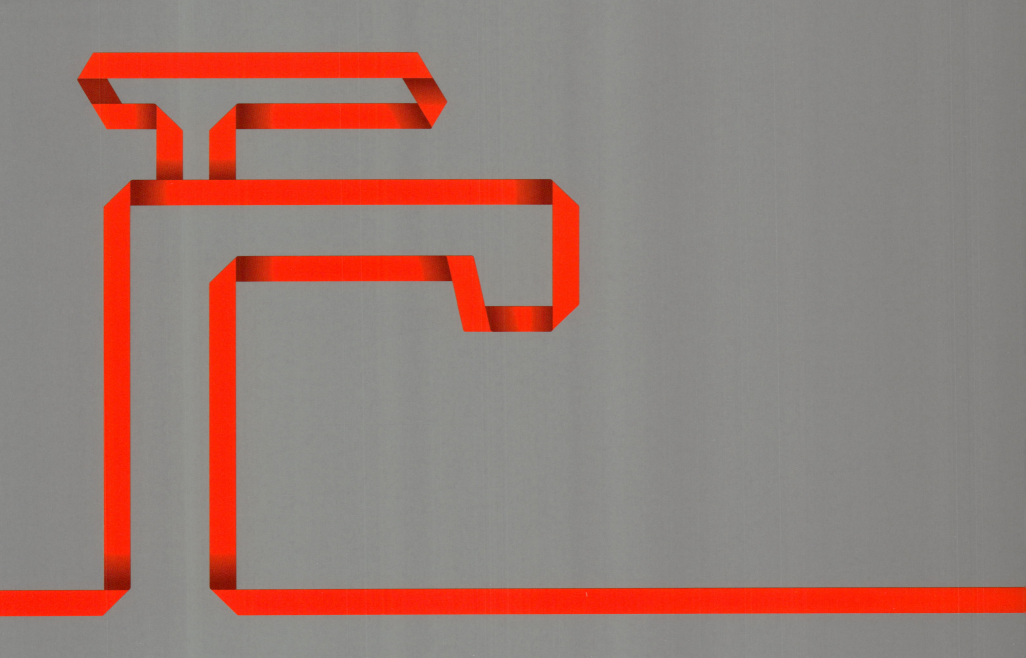

I Bordi
Bathtub
Badewanne

Manufacturer
Teuco SpA,
Montelupone, Italy

Design
Carlo Colombo,
Milan, Italy

Web
www.teuco.com
www.carlocolombo.it

reddot award 2015
best of the best

A symbiosis of design and material

When in the mountains, water collects among the rocks, an observer senses the unity of these two archaic elements. The I Bordi bathtub successfully conveys the primeval strength of a material in a new form language. Regardless of whether it is made from natural raw materials such as valuable Carrara marble or Grey Stone, or easily cleanable white Duralight, the well-balanced oval shape attracts attention. This bathtub has soft and sensual lines with contrasting geometric and architectural elements full of tension. The result is a dynamic balance which lends a bathroom an expressive character. The lines of the edges play a central role here. They stand out asymmetrically from the material in an apparently casual motion which adds a fascinating aspect to the water. The bathtub is available in a free-standing version as well as one that is built-in recessed with a protruding rim. The Duralight version is particularly interesting owing to the functionality and ease of cleaning of the material. This patented solid-surface composite is not only highly durable, but has an aesthetic appeal. The very successful symbiosis of material and design of the I Bordi bathtub creates a sculptural object whose dynamics and lines are appealing and which redefines the limits of a form.

Symbiose von Design und Material

Sammelt sich im Gebirge das Wasser im Gestein, spürt der Betrachter die besondere Einheit dieser archaischen Elemente. I Bordi gelingt es, die Urkraft der Materie in eine Badewanne mit einer neuen Formensprache zu überführen. Gestaltet aus Naturmaterialien wie wertvollem Carrara-Marmor und Stone Grey oder aber dem reinigungsfreundlichen Material Duralight in Weiß, fällt sogleich ihre ausgewogene ovale Form ins Auge. Diese Badewanne zeigt eine weich und sinnlich anmutende Linienführung, bei der sich spannungsreich geometrische und architektonische Elemente abwechseln. Es entsteht eine dynamisch wirkende Balance, die dem Badezimmer einen ausdrucksstarken Charakter verleiht. Eine zentrale Bedeutung haben dabei die Linien der Ränder und Kanten. Sie sind derart gestaltet, dass sie sich asymmetrisch ins Material „schneiden" – in der scheinbar zufälligen Bewegung, durch die auch das Wasser fasziniert. Erhältlich ist diese Badewanne freistehend sowie als Einbauversion mit Außenrand. In der Duralight-Variante beeindruckt insbesondere die Funktionalität und Reinigungsfreundlichkeit des Materials. Als patentiertes Solid Surface-Verbundmaterial bietet es ein hohes Maß an Langlebigkeit und Ästhetik. In einer sehr gelungenen Symbiose von Material und Design entstand mit der Badewanne I Bordi ein skulpturales Objekt das mit der Dynamik seiner Linien begeistert und die Grenzen einer Form neu definiert.

Statement by the jury

The unity of form and material of I Bordi creates a bathtub with notable expressiveness and visual intensity. The exciting, flowing lines and soft curves lend it a particular dynamism. The subtle effect delights the eye and the form instantly reveals its natural simplicity. The design of the I Bordi bathtub has resulted in a product that seems perfect in itself.

Begründung der Jury

Im Einklang von Form und Material zeigt sich mit I Bordi eine Badewanne von bemerkenswerter Ausdruckskraft und visueller Intensität. Mit spannungsvoll fließenden Linien sowie einer weichen Kurvatur erzeugt sie eine besondere Dynamik. Ihre subtile Wirkung zieht den Betrachter sofort in den Bann, wobei sich ihm die Form unmittelbar in ihrer natürlichen Einfachheit erschließt. Die Gestaltung der Badewanne I Bordi führt so zu einem Produkt, das in sich vollkommen erscheint.

Designer portrait
See page 50
Siehe Seite 50

Cape Cod
Bathtub
Badewanne

Manufacturer
Duravit AG, Hornberg, Germany
Design
Philippe Starck, Paris, France
Web
www.duravit.de
www.starck.com

Cape Cod offers a concisely shaped bathtub that has been designed with relaxation in mind: an ergonomic headrest provides a bathing experience that is particularly serene and produces a feeling of lightness. The tub has been designed as a monolith and made of the innovative material DuraSolid A, which provides a pleasantly warm feel and high quality matt look. It is available as a freestanding model and as a back-to-wall and corner version for left or right. The bathtub can be equipped with an unobtrusive air whirl system and optionally an integrated sound system that is operated by Bluetooth-compatible devices.

Bei der prägnant geformten Badewanne von Cape Cod steht die Entspannung im Vordergrund. Eine sanft geformte Ablage für den Kopf ermöglicht ein ruhevolles und von Leichtigkeit geprägtes Badeerlebnis. Die Wanne ist als Monolith gestaltet und aus dem innovativen Material DuraSolid A mit angenehm warmer Haptik und hochwertig matter Optik gefertigt. Zur Auswahl stehen ein freistehendes Modell sowie eine Vorwand- und Eckversion für links oder rechts. Die Wanne wird auch mit einem unauffällig gearbeiteten Air-Whirl-System mit oder ohne integriertem Soundsystem, das über Bluetooth-kompatible Geräte bedient wird, angeboten.

Statement by the jury
A bath in the bathtub Cape Cod appeals to many senses, from grasping the tub edge to relaxed leaning back in the organically shaped headrest.

Begründung der Jury
Ein Bad in der Wanne Cape Cod erweist sich als sinnlicher Genuss, vom Ergreifen des Wannenrandes bis zum entspannten Zurücklehnen in der organisch gestalteten Kopfstütze.

Amiata
Bathtub and Washbasin
Badewanne und Waschbecken

Manufacturer
Victoria + Albert Baths,
Telford, Great Britain
Design
Meneghello Paolelli Associati
(Sandro Meneghello, Marco Paolelli),
Milan, Italy
Web
www.vandabaths.com
www.meneghellopaolelli.com

In this ensemble made up of bathtub and washbasin, exalted British elegance and Mediterranean easiness converge. The curved design of the bathtub wall extends towards the top – a visually pleasing neo-classical reference that allows a relaxing neck position. The symmetrical, ergonomically designed tub can also act as a free-standing sculptural object in living areas. With only 1,645 mm length, it provides an attractive solution even for rooms with little space. The washbasin mirrors the lines of the tub.

Gehobene britische Eleganz und mediterrane Leichtigkeit fließen in diesem Ensemble aus Badewanne und Waschbecken gestalterisch ineinander. Die Kurvenführung der Wannenwand erweitert sich nach oben hin – ein optisch ansprechendes neoklassizistisches Zitat, das eine entspannende Nackenposition ermöglicht. Die symmetrisch geschwungene Wanne kann auch als Solitär im Wohnbereich bestehen. Bei nur 1.645 mm Länge stellt sie selbst für Räume mit weniger Platz eine attraktive Lösung dar. Das Waschbecken nimmt die Linienführung der Wanne auf.

Statement by the jury
The Amiata collection skilfully combines Mediterranean charm and British elegance. Its subtle design stands for a highly developed bathroom culture.

Begründung der Jury
Gekonnt vereint die Kollektion Amiata mediterrane Leichtigkeit und britische Eleganz. Ihre feinsinnige Gestaltung steht für eine hoch entwickelte Badezimmerkultur.

Meisterstück Incava
Bathtub
Badewanne

Manufacturer
Franz Kaldewei GmbH & Co. KG,
Ahlen, Germany
Design
Anke Salomon Product Design
(Anke Salomon),
Potsdam, Germany
Web
www.kaldewei.de
www.ankesalomon.com

The free-standing bathtub Meisterstück Incava impresses through the subtle tension between a soft lined inner form and a sensual geometric exterior. The reduced rim and the conically shaped panel of the bathtub create the impression of lightness. The slightly inclined backrest ensures a highly comfortable lying position. The design is complemented through the flush-fitting enamelled waste cover and the also enamelled overflow that underscores the symmetry of the object.

Die frei stehende Badewanne Meisterstück Incava besticht durch die subtile Spannung zwischen der weich gezeichneten Innenform und dem sinnlich-geometrischen Äußeren. Der reduzierte Rand und die konisch geformte Verkleidung sorgen für einen Eindruck von Leichtigkeit. Die sanfte Rückenschräge erlaubt hohen Liegekomfort. Komplettiert wird die Gestaltung von dem bündig eingepassten und emaillierten Ablaufdeckel sowie einem ebenfalls emaillierten Überlauf, der die Symmetrie des Objekts unterstreicht.

Meisterstück Emerso
Bathtub
Badewanne

Manufacturer
Franz Kaldewei GmbH & Co. KG,
Ahlen, Germany
Design
Arik Levy Studio (Arik Levy), Paris, France
Web
www.kaldewei.de
www.ariklevy.fr

With its high-rise backrest and long base area, this bathtub, designed for one person, offers an ergonomic, modern interpretation of historical bathtub shapes. The central characterising element is the rising twist rim of the bathtub, which overcomes the horizontal plane, thereby shaping a comfortable head area.
The tapered panel of high-quality steel enamel emphasises the sculptural character and gives the bathtub the impression of a modern art object. A flush-fitting waste cover and a discreet overflow made of steel enamel round off the sculpturesque overall appearance.

Mit hoher Rückenschräge und langem Bodenbereich bietet diese Wanne für eine Person die ergonomische, moderne Interpretation historischer Wannenformen. Zentrales charakterisierendes Element ist jedoch der ansteigende in sich gedrehte Wannenrand, welcher die Horizontale verlässt. Zugleich entsteht durch diese Formgebung ein komfortabler Kopfbereich. Die konisch zulaufende Verkleidung aus hochwertigem Stahl-Email betont den skulpturalen Charakter und verleiht der Badewanne die Anmutung eines Kunstobjekts. Ein bündiger Ablaufdeckel und ein dezenter Überlauf aus Stahl-Email ergänzen das skulpturhafte Gesamtbild.

Vavee
Bathtub
Badewanne

Manufacturer
Bathroom Design Co., Ltd.,
Bangkok, Thailand
In-house design
Web
www.bathroomtomorrow.com

The harmonious blend of a surface inspired by nature and water gently cascading down into the basin like a small waterfall: soaking in this bathtub creates a therapeutical spa atmosphere, as it were. Hydro and air bubble jets provide for a beneficial massage. A storage shelf features a structured, anti-slip surface. It is easily shifted by means of a special magnetic mechanism. Light in different shades shines from under it.

Die harmonische Verbindung aus einer von der Natur inspirierten Oberfläche mit Wasser, das sich wie ein behutsamer Wasserfall in Kaskaden in das Becken ergießt: Hier lässt das Liegen in der Badewanne gleichsam therapeutische Spa-Atmosphäre aufkommen. Für eine wohltuende Massage sorgen Wasser- und Luftdüsen. Als sehr praktisch erweist sich die Ablagefläche mit strukturierter rutschfester Oberfläche, die sich durch einen speziellen Magnetmechanismus einfach verschieben lässt. Darunter erstrahlt Licht in verschiedenen Farbtönen.

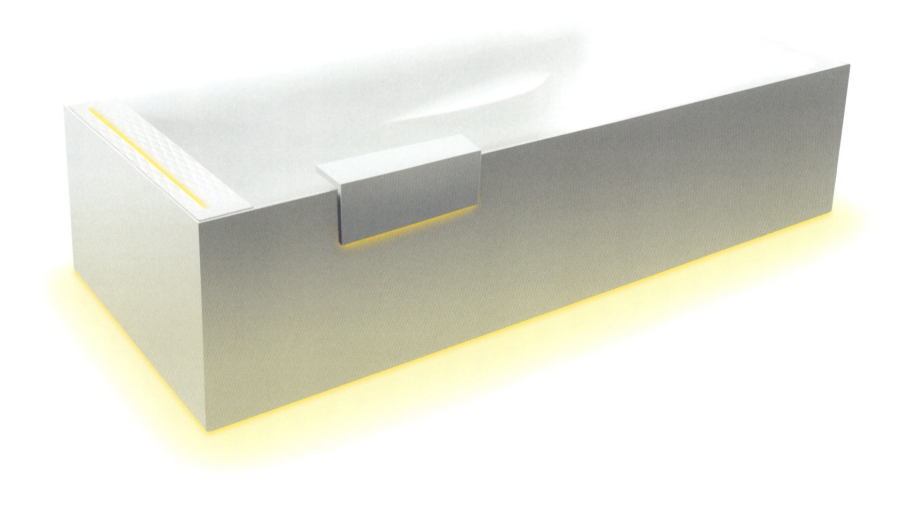

Paiova 5
Bathtub
Badewanne

Manufacturer
Duravit AG, Hornberg, Germany
Design
EOOS, Vienna, Austria
Web
www.duravit.de
www.eoos.com

The symbiosis of a corner bathtub and a free-standing bathtub with differently sloping backrests offers a choice between an upright seating position and relaxed reclining. The pentagonal shape is created by turning a wall-hugging version into the room, which opens up towards the user, as it were. The bathtub entry is slightly lower with an overall height of 580 mm compared to the customary 600 mm. The narrow bathtub rim is well suited to hold on. On the wall side a broad rim provides convenient shelf space. Owing to the different side lengths both tall and small people can support themselves by propping up their feet against the wall of the bathtub.

Diese Symbiose einer Eckwanne und einer frei stehenden Wanne bietet mit unterschiedlichen Rückenschrägen die Wahl zwischen einer aufrechten Sitzposition und entspanntem Liegen. Die pentagonale Formgebung erfolgte durch Drehung einer wandgebundenen Version in den Raum hinein, die sich gleichsam zum Benutzer hin öffnet. Der Wanneneinstieg ist mit einer Gesamthöhe von 580 mm im Vergleich zu den üblichen 600 mm etwas niedriger. Der schmale Wannenrand ist gut zum Festhalten geeignet. Wandseitig bietet ein breiter Rand komfortable Ablagefläche. Dank der unterschiedlichen Seitenlängen können sich große wie kleinere Personen gut mit den Füßen am Wannenende abstützen.

Rainmaker Select 580 3jet
Overhead Shower
Kopfbrause

Manufacturer
Hansgrohe SE, Schiltach, Germany
Design
Phoenix Design GmbH & Co. KG,
Stuttgart, Germany
Web
www.hansgrohe.com
www.phoenixdesign.com

The overhead shower Rainmaker Select 580 3jet places the optimised physical and aesthetic shower experience in the foreground. Three selectable spray modes offer individual showering comfort: Rain-XL thoroughly rinses the hair; RainFlow provides a comforting massage; Mono is an invitation to relax. Glass and a removable spray tray make the shower not only easy to clean but also appealing in any living environment.

Statement by the jury
With its generous tray made of high-quality glass and mounted directly to the wall, the shower immediately conveys what this is all about: optimum comfort.

Bei der Kopfbrause Rainmaker Select 580 3jet stehen das optimierte körperliche wie ästhetische Duscherlebnis ganz im Vordergrund. Drei anwählbare Strahlarten bieten individuellen Komfort: Rain-XL wäscht die Haarpflege gründlich aus, RainFlow bietet wohlige Massage und Mono lädt zum Entspannen ein. Durch den Einsatz von Glas und die abnehmbare Strahlplatte ist die Brause nicht nur leicht zu reinigen, sondern auch ansprechend in jedem Wohnambiente.

Begründung der Jury
Mit einer direkt an der Wand montierten großzügigen Duschplatte aus hochwertigem Glas macht die Brause sofort klar, worum es geht: optimalen Komfort unter der Dusche.

Hansgrohe
Rainmaker Select 460 1jet
Shower Set
Duschenset

Manufacturer
Hansgrohe SE, Sch Itach, Germany
Design
Phoenix Design GmbH & Co. KG,
Stuttgart, Germany
Web
www.hansgrohe.com
www.phoenixdesign.com

The shower set includes the large surface overhead shower Rainmaker Select 460, the ShowerTablet Select 700 and a hand shower. The ShowerTablet is made of high-quality glass and serves for placing bathroom accessories. At the simple push of a button, it selects either overhead or hand shower. The temperature is preselected on the control dial. The surface of the overhead shower is also made of glass. An EcoSmart version ensures energy-efficient water consumption.

Statement by the jury
A well thought-out all-round solution: Bathrooms are fully equipped by installing the Rainmaker Select 460 1jet product family.

Das Duschenset besteht aus der großflächigen Kopfbrause Rainmaker Select 460, dem ShowerTablet Select 700 und einer Handbrause. Das ShowerTablet aus hochwertigem Glas, auf dem auch Badutensilien abgestellt werden können, wählt durch einfachen Knopfdruck Kopf- oder Handbrause an. Die Temperatur wird am Drehregler voreingestellt. Auch die Oberfläche der Kopfbrause besteht aus Glas. Eine EcoSmart-Variante sorgt für ökonomischen Wasserverbrauch.

Begründung der Jury
Eine wohldurchdachte Allround-Lösung: Mit der Installation der Produktfamilie Rainmaker Select 460 1jet ist der Duschbereich komplett ausgestattet.

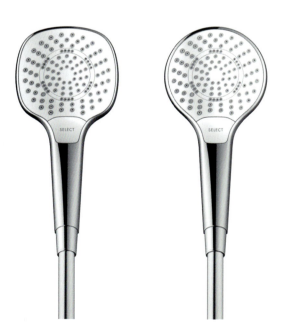

Croma Select
Hand Showers
Handbrausen

Manufacturer
Hansgrohe SE, Schiltach, Germany
Design
Phoenix Design GmbH & Co. KG,
Stuttgart, Germany
Web
www.hansgrohe.com
www.phoenixdesign.com

Croma Select hand showers are available in E and S versions, each with different spray modes. The Croma Select Multi hand shower offers individual showering experiences with three different spray modes: from SoftRain via IntenseRain to the massage spray. In the Croma Select Vario shower the spray intensity can be selected from three different settings, from the soft Rain Spray to the strong Turbo Rain. All showers are available in modern white-chrome with a white spray disk and handle. They are available with EcoSmart technology for reduced water consumption.

Statement by the jury
The hand showers offer a convenient shower experience for any mood: from a pleasant, gentle shower to a powerful massage.

Die Handbrausen Croma Select gibt es in den Versionen E und S mit jeweils verschiedenen Strahlarten. Die Handbrause Croma Select Multi bietet individuelle Duscherlebnisse mit drei Strahlarten: von SoftRain über IntenseRain bis zum Massagestrahl. Bei der Brause Croma Select Vario lässt sich die Strahlintensität in drei Einstellungen, vom sanften Rain-Strahl bis zum kräftigen Turbo-Rain, regulieren. Alle Brausen zeigen sich in modernem Weiß-Chrom mit weißer Strahlscheibe und Griff. Sie sind mit der EcoSmart-Technologie für reduzierten Wasserverbrauch erhältlich.

Begründung der Jury
Mit den Handbrausen Croma Select findet man in jeder Stimmung zum passenden Duscherlebnis: vom wohltuend sanften Schauer bis zur kräftigen Massage.

Inspire
Three-Hole Bathtub Mixer
Drei-Loch-Wannenarmatur

Manufacturer
AM.PM Produktionsgesellschaft S&T mbH,
Berlin, Germany
Design
Design3 GmbH, Hamburg, Germany
Web
www.ampm-germany.com
www.design3.de

The three-hole bathtub mixer of the Inspire collection is characterised by the combination of organic softness and geometric rigor. Generous plane surfaces, clear lines and controlled three-dimensionality define the harmonious overall picture. The integrated water- and energy-saving functions provide for economic consumption.

Statement by the jury
Clarity is the key design element of this three-hole bathtub mixer of the Inspire collection. Modern water- and energy-saving functions make for an environment-friendly solution.

Die Drei-Loch-Wannenarmatur der Kollektion Inspire zeichnet sich durch die Verbindung von organischer Weichheit und geometrischer Strenge aus. Großzügige Flächen, klare Linien und eine kontrollierte Dreidimensionalität prägen das harmonische Gesamtbild. Für einen ökonomischen Verbrauch sorgen die integrierten Wasser- und Energiesparfunktionen.

Begründung der Jury
Klarheit ist das gestalterische Grundelement dieser Drei-Loch-Wannenarmatur aus der Kollektion Inspire. Moderne Wasser- und Energiesparfunktionen nehmen Rücksicht auf die Umwelt.

Axor Citterio E
Shower Set
Duschenset

Manufacturer
Hansgrohe SE, Schiltach, Germany
Design
Antonio Citterio Patricia Viel and Partners,
Milan, Italy
Web
www.hansgrohe.com
www.antoniocitterioandpartners.it

The elegant hand shower set radiates timeless elegance. The angle of the water jet can be individually adjusted by the slide, which runs parallel to the wall bar, as well as by the discreet hinge on the hand shower. The bar also serves as a safety handle under the shower. The clear, precisely shaped surfaces make the object easy to clean.

Statement by the jury
The generously dimensioned slide bar is an expedient component and allows individual adjustment of the jet angle: comfort in the shower for children and tall people alike.

Das elegant anmutende Duschenset zeugt von zeitloser Eleganz. Über den Schieber, der parallel zur Wandstange verläuft, sowie über das dezente Scharnier an der Handbrause selbst, wird der Strahlwinkel des Wassers individuell angepasst. Zugleich kann der Benutzer die Stange als Sicherheitsgriff unter der Dusche verwenden. Dank der klaren, präzise geformten Flächen ist das Objekt leicht zu reinigen.

Begründung der Jury
Als funktionell erweist sich die großzügige Schiebestange, die den Strahlwinkel individuell anpasst: Duschkomfort für Kinder wie auch sehr große Personen.

Axor Citterio E
Floor-Standing Bathtub Thermostat
Bodenstehender Wannenthermostat

Manufacturer
Hansgrohe SE, Schiltach, Germany
Design
Antonio Citterio Patricia Viel and Partners,
Milan, Italy
Web
www.hansgrohe.com
www.antoniocitterioandpartners.it

The floor-standing bathtub thermostat of the collection Axor Citterio E provides classic elegance in bathrooms. Both bathtub and tap present themselves as exclusive independent units. A self-explanatory pull knob allows changing between shower holder and water inlet. The holder for the shower rod is discreetly mounted behind the water column, which thus seems to float.

Statement by the jury
The floor-standing bathtub thermostat presents an aesthetic unit in its own right, yet merges convincingly with the bathtub to create a classic overall impression.

Der bodenstehende Wannenthermostat aus der Kollektion Axor Citterio E verleiht dem Badezimmer klassische Eleganz. Wanne und Mischer präsentieren sich als jeweils allein stehende, edle Einheit. Zwischen Brausehalter und Wanneneinlass wird selbsterklärend per Zugknopf gewechselt. Die Halterung für den Brausestab ist dezent hinter der Wassersäule montiert, dadurch scheint dieser gleichsam zu schweben.

Begründung der Jury
Der bodenstehende Wannenthermostat stellt eine ästhetisch separate Einheit dar und verschmilzt mit der Wanne dennoch überzeugend zu einem klassischen Gesamtbild.

Classic Shower Mixer
Duschenmischer

Manufacturer
Primy Corporation Ltd., Zhuhai, China
In-house design
Yingying Cao, Venkat Tirunagaru
Web
www.primyonline.com

This recyclable shower mixer draws its inspiration from architecture and timeless design. The silicone-coated handle in chilli red provides comfort and a pleasant ambiance. The bathtub spout is infinitely variable from 0 to 180 degrees in vertical direction – a unique function in Classic Shower Mixer. The hand shower is infinitely variable both vertically and horizontally. The mixing tap features three settings for bath, hand and head shower respectively. An aerator in the nozzles reduces the water volume by up to 30 per cent. A special surface finish prevents bacterial growth. All materials used are completely lead-free and very durable.

Statement by the jury
Recyclable materials and a water-saving nozzle function turn the classic-style shower mixer into a contemporary element in the bathroom.

Dieser recycelbare Duschenmischer ist von Architektur und zeitloser Gestaltung inspiriert. Für Komfort und freundliches Ambiente sorgt der silikonbeschichtete chilirote Griff. Der Badewanneneinlauf ist in vertikaler Richtung zwischen 0 und 180 Grad stufenlos schwenkbar – ein ganz besonderes Merkmal bei dem Duschenmischer Classic Shower Mixer. Die Handbrause ist vertikal wie horizontal stufenlos verstellbar, die Mischbatterie erlaubt drei Einstellungen jeweils für Badewanne, Hand- und Kopfbrause. Ein Lüfter in den Düsen reduziert die Wassermenge um bis zu 30 Prozent. Bakterienbildung wird durch eine spezielle Oberflächengestaltung entgegengewirkt. Alle verwendeten Materialien sind komplett bleifrei und besonders langlebig.

Begründung der Jury
Recycelbare Materialien und wassersparende Düsenfunktionen machen den klassisch gestalteten Duschenmischer zu einem zeitgemäßen Element im Badezimmer.

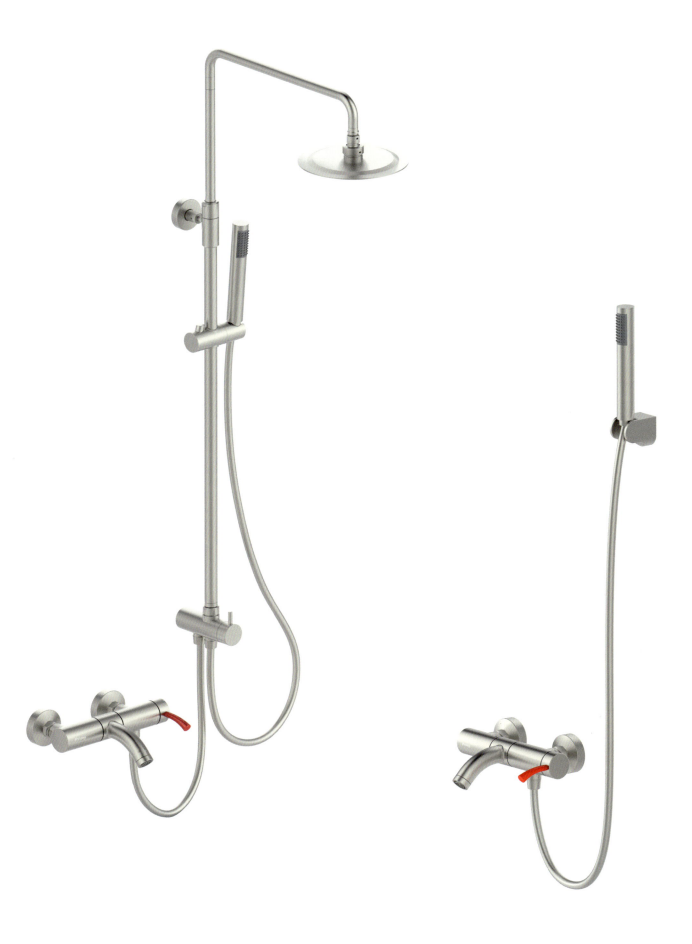

Obliqua
Shower Column
Duschsäule

Manufacturer
Rubinetterie Zazzeri S.p.A.,
Incisa in Val d'Arno (Firenze), Italy
Design
Roberto Innocenti, Florence, Italy
Web
www.zazzeri.it

Thanks to the Flyfall Rain System the water seems to flow directly from the wall. All necessary water circulations required for an even filling of the shower head are unobtrusively integrated in the wall profile, which only measures 12.5 mm. The flow keeps the water sufficiently close to the wall. The design is available in three sizes, also as backlit versions. It is made of polished or brushed stainless steel and conveys intimacy and harmony.

Statement by the jury
As if water was flowing directly from the wall: the reduced appearance of the shower column demonstrates an innovative design consciousness.

Das Flyfall Rain System vermittelt die Idee, dass das Wasser direkt aus den Wänden fließt. In dem nur 12,5 mm starken Wandprofil sind alle notwendigen Wasserkreisläufe für ein gleichmäßiges Befüllen des Duschkopfes optisch dezent integriert. Zudem hält der Durchfluss das Wasser ausreichend nah an der Wand. Das Design ist in drei Größen jeweils mit Gegenlicht-Version verfügbar, welche Intimität und Harmonie schafft, sowie in glänzendem oder gebürstetem Stahl.

Begründung der Jury
Als würde das Wasser direkt aus der Wand strahlen: In ihrer zurückhaltenden Anmutung beweist die Duschsäule innovatives Gestaltungsbewusstsein.

RS 200 AquaSwitch Thermostat
Shower Set
Duscharmatur

Manufacturer
HSK Duschkabinenbau KG,
Olsberg, Germany
In-house design
Web
www.hsk.de

The toggle device of the shower set RS 200 AquaSwitch thermostat makes switching between head and hand shower an obvious matter. Exhaustive pulling or pushing or disagreeable temperatures are things of the past. The 38-degree-centigrade lock, an automatic stop for cold water and the SafeTouch construction that prevents the fitting to heat up ensure safety. The elegant real glass shelf comes in black or white or mirror glass. It provides room for shower items exactly where they are needed.

Statement by the jury
Distinct symbols immediately strike the eye – the toggle switch of the tap body reliably switches between head and hand shower, thus demonstrating high functionality.

Das Umschalten zwischen Kopf- und Handbrause ist mit dem Kippschalter der Duscharmatur RS 200 AquaSwitch Thermostat absolut eindeutig. Kraftaufwändiges Ziehen oder Drücken sowie unliebsame Temperaturen sind damit Geschichte. Für Sicherheit sorgen auch die 38°C-Sperre, die Kaltwasser-Stoppautomatik sowie die SafeTouch-Konstruktionsweise, die ein Aufheizen des Armaturenkörpers verhindert. Die edle Echtglasablage in den Ausführungen Schwarz, Weiß oder Spiegelglas bietet Platz für Duschutensilien direkt dort, wo sie gebraucht werden.

Begründung der Jury
Eindeutige Symbole springen sofort ins Auge – der Kippschalter des Armaturenkörpers wechselt zuverlässig zwischen Kopf- und Handbrause und beweist dadurch hohe Funktionalität.

Lyric
Showerhead
Duschkopf

Manufacturer
American Standarc
In-house design
Web
www.lixil.co.th

The Lyric pairs up to four Bluetooth devices for family members to enjoy their music preferences on one showerhead. It includes over 50 nozzles and a 3-function spray pattern for a pleasurable showering experience. The speaker can be easily removed for charging and used indoors and outdoors up to ten metres away. The built-in rechargeable Lithium-ion battery provides seven hours of run time.

Statement by the jury
The Bluetooth-enabled showerhead Lyric enthuses through its innovative combination of excellent sound and shower experience. In addition, it is easy to handle.

Lyric kann sich mit bis zu vier bluetooth-fähigen Geräten verbinden, so genießt jedes Familienmitglied unter der Dusche seine spezielle Lieblingsmusik. Über 50 Düsen und drei verschiedene Sprüheinstellungen sorgen für ein angenehmes Duscherlebnis. Der Lautsprecher kann zum Laden einfach herausgenommen werden, zudem ist seine Verwendung indoor wie outdoor in einer Entfernung von bis zu zehn Metern möglich. Die aufladbare Lithium-Ionen-Batterie läuft bis zu sieben Stunden.

Begründung der Jury
Der bluetoothfähige Duschkopf Lyric begeistert durch die innovative Kombination von hervorragendem Sound und Duscherlebnis und erlaubt zudem einfaches Handling.

ShowerSelect Glass
Built-in Thermostats
Unterputzthermostate

Manufacturer
Hansgrohe SE, Schiltach, Germany
Design
Phoenix Design GmbH & Co. KG,
Stuttgart, Germany
Web
www.hansgrohe.com
www.phoenixdesign.com

These ShowerSelect Glass thermostats are based on a flush-mounted operating concept featuring a glass surface. The high-quality material embodies the aesthetics of distinctness and is easy to clean. At the push of a button, different water sources can be selected. Their symbols are grasped intuitively and are particularly suitable for older people, children and physically impaired persons.

Statement by the jury
Glass stands for clarity and purity. This flush-mounted operating concept blends into any bathroom in a particularly harmonious and highly functional way.

Bei den Thermostaten ShowerSelect Glass handelt es sich um ein flächenbündiges Unterputz-Bedienkonzept mit Glasoberfläche. Das hochwertige Material vermittelt klare Ästhetik und lässt sich leicht reinigen. Die Symbole auf den Knöpfen, mit denen verschiedene Wasserquellen angewählt werden, sind intuitiv zu verstehen und kommen älteren Menschen, Kindern oder körperlich eingeschränkten Personen besonders entgegen.

Begründung der Jury
Glas steht für Klarheit und Reinheit, wodurch sich das Unterputz-Bedienkonzept besonders harmonisch und hoch funktional in jedes Badezimmer integriert.

Axor Citterio E
Built-in Products
Unterputzprodukte

Manufacturer
Hansgrohe SE, Schiltach, Germany
Design
Antonio Citterio Patricia Viel and Partners,
Milan, Italy
Web
www.hansgrohe.com
www.antoniocitterioandpartners.it

The built-in products of the series Axor Citterio E impress through the intuitive handling they ensure in the shower. Due to their archetypal form, the various handle designs of the thermostat modules help to grasp their respective functions in no time at all: the cross handle signalises the regulation of the water volume, the cylindrical thermostat handle the temperature setting. A two-way-diverter indicates which shower is activated. Horizontally and vertically aligned, the modules present a harmonious overall impression.

Statement by the jury
Thanks to their conclusive design, all modules of the built-in products can be intuitively understood. This leaves no doubt under the shower.

In der Dusche überzeugen die Unterputzprodukte der Kollektion Axor Citterio E durch ihre intuitive Bedienbarkeit. Mit ihrer archetypischen Form helfen die unterschiedlichen Grifflösungen der Thermostatmodule, ihre Funktionen jeweils rasch zu erfassen: Der Kreuzgriff signalisiert die Regulierung der Wassermenge, der zylindrische Thermostatgriff die Temperatureinstellung. Ein Zwei-Wege-Umsteller zeigt, welche Brause gerade angesteuert wird. Horizontal wie vertikal ausgerichtet ergeben die Module ein ausgewogenes Gesamtbild.

Begründung der Jury
Aufgrund ihrer schlüssigen Gestaltung sind alle Module der Unterputzprodukte intuitiv erfassbar und lassen keine Zweifel unter der Dusche aufkommen.

HÜPPE Xtensa pure
Shower Enclosure
Duschabtrennung

Manufacturer
Hüppe GmbH, Bad Zwischenahn, Germany
Design
Phoenix Design GmbH & Co. KG,
Stuttgart, Germany
Web
www.hueppe.com
www.phoenixdesign.com

A highly transparent real glass sliding door design for a shower enclosure with floor-level comfort walk-in provides more safety than a swinging construction. It also saves useful space. The highly transparent glass door disappears almost completely behind the fixed segment. The construction dispenses with the upper guide rail, which enhances the high-quality overall appearance and still ensures overall splash protection. The puristic door handle is mounted on the very edge of the glass surface. Thus, the door can be opened and closed easily both from inside and outside. The optional anti-plaque glass treatment allows easy cleaning.

Das Konzept einer Gleittür bei einer Duschabtrennung mit bodenebenem Komfort-Walk-in bietet höhere Sicherheit als eine ausschwingende Anlage und es spart wertvollen Platz. Die hoch transparente Echtglas-Gleittür verschwindet fast völlig hinter dem festen Segment. Die Konstruktion verzichtet auf die obere Führungsschiene, was den hochwertigen Gesamteindruck verstärkt und dennoch umfassenden Spritzschutz gewährleistet. Der puristische Türgriff ist ganz an der Kante der Glasfläche montiert, so lässt sich die Tür von innen wie außen problemlos öffnen und schließen. Eine optionale Anti-Plaque-Glasveredelung ermöglicht einfache Reinigung.

Statement by the jury
HÜPPE Xtensa pure combines the advantages of a comfortable entry with the increased protection of a sliding door system – a highly expedient, well thought-out design solution.

Begründung der Jury
HÜPPE Xtensa pure verbindet die Vorteile des Komforteinstiegs mit dem höheren Schutz einer Gleittür – eine sehr praktikable, formal gut ausgearbeitete Lösung.

Liga
Shower Enclosure
Duschkabine

Manufacturer
Kermi GmbH, Plattling, Germany
In-house design
Web
www.kermi.de

Liga offers an expedient, easy-to-clean shower solution where it is necessary to save space. Two doors made of 5-mm-thick single-pane safety glass keep the steam in the shower enclosure. They are folded inwardly by means of high-quality joints and elegantly placed against the wall to save space. The raising/lowering mechanism integrated in all hinges as well as ergonomically bow-shaped handles facilitate handling.

Statement by the jury
Liga is a well-devised solution for a beneficial shower experience in small bathrooms.

Liga bietet eine praktische, pflegeleichte Duschlösung, wenn Platz gespart werden muss. Zwei Türen aus 5 mm starkem Einscheiben-Sicherheitsglas halten den Dampf in der Duschkabine. Anschließend werden sie mittels hochwertig verchromter Gelenke in sich gefaltet sowie ökonomisch und elegant an der Wand platziert. Ein in allen Gelenken integrierter Hebe-Senk-Mechanismus und ergonomisch geformte Bügelgriffe erleichtern die Handhabung.

Begründung der Jury
Liga bietet eine gut durchdachte Lösung für ein wohliges Duscherlebnis in kleinen Badezimmern.

RenoDeco
Wall Cladding System
Wandverkleidungssystem

Manufacturer
HSK Duschkabinenbau KG,
Olsberg, Germany
In-house design
Web
www.hsk.de

With the wall cladding system RenoDeco partial renovation of the bathroom is a clean and easy matter. In wet areas like showers and behind washstands and toilets, the 3-mm-thick aluminium composite panels in stone, wood or metal look cover the old or missing tiling. Thus, it is easy to replace a high shower tray or a bathtub by a barrier-free shower area. With a scratchproof, UV-resistant and easy to clean surface finish the decor remains beautiful all the time.

Statement by the jury
RenoDeco impresses with natural materials in humid areas with high hygienic requirements. The renovation succeeds largely free of dirt, and the bathroom can be used again before long.

Mit dem Wandverkleidungssystem RenoDeco gelingt eine einfache und saubere Teilsanierung des Badezimmers. In Feuchtbereichen wie der Dusche und hinter Waschtisch und WC verdecken die 3 mm starken Dekorplatten in Stein-, Holz- und Metalloptik einfach die alte oder fehlende Verfliesung. Damit kann eine hohe Duschtasse oder Wanne leicht durch eine barrierefreie Dusche ersetzt werden. Eine kratzfeste, UV-beständige und pflegeleichte Oberflächenverede ung hält das Dekor dauerhaft schön.

Begründung der Jury
RenoDeco überzeugt mit natürlichen Materialien in Bereichen mit Feuchtigkeit und hohem Hygienebedarf. Die Renovierung gelingt damit weitgehend schmutzfrei und ohne langen Verzicht auf das Bad.

Advantix Vario
Wall Drain
Wandablauf

Manufacturer
Viega GmbH & Co. KG,
Attendorn, Germany
Design
ARTEFAKT industriekultur,
Darmstadt, Germany
Web
www.viega.com
www.artefakt.de

Due to its flat mounting depth of only 25 mm, the wall drainage system allows installation on the surface, in lightweight or exposed walls without damaging the masonry. The floor covering can be laid without interruptions and sloping sections. As the drain profile and the slotted grating can be cut to length to the millimetre, they can be used in lengths of between 300 and 1,200 mm. The simple slotted grating made of stainless steel runs centrally along a 20 mm wide drain joint. It ends with a discreet cover plate on both sides. The drain fitting is self-cleaning, which makes removal and maintenance of the odour trap unnecessary.

Das Wandablaufsystem ermöglicht durch seine flache Einbautiefe von nur 25 mm die Installation im Putz, in der Leichtbau- oder Vorwand, ohne das Mauerwerk zu beschädigen. Auch der Bodenbelag kann ohne Unterbrechungen und Gefälleschnitte verlegt werden. Ablaufprofil und Stegrost sind millimetergenau ablängbar und damit zwischen 300 und 1.200 mm Länge einzusetzen. Der schlichte Stegrost aus Edelstahl verläuft mittig entlang einer 20 mm breiten Ablauffuge und findet beidseitig, mit je einem dezenten Abdeckplättchen, seinen Abschluss. Die Ablaufgarnitur ist selbstreinigend, wodurch sich die Entnahme und Wartung des Geruchsverschlusses erübrigt.

CeraWall P
Wall Drainage System
Wandablaufsystem

Manufacturer
Dallmer GmbH + Co. KG,
Arnsberg, Germany
In-house design
Johannes Dallmer
Web
www.dallmer.de

The wall drainage system CeraWall P is placed directly in the wall. Thus it remains almost invisible. The system fully integrates into the bathroom architecture becoming a part of it. Technically, CeraWall P presents a combination of a floor drain and a concealed channel with an integrated slope. The stainless steel profile is entirely above the bonding seal. The ease of access to the drainage area makes it quick and thorough to clean.

Statement by the jury
With CeraWall P water seems to disappear in an invisible gap on the shower wall – skilful purism that makes any bathroom seem more present.

Das Wandablaufsystem CeraWall P wird direkt in der Wand platziert und bleibt so nahezu unsichtbar. Das System fügt sich gänzlich in die Badezimmerarchitektur ein und wird so ein Teil von ihr. Technisch ist CeraWall P eine Kombination aus Punktablauf und Ablaufprofil mit integriertem Gefälle. Das Edelstahlprofil liegt komplett oberhalb der Verbundabdichtung und ist durch den leichten Zugang zu den Ablaufbereichen schnell und gründlich zu reinigen.

Begründung der Jury
Mit CeraWall P verschwindet das Wasser gleichsam in einen unsichtbaren Spalt an der Duschwand – gekonnter Purismus, der jedes Badezimmer umso präsenter wirken lässt.

CeraWall S
Wall Drainage System
Wandablaufsystem

Manufacturer
Dallmer GmbH + Co. KG,
Arnsberg, Germany
In-house design
Johannes Dallmer
Web
www.dallmer.de

The gentle slope of this wall drainage system is milled. Thus, the visible stainless steel strip with a width of only 54 mm presents itself as both unobtrusive and clearly visible thanks to its precision craftsmanship. The drainagestrip is always entirely located above the bonding seal: flush-mounted, it provides for simple and hygienic cleaning. The drain body is sealed using a simple and secure connection to the wall and floor sealing.

Statement by the jury
The slope of CeraWall S is milled in metal and attests to high quality and longevity. The minimalist strip imparts elegant modernity.

Das sanfte Gefälle dieses Wandablaufsystems ist gefräst. So wirkt die sichtbare Edelstahlschiene mit nur 54 mm Breite zurückhaltend und durch ihre handwerkliche Präzision präsent zugleich. Die Ablaufschiene liegt komplett oberhalb der Verbundabdichtung und gewährleistet durch die flächenbündige Konstruktion eine einfache und hygienische Reinigung. Die Abdichtung des Ablaufgehäuses erfolgt durch eine ebenso einfache wie sichere Anbindung an die Wand- und Flächenabdichtung.

Begründung der Jury
Das in Metall gefräste Gefälle von CeraWall S zeugt von Hochwertigkeit und Langlebigkeit. Die minimalistische Schiene verleiht der Dusche edle Modernität.

ACO Walk-in
Floor-Level Shower Drain
Bodenebener Duschabfluss

Manufacturer
ACO Passavant GmbH,
Stadtlengsfeld, Germany
In-house design
Web
www.aco-haustechnik.de

ACO Walk-in allows floor-level shower-ing without causing impounding water. The water runs entirely off through the grating into an electro-polished stainless steel tray. The cover is resistant towards detergents and soapy water. The shower always stays hygienic as it is easy to clean and standing water is avoided. The grating has a woody, warm feel. Different sizes, colours and slope situations permit versatile scope for design.

Statement by the jury
A functionally sophisticated solution, ACO Walk-in allows extensive showers in floor-level systems without causing puddles.

ACO Walk-in ermöglicht bodenebenes Duschen ohne aufstauendes Wasser. Das Wasser fließt durch den Rost in eine elektropolierte Edelstahlwanne. Das Pro-dukt ist unempfindlich gegenüber Reini-gungsmitteln und Seifenwasser. Durch einfache Reinigung und Vermeidung von stehendem Wasser bleibt die Dusche stets hygienisch. Die Abdeckung bietet eine holzartige, warme Haptik. Verschiedene Größen, Farben und Gefällesituationen schaffen vielfältige Gestaltungsmöglich-keiten.

Begründung der Jury
Eine funktional durchdachte Lösung macht mit ACO Walk-in ausgiebiges Duschen ohne Pfützenbildung in boden-ebenen Anlagen möglich.

CleanLine
Shower Channel
Duschrinne

Manufacturer
Geberit International AG,
Jona, Switzerland
Design
Tribecraft AG, Zürich, Switzerland
Web
www.geberit.com
www.tribecraft.ch

In this shower channel with stainless steel profile, the water flows straight from the surface to the drain. There are no hidden areas that could become soiled. Thus, the channel and the shower floor can be wiped in one go. An easy to remove comb insert replaces the standard outlet sieve, and impurities can be easily stripped off. Infinitely variable trim-ming to length allows for flush installa-tion matching the width of any shower space. The factory-injected sealing foil facilitates easy installation and increases sealing reliability.

Statement by the jury
Owing to its small profile and extra slim look, this shower channel merges with the floor to become a harmonious unit. The comb insert ensures hygienic clean-ing.

Bei dieser Duschrinne mit Edelstahlprofil wird das Wasser direkt von der Oberfläche in den Ablauf geführt, ohne unsichtbare Ebene, die verschmutzen könnte. Die Rinne lässt sich so im selben Wischvorgang mit dem Boden reinigen. Ein leicht entnehm-barer Kammeinsatz ersetzt das Ablaufsieb – Verunreinigungen können problemlos abgestreift werden. Stufenloses Ablängen lässt einen bündigen Einbau in jeder Dusch-platzbreite zu. Die werkseitig eingespritzte Dichtfolie unterstützt eine einfache Instal-lation und erhöht die Dichtsicherheit.

Begründung der Jury
Dank des schmalen Profils und der beson-ders schlanken Optik verschmelzen Dusch-rinne und Boden zu einer harmonischen Einheit. Der Kammeinsatz gewährleistet hygienische Reinigung.

F&C Drain Assembly
Drainage System
Abflussanlage

Manufacturer
Taizhou Haitian Brass Manufacture Co., Ltd.,
Taizhou, China
In-house design
Qiquan Bao
Web
www.neverback.net

The drainage system allows water to drain with 30 litres per minute, while a cellular filter reliably collects hair and other soiling from the bathtub. A magnet-structure inner core allows the interior to open when the water flows; when there is no water, it automatically snaps back. The silicone seal ring at the bottom of the magnet is non-toxic, non-stick and chemically stable and helps to prevent malodour and bacteria inside the drainage system. The drain pipe features a universal joint with a rotation angle of 270 degrees. This allows installation in very differing wall conditions. The drainage pipe is made of copper, which also prevents the formation of bacteria.

Die Abflussanlage lässt das Wasser mit 30 Litern pro Minute ablaufen, während ein zellenartig aufgebauter Filter Haare und andere Verschmutzungen aus der Wanne zuverlässig auffängt. Das magnetische Innere des Systems öffnet sich, wenn Wasser fließt, andernfalls schnappt es automatisch zurück. Der Silikonring an der Unterseite des Magneten ist ungiftig, haftabweisend und chemisch beständig, er verhindert schlechten Geruch und die Bildung von Bakterien innerhalb des Rohres. Das Abflussrohr besteht aus einem Kardangelenk mit einem Rotationswinkel von 270 Grad, was die Installation in sehr unterschiedlichen Wandgegebenheiten erlaubt. Die Kupferbeschichtung im Inneren verhindert ebenfalls die Bildung von Bakterien.

Statement by the jury
F&C Drain Assembly meets all the requirements of a modern drainage system: the water runs off quickly and residues are reliably caught. The material has an antibacterial effect.

Begründung der Jury
F&C Drain Assembly erfüllt alle Ansprüche an eine moderne Abflussanlage: Das Wasser läuft rasch ab, Rückstände werden verlässlich zurückgehalten, das Material wirkt antibakteriell.

Neorest
Washbasin
Waschbecken

Manufacturer
Toto Ltd., Fukuoka, Japan
In-house design
Yuji Yoshioka
Web
www.toto.co.jp/en/index.htm

The washbasin Neorest impresses through its clear lines and aesthetical surface finish. The design is mainly characterized by parallel front and back edges as well as the curved bowl interior. The side which faces the user slopes more gently than the opposite side. The drain in the same colour as the basin enhances the modern appearance of the solution. The basin looks very flat, and thus as elegant as stylish. Since it is embedded in the washstand, it is deep enough to prevent water splashing.

Das Waschbecken Neorest überzeugt durch seine klare Linienführung und eine ästhetisch ansprechende Oberflächenbeschaffenheit. Prägende Elemente der Formgebung sind die parallel verlaufenden vorderen und hinteren Kanten sowie das geschwungene Beckeninnere, das an der dem Benutzer zugewandten Seite sanfter abfällt als an der gegenüberliegenden Seite. Der Abfluss in gleicher Farbe wie das Becken verstärkt die moderne Anmutung. Das Bassin wirkt sehr flach und dadurch elegant wie modern, da es aber in den Waschtisch eingelassen wird, hat es ausreichend Tiefe, sodass kein Spritzwasser entsteht.

Statement by the jury
The thin edge and straight lines demonstrate a creative independence; the curved surface of the basin matches both urban and natural environments.

Begründung der Jury
Der dünne Rand und gerade Linien beweisen gestalterische Eigenständigkeit, die geschwungene Beckenoberfläche passt gut in eine urbane wie natürliche Umgebung.

Serel Purity
Washbasin
Waschbecken

Manufacturer
Serel Sanitary Factory, Manisa, Turkey
In-house design
Ali Yıldız, Aldonat Sunar, Didem Durmaz,
Metin Murat Elbeyli, Zafer Doğan
Web
www.serelseramik.com.tr

With Serel Purity, the water streams in like a gushing waterfall and disappears unnoticed. A drain, which is invisible to users, makes this possible. The organically shaped basin merges into two convenient deposit shelves on both sides. A special surface technology prevents dirt from sticking and enables easy cleaning.

Bei Serel Purity verschwindet das zuerst in einem überschwänglichen Wasserfall einströmende Wasser ganz unbemerkt. Möglich macht dies ein dem Benutzer nicht einsehbarer Abfluss. Das organisch geformte Becken geht fließend in eine komfortable Ablagefläche an beiden Seiten über. Eine spezielle Oberflächentechnologie verhindert das Anhaften von Schmutz und erlaubt einfache Reinigung.

Statement by the jury
The extraordinary design of Serel Purity unites two facets of water: it pours lively into the basin and imperceptibly escapes.

Begründung der Jury
In seiner außergewöhnlichen Formgebung vereint Serel Purity zwei Facetten des Wassers: Ausgelassen strömt es ins Becken, ganz unauffällig fließt es ab.

Cape Cod
Wash Bowl with Console
Waschschale mit Konsole

Manufacturer
Duravit AG, Hornberg, Germany
Design
Philippe Starck, Paris, France
Web
www.duravit.de
www.starck.com

Iconic and softly curved forms are the basis of the Cape Cod range by Philippe Starck. The exclusive basins are produced from the specially developed high-strength ceramic material DuraCeram with an elegant finish. This material makes it possible to create countertop basins with a rim thickness of only 5 mm that are still very robust, impact-resistant and easy to maintain. For individual bathroom designs, the wash bowls are available in three different shapes: round, square and tri-oval. The consoles with shelves make clever use of the contrast between the floor-standing console in cool, smooth chrome and exclusive wood finishes.

Sanft geschwungene Formen bilden die Basis der Serie Cape Cod von Philippe Starck. Die Becken in exklusiver Anmutung sind aus der speziell entwickelten Keramikmasse DuraCeram gefertigt, die eine hohe Festigkeit aufweist und ein edles Finish ermöglicht. Dieser Werkstoff erlaubt es zudem, Aufsatzbecken mit einer Randstärke von gerade einmal 5 mm zu formen, die dennoch pflegeleicht, ausgesprochen stabil und schlagfest bleiben. Für eine individuelle Badraumgestaltung sind die Waschschalen in drei Formen erhältlich: kreisrund, quadratisch und trioval. Die Konsole mit Ablage spielt mit dem Kontrast aus dem bodenstehenden Gestell in kühlem glatten Chrom und der erlesenen Holzoberfläche.

Fuse
Washbasin
Waschbecken

Manufacturer
NotOnlyWhite B.V.,
Amsterdam, Netherlands
In-house design
Marike Andeweg
Web
www.notonlywhite.com

The washbasin Fuse seems to float in the air. This effect is caused by its stepped backside, which also serves as an expedient deposit space for tall accessories or bathroom decoration. The combination of distinct, straight lines and organically curved shapes accounts for the washbasin's appearance. The bottle trap, hidden to the viewer, and the push-open drain, which forms a unit with the white basin made of Cristalplant, make for a discreet and classy look.

Statement by the jury
The combination of distinct, floating lines and organic shapes manifest in Fuse give artful expression to the clarity of water.

Das Waschbecken Fuse schwebt förmlich in der Luft. Diesen Effekt erzeugt die abgesetzte Rückseite, die zugleich als praktische Abstellfläche für höhere Accessoires oder Badschmuck dient. Charakterisierend ist die Kombination von klaren geraden Linien und organisch geschwungenen Formen. Dezent und edel wirken der dem Betrachter verborgene Überlauf und der Push-Open-Ablauf, der eine Einheit mit dem weißen Cristalplant des Beckens darstellt.

Begründung der Jury
Klare, schwebende Linienführung in Kombination mit organischen Formen bringen in Fuse gekonnt die Reinheit des Wassers zum Ausdruck.

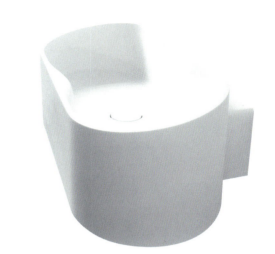

Moment
Washbasin
Waschbecken

Manufacturer
ZeVa GmbH,
Griesheim/Darmstadt, Germany
In-house design
Web
www.zevalife.com

The waveform of the washbasin reflects the smooth rhythm of water. Thanks to its round shape, users get a sense of balance and stability. Without any superfluous design elements, Moment provides prime functions and an immediate sensory experience. The material is rippled on both sides suggesting the natural play of the water surface; the hidden overflow induces a feeling of simplicity. The compact form makes the washbasin also suited for smaller bathrooms.

Statement by the jury
The round contours of the basin expressively make reference to the safety and comfort of the gentle rhythm of water.

Die Wellenform des Waschbeckens verleiht dem sanften Rhythmus des Wassers Ausdruck. Gemeinsam mit der runden Form wird so ein Eindruck von Balance und Stabilität erzeugt. Ganz ohne überflüssige Designelemente bietet Moment beste Funktionen sowie eine unmittelbare Sinneserfahrung. Das beidseitig geriffelte Material suggeriert das natürliche Spiel der Wasseroberfläche. Schlichtheit drückt der verdeckte Ablauf aus, die kompakte Form des Beckens ermöglicht die Nutzung auch im kleinen Badezimmer.

Begründung der Jury
Die runden Konturen des Beckens bringen die Sicherheit und Behaglichkeit des sanften Wasserrhythmus gekonnt zum Ausdruck.

Spirit
Washbasin
Waschbecken

Manufacturer
AM.PM Produktionsgesellschaft S&T mbH,
Berlin, Germany
Design
Design3 GmbH, Hamburg, Germany
Web
www.ampm-germany.com
www.design3.de

The washbasin of the Spirit collection combines clear contours with pleasant simplicity, which makes it a timeless object in the bathroom. Thanks to its distinct shape, it can be installed either as wall-hung or tabletop variant. The low-key design of the washbasin creates a characteristic appearance. Ergonomically high standards are met.

Statement by the jury
The washbasin impresses by its reduced design, which also demonstrates functionality: it can be installed on the wall and also as tabletop variant.

Das Waschbecken der Kollektion Spirit verbindet klare Konturen mit angenehmer Schlichtheit und wird somit zu einem zeitlosen Objekt im Badezimmer. Durch seine eigenständige Formgebung kann es sowohl wandhängend als auch in der Table-Top-Variante installiert werden. Die zurückhaltende Gestaltung verleiht dem Becken ein charakteristisches Erscheinungsbild. Ergonomisch werden hohe Komfortansprüche erfüllt.

Begründung der Jury
Das Waschbecken überzeugt durch reduzierte Gestaltung, die zugleich Funktionalität beweist, indem es an der Wand als auch in der Table-Top-Variante installiert werden kann.

Green
Washbasin Collection
Waschbecken-Kollektion

Manufacturer
Ceramica Catalano Srl,
Fabrica di Roma (Viterbo), Italy
Design
Carlo Martino e Partners Srl con socio unico,
Rome, Italy
Web
www.catalano.it

The entire life cycle of the washbasin collection Green is focused on sustainability – from the raw materials to the manufacturing process to recycling. The extra thin rims are a characteristic feature. They allow for a high capacity since the entire width of the basin is used. Curved and continuous surfaces support the natural discharge of the water, which ensures hygiene and cleanliness. The collection comes in seven versions for full or semi-pedestal or wall-mount installation.

Bei der Waschbecken-Kollektion Green steht Nachhaltigkeit im gesamten Lebenszyklus im Zentrum – von den Ausgangsmaterialien über den Herstellungsprozess bis zum Recycling. Charakterisierend sind die extrem dünnen Ränder, die ein hohes Fassungsvermögen erlauben, da praktisch die gesamte Breite des Beckens genutzt wird. Geschwungene und durchgehende Oberflächen unterstützen das natürliche Abfließen des Wassers, was den Umgang mit Hygiene und Sauberkeit erleichtert. Die Kollektion steht in sieben Varianten für Wand- oder Säuleninstallation oder als Aufbau zur Verfügung.

Nature
Washbasin
Waschbecken

Manufacturer
AMA Design by Materfut S.A.,
V.N. Gaia, Portugal
In-house design
Alexandre Sousa, Hugo Vaz
Web
www.amadesign.pt

The washbasin Nature has the shape of a column and combines two materials. The washbasin with an unobtrusive drain at the inner edge is made of the solid surface material Corian, which has anti-bacterial properties and is easy to clean. The natural cork material of the column is humidity-resistant and air-imperme-able. The whole column is lightweight; nevertheless, it allows for very good thermal and acoustic insulation. It can be installed in one piece.

Statement by the jury
Easy installation in one piece demonstrates sophisticated functionality. The combination of the mineral basin with organic cork is an innovative feature.

Das Waschbecken Nature in Säulenform vereint zwei Materialien: So besteht das Becken mit dezentem Ablauf am Rand aus dem Mineralwerkstoff Corian. Dieser hat antibakterielle Eigenschaften und kann problemlos gereinigt werden. Das Natur-material Kork der Säule ist feuchtigkeits- und luftundurchlässig. Die gesamte Säule ist leicht und ermöglicht dennoch eine sehr gute Wärme- und Schalldämmung; sie kann in einem Stück installiert werden.

Begründung der Jury
Die einfache Installation in nur einem Stück beweist durchdachte Funktionalität. Innovativ ist die Kombination des mineralischen Beckens mit organischem Kork.

SaniQ Kristall
Washbasin
Waschbecken

Manufacturer
Acrysil Limited, Bhavnagar, India
Design
Emamidesign, Berlin, Germany
Web
www.sternhagen.com
www.emamidesign.de
Honourable Mention

The washbasin SaniQ Kristall was inspired by the multifaceted light reflection of a pure, natural crystal. For easy cleaning all edges of the basin are inconspicuously rounded. The look of freshly hewn splinters is maintained. Just as no crystal is like another in nature, this object leaves different impressions depending on the perspective. The quartz surface has an antibacterial effect; it is harder and more heat-resistant than ceramic.

Statement by the jury
Looking like pure crystal, SaniQ Kristall conveys the impression of durability and purity. The high-quality version in quartz clearly comes up to this requirement.

Die Idee hinter dem Waschbecken SaniQ Kristall ist das facettenreich reflektierte Licht eines reinen, natürlichen Kristalls. Für die einfache Reinigung sind alle Kanten des Beckens unsichtbar abgerundet, die Optik frisch gehauener Splitter bleibt dabei erhalten. So wie in der Natur kein Kristall dem anderen gleicht, erscheint dieses Objekt aus jedem Blickwinkel anders. Die Oberfläche aus Quarz wirkt antibakteriell, sie ist härter und hitzebeständiger als Keramik.

Begründung der Jury
Durch die Optik des reinen Kristalls vermittel: SaniQ Kristall Langlebigkeit und Reinheit. Die hochwertige Ausführung in Quarz erfüllt diese Vorgabe sehr gut.

Essence Plus
Faucet Collection
Armaturen-Kollektion

Manufacturer
Grohe AG, Düsseldorf, Germany
In-house design
Web
www.grohe.com

Purism and well-balanced proportions determine the charm of the faucet collection Essence Plus. The dominant visual aspects of this series are cylinder, straight line and right angle. Rounded transitions, however, transform this strictly geometric language into a flowing, organic profile that is inspired by nature. The striking component of each element is its U-shaped outlet. With its characteristic features, the series conveys a self-confident statement of an independent aesthetics without being obtrusive or mannerist.

Purismus und ausbalancierte Proportionen bestimmen den Charme der Armaturen-Kollektion Essence Plus. Vorherrschende visuelle Elemente bei der Gestaltung dieser Serie sind der Zylinder, die Gerade sowie der rechte Winkel. Diese streng geometrische Sprache erfährt jedoch durch abgerundete Übergänge ein fließendes, von der Natur inspiriertes organisches Profil. Markant in jedem Element ist jeweils der u-förmige Auslauf. In ihrer Charakteristik gibt die Serie ein selbstbewusstes Statement für eine eigenständige Ästhetik ab, ohne dabei aufdringlich oder manieriert zu wirken.

Eurosmart New
Faucet Collection
Armaturen-Kollektion

Manufacturer
Grohe AG, Düsseldorf, Germany
In-house design
Web
www.grohe.com

The design of the faucet collection Eurosmart New focuses on high comfort for washing one's hands. The series displays a self-confident appearance with a rather high-positioned water outlet. It is not only an eye-catcher but also allows excellent free moving space without running the risk of getting into unnecessary contact with the material. In addition, a high quality chrome finish ensures ease of cleaning. The energy-saving cold start adapts to the changing needs of users for more sustainability.

Bei der Armaturen-Kollektion Eurosmart New stand hoher Komfort beim Händewaschen im Vordergrund der Gestaltung. Der relativ hohe Auslaufstutzen ist daher nicht nur optisch ein Blickfang, mit dem die Serie ein selbstbewusstes Auftreten demonstriert. Er ermöglicht dem Benutzer auch ausgezeichnete Bewegungsfreiheit beim Waschen der Hände, ohne Gefahr zu laufen, unnötig mit dem Material in Kontakt zu kommen. Eine hochwertige Chromoberfläche sorgt zusätzlich für Reinigungskomfort. Der energiesparende Kaltstart passt sich den veränderten Bedürfnissen der Verbraucher nach mehr Nachhaltigkeit an.

Statement by the jury
Eurosmart New enables comfortable and almost non-contact hand washing. The cold start function demonstrates environmental awareness.

Begründung der Jury
Eurosmart New macht komfortables und weitgehend kontaktfreies Händewaschen möglich. Die Kaltstart-Funktion beweist Umweltbewusstsein.

Eurocube Joy
Faucet Collection
Armaturen-Kollektion

Manufacturer
Grohe AG, Düsseldorf, Germany
In-house design
Web
www.grohe.com

Inspired by the contemporary architecture of European metropolises, Eurocube Joy proves to be committed to a puristic-cubist design vocabulary. Despite the clarity of lines, the collection surprises by playful details presenting the cuboid in different ways and making it the centre point. The joystick lever is an outstanding design element: reduced and unconventional at the same time, it takes up the minimalist claim in a sophisticated manner and allows precise and comfortable water flow.

Inspiriert von der zeitgemäßen Architektur europäischer Metropolen zeigt sich Eurocube Joy einer puristisch-kubistischen Formensprache verpflichtet. Trotz aller Klarheit in der Linienführung überrascht die Kollektion durch verspielte Details, die den Quader immer wieder anders inszenieren und ihn zum Mittelpunkt erheben. Der Joystick-Hebel ist dabei herausragendes Gestaltungselement, reduziert und unkonventionell zugleich. Er greift den minimalistischen Anspruch raffiniert auf und ermöglicht eine präzise und komfortable Wasserführung.

Kovera
Basin Mixer
Waschtischmischer

Manufacturer
Vado, Cheddar,
Somerset, Great Britain
In-house design
Andrew Breeds
Web
www.vado-uk.com

Flat and calm like a stream, the handle shape flows towards the bowl. Reminiscent of a waterfall, the form slopes towards the basin. The concave rim sweeps from the outlet, before curving down the front of the body. The design offers a solution that blends well with both round and oval basins. Flow regulators, fitted as standard, help to reduce water usage.

Statement by the jury
The handle inclined into the basin resembles a stream that pours down into the depths like a waterfall – a truly apt design for any bathroom.

Glatt und ruhig wie ein Strom fließt der Einstellhebel dem Waschbecken entgegen. An einen Wasserfall erinnernd neigt er sich zum Becken hin. Die Wölbung des Hahns zieht sich den Wasserauslass entlang und fällt schließlich nach unten hin ab. Die Gestaltung bietet eine Lösung, die sich gut in runde wie ovale Waschbecken einfügt. Der serienmäßig integrierte Durchflussregler trägt dazu bei, den Wasserverbrauch zu senken.

Begründung der Jury
Der ins Becken geneigte Griff gleicht einem Strom, der sich als Wasserfall in die Tiefe ergießt – eine wirklich treffende Gestaltung, die in jedes Badezimmer passt.

Sydney
Basin Mixer
Waschtischmischer

Manufacturer
AWA Faucet Ltd., Taipei, Taiwan
In-house design
Christian Laudet
Web
www.awafaucet.com

The Sydney Opera House was the source of inspiration for the dynamic, open angles and contours of the tap. An aerator reduces the water volume by 40 per cent and increases the soft feeling of the jet on the skin. By simply unscrewing it, limescale is quickly removed. Setting the two-step cartridge to a particular position reduces the water flow by another 50 per cent. In the centre setting, only cold water is supplied. The interior is made of high-quality, recyclable brass.

Statement by the jury
A major issue in the basin mixer Sydney is the conservation of resources: the material is recyclable; water flow and temperature are kept as low as possible.

Die dynamischen, offenen Winkel und Linien der Armatur sind vom Opernhaus in Sydney inspiriert. Eine Luftzufuhr verringert die Wassermenge um 40 Prozent, zudem verstärkt sie das weiche Gefühl des Strahls auf der Haut. Durch einfaches Abschrauben des Luftsprudlers wird hier Kalk rasch entfernt. Eine Zwei-Stufen-Kartusche verringert den Wasserdurchfluss in einer Einstellung um weitere 50 Prozent, in der Mittelstellung fließt nur kaltes Wasser. Das Innere besteht aus hochwertigem recyclebarem Messing.

Begründung der Jury
Der Waschtischmischer Sydney steht ganz im Zeichen der Ressourcenschonung: Das Material ist recycelbar, Wassermenge und -temperatur werden möglichst niedrig gehalten.

Inspire
Single-Lever Mixer
Einhandmischer

Manufacturer
AM.PM Produktionsgesellschaft S&T mbH,
Berlin, Germany
Design
Design3 GmbH, Hamburg, Germany
Web
www.ampm-germany.com
www.design3.de

The single-lever mixer of the Inspire collection is characterised by a skilful combination of soft and geometric forms. Generous plane surfaces, distinct lines and a controlled three-dimensional structure create a harmonious overall impression. An integrated water and energy saving function guarantees economic consumption.

Statement by the jury
The single-lever mixer stands out by its organic and at the same time strictly geometric design. Water and energy saving functions meet high economic standards.

Der Einhandmischer der Kollektion Inspire zeichnet sich durch die gekonnte Verbindung von weichen und geometrischen Formen aus. Großzügige Flächen, klare Linien und eine kontrollierte Dreidimensionalität prägen das harmonische Gesamtbild. Die integrierte Wasser- und Energiesparfunktion sorgt für einen ökonomischen Verbrauch.

Begründung der Jury
Der Einhandmischer fällt auf durch organische und zugleich streng geometrische Gestaltung. Wasser- und Energiesparfunktionen erfüllen hohe ökonomische Standards.

Colan
Single-Lever Mixer
Einhandmischer

Manufacturer
Bonke Kitchen & Sanitary Industrial Co., Ltd.,
Comdr, Foshan, China
Design
Weicong Luo, Foshan, China
Web
www.comdr.cn
www.ypss365.com

A special technology allows lending the stainless steel of the single-lever mixer Colan a strongly curving line despite the material's hardness. The treatment of the surface harks back to the wire drawing technique and ensures a very indurate metallic texture. Since stainless steel has antioxidant properties, the surface need not be galvanised. Thus, the manufacturing process is comparatively eco-friendly. No patina forms on the easy-to-clean material.

Statement by the jury
Colan appeals through its unusually strong curvature and its distinct design that relies on solid stainless steel.

Eine spezielle Technologie macht es beim Einhandmischer Colan möglich, dass hier Edelstahl trotz seiner Härte zu einem weiten Bogen geformt wird. Bei der Oberflächenbehandlung wurde auf die Technik des Drahtziehens zurückgegriffen, was der Armatur eine sehr harte metallische Textur verleiht. Der Edelstahl hat antioxidative Eigenschaften, dadurch muss die Oberfläche nicht galvanisiert werden. Der Herstellungsprozess verläuft deshalb vergleichsweise umweltschonend. Das Material setzt zudem keine Patina an und kann leicht gereinigt werden.

Begründung der Jury
Colan besticht mit seiner klaren Gestaltung aus massivem Edelstahl und der ungewöhnlich weiten Bogenform.

Axor Citterio E
Basin Mixers
Waschtischmischer

Manufacturer
Hansgrohe SE, Schiltach, Germany
Design
Antonio Citterio Patricia Viel and Partners,
Milan, Italy
Web
www.hansgrohe.com
www.antoniocitterioandpartners.it

It is the mixture of the familiar and the new that constitutes the charm of Axor Citterio E. The collection combines timeless premium quality with subtle elegance. Precisely shaped surfaces and edges merge with smooth transitions. The slim, high-rise design of the single-lever tap that works like a modern joystick is impressive. It snaps into a special cartridge in the off-position, thus increasing ease of use.

Statement by the jury
The basin mixers match many bathroom styles due to a well thought-out combination of different design elements. Precise tap control provides high user comfort.

Der Charme von Axor Citterio E steckt in der Kombination aus Vertrautem und Neuem. Die Kollektion verbindet zeitlose Hochwertigkeit mit feiner Eleganz, präzise geformte Flächen und Kanten verschmelzen mit weichen Übergängen. Gestalterisch einprägsam ist jeweils die schlanke, hoch aufragende Einhebelarmatur, die wie ein moderner Joystick funktioniert. In der Aus-Position rastet sie in einer speziellen Kartusche ein, was zum Nutzungskomfort beiträgt.

Begründung der Jury
Überlegte Kombination unterschiedlicher Gestaltungselemente macht die Waschtischmischer mit vielen Badstilen kompatibel. Die präzise Führung der Armaturen bietet hohen Nutzerkomfort.

Axor Starck Organic
Basin Mixer
Waschtischmischer

Manufacturer
Hansgrohe SE, Sch Itach, Germany
Design
Starck Network, Paris, France
Web
www.hansgrohe.com
www.starck.com

The sensor-controlled electronic version of the basin mixer Axor Starck Organic impresses through its organic minimalist solution and a generous shower spray. Hands are extensively and thus efficiently wetted. This enhancement of functionality and water experience is accompanied by a low water flow rate of 3.5 litres per minute only. The thermostat, which indicates the specifically selected temperature, visually melts into the tap body.

Statement by the jury
Sensor controls and a generous shower spray enable efficiency and hygiene. The union of organic structures and minimalism is artfully implemented.

Die sensorgesteuerte Elektronikvariante des Waschtischmischers Axor Starck Organic besticht durch organisch-minimalistische Formgebung und einen großzügigen Brausestrahl. Die Hände werden besonders großflächig und dadurch effizient benetzt. Dieses Mehr an Funktionalität und an Wassererlebnis geht mit einem geringen Wasserdurchfluss von nur 3,5 Litern pro Minute einher. Der Thermostat zeigt die jeweils gewählte Temperatur an und verschmilzt visuell mit dem Armaturenkörper.

Begründung der Jury
Sensorsteuerung und ein großzügiger Brausestrahl gewährleisten Effizienz und Hygiene. Gestalterisch überzeugend umgesetzt ist die Vereinigung aus Organischem und Minimalismus.

Talis Select S
Basin Mixers
Waschtischmischer

Manufacturer
Hansgrohe SE, Schiltach, Germany
Design
Phoenix Design GmbH & Co. KG,
Stuttgart, Germany
Web
www.hansgrohe.com
www.phoenixdesign.com

The tapering spout of the product family Talis Select S, available in three sizes, recalls traditional organic materials. The special Select technology permits the user to switch the water flow on and off by pressing the handle top. A cartridge adjusts the water volume. It makes sense to intuitively press the arm down on the tap allowing especially hygienic use of the tap. Turning the handle controls the temperature.

Statement by the jury
Thanks to Select technology water flows without the need for using one's hands – a functional solution for high hygienic standards.

Die konisch verlaufenden Grundkörper der Produktfamilie Talis Select S in drei Höhen erinnern an traditionelle organische Materialien. Die spezielle Select-Technologie erlaubt es dem Nutzer, mittels Druck auf die Griffoberfläche den Wasserfluss ein- und auszuschalten, eine Kartusche regelt die Wassermenge. Ein intuitives Anpressen mit dem Arm bietet sich an, was einen besonders hygienischen Gebrauch des Hahns zulässt. Das Drehen des Griffes reguliert die Temperatur.

Begründung der Jury
Dank Select-Technologie fließt das Wasser, ohne dass die Hände ins Spiel gebracht werden müssten – eine funktionale Lösung für hohe Hygienestandards.

Axor Citterio E
Three-Hole basin mixers
Drei-Loch-Waschtischmischer

Manufacturer
Hansgrohe SE, Schiltach, Germany
Design
Antonio Citterio Patricia Viel and Partners,
Milan, Italy
Web
www.hansgrohe.com
www.antoniocitterioandpartners.it

The classic cross handles of the tap are characteristic features of the three-hole washbasin mixer. With their slim, rounded design, they create a link between nostalgia and modernity. They imply timeless quality and simple elegance. Modern functionality harmonises with a pleasant feel. The collection's soft vocabulary of forms fits in a wide variety of styles.

Statement by the jury
With their appealing design the cross handles of the three-hole basin mixer collection Axor Citterio E create a visual bridge between familiar classicism and functional modernity.

Die klassischen Kreuzgriffe an der Armatur sind prägende Gestaltungsmerkmale dieser Drei-Loch-Waschtischmischer. Durch ihre abgerundete Form schaffen sie eine Verbindung zwischen Nostalgie und Moderne, sie vermitteln Hochwertigkeit mit schlichter Eleganz. Zeitgemäße Funktionalität harmoniert mit angenehmer Haptik. Die sanfte Formensprache der Kollektion fügt sich in ganz unterschiedliche Stilrichtungen ein.

Begründung der Jury
Bei den Drei-Loch-Waschtischmischern der Kollektion Axor Citterio E überzeugt die Gestaltung der Kreuzgriffe, die eine optische Brücke zwischen vertrauter Klassik und funktionaler Moderne schlagen.

Composed
Single-Lever Mixer
Einhandmischer

Manufacturer
Kohler Company, Kohler, Wisconsin, USA
In-house design
Rafael Rexach
Web
www.kohler.com

The long-drawn outlet of the single-lever mixer reaching out to the user makes an almost cheeky and experimental impression. Timeless in character, the otherwise uncompromising cubic shape fits into any environment and still looks minimalist and modern. The side-mounted control consistently continues the design concept and allows easy control of flow rate and temperature.

Statement by the jury
The cubic shape lends the single-lever mixer an uncompromising touch. Its easy handling makes for a convincing solution.

Nahezu vorwitzig und experimentell wirkt der lang gezogene, dem Benutzer sich entgegenstreckende Auslauf dieses Einhandmischers. Die ansonsten kompromisslos kubische Gestaltung fügt sich in ihrer Zeitlosigkeit in jedes Ambiente ein und wirkt dennoch minimalistisch-modern. Die seitlich montierte Steuerung setzt das formgebende Konzept konsequent fort und ermöglicht einfache Regulation von Durchflussmenge und Temperatur.

Begründung der Jury
Der Einhandmischer zeigt sich in seiner kubischen Formgebung kompromisslos. Auch die einfache Bedienbarkeit überzeugt.

Avid
Single-Lever Mixer
Einhandmischer

Manufacturer
Kohler Company, Kohler, Wisconsin, USA
In-house design
Zheyan Hong
Web
www.kohler.com

The graceful design of the single-lever mixer Avid is reminiscent of liquid metal. This particularly shows in the drop-shaped drain as well as in the control lever. The connection between lever and mixing tap is arranged as far back as possible to represent a flowing unit in the 90-degree rotation as well. Due to its low-key language of forms, the tap fits into any modern environment.

Statement by the jury
The single-lever mixer blends in well with any modern ambience, notwithstanding that its flowing forms give it a self-contained appearance.

In seiner grazilen Formgebung erinnert der Einhandmischer Avid an flüssiges Metall. Insbesondere zeigt sich dies in dem tropfenförmigen Auslauf sowie im Mischhebel. Die Verbindung zwischen Hebel und Hahn ist weitestmöglich nach hinten verlegt, um auch in der 90-Grad-Drehung eine fließende Einheit darzustellen. Durch seine zurückhaltende Formensprache fügt sich der Hahn in jede moderne Umgebung ein.

Begründung der Jury
Der Einhandmischer passt sich in jedes moderne Ambiente gut ein, tritt mit seinen fließenden Formen gestalterisch dennoch eigenständig auf.

Modus
Faucet
Wasserhahn

Manufacturer
VSI Industrial S.A.C., Lima, Peru
In-house design
Stefano Camaiora
Web
www.vainsa.com

The combination of a simple cylinder and a flat plate in the water tap Modus represents a very plain solution. The flat lever regulates the water flow, which allows intuitive and very easy handling. Its shape reflects expressive elegance. The unit has been designed for a regional market that asks for solely cold-water taps at an affordable price.

Statement by the jury
There could hardly be a more simple and distinct tap design. In addition, Modus meets the demand for an energy-efficient cold-water tap.

Ganz besonders schlicht ist der Wasserhahn Modus in seiner Kombination eines einfachen Zylinders und einer flachen Platte ausgeführt. Der in den Auslauf integrierte flache Hebel regelt den Wasserdurchfluss. Das macht eine intuitive und sehr leichte Handhabung möglich, zugleich zeugt diese Gestaltung von charaktervoller Eleganz. Die Einheit ist für einen regionalen Markt konzipiert, der reine Kaltwasserarmaturen zu einem erschwinglichen Preis verlangt.

Begründung der Jury
Klarer und einfacher lässt sich ein Wasserhahn kaum gestalten. Zudem wird Modus der Nachfrage nach einer ökonomischen Kaltwasserarmatur gerecht.

E-Quick
Multifunctional Basin Mixer
Multifunktionaler Waschtischmischer

Manufacturer
Xiamen Water Nymph Sanitary Technology Co., Ltd., Xiamen, China
Design
Sprgo Design (Jianquan Chen, Qideng Huang, Zengyi Cai, Junwen Huang, Shiliang Yang, Feng Chen, Yaoxiar Lai, Jianming Huang, Kai Lin, Xinhui Hong), Xiamen, Fujian, China
Web
www.waternymph.cn
www.sprgo.cn

E-Quick stands for comfort and ease of use. A special aerator switches between spray and bubble function, which reduces water consumption significantly. Blue, anti-slip rings enable easy change. With two simple steps, the pull-out tap allows different spray options. E-Quick is made of high-quality stainless steel.

Statement by the jury
Well-devised attachments turn the E-Quick into a multifunctional basin mixer that stands out by easy handling.

E-Quick steht für Komfort und Benutzerfreundlichkeit. Er hat einen speziellen Strahlregler für den Wechsel zwischen Sprüh- und Sprudelfunktion, was den Wasserverbrauch deutlich reduziert. Blaue, rutschfeste Ringe erlauben einen mühelosen Wechsel. Die ausziehbare Armatur ermöglicht mit zwei einfachen Handgriffen verschiedene Sprühoptionen. Der Hahn ist aus hochwertigem Edelstahl gefertigt.

Begründung der Jury
Durchdachte Zusätze machen aus dem E-Quick einen multifunktionalen Waschtischmischer, der durch einfache Handhabbarkeit punktet.

POP

Basin Mixer
Waschtischmischer

Manufacturer
Rubinetterie Zazzeri S.p.A.,
Incisa in Val d'Arno (Firenze), Italy
Design
Studio Batoni Architettura+Design
(Fabrizio Batoni),
Colle di Val d'Elsa (Siena), Italy
Web
www.zazzeri.it
www.studiobatoni.com

The archetypal shape of the tap served as the model for this solution. The fusion of the simple, clean lines of metal with the softness of silicone creates a visually cheerful, bright-coloured mix that is pleasant to touch. The metal part of the device is available in chrome, brushed steel or black and white lacquered finish. The combinable coloured silicone comes in several designs including a phospho-rescent version. An aerator mixes water and air and ensures energy efficiency.

Die Urform des Wasserhahns stand hier Pate. Aus der Vermählung der schlichten, klaren Linien des Metalls mit der Weichheit des Silikons erwächst optisch ein spieler-ischer, farbenfroher Mix, der angenehm anzufassen ist. Der metallische Teil der Vorrichtung ist in Chrom, gebürstetem Stahl sowie weißem oder schwarzem Lack verfügbar, der farbige Silikonauslauf in mehreren, auch phosphoreszierenden Aus-führungen dazu kombinierbar. Ein Strahl-regler mischt Luft und Wasser und sorgt für Energieeffizienz.

Statement by the jury
Brilliantly simple, the basin mixer POP combines an archetypal shape with play-ful and colourful ideas.

Begründung der Jury
Der Waschtischmischer POP kombiniert bestechend einfach archetypische Form-gebung mit spielerischen und farbenfrohen Möglichkeiten.

Affability Faucet
Basin Mixer
Waschtischmischer

Manufacturer
Bravat China GmbH, Guangzhou, China
Design
Sprgo Design (Jianquan Chen, Qideng Huang,
Zengyi Cai, Junwen Huang, Shiliang Yang,
Feng Chen, Yaoxian Lai, Jianming Huang, Kai Lin),
Xiamen, Fujian, China
Web
www.bravat.com
www.sprgo.cn

The concise geometric column is the core design element of the basin mixer Affability Faucet. The single functional parts feature a transition from hardness to softness reflecting the nature of the hydrological cycle. At the gentle touch of a key, the air mixer changes between a soft spray jet and a rippling watercourse. A simple pulling movement extends the water outlet allowing to spray plants set up nearby, for example.

Wichtigstes gestalterisches Element des Waschtischmischers Affability Faucet ist die konzise geometrische Säule. Die einzelnen funktionalen Teile zeigen einen Übergang von Härte zu Weichheit, was die Natur des Wasserkreislaufes widerspiegelt. Auf Tastendruck wechselt der Luftmischer zwischen einem sanften Sprühstrahl und plätscherndem Wasserlauf. Mit einer einfachen Zugbewegung wird der Wasserauslass verlängert, wodurch es möglich wird, etwa in der Nähe aufgestellte Pflanzen zu besprühen.

Statement by the jury
The basin mixer Affability Faucet impresses with the timeless shape of a classic water column. Its pull-out solution provides multifunctionality in bathrooms.

Begründung der Jury
Der Waschtischmischer Affability Faucet überzeugt in seiner zeitlosen Gestalt nach Art einer klassischen Wassersäule. Der ausziehbare Hahn macht Multifunktionalität im Badezimmer möglich.

DOUHO
Water Tap Set
Wasserhahnset

Manufacturer
JOMOO Kitchen & Bath Co., Ltd., Xiamen, China
In-house design
Xiaofa Lin, Shan Lin, Linda Dong, Faxing Miao,
Chengmin Huang, Jiqiao Lin, Zhiqin Lin,
Shaohang Zhang, Xiaoming Lv
Web
www.jomoo.com.cn

Inlet and outlet of the water tap set
DOUHO merge to become a pair of
opposites: the colours black and white
and cold and warm materials highlight
their relatedness. The particularly long
lever in the form of a simple comma,
which facilitates handling, is a distinctive feature. Each part of the lever ends
flush in a valve block. This gives the
tap a modern, holistic appearance and
allows easy cleaning.

In dem Wasserhahnset DOUHO verbinden
sich Zulauf und Abfluss zu einem optischen Gegensatzpaar: Die Farben Schwarz
und Weiß, kalte und warme Materialien
verweisen auf ihre Verbundenheit. Auffallend ist der besonders lange Griff in Form
eines schlichten Kommas, der eine einfache
Handhabung ermöglicht. Jedes Griffteil
mündet plan in einen Ventilblock. Das
verleiht dem Hahn ein modernes, ganzheit-
liches Erscheinungsbild, zudem erlaubt es
eine leichte Reinigung.

10°
Bathroom Series
Badserie

Manufacturer
RAVAK a.s., Pribram, Czech Republic
Design
Nosal Design Studio (Krystof Nosal),
Cernosice, Czech Republic
Web
www.ravak.com
www.nosaldesign.cz

The principle of this bathroom series is based on the slight rotation of individual elements, each by ten degrees. The bathtub, the washbasin and the water taps create the impression of turning to the user. The oblique position of the rectangular bathtub entails an asymmetry, which produces an additional surface for seating or placing bathroom accessories. The slightly turned position of the washbasin offers a lot of storage space and comfortable access to the basin.

Das Prinzip dieser Badserie beruht auf einer leichten Rotation der einzelnen Elemente um jeweils zehn Grad. Die Badewanne, das Waschbecken sowie die Armaturen erwecken auf diese Weise den Eindruck, als würden sie sich dem Benutzer zuwenden. Durch die Schrägstellung der rechteckigen Badewanne entsteht mit der Asymmetrie eine zusätzliche Fläche, auf der man Badutensilien abstellen oder auch sitzen kann. Viel Ablagefläche und einen komfortableren Zugriff auf das Bassin ermöglicht zudem die leicht gedrehte Positionierung des Waschbeckens.

Neorest RH
Shower Toilet
Dusch-WC

Manufacturer
Toto Ltd., Fukuoka, Japan
In-house design
Minoru Tani, Yuji Yoshioka,
Hirotaka Nakabayashi
Web
www.toto.co.jp/en/index.htm

The shower toilet features a warm-water massage and cleansing system. It also disperses electrolysed tap water, which reliably disinfects, deodorises and also removes dirt on the bowl and in the nozzles. A pump-driven water tank, direct pressure from the water pipeline as well as a two-flush system of 3.8 and 3.3 liters provide for environment-friendly usage. The rounded lid and bowl with gradually tapering edges convey a feeling of coherence and an unobtrusive familiarity.

Das Dusch-WC mit Warmwassermassage und -reinigung versprüht zusätzlich elektrolysiertes Leitungswasser, welches Verunreinigungen auf der Schüssel und in den Düsen zuverlässig beseitigt sowie desinfiziert und desodoriert. Ein pumpengetriebener Wassertank und direkter Druck von der Wasserleitung sowie zwei verschieden starke Spülungen von nur 3,8 und 3,3 Liter sorgen für Umweltfreundlichkeit. Die Rundungen von Deckel und Schüssel mit sich allmählich verjüngenden Kanten vermitteln Stimmigkeit und eine unaufdringliche Vertrautheit.

Admire
Close Coupled Toilet
Stand-WC

Manufacturer
AM.PM Produktionsgesellschaft S&T mbH,
Berlin, Germany
Design
Gneiss Group (Torben S. Jorgensen),
Copenhagen, Denmark
Web
www.ampm-germany.com
www.gneissgroup.com

The close coupled toilet of the Admire collection impresses through its elegant and unobtrusive simplicity. All details like hinges and transitions are precisely finished. Combined with elaborate technology like the hidden wall mount or the toilet seat with soft close lid, the toilet ensures timeless and stylish quality.

Statement by the jury
The close coupled toilet of the Admire collection meets high quality standards. Each detail is meticulously finished; the timeless design underscores its longevity.

Das Stand-WC der Kollektion Admire besticht durch elegante und zurückhaltende Reduktion. Alle Details wie Scharniere und Übergänge sind präzise verarbeitet. In Kombination mit wohldurchdachter Technik wie der verdeckten Wandaufhängung oder dem WC-Sitz mit Absenkautomatik gewährleistet die Toilette zeitlose und stilvolle Qualität.

Begründung der Jury
Das Stand-WC der Kollektion Admire erfüllt hohe Qualitätsansprüche. Jedes Detail ist exakt verarbeitet, die zeitlose Gestaltung steht im Zeichen der Langlebigkeit.

Serenity
Close Coupled Toilet
Stand-WC

Manufacturer
ZeVa GmbH, Griesheim/Darmstadt, Germany
In-house design
Web
www.zevalife.com

Its design makes this toilet an organic unit. The circumferential plain lines inhibit the formation of bacteria and support easy cleaning. The shape takes the human posture into consideration: the lower part runs towards the wall; thus, direct contact with the user's feet is avoided and high ease of use is ensured.

Statement by the jury
Its distinct design turns Serenity into a visual haven of tranquillity. In addition, ergonomics and hygienic properties have been taken into account.

Die Toilette erscheint in ihrer Gestaltung als organische Einheit. Die umlaufend einfache Linienführung hemmt zudem die Bildung von Bakterien und unterstützt eine leichte Reinigung. Die Form nimmt Rücksicht auf die menschliche Körperhaltung: Der untere Teil verläuft in Richtung Wand, so wird direkter Kontakt mit den Füßen des Nutzers vermieden und ein hohes Maß an Komfort gewährleistet.

Begründung der Jury
Ihre klare Gestaltung macht Serenity zum optischen Ruhepol, zudem wurde auf Ergonomie und gute Hygieneeigenschaften geachtet.

Geberit AquaClean Mera
Shower Toilet
Dusch-WC

Manufacturer
Geberit International AG,
Jona, Switzerland
Design
Christoph Behling Design Ltd.,
London, Great Britain
Web
www.geberit.com
www.christophbehlingdesign.com

The patented WhirlSpray shower technology ensures particularly thorough cleaning. A pulsating shower spray provides immediately and continuously tempered water, while infused air refines the water jet. The entire asymmetric internal geometry of the rimless ceramics is flushed very carefully and gently. Additional comfort functions include an intuitive remote control, a discreet light for orientation at night, an automatic lid, heating, dryer, and an integrated odour extraction.

Statement by the jury
The Geberit AquaClean Mera shower toilet meets all hygienic requirements in a compelling way. The design holds out against visual and technical upscale standards.

Die patentierte WhirlSpray-Duschtechnologie ermöglicht eine besonders gründliche Reinigung. Ein pulsierender Duschstrahl aus sofort und anhaltend temperiertem Wasser wird mittels Luftbeimischung verfeinert. Sorgfältig und leise erfolgt die Ausspülung der gesamten asymmetrischen Innengeometrie der spülrandlosen Keramik. Weitere Komfortfunktionen sind eine intuitive Fernbedienung, dezentes Licht zur nächtlichen Orientierung, ein automatischer Deckel, Heizung, Trocknungsfunktion sowie eine integrierte Geruchsabsaugung.

Begründung der Jury
Geberit AquaClean Mera entspricht überzeugend sämtlichen Hygieneerfordernissen. Die Gestaltung hält gehobenen optischen wie auch technischen Standards stand.

Axent One
Shower Toilet
Dusch-WC

Manufacturer
Axent Switzerland AG,
Rapperswil-Jona, Switzerland
Design
Matteo Thun & Partners
(Matteo Thun, Antonio Rodriquez),
Milan, Italy
Web
www.axentbath.com
www.matteothun.com

The very name of the shower toilet Axent One suggests that it can be controlled by a single multifunction button. In the basic setting, a clockwise rotation releases the ladies shower. A counter-clockwise rotation activates the anal shower function. Water flow and temperature can be adjusted in both positions. A subtle clicking sound supports finding the right setting acoustically. For repeated use by one person, a quick function remembers the last setting.

Statement by the jury
A shower toilet has to meet manifold hygienic standards. Axent One fully matches them with one single, easy to operate multifunction button.

Der Name des Dusch-WCs Axent One weist bereits darauf hin, dass es mittels eines einzigen Multifunktionsknopfs gesteuert werden kann. In dessen Grundeinstellung löst eine Drehung im Uhrzeigersinn die Lady-Dusche aus. Bei einer Drehung entgegen dem Uhrzeigersinn wird die Anal-Dusche in Funktion gesetzt. In der jeweiligen Nutzungsposition können Wassermenge und -temperatur reguliert werden. Ein dezenter Klickton unterstützt das Finden der richtigen Einstellung akustisch. Für die mehrmalige Nutzung durch eine Person findet eine Quick-Funktion zur jeweils letzten Einstellung zurück.

Begründung der Jury
Ein Dusch-WC muss sehr vielfältigen Hygieneansprüchen entsprechen. Axent One wird dem mit einem einzigen leicht bedienbaren Multifunktionsknopf voll gerecht.

Tonic II
Wall-Hung Toilet
Wand-WC

Manufacturer
Ideal Standard International BVBA,
Zaventem, Belgium
Design
ARTEFAKT industriekultur,
Darmstadt, Germany
Web
www.idealstandard.com
www.artefakt.de

Tonic II achieves a contemporary yet timeless and elegant appearance. The patented AquaBlade technology gives optimum water flow from the top of the bowl on every flush and covers every inch of the ceramic with its powerful water jet. Thanks to its smooth gently curved design with no overhanding flushing rim, the toilet is very hygienic and easy to clean. As the water jet flows along the bowl, rinsing is carried out very quietly and splashing is greatly reduced.

Statement by the jury
Thanks to its innovative AquaBlade technology, an unbroken current of water cleans the complete inner surface of the toilet bowl – a compelling easy-to-maintain solution that is also aesthetically pleasing.

Tonic II zeigt ein aktuelles und dennoch zeitloses wie elegantes Erscheinungsbild. Die spezielle AquaBlade-Technologie ermöglicht stets die äußerst gründliche Ausspülung des WC-Beckens mit einem starken, von oben umlaufenden Wasserstrahl, der jede Stelle der Keramik erfasst. Durch die glatte, nur leicht gewölbte Form ohne klassischen Spülrand ist das WC sehr hygienisch und einfach zu reinigen. Der Wasserstrom fließt unmittelbar an der Wandung entlang, damit werden Spülgeräusche reduziert und es entsteht weniger Spritzwasser.

Begründung der Jury
Dank seiner innovativen AquaBlade-Technologie reinigt ein lückenloser Wasserstrom das komplette Beckeninnere des WCs – eine überzeugend wartungsfreundliche wie formschöne Lösung.

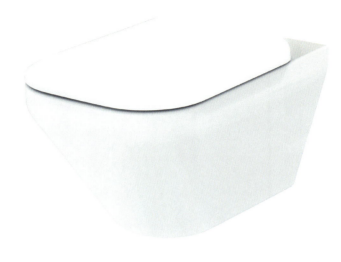

Wall-Hung Toilet
Wand-WC

Manufacturer
Bien Yapi Ürünleri Sanayi Turizm
Ve Ticaret A.Ş., Bilecik, Turkey
In-house design
Tolga Berkay
Web
www.bienseramik.com.tr

The wall-hung toilet made of high-quality vitreous porcelain creates an independent definition of the term "organic" as its shape elegantly recalls the anatomy of human bones. With a flush volume of 4 respectively 2.5 litres, the system works rather economically. A very slim lid and the concealed installation system enhance the positive aesthetic overall impression.

Statement by the jury
A very unconventional design concept was aesthetically implemented in a masterly manner. High-quality vitreous porcelain emphasises the stylish overall appearance.

Das Wand-WC aus hochwertigem Glasporzellan schafft eine eigenständige Definition des Begriffs „organisch", indem seine Form auf elegante Weise an die Anatomie menschlicher Knochen erinnert. Mit einer Spülmenge von 4 bzw. 2,5 Litern arbeitet das System vergleichsweise ökonomisch. Ein sehr dünner Deckel und das versteckte Installationssystem verstärken den positiven ästhetischen Gesamteindruck.

Begründung der Jury
Ein sehr eigenwilliges Formkonzept wurde gekonnt ästhetisch umgesetzt. Hochwertiges Glasporzellan verstärkt den edlen Gesamteindruck.

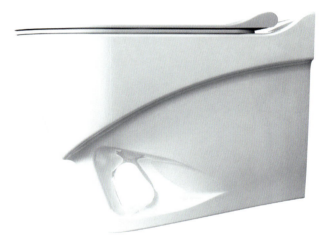

Sott'Aqua
Wall-Hung Toilet
Wand-WC

Manufacturer
Isvea Eurasia Yapı Malz.San.ve Tic. A.Ş.,
Turkey
Design
Nativita Design House, Milan, Italy
Web
www.isveabagno.it
www.nativitamilano.com

With a water flow of 1.5 or 2.7 litres only and a high-performance flush, Sott'Aqua is one of the leading models that have been designed with regard to the conservation of resources. The protruding shape of the toilet bowl is balanced by a slim toilet lid. A concealed fixing system enhances the well-proportioned appearance and enables easy installation.

Statement by the jury
Due to a special high-performance flush, the toilet requires a water flow of 1.5 or 2.7 litres only – a significant contribution to the conservation of resources.

Mit einer Wassermenge von nur 1,5 bzw. 2,7 Litern und einer speziellen Hochleistungsspülung gehört Sott'Aqua zu den führenden Modellen, die unter dem Aspekt der Ressourcenschonung entwickelt wurden. Das hervortretende WC-Becken wird optisch durch einen schmalen Deckel ausgeglichen. Ein verstecktes Fixierungssystem verstärkt das ebenmäßige Erscheinungsbild und ermöglicht eine einfache Installation.

Begründung der Jury
Durch seine spezielle Hochleistungsspülung benötigt das WC eine Wassermenge von nur 1,5 bzw. 2,7 Litern – ein beachtlicher Beitrag zur Ressourcenschonung.

SensoWash Slim
Shower Toilet
Dusch-WC

Manufacturer
Duravit AG, Hornberg, Germany
In-house design
Web
www.duravit.de

SensoWash Slim is the purist version of a shower toilet. Thanks to harmonious proportions and a decidedly flat seat it looks like a classic toilet. Simple handling and functionality make it a convenient solution for private and semi-public space. A slim remote control navigates between rear, comfort and lady wash. Intensity and position of the spray as well as the water temperature are set via illuminated icons. The entire seating can be removed and replaced with effortless ease thus allowing quick and easy cleaning.

SensoWash Slim ist die puristische Version eines Dusch-WCs. Dank ausgewogener Proportionen und betont flachem Sitz wirkt es wie eine klassische Toilette. Einfache Bedienbarkeit und Funktionalität machen es zur geeigneten Lösung für den privaten oder auch halböffentlichen Raum. Eine schlanke Fernbedienung steuert Gesäß-, Komfort- und Ladydusche. Intensität und Ausrichtung des Dusch-strahls sowie die Wassertemperatur sind über illuminierte Symbole einstellbar. Die komplette Sitzfläche lässt sich mit nur einem Handgriff abnehmen und wieder aufsetzen, was eine schnelle und leichte Reinigung ermöglicht.

Statement by the jury
SensoWash Slim meets all requirements on a shower toilet at a high level: ease of use thanks to clearly graspable function-ality as well as easy maintenance.

Begründung der Jury
SensoWash Slim erfüllt auf hohem Niveau alle Anforderungen an ein Dusch-WC: einfache Bedienbarkeit dank klar erfass-barer Funktionalität sowie unkomplizierte Wartung.

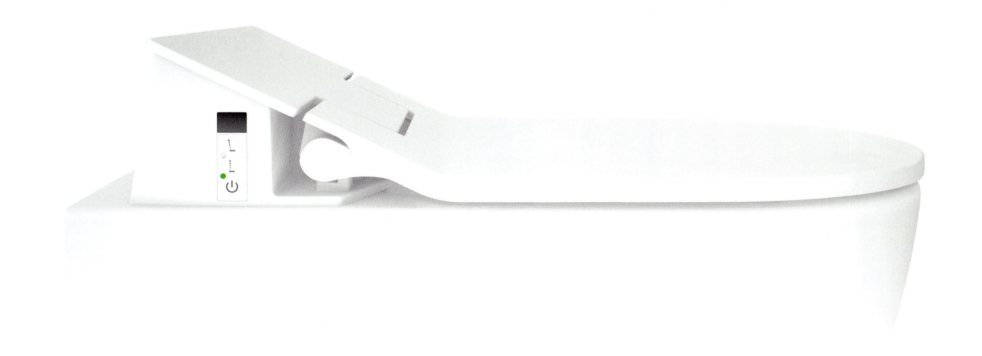

Evolution
Toilet Seats
WC-Sitz-Serie

Manufacturer
Hamberger Sanitary GmbH,
Rohrdorf, Germany
In-house design
Maciej Panas
Design
Staudenmayer design development,
Munich, Germany
Web
www.hamberger-sanitary.com
www.m-staudenmayer.de

Both models of the series Evolution captivate at first sight because of their sensuous minimalism. The model Evo shows playful lightness. Its gently curved silhouette dissolves the tension of the extremely reduced form in an expressive way. Era represents a reinterpretation of the classic form with clear surfaces and precisely drawn lines. Both models hide mechanic details in a masterly way; they correspond to high hygienic standards.

Statement by the jury
The extremely reduced design of the series Evolution imposingly conveys sensuous lightness and impresses with advanced technical details.

Beide Modelle der Serie Evolution bestechen auf den ersten Blick durch ihren sinnlichen Minimalismus. Das Modell Evo vermittelt spielerische Leichtigkeit. Seine sanft geschwungene Silhouette löst die Spannung der extrem reduzierten Form ausdrucksstark auf. Era zeigt mit klaren Flächen und exakt gezogenen Linien eine Neuinterpretation der klassischen Form. Beide Modelle verbergen sehr gut mechanische Details und entsprechen hohen Hygieneerfordernissen.

Begründung der Jury
Die extrem reduzierte Gestaltung der Serie Evolution vermittelt gekonnt sinnliche Leichtigkeit und beeindruckt mit ausgeklügelten technischen Details.

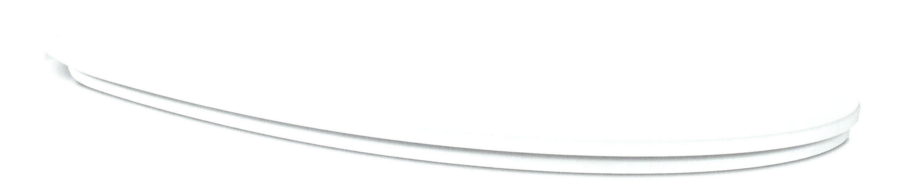

Geberit
Remote Flush Actuation
Type 70
Geberit Fernbetätigung Typ 70

Manufacturer
Geberit International AG, Jona, Switzerland
Design
Christoph Behling Design Ltd.,
London, Great Britain
Web
www.geberit.com
www.christophbehlingdesign.com

This remote flush actuation for toilets can be placed freely up to two metres from the tank. This offers a maximum of flexibility when designing bathrooms. With dimensions of only 50 × 112 mm it is extremely small. Two visually separated operational areas on the plate, which is made of either glass or stainless steel, indicate how to use the water-saving dual flush function. The surface allows thorough cleaning, since there are no gaps or movable parts where dirt or bacteria could accumulate.

Statement by the jury
The Geberit remote flush actuation type 70 features no buttons and meets high standards of aesthetics and cleanliness. As it can be placed almost anywhere, there is much creative design flexibility.

Die WC-Fernauslösung ist bis zu zwei Meter vom Spülkasten frei platzierbar und bietet so absolute Flexibilität in der Badgestaltung. Mit ihren Maßen von 50 × 112 mm ist sie zudem extrem klein. Zwei optisch getrennte Betätigungsfelder der aus Glas oder Edelstahl geformten Platte weisen den Benutzer auf die wassersparende 2-Mengen-Spülung hin. Die Oberfläche lässt ein gründliches Reinigen zu, da sie keine beweglichen Teile oder Spalten aufweist, wo sich Schmutz und Bakterien ansammeln könnten.

Begründung der Jury
Die tastenlose Geberit Fernbetätigung Typ 70 genügt hohen Ansprüchen an Ästhetik wie Sauberkeit. Weitgehend freie Platzierbarkeit ermöglicht eine flexible Raumplanung.

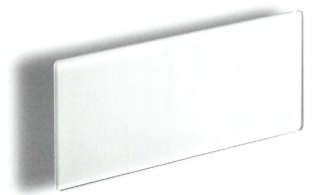

CI

Bathroom Collection
Badezimmer-Kollektion

Manufacturer
Toto Ltd., Fukuoka, Japan
In-house design
Shingo Kawakami, Takeaki Miyamoto
Web
www.toto.co.jp/en/index.htm

The bathroom collection CI comprises a washbasin with tap and vanity unit, a mirror cabinet and a stand-alone element. It features minimalist lines without any unnecessary parts. Nevertheless, it offers a range of user-friendly details and plenty of storage space. The washbasin is generously sized. The tap with eco-cap saves up to 25 per cent of water. The mirror cabinet is furnished with a socket inside; the mirror is well illuminated by LEDs at its top and bottom. The rear side of the door of the high cabinet is mirrored. The drawers offer a variety of storage options.

Die Badezimmer-Kollektion CI besteht aus Waschbecken mit Hahn und Schrank darunter, Spiegelschrank sowie einem frei stehenden Element. Sie zeigt eine minimalistische Linienführung ohne überflüssige Zusätze, wartet aber mit praktischen Details und viel Stauraum auf. Das Waschbecken ist großzügig dimensioniert. Der Hahn mit Eco-Cap spart bis zu 25 Prozent Wasser. Der Spiegelschrank ist innen mit einer Steckdose ausgestattet, LEDs oben und unten leuchten den Spiegel gut aus. Die Rückseite der Tür des hohen Schranks ist verspiegelt, die Schubladen bieten eine Vielzahl von Aufbewahrungsmöglichkeiten.

Statement by the jury
The design of the bathroom collection CI scores with minimalism and yet holds numerous functional and useful details in store.

Begründung der Jury
Die Badezimmer-Kollektion CI überzeugt mit gestalterischem Minimalismus und hält dennoch viele praktische, funktionale Details bereit.

Fluent
Bathroom Collection
Badezimmer-Kollektion

Manufacturer
Inbani, Alicante, Spain
Design
Arik Levy
Web
www.inbani.com
www.ariklevy.fr

The name Fluent alludes to the character of the ensemble since it stands for smooth transitions: between round and angular, broad and narrow, vertical and horizontal. The different elements of the bathroom collection match in style, visually well thought out, and form a balanced whole. All shapes of the collection are soft and elegant and create a calm effect in every bathroom. Cabinets, mirrors and washbasins featuring a subtly hidden drain are adaptable for classical as well as for contemporary interiors.

Der Name Fluent spricht bereits den Charakter des Ensembles an, denn er steht für fließende Übergänge: zwischen rund und eckig, breit und schmal, vertikal und horizontal. Die verschiedenen Elemente der Badezimmer-Kollektion sind optisch sehr überlegt aufeinander abgestimmt und bilden ein ausgewogenes Ganzes. Alle Formen der Kollektion wirken weich und elegant und verbreiten eine ruhige Wirkung in jedem Badezimmer. Schränke, Spiegel und mit einem raffiniert versteckten Ablauf versehene Waschbecken sind für klassische wie für zeitgenössische Innenräume adaptierbar.

Statement by the jury
Fluent impresses through a design concept in which harmony is in the fore – visually balanced forms combine to create an accomplished calm and expressive overall appearance.

Begründung der Jury
Fluent beeindruckt durch ein Gestaltungskonzept, in dem Harmonie ganz im Vordergrund steht – optisch ausgewogene Formen vereinen sich zu einem gekonnt ruhigen wie ausdrucksstarken Gesamtbild.

LOOQ
Functional Toilet Wall
Toiletten-Funktionswand

Manufacturer
Architekten Spiekermann, Beelen, Germany
In-house design
Oliver Spiekermann
Web
www.architekten-spiekermann.de
Honourable Mention

The functional toilet wall LOOQ allows to inconspicuously hide all the accessories of a toilet facility. All closet surfaces are opened by gentle hand pressure. The middle compartment keeps toilet paper at hand; only the beginning of the toilet roll protrudes. Above, spare rolls are stored; beneath, there is a toilet brush, which can be removed for cleaning. In the large compartment beside, there are racks for magazines and newspapers.

Statement by the jury
Purism in the toilet: with LOOQ all the accessories of a toilet facility discreetly disappear in the wall.

Die Toiletten-Funktionswand LOOQ lässt alle Accessoires einer WC-Anlage dezent verschwinden. Alle Schrankflächen werden durch sanften Handdruck geöffnet. Das mittlere Fach hält Toilettenpapier bereit, nur der Anfang der Rolle blickt hervor. Darüber werden Ersatzrollen gelagert, darunter eine Reinigungsbürste, die zur Säuberung herausgenommen werden kann. Im größeren Fach daneben befinden sich Halterungen für Magazine und Zeitungen.

Begründung der Jury
Purismus in der Toilette: Mit LOOQ verschwinden diskret alle Accessoires einer WC-Anlage in der Wand.

Axor Universal Accessories
Bathroom Accessories
Badezimmer-Accessoires

Manufacturer
Hansgrohe SE, Schiltach, Germany
Design
Antonio Citterio Patricia Viel and Partners, Milan, Italy
Web
www.hansgrohe.com
www.antoniocitterioandpartners.it

Axor Universal Accessories is an equipment system suitable for the whole collection. It includes railings, storage shelves and handles, which can be used for individual or comprehensive solutions in bathrooms and to some extent in kitchens. Premium single components ensure harmonious overall accessory combinations that are both coherent and visually pleasing. If required, they can be rearranged or complemented with additional elements. This makes it much easier to customise room designs.

Statement by the jury
Axor Universal Accessories impress by their functionality: a modular, versatile and all-purpose system to be used in bathrooms and in kitchens.

Die Axor Universal Accessories sind ein kollektionsübergreifendes Zubehörsystem, das aus Relings, Ablageelementen und Halterungen besteht. Es kann als Einzel- oder als Systemlösung im Bad und teilweise in der Küche eingesetzt werden. Aus hochwertigen Einzelkomponenten entstehen visuell ansprechende und harmonische Accessoire-Gesamtlösungen. Bei Bedarf können sie neu ausgerichtet oder um weitere Elemente ergänzt werden. Die individuelle Gestaltung des Raumes wird somit entscheidend erleichtert.

Begründung der Jury
Die Axor Universal Accessories überzeugen durch Funktionalität, denn sie sind modular, vielseitig und universell in Küche oder auch Badezimmer einsetzbar.

Admire
Towel Holder
Handtuchhalter

Manufacturer
AM.PM Produktionsgesellschaft S&T mbH, Berlin, Germany
Design
Gneiss Group (Torben S. Jorgensen), Copenhagen, Denmark
Web
www.ampm-germany.com
www.gneissgroup.com

The design of the towel holder of the Admire Accessory collection combines formal rigor with the elegance of a bathroom for superior demands. The mechanics of the swivelling accessory bracket stays hidden under the stable master arm. Due to its timeless purist design language, the object easily fits in modern as well as classic styles.

Statement by the jury
With its geometric simplicity, the towel holder of the Admire Accessory collection becomes an artistic all-rounder in every bathroom.

Der Handtuchhalter aus der Kollektion Admire Accessory verbindet in seiner Gestaltung formale Strenge und die Eleganz eines Badezimmers für gehobene Ansprüche. Die Mechanik des ausschwenkbaren Zusatzbügels bleibt unter dem stabilen Hauptarm verborgen. In seiner zeitlosen, puristischen Formgebung fügt sich das Objekt in moderne wie klassische Stile problemlos ein.

Begründung der Jury
In seiner geometrischen Schlichtheit wird der Handtuchhalter aus der Kollektion Admire Accessory zum gestalterischen Allrounder in jedem Badezimmer.

Björk
Hand Dryer
Händetrockner

Manufacturer
Dan Dryer A/S, Randers, Denmark
Design
VE2 (Hugo Dines Schmidt), Aarhus, Denmark
Web
www.dandryer.com
www.ve2.dk

The cylindrical form of the hand dryer Björk presents a design solution beyond the usual. The almost unbroken dome of the body forms an entity with the wall. This also offers functional benefits as the cylinder softens the volume of the flowing air jet. The upper part is available in different colours and materials and can thus be customised to match the environment.

Statement by the jury
Thanks to its cylindrical shape, the hand dryer Björk is treading new ground not only in bathroom aesthetics, but also dampens the sound of escaping air.

Die zylindrische Form des Händetrockners Björk zeigt eine gestalterische Lösung abseits des Gewohnten. Die fast ungebrochene Rundung des Körpers bildet eine Einheit mit der Wand. Zudem bietet diese Formgebung funktionelle Vorteile, denn der Zylinder dämpft die Lautstärke des austretenden Luftstrahls. Der obere Teil wird in verschiedenen Farben und Materialien angeboten und kann so der Umgebung individuell angepasst werden.

Begründung der Jury
In seiner zylindrischen Form beschreitet der Händetrockner Björk nicht nur neue Wege in der Badezimmerästhetik, er dämpft auch das Geräusch der austretenden Luft.

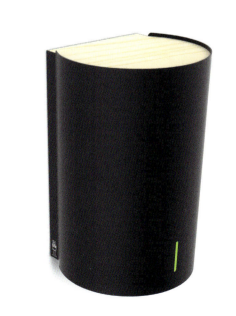

ghd Aura
Hairdryer
Haartrockner

Manufacturer
ghd, London, Great Britain
In-house design
Web
www.ghdhair.com

The hairdryer focuses on innovation, functionality and premium-level quality and has a stylish, ergonomic appearance. A special Laminair technology delivers a concentrated, non-turbulent stream of air. Thus, while working on one section of hair, other sections of hair remain undisturbed. This results in precision styling. The Cool-Wall technology keeps the outer casing and nozzle cool to touch, creating a ring of cool air surrounding the hot air stream.

Der Haartrockner legt den Fokus auf Innovation, Funktionalität sowie hohe Qualität, dabei bietet er ein formschönes ergonomisches Erscheinungsbild. Eine spezielle Laminair-Technologie erzeugt einen konzentrierten, nicht verwirbelnden Luftstrom. So werden die anderen Haarpartien während der Arbeit an einer Strähne nicht durcheinandergebracht und das Haar kann kontrolliert gestylt werden. Die Cool-Wall-Technologie hält das Außengehäuse und die Düse so kühl, dass es berührt werden kann, indem sich um die heiße Luft herum ein Ring kalter Luft bildet.

Statement by the jury
Widespread problems with hair styling are solved: a concentrated stream of air works without destroying other sections of hair. During operation, the outer casing of the device remains cool.

Begründung der Jury
Hier werden verbreitete Probleme beim Haarstyling gelöst: Ein konzentrierter Luftstrom arbeitet, ohne andere Partien zu zerstören. Das Gerät wird im Betrieb außen nicht heiß.

EssentialCare
Hairdryer
Haartrockner

Manufacturer
Royal Philips, Eindhoven, Netherlands
In-house design
Web
www.philips.com

The hairdryer EssentialCare was especially designed for the needs of young and modern Chinese women. It is available in the colour versions pink and fuchsia. Its lily design makes it look natural. Three levels to control heat and speed, an extra cooling function for the premium model, diffuser and nozzle enable versatile hair styling.

Statement by the jury
Making hair styling easy for young Chinese women, the hairdryer proves to be a functional solution in fast-paced everyday lives.

Der EssentialCare Haartrockner wurde speziell für die Bedürfnisse junger, moderner chinesischer Frauen entworfen. Er ist in den farblichen Ausführungen Pink und Fuchsia verfügbar. Seine Lilienform verleiht dem Gerät eine natürliche Anmutung. Drei Stufen zur Wärme- und Geschwindigkeitskontrolle, mit zusätzlicher Kühlung beim Spitzenmodell, Diffuser und Düse ermöglichen ein vielseitiges Haarstyling.

Begründung der Jury
Der Haartrockner erleichtert jungen chinesischen Frauen das Haarstyling und erweist sich daher als funktionales Tool in einem schnelllebigen Alltag.

ProCare Dryer
Hairdryer
Haartrockner

Manufacturer
Royal Philips, Eindhoven, Netherlands
In-house design
Web
www.philips.com

The hairdryer ProCare serves the needs for professional and quick styling outside a hairdresser's shop. A special Style & Protect concentrator delivers excellent drying performance with 25 per cent less damage due to overheating. In addition, drying the hair is cut down to half of the usual time. Robust materials and an ergonomic design complete the concept.

Statement by the jury
The hairdryer ProCare demonstrates innovative and functional features: it allows drying the hair both quickly and gently.

Der Haartrockner ProCare bedient die Bedürfnisse nach professionellem und schnellem Styling außerhalb des Friseursalons. Ein spezieller Style-&-Protect-Konzentrator bietet eine sehr gute Trocknungsleistung mit 25 Prozent weniger Schädigung aufgrund von Überhitzen. Zudem wird das Trocknen der Haare auf die Hälfte der üblichen Trocknungszeit verkürzt. Robuste Materialien und eine ergonomische Formgebung runden das Konzept ab.

Begründung der Jury
Der Haartrockner ProCare beweist Innovation und Funktionalität, indem er schnelle Trocknung bei schonender Behandlung des Haars ermöglicht.

Style to Go
Hairstyling Set

Manufacturer
Robert Bosch Hausgeräte GmbH,
Munich, Germany
In-house design
Helmut Kaiser, Yvonne Janet Weisbarth
Design
Stascha Offenbeck, Ottobrunn, Germany
Web
www.bosch-hausgeraete.de
www.offenbeck.eu

Its compact, handy-size design makes this hairstyling set an expedient accessory for travel. Thanks to the option to switch to different voltages, dryer, straightener and curling tongs can be used throughout the world. Even weight distribution and ergonomic handle shapes with partial texture surfaces make each device ie comfortably and safely in the hand. Important control elements are highlighted by colour accents. They click precisely into place, which facilitates handling.

Das kompakte und handliche Design macht dieses Haarstylingset zum idealen Reisebegleiter. Dank der Möglichkeit, auf unterschiedliche Stromspannungen umzuschalten, sind Haartrockner, Haarglätter und Lockenstab weltweit einsetzbar. Durch die gut austarierte Gewichtsverteilung und ergonomische Griffgestaltung mit partieller Textur-Oberfläche liegt jedes Gerät angenehm und sicher in der Hand. Wichtige Bedienelemente sind durch farbliche Akzente hervorgehoben und zeichnen sich durch präzise Rastung aus, was die Handhabung erleichtert.

ghd Curve Range
Hair Curlers
Lockenstäbe

Manufacturer
ghd, London, Great Britain
In-house design
Web
www.ghdhair.com

Each of the four hair curlers creates a different look. The patented tri-zone technology featuring six highly sensitive sensors constantly and evenly maintains the optimum styling temperature of 185 degrees centigrade along the barrel. To allow styling of as much hair as possible at one time, the clamp/turn feature was removed from the barrel and placed on the handle. The protective cool tip safeguards the fingers when fixing the hair. A safety stand allows placing the curlers down without damaging surfaces. The devices go into sleep mode when not in use for 30 minutes.

Jeder der vier Lockenstäbe kreiert einen anderen Look. Die patentierte Tri-Zone-Technologie mit sechs hochsensiblen Sensoren hält die optimale Stylingtemperatur von 185 Grad Celsius gleichmäßig und konstant entlang des Stabes. Um das Stylen von möglichst viel Haar auf einmal zu ermöglichen, wurde die Klemm-/Drehfunktion vom Stab auf den Griff verlegt. Die schützende Cool-Tip-Spitze schützt die Finger beim Fixieren der Haare. Dank des Sicherheitsständers lässt sich der Stab ablegen, ohne Oberflächen zu beschädigen. Die Geräte fallen nach 30 Minuten in einen Schlafmodus, wenn sie nicht benutzt werden.

Statement by the jury
The hair curlers of the ghd Curve Range offer a convincing solution: they allow for quick styling and long-lasting curls combined with high ease of use.

Begründung der Jury
Die Lockenstäbe der ghd Curve Range überzeugen, denn sie ermöglichen das schnelle Styling lang haltbarer Locken bei hoher Benutzerfreundlichkeit.

RB844
Hot Rotating Silicone Brush
Elektrische Rundbürste

Manufacturer
Kenford Industrial Company Limited,
Hong Kong
In-house design
Michael Keong
Web
www.kenford.com.hk

The hot rotating brush achieves professional hair styling at home. The curling tongs heat up without loss of time. Their rotating brush performs three steps in one go: brushing and drying the hair as well as styling waves. The hairbrush is made of silicone, which prevents the hair from entangling. In addition, it gets smooth rather quickly.

Statement by the jury
With a rotating brush the hair is dried, dressed, smoothed and styled in one go – an impressively expedient solution.

Die Rundbürste macht professionelles Haarstyling zu Hause möglich. Der Lockenstab erhitzt ohne Zeitverlust und seine rotierende Bürste erledigt gleich drei Arbeitsdurchgänge in einem, nämlich das Frisieren und Trocknen der Haare sowie das Stylen der Wellen. Die Borsten bestehen aus Silikon, so kann verhindert werden, dass sich die Haare verwirren, und sie werden schneller geschmeidig.

Begründung der Jury
Mit dem rotierenden Bürstenkopf wird das Haar in einem Durchgang getrocknet, frisiert, geschmeidig gemacht und gestylt – eine überzeugend funktionale Lösung.

Magic Hair Styler
Hair Curling Device
Lockenstab

Manufacturer
Kenford Industrial Company Limited,
Hong Kong
In-house design
Michael Keong
Web
www.kenford.com.hk

The Magic Hair Styler ensures stable curly hair without causing unnecessary damage. The heating element catches and wraps a strand of hair; it is then guided like a brush towards the hair tips. Thus, the keratin chains in the hair are heated only during the styling process. Since the rod only heats up inside, the risk that the scalp or the fingers are hurt is minimised.

Statement by the jury
Intensive styling with a curling iron often leads to damage to the hair structure. The Magic Hair Styler offers a sophisticated solution to this problem.

Der Magic Hair Styler sorgt für stabil gelocktes Haar, ohne dieses unnötig zu schädigen. Das Heizelement erfasst und umwickelt eine Strähne und wird anschließend wie eine Bürste in Richtung Haarspitzen geführt. Dadurch werden die Keratinketten im Haar nur für die Dauer des Stylingvorgangs erhitzt, nicht aber darüber hinaus. Da sich der Stab nur im Inneren erhitzt, minimiert sich auch die Gefahr, dass Finger oder Kopfhaut verletzt werden.

Begründung der Jury
Intensives Styling mit dem Lockenstab führt häufig zur Schädigung der Haarstruktur – der Magic Hair Styler schafft dafür auf durchdachte Weise Abhilfe.

MoistureProtect
Hair Straightener
Haarglätter

Manufacturer
Royal Philips, Eindhoven, Netherlands
In-house design
Web
www.philips.com

The hair straightener MoistureProtect ensures smooth styling without drying out the hair unnecessarily. An innovative sensor controls the heat and thus maintains the natural hair moisture. Elastic ceramic plates apply slight pressure reducing the risk of hair breakage. An ionisation feature prevents electrostatic charge. Its reduced design with a hidden hinge and self-explanatory functions makes the device visually appealing.

Statement by the jury
MoistureProtect allows giving hair a fashionable, smooth look without drying it out – an innovative solution for a common problem.

Der Haarglätter MoistureProtect sorgt für glattes Styling, ohne das Haar unnötig auszutrocknen. Ein innovativer Sensor kontrolliert die Wärme und erhält so die natürliche Feuchtigkeit der Haare. Federnde Keramikplatten üben geringen Druck aus, was die Gefahr des Haarbruchs reduziert. Eine Ionisierungsfunktion verhindert statische Aufladung. Optisch besticht das Gerät durch reduzierte Formgebung mit verstecktem Gelenk und selbsterklärenden Funktionstasten.

Begründung der Jury
MoistureProtect stylt das Haar zu einem modisch glatten Look, ohne es auszutrocknen – hier wurde die innovative Lösung für ein weit verbreitetes Problem gefunden.

Series 9000
Hairclipper
Haarschneider

Manufacturer
Royal Philips, Eindhoven, Netherlands
In-house design
Web
www.philips.com

This hairclipper immediately attracts attention owing to its ergonomic construction. The digital display in subtle blue is both user-friendly and modern. The desired length is selected from 400 options via a digital swipe feature. An intelligent memory recalls the last cutting length and three other settings. Cutting elements made of stainless steel ensure reliable results and make the product very durable.

Statement by the jury
With its ergonomically advanced design the hairclipper proves to be a precisely working and visually pleasing accessory in bathrooms.

Der Haarschneider fällt sofort wegen seiner ergonomischen Bauweise ins Auge. Benutzerfreundlich und modern zugleich wirkt die digitale Anzeige in dezentem Blau. Über eine Digital-Swipe-Funktion wird aus 400 Möglichkeiten die gewünschte Länge gewählt. Ein intelligenter Speicher merkt sich die letzte Schnittlänge sowie drei weitere Einstellungen. Schneideelemente aus Edelstahl sorgen für zuverlässige Ergebnisse und machen das Produkt besonders langlebig.

Begründung der Jury
In seiner ergonomisch ausgereiften Bauweise erweist sich der Haarschneider als präzise funktionierendes und gestalterisch ansprechendes Accessoire im Badezimmer.

Povos PS508
Razor
Rasierapparat

Manufacturer
Loe Design, Shanghai, China
In-house design
Web
www.loedesign.com

With its slim silhouette of only 18 mm, the minimalist design of the razor catches the eye. The reduced form of a slender cuboid with rounded edges suggests a modern communication medium. The aluminium of the all-metal body originates from aviation technology and is particularly wear-resistant. The round control button with smart-click function and subtle, blue background lighting underlines the razor's elegant technological appearance.

Statement by the jury
High-quality metal from the aircraft industry and a slim-shaped silhouette emphasise the technological qualities of Povos PS508.

Der minimalistisch gestaltete Rasierapparat fällt durch seine schmale Silhouette von nur 18 mm ins Auge. Die reduzierte Form eines schlanken Quaders mit abgerundeten Kanten erinnert an ein modernes Kommunikationsmittel. Das Aluminium des Ganzmetallkörpers entstammt der Flugzeugtechnologie und ist dadurch besonders verschleißfest. Der runde Schaltknopf mit Smart-Click-Funktion und dezenter blauer Hintergrundbeleuchtung betont die elegante technologische Anmutung.

Begründung der Jury
Hochwertiges Metall aus der Flugzeugindustrie und eine schmal gestaltete Silhouette unterstreichen die technologisch ausgefeilten Qualitäten von Povos PS508.

Lumea Essential
IPL Hair Removal System
IPL-Haarentfernungssystem

Manufacturer
Royal Philips, Eindhoven, Netherlands
In-house design
Web
www.philips.com

Lumea Essential makes for hair free skin with the IPL (Intensed Pulsed Light) hair removal system. After four treatments at intervals of two weeks, hair is reduced by 75 per cent. The unit works gently and pain-free with five intensity levels depending on the type of skin. A skin tone sensor suppresses the flash if the skin is very dark, thus guaranteeing high safety. There are several attachments for different parts of the body.

Statement by the jury
Lumea Essential reliably provides for hair reduction by up to 75 per cent and ensures smooth and safe operation.

Lumea Essential sorgt für haarfreie Haut mittels IPL-Haarentfernungssystem. Es wird eine Haarreduzierung von 75 Prozent nach vier Behandlungen im Abstand von zwei Wochen erreicht. Das Gerät arbeitet besonders schmerzfrei und sanft in fünf Intensitätsstufen je nach Hauttyp. Für hohe Sicherheit unterbindet ein Hauttonsensor den Lichtblitz bei zu dunkler Haut. Es gibt mehrere Aufsätze für unterschiedliche Körperpartien.

Begründung der Jury
Lumea Essential sorgt zuverlässig für Haarreduzierung um bis zu 75 Prozent und gewährleistet ein sanftes und sicheres Arbeiten.

VisaCare SC6240
Microdermabrasion Device
Mikrodermabrasionsgerät

Manufacturer
Royal Philips, Eindhoven, Netherlands
In-house design
Web
www.philips.com

VisaCare SC6240 brings skin rejuvenation treatment in beauty parlour quality into one's own home. The microdermabrasion device combines the dual action feature exfoliating and vacuum massage to remove dead skin cells. With its high-quality materials and a haptically fine surface, it is a very feminine product that creates an elegant and sophisticated impression in cosmetic shelves.

Statement by the jury
The microdermabrasin device impresses through both its feminine design and its functionality.

Mit VisaCare SC6240 ist die Hautverjüngungsbehandlung in Salonqualität im eigenen Zuhause möglich. Das Mikrodermabrasionsgerät kombiniert die Doppelfunktion Peeling und Vakuummassage, um abgestorbene Hautzellen zu entfernen. Mit seinen hochwertigen Materialien und einer haptisch feinen Oberfläche handelt es sich um ein sehr feminines Produkt, das im Kosmetikregal einen eleganten und qualitätvollen Eindruck verbreitet.

Begründung der Jury
Das Mikrodermabrasionsgerät überzeugt sowohl durch seine feminin inspirierte Ausgestaltung als auch durch seine Funktionalität.

VisaPure Advanced SC5370
Home Facial Device
Gesichtspflegegerät

Manufacturer
Royal Philips, Eindhoven, Netherlands
In-house design
Web
www.philips.com

VisaPure Advanced SC5370 unites the experience of dermatology and Japanese facial massage in a tool to clean and care for the facial skin. The 3-in-1 device combines a cleansing brush, a Fresh Eyes head and a revitalising massage head. Due to an intelligent brush head recognition technology via NFC tag, VisaPure Advanced controls a Dual-Motion program with specific rotation, vibration and program duration that is adapted to each attachment.

Statement by the jury
VisaPure Advanced SC5370 combines the knowledge of dermatology and Japanese facial massage in a single unit that stands out by innovative and functional features.

VisaPure Advanced SC5370 vereint Erfahrungen der Dermatologie und der japanischen Gesichtsmassage zu einem Instrument für die Reinigung und Pflege der Gesichtshaut. Das 3-in-1-Gerät kombiniert eine Reinigungsbürste, einen Fresh-Eyes- und einen belebenden Massageaufsatz. Dank intelligenter Bürstenkopferkennung mittels NFC-Tag steuert das Gerät ein für jeden Aufsatz angepasstes DualMotion-Programm mit spezieller Rotation, Vibration und Programmdauer.

Begründung der Jury
VisaPure Advanced SC5370 vereint Erkenntnisse der Dermatologie und der japanischen Gesichtsmassage in einem Gerät, das sich durch Innovation wie Funktionalität auszeichnet.

Breo iScalp
Scalp Massager
Kopfhaut-Massagegerät

Manufacturer
Shenzhen Breo Technology Co., Ltd.,
Shenzhen, China
In-house design
Paul Cohen
Web
www.breo.com.cn

The ergonomic massaging device features six techniques of the traditional Chinese massage. The unit promotes blood circulation, drives off fatigue, alleviates itching and creates the pleasant and relaxing feeling of a professional treatment. The four attachments made of silicone can be removed and cleaned or replaced by other elements. The waterproof massager can be used in the bathtub, in the shower or in combination with massage oil and other wellness products.

Das ergonomische Gerät bietet sechs Techniken der traditionellen chinesischen Massage. Es fördert die Blutzirkulation, vertreibt Müdigkeit, lindert Juckreiz und bereitet das entspannende und komfortable Gefühl einer professionellen Behandlung. Die vier Aufsätze aus Silikon können abgenommen und gereinigt oder durch andere Elemente ersetzt werden. Die wasserfeste Ausführung erlaubt eine Anwendung in der Wanne, unter der Dusche oder zusammen mit Massageöl oder anderen Pflegeprodukten.

Statement by the jury
With its simple and ergonomically comfortable design, Breo iScalp allows to experience the soothing techniques of traditional Chinese massage at home.

Begründung der Jury
In schlichter und ergonomisch komfortabler Gestaltung macht Breo iScalp wohltuende Techniken der chinesischen Massage zu Hause erlebbar.

309

Braun Face
Facial Epilator and Cleansing Brush
Gesichtsepilierer und Reinigungsbürste

Manufacturer
Procter & Gamble Manufacturing GmbH,
Kronberg im Taunus, Germany
In-house design
Braun Design Team
Web
www.braun.com

The facial epilator Braun Face gently removes very fine hair by the root on upper lip and eyebrows. Its sleek and compact design allows handling it like mascara. The extra-slim, precise head is arranged in a free and perpendicular position to the device so that the area to be epilated is clearly visible. By clockwise and counter-clockwise rotation the hair is easily captivated and removed. An additional brush head cleanses and exfoliates the skin.

Der Braun Face Gesichtsepilierer entfernt feinste Haare an Oberlippe und Augenbrauen sanft an der Wurzel. Durch sein schlankes und kompaktes Design lässt er sich ähnlich wie Mascara handhaben. Der besonders kleine, präzise Kopf steht frei und senkrecht zum Gerät, was die Sicht auf den Epilationsbereich freigibt. Durch Rechts- und Linkslauf werden die Haare optimal erfasst und entfernt. Ein zusätzlicher Bürstenkopf reinigt und peelt die Haut.

Statement by the jury
Braun Face easily and reliably exfoliates the facial area. Thanks to its slim and compact design it disappears discreetly in a purse.

Begründung der Jury
Braun Face epiliert den Gesichtsbereich zuverlässig und unkompliziert. Durch seine schlanke und kompakte Form verschwindet er diskret in jeder Handtasche.

Pureo Deep Cleansing FC 95
Facial Cleansing Brush
Gesichtsreinigungsbürste

Manufacturer
Beurer GmbH, Ulm, Germany
Design
Pulse Design (Matthias Kolb),
Munich, Germany
Web
www.beurer.com
www.pulse-design.de

The facial cleansing brush has a handy size and is used for daily gentle and pore refining cleansing. The result outperforms any manual care. At the same time, blood circulation is stimulated to make the skin smooth and tender. The device being waterproof, it is suitable for use in the shower or bathtub. Four different brush attachments are available with either circular or oscillating rotation, each with three speed settings.

Statement by the jury
Pureo Deep Cleansing FC 95 proves to be innovative and functional: it leaves the facial skin deep cleaned and glowing as in a beauty parlour – achieved in the private bathroom.

Die Gesichtsreinigungsbürste im handlichen Format dient der täglichen sanften und porentiefen Reinigung. Das Ergebnis fällt gründlicher aus als bei manueller Pflege. Gleichzeitig wird die Durchblutung angeregt, so soll die Haut geschmeidig und zart werden. Das Gerät ist wasserfest und kann daher unter der Dusche und in der Badewanne verwendet werden. Es stehen vier unterschiedliche Bürstenaufsätze zur Verfügung, die jeweils kreisförmig oder oszillierend in drei Geschwindigkeiten rotieren.

Begründung der Jury
Pureo Deep Cleansing FC 95 beweist sich als innovativ und funktional: Porentief gereinigte, strahlende Gesichtshaut wie vom Kosmetikinstitut wird im eigenen Bad erreicht.

Pureo Derma Peel FC 100
Microdermabrasion Device
Mikrodermabrasionsgerät

Manufacturer
Beurer GmbH, Ulm, Germany
Design
designflow (Daniel Nusser),
Aalen, Germany
Web
www.beurer.com
www.designflow.de

Pureo Derma Peel FC 100 permits microdermabrasion at home. The round, flat case opens by pressing an illuminated button. In addition to the material properties, the generous arrangement of the components inside enhances the high-quality impression. Three sapphire-coated top parts are available. An integrated mirror, an indicator lamp for speed indication and replacement filters complete the set.

Statement by the jury
Professional microdermabrasion made possible at home. After use, a discreetly designed case hides all high-quality components.

Pureo Derma Peel FC 100 ermöglicht die Mikrodermabrasion zu Hause. Das runde, abgeflachte Etui wird durch Druck auf einen beleuchteten Knopf geöffnet. Die großzügige Anordnung der Bestandteile im Inneren unterstützt neben der Materialbeschaffenheit den hochwertigen Eindruck; es stehen drei saphirbeschichtete Aufsätze zur Verfügung. Ein integrierter Spiegel, eine Kontrollleuchte zur Geschwindigkeitsanzeige und Ersatzfilter komplettieren das Set.

Begründung der Jury
Professionelle Mikrodermabrasion wird zuhause möglich. Ein dezent gestaltetes Etui verbirgt danach gekonnt alle hochwertig verarbeiteten Bestandteile.

Sonic Black Whitening
Sonic Toothbrush
Schallzahnbürste

Manufacturer
megasmile AG, Teufen, Switzerland
In-house design
Dr. Roland Zettel
Design
Fabienne Meyer/AIM studio GmbH,
Zürich, Switzerland
Web
www.megasmile.com

The Sonic Black Whitening removes
dental discolouration and plaque. The
surface of the teeth is polished very
thoroughly by black active charcoal
particles, which are released from the
brush head with up to 45,000 vibra-
tions per minute. Thanks to its memory
function, the toothbrush "remembers"
the last selected setting. The hand piece
consists of soft, slip-proof material
and has an luscious jet-black finish.
It features a lithium-ion battery and
a USB port.

Statement by the jury
Due to an innovative technology, the
functionality of Sonic Black Whitening
is impressive. Its dark black appearance
complies with high aesthetic demands.

Die Sonic Black Whitening entfernt Zahn-
verfärbungen sowie Belag und poliert
Zähne mittels schwarzer Aktivkohle-
partikel, die der Bürstenkopf mit bis zu
45.000 Schwingungen pro Minute
freisetzt. Diese polieren die Oberflächen
zusätzlich besonders gründlich. Dank
Memoryfunktion „erinnert" sich die Zahn-
bürste an die zuletzt gewählte Anwendung.
Das Handstück besteht aus weichem,
rutschfestem Material. Das Gerät ist in
sattem Nachtschwarz gehalten und mit
Lithium-Ionen-Akku sowie einem USB-
Anschluss ausgestattet.

Begründung der Jury
Sonic Black Whitening erweist sich durch
innovative Technik als überzeugend funktionell. Das nachtschwarze Erscheinungsbild
genügt hohen ästhetischen Ansprüchen.

SHAPL
Smart Travel Containers
Reiseset für Körperpflegeprodukte

Manufacturer
SHAPL, Seoul, South Korea
In-house design
Changsoo Jin
Web
www.shapl.com

The travel set for body care products provides a very functional solution for unfamiliar bathrooms. One's own products are filled in transparent bottles made of recyclable material. Their opening reliably seals and faces downwards so that even very small quantities can be used. The individual containers can be attached to each other. Thus, no toiletry bag is required to carry them into a bathroom or a lavatory. The containers need not be stored on the floor, in shelves or on a seat while caring for the body, but are simply and hygienically attached to the wall by means of suction cups.

Dieses Reiseset für Körperpflegeprodukte ist sehr praktisch beim Benutzen fremder Badezimmer. Die eigenen Produkte werden in transparente Behälter aus recyclebaren Materialien gefüllt. Deren zuverlässig dichtende Öffnung ist nach unten gerichtet, so kann selbst auf sehr kleine Mengen zugegriffen werden. Die einzelnen Behälter können aneinandergeheftet und somit ohne Kulturbeutel in ein Badezimmer oder einen Waschraum getragen werden. Die Behältnisse müssen während der Körperpflege nicht auf dem Boden, in Regalen oder auf Sitzvorrichtungen gelagert werden, sondern bleiben mittels Saugknopf einfach und hygienisch an der Wand befestigt.

Statement by the jury
SHAPL puts an end to well-known inconveniences in unfamiliar bathrooms; the set provides hygienic solutions even in less comfortable surroundings.

Begründung der Jury
SHAPL macht Schluss mit wohlbekannten Unannehmlichkeiten in fremden Badezimmern und bietet Hygiene auch in weniger komfortabler Umgebung.

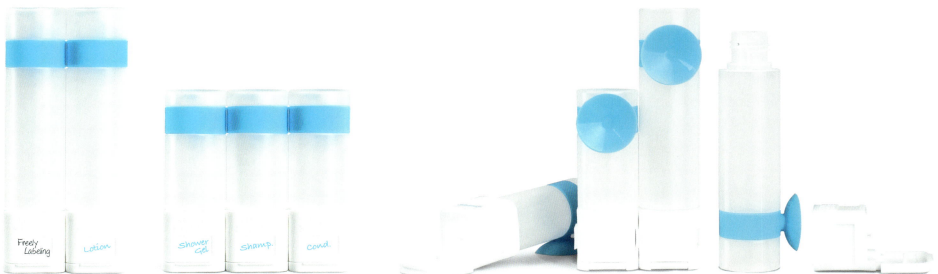

PPW10
Scale
Waage

Manufacturer
Robert Bosch Hausgeräte GmbH,
Munich, Germany
In-house design
Helmut Kaiser, Yvonne Janet Weisbarth
Web
www.bosch-hausgeraete.de

Clean lines, an elegant colour scheme and the use of high-quality Swarovski crystals lend the scales exclusivity in bathroom or bedroom. A light touch activates the patent-pending display solution, making it an eye-catcher. Blue LEDs backlight selected crystal elements and put an atmospheric spotlight on the measured values. A surface made of premium shockproof glass protects the display element and guarantees perfect hygienic conditions.

Mit einer klaren Linienführung, eleganter Farbgebung und durch die Verwendung von qualitätvollen Swarovski-Kristallen vermittelt die Waage Exklusivität in Bad oder auch Schlafzimmer. Durch simple Berührung aktiviert, wird die zum Patent angemeldete Anzeigelösung zum Blickfang. Blaue LEDs hinterleuchten ausgewählte Kristallelemente und setzen Messwerte atmosphärisch in Szene. Eine Oberfläche aus hochwertigem stoßfestem Glas schützt das Anzeigeelement und garantiert darüber hinaus optimale Hygiene- und Reinigungseigenschaften.

Statement by the jury
High-quality material and sophisticated technology make the scale a stylish home accessory, also outside of the bathroom.

Begründung der Jury
Hochwertige Ausführung und ausgeklügelte Technik machen die Waage zum edlen Wohnaccessoire auch außerhalb des Badezimmers.

BORK N701
Scale
Waage

Manufacturer
Bork-Import LLC, Moscow, Russia
In-house design
Web
www.bork.com

The highly robust appearance of Bork N701 promises to bear any weight without changing its form. The square surface of black anodised aluminium constitutes a visually compact unit and is smooth to touch. The top is only interrupted by a discreet bar at the upper edge, into which an unobtrusive stylish display with blue LEDs is incorporated.

Statement by the jury
The elegant square design imposingly shows that Bork N701 provides high-capacity weighing.

In ihrem äußerst soliden Erscheinungs-bild verspricht Bork N701 jedem Gewicht standzuhalten, ohne sich zu verbiegen. Die quadratische Oberfläche aus schwarz eloxiertem Aluminium wirkt als optisch kompakte Einheit, die auch eine angeneh-me Haptik bietet. Unterbrochen wird die Fläche nur durch eine unauffällige Leiste an der Oberkante, in die ein modisch-dezentes Display aus blauen LEDs eingearbeitet ist.

Begründung der Jury
Bork N701 unterstreicht in ihrer edlen quadratischen Gestaltung überzeugend, dass sie auch sehr hoher Belastung standhält.

ITO
Health Index Scale
Gesundheitswaage

Manufacturer
Shenzhen HCOOR Technology Co., Ltd., Shenzhen, China
Design
Shenzhen Bifoxs Industrial Design Co., Ltd. (Yuebing Chen, Baomin Wang), Shenzhen, China
Web
www.hcoor.com
www.bifoxs.com

ITO is used like common bathroom scales. By analysing the soles of the feet by means of a conductive surface, the device evaluates different indicators. This allows users to draw simple conclusions about their state of health. About 20 values, such as body fat, water, and muscle and bone mass, are measured. An app processes all information into an impedance analysis and provides for a better health management.

Statement by the jury
The whole family participates in health management in a very straightforward manner. Thanks to a modern app, all data are clearly evaluated.

ITO wird wie eine normale Körperwaage verwendet. Durch Analyse der Fußsohlen mittels einer leitfähigen Oberfläche wertet das Gerät verschiedene Indikatoren aus, die dem Benutzer einfache Rückschlüsse auf seinen Gesundheitszustand ermöglichen. Gemessen werden rund 20 Werte, etwa der Anteil an Körperfett und Wasser, Muskel- und Knochenmasse. Eine App verarbeitet alle Informationen zu einer Impedanzana-lyse und ermöglicht ein besseres Gesund-heitsmanagement.

Begründung der Jury
Gesundheitsmanagement für die ganze Familie funktioniert hier auf unkomplizierte Weise. Dank innovativer App werden alle Daten übersichtlich ausgewertet.

yutanpo
Hot-Water Bottle
Wärmflasche

Manufacturer
ceramic japan Co., Ltd.,
Seto City, Aichi Prefecture, Japan
Design
Masahiro Minami Design (Masahiro Minami),
Shiga, Japan
Web
www.ceramic-japan.co.jp
www.masahiro-minami.com

The shape of the ceramic hot-water bottle meets high aesthetic and functional requirements. The flat design without any unevenness nestles comfortably to the body. The lid is inserted in the middle of the object. The material ensures that the skin of the user is not dried out. The hot-water bottle holds the temperature well so that the heat fades out only in the morning hours. The same water can be reheated in the form after removing the lid in the microwave. The sustainable option takes into account the shortage of resources in Japan after the earthquake of 2011.

Die Wärmflasche aus Keramik genügt in ihrer Formgebung hohen ästhetischen wie funktionalen Ansprüchen. Die flache Gestaltung ohne Unebenheiten schmiegt sich angenehm an den Körper. Der Deckel ist in der Mitte des Objekts eingelassen. Das verwendete Material sorgt dafür, dass die Haut des Benutzers nicht austrocknet. Die Wärmflasche hält auch die Temperatur gut, sodass die Wärme erst in den Morgenstunden abklingt. Dasselbe Wasser lässt sich in dem Gefäß mit geöffnetem Deckel in der Mikrowelle wieder erhitzen. Diese nachhaltige Option berücksichtigt die Ressourcenknappheit in Japan nach dem Erdbeben von 2011.

Statement by the jury
The hot-water bottle yutanpo makes an efficient contribution to protecting Japanese electricity and water resources, thus demonstrating an ecologically well thought-out design concept.

Begründung der Jury
Die Wärmflasche yutanpo leistet einen effizienten Beitrag zur Schonung japanischer Strom- und Wasserressourcen und lässt damit ein ökologisch durchdachtes Gestaltungskonzept erkennen.

Auxiliary Evaporator R
Zusatzverdampfer R

Manufacturer
sentiotec GmbH, Regau, Austria
Design
dbgp (Peter Groiss), Laakirchen, Austria
Web
www.sentiotec.com
www.dbgp.at

The design of the auxiliary evaporator R is characterised by a high-quality appearance, with an exterior consisting of brushed and powder-coated stainless steel. The cover shell for fragrance oils and herbs is made of dark glazed ceramic and is suitable for dishwasher use. The device has an evaporator capacity of 2.5 kW, contributing to a comfortable climate in the sauna. The auxiliary evaporator can be used alone or with an extension module.

Statement by the jury
The auxiliary evaporator R convinces thanks to the use of high-grade materials. Whether all by itself or as an auxiliary device, it cuts a good figure.

Wertigkeit prägt die Gestaltung des Zusatzverdampfers R. Seine Außenteile bestehen aus gebürstetem sowie pulverbeschichtetem Edelstahl. Die Abdeckschale für Duftöle und Kräuter ist aus dunkel glasierter Keramik gefertigt und spülmaschinengeeignet. Das Gerät hat eine Verdampferleistung von 2,5 kW, dies sorgt in der Sauna für ein angenehmes Klima. Der Zusatzverdampfer kann sowohl allein als auch als Erweiterungsmodul verwendet werden.

Begründung der Jury
Der Zusatzverdampfer R überzeugt durch die Verarbeitung edler Materialien und macht eine gute Figur, ob allein verwendet oder als Zusatzgerät.

Huum Drop
Sauna Stove
Saunaofen

Manufacturer
Huum OÜ, Tartu, Estonia
Design
Mihkel Masso (Joon DB), Tallinn, Estonia
Web
www.huum.eu
www.masso.format.com
Honourable Mention

The sauna stove with its modern Nordic design is heated up by a high-tech, app-controlled electronic system ensuring constant high humidity. Thanks to its minimalist construction, there is much room for the organically shaped stones. The oval form of the metal coat is inspired by the idea of an internally heated rib cage or arms that gently embrace a warm body. The stove is suitable for saunas up to 20 cbm.

Statement by the jury
The app-controlled, cutting-edge technology of this stove makes a beneficial sauna experience a simple matter.

Der Saunaofen in moderner nordischer Gestaltung heizt mit einem hochtechnologischen elektronischen System samt Steuerung via App, das stets hohe Luftfeuchtigkeit garantiert. Dank seiner minimalistischen Bauweise bleibt viel Platz für die organisch geformten Steine. Die ovale Form des Metallmantels ist inspiriert vom innerlich erwärmten Brustkorb bzw. von Armen, die sanft einen warmen Körper umfangen. Der Ofen ist geeignet für Saunen bis zu 20 cbm.

Begründung der Jury
Modernste Technologie, über eine App steuerbar, macht das wohltuende Saunaerlebnis zu einer ganz unkomplizierten Angelegenheit.

Bedroom lamps	Außenleuchten
Built-in lighting	Deckenleuchten
Ceiling lighting	Downlights
Desk lamps	Einbauleuchten
Downlights	LED-Leuchten
Floor lamps	Leseleuchten
LED lamps	Leuchtmittel
Light bulbs and tubes	Lichtsysteme
Lighting systems	OLED-Leuchten
Living room lamps	Pendelleuchten
OLED lamps	Schlafraumleuchten
Outdoor lighting	Schreibtischleuchten
Pendant luminaires	Stehleuchten
Reading lamps	Strahler
Spotlights	Taschenlampen
Table lamps	Tischleuchten
Torches	Wandleuchten
Wall lamps	Wohnraumleuchten

Lighting and lamps
Licht und Leuchten

LumiStreet
LED Street Light
LED-Straßenleuchte

Manufacturer
Royal Philips,
Eindhoven, Netherlands

In-house design
Royal Philips

Web
www.philips.com

reddot award 2015
best of the best

In a new light

Street lights represent an enormous economic burden for communities. Yet, retrofitting street lights to much more energy-efficient LED lights is an economically as well as environmentally friendly option. Since the LumiStreet street light was developed with communities particularly in mind, the concept well reflects their unique needs. The aim is to transform open spaces by adding a modern and energy-efficient lighting system. The street light therefore features a compact design that showcases a purism of convincing and attractive appearance. Equipped with highly efficient LED technology, it constitutes an innovative, low-cost solution particularly for roads and streets in residential areas and impresses with a well thought-through functionality. The street light is easy to install and ensures exceptionally low maintenance costs. It is manufactured from high-quality and robust materials that make all elements highly durable and wear resistant. This ensures that the benefits from an initial installing last over the longest possible timeframe. As such, they embody a well thought-through solution for communities to plan their street lighting in a future-oriented manner. Once installed, LumiStreet exudes an air of elegance that gives it a timeless appearance.

In neuem Licht

Die Straßenbeleuchtung stellt für die Kommunen eine enorme wirtschaftliche Belastung dar. Eine Umrüstung auf wesentlich energieeffizientere LED-Leuchten ist dabei eine ökonomische wie auch sehr umweltfreundliche Variante. Da sie nahe an dieser Situation der Kommunen entwickelt wurde, spiegelt das Konzept der Straßenleuchte LumiStreet deren Bedürfnisse gut wider. Das Ziel ist es, die Straßenräume mit einer moderneren und energieeffizienteren Beleuchtungsanlage neu zu gestalten. Diese Straßenleuchte ist deshalb kompakt konzipiert, wobei ihr ein überzeugender und attraktiv anmutender Purismus zu eigen ist. Ausgestattet mit einer hocheffizienten LED-Technologie bietet sie insbesondere für die Straßen in Wohngebieten eine innovative, kostengünstige Lösung, die zudem durch ihre zu Ende gedachte Funktionalität beeindruckt. Sie ist leicht zu installieren und auch die späteren Wartungskosten halten sich in einem ausgesprochen kostengünstigen Rahmen. Gefertigt wird sie aus hochwertigen und robusten Materialien, weshalb alle Elemente sehr langlebig und verschleißarm sind. Auf diese Weise können die mit der Umstellung verbundenen Vorteile über den längstmöglichen Zeitraum garantiert werden. Den Kommunen bietet sie so eine sehr gut durchdachte Möglichkeit, ihre Straßenbeleuchtung zukunftsweisend zu planen. Im Straßenbild wirkt LumiStreet dabei elegant und überaus zeitlos.

Statement by the jury

The LumiStreet street light merges form and technology in perfect balance. Conceived as an effective solution for modern LED light retrofitting, it aestheticises the scenery of public spaces and lends them a strong identity. The LumiStreet is easy and affordable to install, is made of robust materials and highly durable. As a lighting concept it is suitable for communities as a consistent and well-designed option for future planning.

Begründung der Jury

Form und Technologie befinden sich bei der Straßenleuchte LumiStreet in perfekter Balance. Konzipiert als eine effektive Lösung für die zeitgemäße Umrüstung auf LED-Beleuchtung ästhetisiert sie das Straßenbild und verleiht ihm zugleich eine eigene Identität. Die LumiStreet ist einfach und kostengünstig zu installieren, besteht aus robusten Materialien und ist langlebig. Den Kommunen gibt dieses Beleuchtungskonzept eine schlüssige wie auch gut gestaltete Alternative an die Hand.

Designer portrait
See page 52
Siehe Seite 52

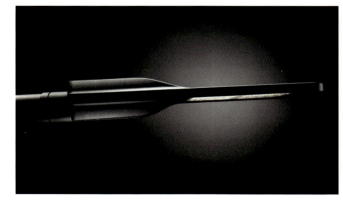

Prona
Smart Street Light
Intelligente Straßenleuchte

Manufacturer
Prona, Shanghai, China
Design
Shanghai Moma
Industrial Product Design Co., Ltd.
(Gao Yichen, Li Yukang), Shanghai, China
Web
www.prona.cc
www.designmoma.com

A multifunctional design concept divides Prona into three functional areas. The top section features a high-performance solar-powered LED light source, which features a photovoltaic system on top of the street light. In the middle section, a display shows multimedia information and advertising. The interactive screen in the lower part of the lamp provides real time information on four different topics: social community networks, business networks, points of interest and industrial areas.

Statement by the jury
This purist street light not only answers to safety aspects but also to modern information needs – a successful implementation.

Ein multifunktionales Gestaltungskonzept unterteilt Prona in drei Funktionsbereiche: Im oberen Teil befindet sich eine leistungsstarke LED-Lichtquelle, die mittels Solarenergie betrieben wird. Ihr Photovoltaikgenerator wurde direkt auf der Leuchte platziert. Im Mittelteil zeigt ein Bildschirm multimediale Informationen und Werbung. Der interaktiv bedienbare Bildschirm im unteren Teil des Mastes informiert hingegen in Echtzeit über vier Themenbereiche: Stadt, Geschäftswelt, Sehenswürdigkeiten sowie Gewerbegebiete.

Begründung der Jury
Diese puristisch gestaltete Straßenleuchte wird Sicherheitsaspekten gerecht und erfüllt zeitgemäße Informationsbedürfnisse – eine überzeugende Umsetzung.

CFT500
Post Top Luminaire Series
Mastaufsatzleuchten-Serie

Manufacturer
WE-EF LEUCHTEN GmbH & Co. KG,
Bispingen, Germany
In-house design
Petra Denst
Web
www.we-ef.com

The CFT500 is a post top luminaire series with a circular, streamlined design made of corrosion resistant die-cast aluminium. Its symmetric or rectangular light distribution is used for lighting public spaces, pedestrian zones and parks. The series uses a unique lens technology, where every single LED illuminates the entire area. In addition, a proprietary optical component increases the efficiency of the LEDs. In combination with the quadrant-shaped LED boards, this achieves a homogeneous light distribution.

Statement by the jury
This functionally sophisticated post top luminaire series offers a high practical value. It owes its objective appearance to a precise geometry.

Die CFT500 ist eine ringförmige, reduziert gestaltete Mastaufsatzleuchten-Serie aus korrosionsbeständigem Aluminium-guss. Ihre symmetrische oder rechteckige Lichtlenkung dient der Beleuchtung von Plätzen, Fußgängerzonen und Parks. Die Serie nutzt eine spezifische Linsentechnik, bei der jede einzelne LED das gesamte Bewertungsfeld ausleuchtet. Zudem erhöht eine herstellereigene Optikkomponente den Wirkungsgrad der LEDs. In Verbindung mit den viertelkreisförmigen LED-Platinen entsteht eine homogene Lichtverteilung.

Begründung der Jury
Einen hohen Gebrauchswert bietet diese funktional ausgereifte Mastaufsatz-leuchten-Serie. Ihre sachliche Anmutung verdankt sie ihrer präzisen Geometrie.

Keen
LED Floodlight
LED-Fluter

Manufacturer
Simes, Corte Franca (Brescia), Italy
In-house design
Web
www.simes.com

Keen is an innovative development of a floodlight for outdoor lighting. A special double joint allows the projector to take infinite positions for different light effects. Thanks to a special anchor base, the luminaires can be mounted on poles of different diameters, 60 mm or more.

Statement by the jury
This LED floodlight combines a high ease of use with a subtle form language – a convincing product solution.

Keen ist eine innovative Weiterentwicklung eines Fluters für die Außenbeleuchtung. Ein speziell entwickeltes Doppelgelenk ermöglicht es, den LED-Fluter in jede beliebige Richtung zu schwenken und somit im individuellen Projekt eine bestmögliche Ausleuchtung zu erreichen. Mittels der speziell geformten Befestigungsplatte können die Keen-Strahler an Masten mit einem Durchmesser von 60 mm und mehr befestigt werden.

Begründung der Jury
Dieser LED-Fluter verbindet einen hohen Bedienkomfort mit einer dezenten Formensprache – eine überzeugende Produktlösung.

Skill Square
Skill Quadratisch
Wall Mounted Luminaire
Wandaufbauleuchte

Manufacturer
Simes, Corte Franca (Brescia), Italy
In-house design
Web
www.simes.com

Skill Square is an outdoor orientation-lighting system with a minimalist form language. The housing, with an exterior height of only 3 cm, harmoniously blends with modern architecture. Alternatively, a special in-wall housing allows for an entirely flush installation. The high light output and uniform light distribution of the LED light guarantees the convenient use in staircases and along paths. Skill Square is part of a variably combinable product range.

Als Orientierungsbeleuchtung für den Außenbereich zeigt Skill Quadratisch eine minimalistische Formensprache. Das Gehäuse mit einer Aufbauhöhe von nur 3 cm fügt sich in eine zeitgemäße Architektur harmonisch ein. Alternativ ermöglicht ein spezielles Wandeinbaugehäuse eine komplett flächenbündige Montage. Eine hohe Lichtausbeute sowie eine gleichmäßige Lichtstreuung der LED-Leuchte stellen eine komfortable Nutzung von Treppen und Wegen sicher. Skill Quadratisch ist Bestandteil eines variabel kombinierbaren Produktprogramms.

Statement by the jury
Plain, geometrical lines underline the sophisticated functionality of this wall mounted luminaire for outdoor application.

Begründung der Jury
Eine schlichte, geometrische Linienführung untermalt die ausgereifte Funktionalität dieser Wandaufbauleuchte für den Outdoor-Bereich.

Slot LED
Recessed Wall Luminaire
Wandeinbauleuchte

Manufacturer
Paulmann Licht GmbH,
Springe-Völksen, Germany
In-house design
Web
www.paulmann.com

Slot LED is an IP44 recessed wall luminaire with a minimalistic design – especially conceived as a functional lighting for outdoor use. The angular application of the LED creates a precise, non-reflecting light effect on the floor. Another characteristic feature is the screwless installation: the mounting box is simply inserted into the façade of the building. The compact luminaire, made of cast aluminium, is optionally available with a matt white or anthracite powder coating.

Statement by the jury
Thanks to its convenient installation and its high level of efficiency, the Slot LED impresses as a functionally well thought-through wall light.

Slot LED ist eine minimalistisch gestaltete IP44 Wandeinbauleuchte, die speziell als Funktionsbeleuchtung für den Außenbereich konzipiert ist. Durch ihren schrägen Einsatz, auf dem die LED sitzt, wird das Licht präzise und blendfrei auf den Boden gelenkt. Eine weitere Besonderheit liegt in der schraubenlosen Montage, wobei die Montagebox einfach in die Außenfassade des Gebäudes eingesetzt wird. Die kompakte Leuchte aus Aluminiumguss ist optional mit mattweißer oder anthrazitfarbener Pulverlackierung erhältlich.

Begründung der Jury
Dank seiner komfortablen Montage und seines hohen Wirkungsgrades überzeugt Slot LED als funktional durchdachte Wandeinbauleuchte.

Plate Out
Wall-Mounted LED Luminaire
LED-Wandleuchte

Manufacturer
Eureka Lighting, Montreal, Canada
In-house design
Web
www.eurekalighting.com

The Plate Out wall-mounted luminaire is suitable for both interior and exterior applications. Its slim body encompasses a diffused 16 watts indirect LED light source. Furthermore, it integrates intelligent controls for dimming and automatic thermal regulation. The housing of the Plate Out is made from solid aluminium. Designed to IP65 standards, this lighting fixture can stand up to extreme weather conditions, as well as physical abuse, and has been designed for an service life of more than 20 years.

Statement by the jury
A purist design characterises this wall-mounted luminaire, the indirect light of which yields an appealing contrast to the geometrical contours.

Die Wandleuchte Plate Out ist sowohl für eine indirekte Innen- als auch Außenbeleuchtung geeignet. Ihr schmaler Korpus umschließt eine weit streuende 16-Watt-LED-Lichtquelle. Zudem wurde eine intelligente Steuerung zum Dimmen und zur automatischen Wärmeregulierung integriert. Das Gehäuse der Plate Out ist aus solidem Aluminium gefertigt. Die nach IP65-Normen gestaltete Leuchte kann extremen Wetterbedingungen sowie Gewaltanwendung standhalten und wurde für eine Lebensdauer von mehr als 20 Jahren konzipiert.

Begründung der Jury
Eine puristische Linienführung zeichnet diese Wandleuchte aus, deren indirektes Licht einen reizvollen Kontrast zur geometrischen Kontur schafft.

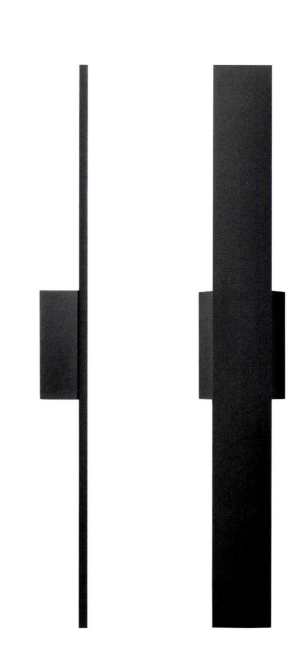

Playbulb Garden
Solar Garden Lamp
Solar-Gartenleuchte

Manufacturer
Shenzhen Baojia Battery Technology Co., Ltd.,
Shenzhen, China
In-house design
Stanley Yeung
Web
www.mipow.com

Playbulb Garden is a wireless solar garden lamp which can be controlled by a smartphone. A free app allows users to switch the lamp on and off, or to control the brightness. In addition, there is a variety of RGBW colours and light effects to choose from as well as a timer function. The group control function allows users to control any number of Playbulbs via only one mobile device. Supplementing the solar energy, a 1,000 mAh battery lengthens the possible operating time. The garden lamp is adjustable in height by stacking either one or two holders on top of each other.

Playbulb Garden ist eine kabellose Solar-Gartenleuchte, die sich per Smartphone steuern lässt. Über eine kostenlose App wird sie an- und ausgeschaltet oder ihre Helligkeit bestimmt. Darüber hinaus kann aus diversen RGBW-Farben und Licht-effekten ausgewählt sowie eine Zeitpro-grammierung vorgenommen werden. Die Gruppensteuerungsfunktion kontrolliert beliebig viele Playbulbs über ein mobiles Steuerungsgerät. Ergänzend zum Solar-betrieb verlängert ein 1.000 mAh starker Akku die Laufzeit. Die Höhe der Garten-leuchte ist variabel, indem optional ein oder zwei Halter übereinander eingesetzt werden.

Statement by the jury
Thanks to its innovative design, this solar garden lamp allows for a variety of aesthetically-pleasing light effects.

Begründung der Jury
Aufgrund ihrer innovativen Gestaltung ermöglicht diese Solar-Gartenleuchte eine Vielzahl ästhetischer Lichteffekte.

Joe
LED Outdoor Lighting
LED-Außenleuchte

Manufacturer
Linea Light Group, Treviso, Italy
In-house design
Web
www.linealight.com

Joe is a robust and reliable path marker that is available in three different heights. At any of these heights, the LED outdoor lighting produces a grazing light that is directed to the ground, providing pedestrians and car drivers with comfortable orientation. The optical unit is encased in an extruded aluminium frame while a weather resistant polyester powder coating with protective base coat ensures long-term preservation from wear and tear.

Statement by the jury
An impressive lighting technology of high functionality is hidden inside the plain housing of the Joe LED outdoor lighting.

Joe ist eine robuste und zuverlässige Wegleuchte, die in drei unterschiedlichen Höhen erhältlich ist. In allen Ausführungen liefert die LED-Lichtquelle ein auf den Boden gerichtetes Streiflicht, das Passanten und Autofahrern eine komfortable Orientierung ermöglicht. Die optische Einheit ist in das Gehäuse aus Aluminiumspritzguss integriert. Eine witterungsbeständige Polyester-Pulverlack-Beschichtung mit schützender Oberflächenbehandlung sorgt dafür, dass das Aussehen der LED-Außenleuchte langfristig erhalten bleibt.

Begründung der Jury
Im schlichten Korpus der LED-Außenleuchte Joe verbirgt sich eine überzeugende Beleuchtungstechnologie von hoher Funktionalität.

Logico Garden
Outdoor Luminaire
Außenleuchte

Manufacturer
Artemide S.p.A., Pregnana Milanese, Italy
Design
architetto Michele De Lucchi S.r.L. (Daniele Moioli), Milan, Italy
Web
www.artemide.com
www.amdl.it

The Logico Garden outdoor luminaire was developed for garden or pedestrian walkway lighting. Its undulating support pole is made of chromed, extruded aluminium while its base, made of electro-zinc-coated and painted steel, can be fixed to the ground by means of screws or plugs. The body of the lamp is made of die-cast aluminium that is refined in three stages. The metallised polycarbonate reflectors have a shock-resistant and weather-proof projection surface.

Statement by the jury
The Logico Garden outdoor luminaire is characterised by a high practical value. Its distinctive contours achieve a powerful appeal.

Die Außenleuchte Logico Garden wurde für die Beleuchtung von Gärten oder Fußgängerwegen konzipiert. Ihre wellenförmige Stützstange besteht aus verchromtem, stranggepresstem Aluminium, wobei sich die Basis aus elektroverzinktem und lackiertem Stahl mittels Schrauben oder Dübeln im Boden befestigen lässt. Der Leuchtenkörper aus Aluminium-Druckguss wurde dreifach veredelt. Die metallisierten Polycarbonat-Reflektoren haben eine stoß- und witterungsfeste Projektionsfläche.

Begründung der Jury
Einen hohen Gebrauchswert zeichnet die Außenleuchte Logico Garden aus. Ihre charakteristische Kontur erzielt eine markante Anmutung.

iHF 3D
Motion Detector
Bewegungsmelder

Manufacturer
Steinel GmbH, Herzebrock-Clarholz, Germany
Design
Kurz Kurz Design (Dorian Kurz), Solingen, Germany
Web
www.steinel.de
www.kurz-kurz-design.de

This outdoor motion detector features an innovative high-frequency technology for precise detection. Its active sensor, with a coverage angle of 160 degrees, detects human walking movements in a radius of eight metres. Its novel 3D aerial technology allows the detection zone to be precisely set on three axes in order to cover the full area. Via signal analysis, the sensor distinguishes between moving persons and moving objects such as bushes or animals precluding the risk of switching errors.

Statement by the jury
The unobtrusive iHF 3D motion detector impresses as an innovative product solution with remarkable functionality.

Dieser Bewegungsmelder für den Außenbereich nutzt eine innovative Hochfrequenz-Technologie für eine exakte Erfassung. Sein aktiver Sensor mit einem Erfassungswinkel von 160 Grad erkennt im Radius von acht Metern menschliche Gehbewegungen. Dank einer neuartigen 3D-Antennentechnik kann der Überwachungsbereich über drei Achsen vollflächig eingestellt werden. Mittels Signalanalyse unterscheidet der Sensor zwischen sich bewegenden Personen und sich bewegenden Sträuchern oder Tieren, was Fehlschaltungen ausschließt.

Begründung der Jury
Als innovative Produktlösung überzeugt der unauffällige Bewegungsmelder iHF 3D, seine ausgereifte Funktionalität ist bemerkenswert.

Lunocs

Lamp
Leuchte

Manufacturer
Degardo, Bad Oeynhausen, Germany
In-house design
Andreas Arndt
Design
Neufeld & Stein Produktdesign
(Michael Stein), Velbert, Germany
Web
www.degardo.de
www.neufeldundstein.de

The Lunocs light objects are designed for both indoor and outdoor use. The stylish, more than two metres high luminaires are extravagant highlights in any private or public environment. Available in either warm white or eight-colour LED versions, and with three different mounting options, the lamps are highly versatile in use. The rotomoulded bodies of the lamps are made of impact-, UV- and weather-resistant polyethylene. The powder coated metal components are made of V2A stainless steel and weather-proof.

Statement by the jury
The appearance of these lamps is defined by their strong focus on functionality. Their sophisticated, complex line arrangement captures the zeitgeist.

Die Leuchtobjekte Lunocs eignen sich sowohl für den Einsatz in Innenräumen als auch zur Außenbeleuchtung. Ihre stilvollen, über zwei Meter hohen Leuchtkörper setzen in jeder privaten sowie öffentlichen Umgebung extravagante Lichtakzente. Wahlweise mit warmweißer oder achtfarbiger LED-Beleuchtung sowie mit drei Befestigungsoptionen erhältlich, lassen sich die Leuchten vielseitig einsetzen. Die Leuchtkörper sind aus schlagfestem, UV-, frost- und wetterbeständigem Polyethylen mittels Rotationsschmelzguss gefertigt. Die Metallelemente bestehen aus V2A-Edelstahl und sind zusätzlich wetterfest pulverbeschichtet.

Begründung der Jury
Der Fokus auf Funktionalität definiert die Anmutung dieser Leuchtobjekte. Ihre variantenreiche Linienführung trifft den Zeitgeist.

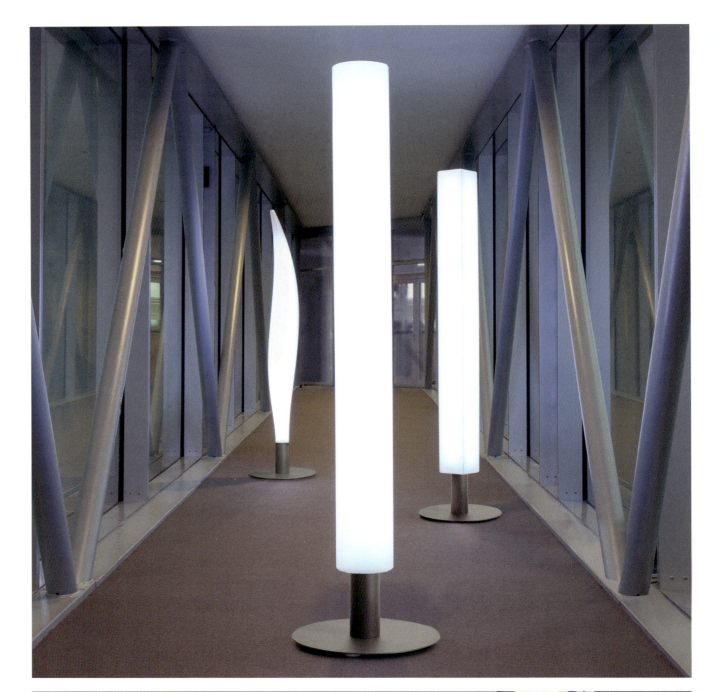

TASK
Floor Luminaire
Stehleuchte

Manufacturer
XAL GmbH, Graz, Austria
In-house design
Web
www.xal.com

At a structural height of only 15 mm, Task combines subtle aesthetics and innovative technology in an understated luminaire. The glare suppression prisms integrated into the cover ensure high suitability for use with computer screens. Intelligent sensors transmit wireless signals to neighbouring luminaires. A self-learning system controlled by presence and daylight sensors regulates the necessary light intensity automatically, or can be programmed for predefined light scenarios.

Statement by the jury
With a high degree of functionality, the Task floor luminaire flexibly adjusts to individual lighting conditions in workplaces.

Mit einer Aufbauhöhe von nur 15 mm vereint Task eine subtile Ästhetik und eine innovative Technologie in einem dezenten Leuchtenkörper. In die Abdeckung integrierte Entblendungsprismen garantieren eine gute Bildschirmtauglichkeit. Eine intelligente Sensorik übermittelt Funksignale an die benachbarten Leuchten. Ein selbstlernendes System, das über den Anwesenheits- und Tageslichtsensor gesteuert wird, reguliert die benötigte Lichtintensität selbsttätig oder wird durch vordefinerte Lichtszenarien programmiert.

Begründung der Jury
Mit einem hohen Maß an Funktionalität passt sich die Stehleuchte Task den jeweiligen Lichtverhältnissen eines Arbeitsplatzes flexibel an.

Solium
Floor Luminaire
Stehleuchte

Manufacturer
Artemide S.p.A., Pregnana Milanese, Italy
Design
Karim Rashid Inc., New York, USA
Web
www.artemide.com
www.karimrashid.com

The design concept of the Solium floor luminaire was inspired by the sensuous silhouette of a screen goddess. The undulated luminaire projects an intensive light onto the wall. The luminous frame is made of fibreglass, a sustainable, very light and long-lasting material. Its smooth surface supports the high performance of the LED lights. Its body serves as a reflector and, thanks to a dedicated optical unit, the luminaire achieves a high light-diffusion efficiency.

Statement by the jury
The dynamic lines of this floor luminaire yield an architectural quality. Its elegant appearance underlines the aesthetic light effect.

In Anlehnung an die sinnliche Silhouette einer Filmdiva entstand das Gestaltungskonzept für die Stehleuchte Solium. Von der wellenförmigen Leuchte wird ein intensives Licht – beispielsweise an eine Wand – reflektiert. Der Leuchtrahmen besteht aus Faserglas, einem nachhaltigen, sehr leichten und dennoch langlebigen Material. Seine glatte Oberfläche unterstützt die hohe LED-Lichtleistung: Der Leuchtkörper selbst dient als Reflektor und erzielt aufgrund einer speziellen Optikeinheit eine hohe Lichtstreuungseffizienz.

Begründung der Jury
Eine architektonische Qualität erreicht die schwungvolle Linienführung dieser Stehleuchte. Ihre elegante Anmutung betont den ästhetischen Lichteffekt.

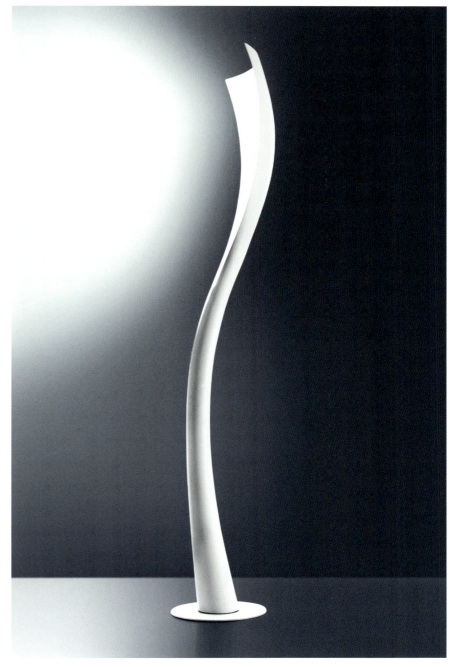

Barrisol – Lovegrove Manta
Lamp
Leuchte

Manufacturer
Barrisol Normalu SAS,
Kembs, France

Design
Ross Lovegrove Ltd,
London, Great Britain

Web
www.barrisol.com
www.rosslovegrove.com

reddot award 2015
best of the best

Form as dialogue

Organic forms occurring in nature, such as the shape of jellyfish floating gently in the water, fascinate observers with their exuberant, almost voluptuous beauty and plasticity. The design of the Manta lamp was inspired by such natural forms, transforming this association into an object with a mysteriously floating appearance. The lamp, by designer Ross Lovegrove, was developed as part of an installation for Barrisol, a manufacturer of lighting and suspended ceilings. As intended by the designer, this installation enters new aesthetic territory somewhere between deep ocean and space, forming a "dialogue for the 21st century between aquatic biomorphism and NASA-like intelligent systems". The unique, organic appearance of the lamp is based on a filigree aluminium frame featuring a stretched translucent membrane, a material that Barrisol otherwise commonly uses for ceiling constructions. Thus given a new form, the material shows its particular qualities: a soft appearance spreads a beautiful and atmospheric light. The lamp is illuminated by LEDs integrated into the curved aluminium profile. The Manta lamp is easy to install and transforms the interior into a cosmos of natural appearance – with structure, material and light entering into a continuous dialogue.

Form als Dialog

Die in der Natur vorkommenden organischen Formen, wie etwa die sich sachte im Wasser bewegende Qualle, beeindrucken durch ihre ausladende, nahezu schwelgerische Plastizität und Schönheit. Die Gestaltung der Leuchte Manta ist inspiriert von solchen natürlichen Formen, wobei sie diese Anleihen in ein geheimnisvoll schwebend anmutendes Objekt verwandelt. Die Leuchte des Designers Ross Lovegrove entstand im Rahmen einer Installation für Barrisol, einen Hersteller für Spann- unc Lichtdecken. Im Sinne des Designers betritt diese Installation ästhetisches Neuland zwischen Tiefsee und Weltall und bildet dabei einen „Dialog für das 21. Jahrhundert aus Meeresbiomorphismus und intelligenten NASA-typischen Systemen". Die einzigartige, organische Anmutung der Leuchte basiert auf einem filigranen Aluminiumrahmen. Dieser ist mit transluziden Folien bespannt, einem Material, welches von Barrisol normalerweise für Deckenkonstruktionen eingesetzt wird. In der neuen Form zeigt es seine besonderen Qualitäten: Es mutet weich an und verbreitet ein schönes und atmosphärisches Licht. Als Lichtquelle dienen LEDs, welche in das geschwungene Aluminiumprofil integriert werden. Die Leuchte Manta ist einfach zu installieren und verwandelt den Raum in einen naturähnlichen Kosmos – Struktur, Material und Licht stehen dabei in einem ständigen Dialog.

Statement by the jury

The Manta lamp, with its organic design, transforms into an object of floating appearance in the room. It realises a unique symbiosis between material and form, embodying a fascinating visual experience. Used in a different context here, the foil material otherwise employed in interiors develops extraordinary qualities. This impressive lamp interacts with the observers, challenging them with its three-dimensional plasticity and sensuousness.

Begründung der Jury

Mit ihrer organischen Gestaltung wird die Leuchte Manta zu einem fließend anmutenden Objekt im Raum. Sie verwirklicht eine einzigartige Symbiose zwischen Material und Form, die etwas faszinierend Neues entstehen lässt. Hier in einem anderen Kontext eingesetzt, zeigt das verwendete Folienmaterial aus dem Interior-Bereich außergewöhnliche Qualitäten. Diese imposante Leuchte interagiert mit dem Betrachter, wobei sie ihn mit ihrer dreidimensionalen Plastizität und Sinnlichkeit herausfordert.

Designer portrait
See page 54
Siehe Seite 54

Kyklos
LED Pendant Luminaire
LED-Hängeleuchte

Manufacturer
Linea Light Group, Treviso, Italy
In-house design
Web
www.linealight.com

Kyklos is a ring-shaped LED pendant luminaire which projects light upwards and thus provides pleasant indirect lighting. An additional ring helps to reflect part of the light downwards. The intensity of the light is individually adjustable. A pendant version provided with three additional LEDs with direct light and a semi-circular contoured wall-mounted luminaire complete the product family.

Kyklos ist eine indirekt strahlende LED-Pendelleuchte in Ringform mit homogener Lichtverteilung. Ein zusätzlicher, individuell einstellbarer Ring sorgt dafür, dass ein Teil des Lichts nach unten reflektiert wird, wobei seine semi-transluzente Beschaffenheit die Schattenbildung an der Decke auf ein Minimum reduziert. Die Intensität des Lichts ist mittels Dimmer individuell einstellbar. Die Serie wird durch eine LED-Pendelleuchte mit drei zusätzlichen LEDs für direkten Lichtaustritt nach unten und eine halbkreisförmige LED-Wandanbauleuchte ergänzt.

Statement by the jury
The form language of this LED pendant luminaire showcases an autonomous dynamism. Moreover, Kyklos achieves attractive ambient light effects.

Begründung der Jury
Eine eigenständige Dynamik zeigt die Formensprache dieser LED-Hängeleuchte. Kyklos erzielt zudem ansprechende Raumlichteffekte.

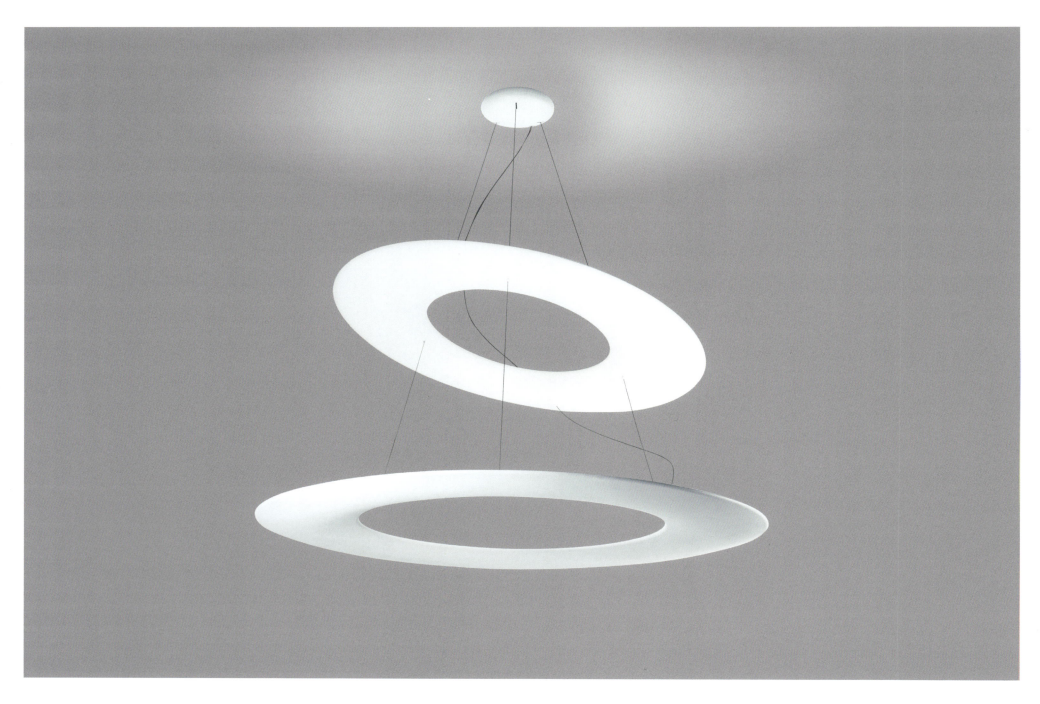

Mr. Magoo
LED Pendant Luminaire
LED-Hängeleuchte

Manufacturer
Linea Light Group, Treviso, Italy
In-house design
Web
www.linealight.com

Mr. Magoo is an LED pendant luminaire with an iconic appeal, which was designed to express the harmony of shapes, light and visually generated emotions. The asymmetric contours of the luminaire are embraced by well-balanced lines. The light is evenly distributed by the housing of the luminaire, resulting in an ambient lighting with a natural appeal. The large-size pendant is also available as a floor lamp or ceiling light.

Mr. Magoo ist eine ikonisch anmutende LED-Hängeleuchte, welche die Harmonie der Formen, des Lichts und der visuell erzeugten Emotionen zum Ausdruck bringen soll. Eine ausgewogene Linienführung schmeichelt den asymmetrischen Konturen des Leuchtkörpers. Das Licht wird gleichmäßig vom Korpus der Hängeleuchte abgestrahlt und erreicht eine natürlich wirkende Raumausleuchtung. Die großformatige Hängeleuchte ist auch in einer Ausführung als Stehleuchte oder Deckenleuchte erhältlich.

Statement by the jury
The design of Mr. Magoo achieves an entirely discrete appearance, while the gently curved contours convey a timeless elegance.

Begründung der Jury
Eine gänzlich eigenständige Anmutung erzielt die Gestaltung von Mr. Magoo, wobei die sanft geschwungenen Konturen eine zeitlose Eleganz vermitteln.

Pirouet
LED Pendant Luminaire
LED-Pendelleuchte

Manufacturer
Ridi Leuchten GmbH, Jungingen, Germany
Design
KDID, Koslowski & Dwalischwili GbR
(Malte Koslowski, Georg Dwalischwili),
Berlin, Germany
Web
www.ridi-group.de
www.kdid.de
Honourable Mention

Like a blossom, the Pirouet pendant luminaire floats above the table. The lightweight design of the body seems to be supported by the satinised, opal plastic material of the lamp shade. Thanks to a simple mechanism, the shade unfolds like a calyx in order to continuously adjust the brightness. Hence the radius of the light is variable. In addition, the intertwining elements of the shade create different surface structures, changing the luminaire's impact on the ambience.

Gleich einer Blüte schwebt die LED-Pendelleuchte Pirouet über dem Tisch. Die Leichtigkeit ihres Korpus wird durch den leicht satinierten, opalen Kunststoff des Leuchtenschirms unterstützt. Um die Lichtstärke individuell anzupassen, lässt sich der Schirm mittels einer simplen Mechanik blumenkelchartig auffächern. Dadurch variiert der Radius des Lichtscheins. Das Ineinandergleiten der Schirmelemente ergibt zudem wechselnde Oberflächenstrukturen und verändert so die Raumwirkung der Leuchte.

Statement by the jury
The precise implementation of this design idea convinces with its dynamic lighting options.

Begründung der Jury
Die präzise Umsetzung dieser Gestaltungsidee überzeugt aufgrund ihrer dynamischen Beleuchtungsmöglichkeiten.

IKEA PS 2014
Pendant Luminaire
Hängeleuchte

Manufacturer
IKEA of Sweden AB, Älmhult, Sweden
In-house design
David Wahl
Web
www.ikea.com
Honourable Mention

The PS 2014 pendant luminaire was inspired by science fiction movies and video games. Its resemblance to a spaceship or an imploding planet yields a particular dynamic appeal. A simple mechanism changes the contour of the luminaire: pulling the strings changes the shape of the lamp shade as well as the intensity of the light. When closed, only a small amount of light finds its way out; when fully open, more light pours out and creates interesting light effects on the walls.

Inspiriert von Science-Fiction-Filmen und Videospielen entstand die Hängeleuchte PS 2014. Ihre Ähnlichkeit mit einem Raumschiff oder einem implodierenden Planeten erreicht eine besondere Dynamik. Denn mittels einer einfachen Mechanik verändert sich die Kontur. Sobald an den Schnüren gezogen wird, verändern sich Schirmform und Lichtintensität: Sind die Klappen geschlossen, dringt nur wenig Licht nach außen. Bei geöffneten Klappen strömt mehr Licht aus und es entstehen interessante Lichteffekte an den Wänden.

Statement by the jury
The creative design of this pendant luminaire surprises with a mechanically simple but effective dimmer function.

Begründung der Jury
Diese kreativ gestaltete Hängeleuchte überrascht mittels einer mechanisch einfachen, aber effektvollen Dimm-Funktion.

Nikau
Pendant Luminaire
Pendelleuchte

Manufacturer
David Trubridge Ltd ,
Hastings, New Zealand
In-house design
David Trubridge
Web
www.davidtrubridge.com

The Nikau pendant luminaire is made of FSC-certified bamboo plywood and sold as a kit of single parts. It can easily be assembled on site via push-in nylon clips. The lamp comes in two versions: the long version works best in the corner of a room and several of them serve as a room divider, whereas the short version shown here is suited for direct illumination, for example of tables. The shadow play of the luminaire yields a dynamic effect. The name Nikau refers to the name of the only indigenous palm tree in New Zealand.

Statement by the jury
The Nikau pendant luminaire showcases a natural appearance with distinctive lines that produce charming effects of light and shade.

Die Pendelleuchte Nikau wird aus FSC-zertifiziertem Bambusholz gefertigt und in Einzelteile zerlegt verschickt. Vor Ort wird sie einfach mittels Einsteck-Klemmen aus Nylon zusammengebaut. Es gibt zwei Varianten: Die lange Version passt gut in die Ecken eines Raumes; mehrere nebeneinander dienen als Raumteiler. Hingegen eignet sich die hier gezeigte Kurzversion für die direkte Beleuchtung, beispielsweise über Tischen. Mit ihren Schatteneffekten erzielt die Leuchte eine dynamische Wirkung. Der Name Nikau bezieht sich auf die einzige einheimische Palmenart Neuseelands.

Begründung der Jury
Eine natürliche Anmutung vermittelt die Pendelleuchte Nikau, deren prägnante Linienführung reizvolle Licht- und Schatteneffekte erzeugt.

Snowflake
Pendant Luminaire
Pendelleuchte

Manufacturer
David Trubridge Ltd.,
Hastings, New Zealand
In-house design
David Trubridge
Web
www.davidtrubridge.com

Snowflake is a pendant luminaire made of FSC-certified bamboo plywood. It is sold as a kit, which reduces the volume for shipping to a minimum. Furthermore, this gives users the fun and satisfaction of having easily assembled their own lighting system. Push-in nylon clips keep the pieces together and ensure that the lamp automatically takes up its shape during assembly. The design of the luminaire, which is available in three different sizes, was inspired by a trip to the Antarctic.

Statement by the jury
The Snowflake pendant luminaire creates an effective shadow play. It is a fully sustainable product from the choice of materials to distribution.

Snowflake ist eine Pendelleuchte aus FSC-zertifiziertem Bambusholz. Da die Leuchte in Einzelteile zerlegt verschickt wird, reduziert sich das Frachtvolumen auf ein Minimum. Zudem wird dem Käufer das Erfolgserlebnis einer leicht zusammenbaubaren Beleuchtung ermöglicht. Einsteck-Klemmen aus Nylon halten die Einzelteile zusammen, wodurch die Leuchte während des Zusammenbaus automatisch ihre Form annimmt. Inspiriert von einer Reise in die Antarktis entstand die in drei Größen erhältliche Leuchte.

Begründung der Jury
Ein effektvolles Schattenspiel erzeugt die Pendelleuchte namens Snowflake. Von der Materialwahl bis zum Versand ein nachhaltiges Produkt.

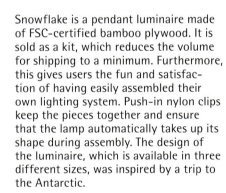

Conbrio

LED Pendant Luminaire

LED-Hängeleuchte

Manufacturer
Royal Philips, Eindhoven, Netherlands
In-house design
Web
www.philips.com

Conbrio offers an elegant, LED-based lighting experience for domestic interior decoration at an affordable price. The complex honeycomb structure of the lamp shade is made possible through the use of silicone. The semi-translucent, natural glow of the material emphasises the shade's extraordinary texture, creating a particularly pleasant light. The LEDs create light effects in a warm colour tone, which can be optimised and specifically targeted according to individual needs.

Statement by the jury
Inspired by a honeycomb structure, this pendant luminaire evokes positive emotions. A formally self-contained yet also functional design.

Als elegantes, LED-basiertes Beleuchtungserlebnis eignet sich Conbrio für die kostengünstige Innenausstattung von Wohnräumen. Eine Fertigung aus Silikon ermöglicht die komplexe Wabenstruktur des Leuchtenschirms. Seine ungewöhnliche Textur wird durch das semitransparente, natürlich glänzende Material visuell betont und sorgt für ein besonders angenehmes Licht. Die LEDs kreieren Lichteffekte in einer warmen Lichttemperatur, die bei Bedarf optimiert und gezielt ausgerichtet werden können.

Begründung der Jury
In Anlehnung an die Struktur von Honigwaben erweckt diese Hängeleuchte positive Emotionen. Ein formal eigenständiger und zugleich funktionaler Entwurf.

Jars

Pendant and Table Light

Hänge- und Tischleuchte

Manufacturer
Royal Philips, Eindhoven, Netherlands
In-house design
Web
www.philips.com

As an affordable range of lights with interchangeable glass shades, Jars especially appeals to a young target group. The shades, available in three different sizes and shapes, can be creatively combined, offering numerous possibilities for customisation. Furthermore, the shades of the luminaires come in a variety of stylish colours. The colour palette was especially designed for retrofit lamps, creating subtle light effects with soft graduated hues.

Statement by the jury
The Jars range of lights follows an original design concept and convinces with a variety of attractive possible combinations.

Als preiswertes Leuchten-Sortiment mit austauschbaren Glas-Lampenschirmen spricht Jars eine junge Zielgruppe an. Die Schirme in drei unterschiedlichen Größen und Formen können kreativ miteinander kombiniert werden und bieten zahlreiche Möglichkeiten zur Personalisierung. Zudem sind die Schirme der Leuchten in unterschiedlichen, trendbewussten Farben erhältlich. Die ausgewählten Farben wurden speziell für LED-Retrofit-Lampen entwickelt, wobei die Farbtöne in Abstufungen für dezente Lichteffekte sorgen.

Begründung der Jury
Das Leuchten-Sortiment Jars folgt einem originellen Gestaltungskonzept und überzeugt mittels einer Vielzahl an attraktiven Kombinationsmöglichkeiten.

Bishop

Pendant Luminaire

Pendelleuchte

Manufacturer
Wever & Ducré BVBA, Roeselare, Belgium
Design
3HDraft, Hong Kong
Web
www.weverducre.com
Honourable Mention

Bishop appears to be a pendant luminaire made of concrete. Due to an innovative manufacturing technique, the problems with some of the properties of this material, such as weight, fragility, discolouration as well as dust abrasion, could be solved. The luminaire, which is cast in one piece, showcases a timeless appearance with gently curved lines. The pleasant colour temperature of 2,200 to 2,700 kelvins is reminiscent of candlelight and bathes the room in a romantic atmosphere of light. The red power cord forms an appealing contrast.

Statement by the jury
This simple, appealing pendant luminaire impresses with its material and provides a well-balanced light quality.

Bishop sieht aus wie eine Pendelleuchte aus Beton. Mithilfe eines innovativen Fertigungsverfahrens konnten die problematischen Eigenschaften des Materials – Gewicht, Zerbrechlichkeit, Verfärbungen sowie Staubabrieb – ausgeschlossen werden. Aus einem Guss entstand eine zeitlos wirkende Leuchte mit einer sanft geschwungenen Linienführung. Die angenehme Farbtemperatur von 2.200 – 2.700 Kelvin erinnert an Kerzenlicht und taucht den Raum in eine romantische Lichtstimmung. Das rote Stromkabel bildet einen reizvollen Kontrast.

Begründung der Jury
Die schlicht anmutende Pendelleuchte überrascht durch ihre Materialität und bietet eine ausgewogene Lichtqualität.

Conversio
LED Pendant Luminaire
LED-Pendelleuchte

Manufacturer
Illuminartis, c/o FL Metalltechnik AG,
Grünen, Switzerland
Design
atelier oï SA, Neuveville, Switzerland
Web
www.illuminartis.ch
www.atelier-oi.ch

The Conversio LED pendant luminaire features a mechanical light control: the interior ring, which contains the circular light source, interacts with the exterior reflector ring via an intelligent mechanism, which directs up to 95 per cent of the warm LED light upwards or downwards, depending on the positioning. At the centre setting, 75 per cent of the light is directed at the table and 25 per cent towards the ceiling. Conversio is available with various anodised layers in different colours, lending the lamp a sculptural appearance even in daylight.

Statement by the jury
Harmonious colours in combination with elegant lines underline the self-contained aesthetics of this pendant luminaire.

Die LED-Pendelleuchte Conversio verfügt über eine mechanische Lichtlenkung: Der innere Ring, welcher die kreisförmige Lichtquelle beinhaltet, interagiert mittels einer raffinierten Mechanik mit dem äußeren Reflektorring. Dieser leitet das warme LED-Licht je nach Position bis zu 95 Prozent nach unten oder oben. In der Mittelstellung fallen 75 Prozent des Lichts auf den Tisch sowie 25 Prozent an die Decke. Conversio ist in Eloxaltönen erhältlich, die der Leuchte auch bei Tageslicht einen skulpturalen Charakter verleihen.

Begründung der Jury
Harmonische Farben in Kombination mit einer eleganten Linienführung unterstreichen die eigenständige Ästhetik dieser Pendelleuchte.

Lateralo Ring
LED Pendant Luminaire
LED-Hängeleuchte

Manufacturer
Trilux GmbH & Co. KG, Arnsberg, Germany
Design
Hartmut S. Engel Design Studio,
Ludwigsburg, Germany
Web
www.trilux.com
www.designer-profile.de/profiles/14-hartmut

As a prestigious circular LED pendant luminaire, the Lateralo Ring provides a homogenous light. Harmoniously balanced direct and indirect light components illuminate the room in all directions exceptionally brightly. Switched off during the daytime, the luminaire appears like a floating ring. It is suspended by three thin length-adjustable wires connected to the ceiling. The transparent panel and wire suspension lend this luminaire, which shows no visible power cable, a delicate appearance.

Statement by the jury
This ring-shaped LED pendant luminaire convinces with its elegant form language and the high quality of its light.

Als repräsentative, kreisrunde LED-Hängeleuchte bietet die Lateralo Ring eine homogene Raumausleuchtung. Harmonisch aufeinander abgestimmte, direkte und indirekte Lichtanteile leuchten den Raum rundum und ungewöhnlich hell aus. Bei Tageslicht ausgeschaltet, wirkt die Leuchte wie ein schwebender Ring. Die Aufhängung erfolgt über drei dünne Seile, die am Baldachin längenverstellbar befestigt werden. Die transparente Scheibe und die Seil-Aufhängung ohne sichtbares Stromkabel erreichen eine filigrane Ästhetik.

Begründung der Jury
Die ringförmige LED-Hängeleuchte überzeugt durch ihre elegante Formensprache und ihre hohe Beleuchtungsqualität.

Sidelite Round FerroMurano
LED Wall and Pendant Luminaire
LED-Wand- und Pendelleuchte

Manufacturer
RZB Rudolf Zimmermann, Bamberg GmbH,
Bamberg, Germany
In-house design
Helmut Heinrich
Web
www.rzb.de

The Sidelite Round FerroMurano wall and pendant luminaire combines the Venetian art of glassmaking with modern LED technology. Its centrepiece is an extremely thin Murano glass panel, the unique design of which turns each lamp into an individually unique item. Due to the lateral light coupling, the Sidelite technology, the luminaire provides a homogeneous and powerful yet mild light. The power is supplied directly through the suspension cable, which makes the luminaire look even more delicate and weightless.

Die Wand- und Pendelleuchte Sidelite Round FerroMurano verbindet die venezianische Glasmacherkunst mit einer zeitgemäßen LED-Technologie. Ihr Blickfang ist eine hauchdünne Murano-Glasscheibe, deren Zeichnung jede Leuchte zu einem Einzelstück macht. Aufgrund der seitlichen Lichtankopplung, der Sidelite-Technologie, entsteht ein homogenes, leistungsstarkes und zugleich sanftes Licht. Die Stromzufuhr erfolgt direkt über die Abhängeseile, was die Leuchte noch filigraner und schwereloser erscheinen lässt.

Statement by the jury
This wall and pendant luminaire yields a characteristic overall appearance. Its use of materials ensures a high degree of attention and light quality.

Begründung der Jury
Diese Wand- und Pendelleuchte erreicht ein charakteristisches Gesamtbild. Ihre Materialität sorgt für eine hohe Aufmerksamkeit und Lichtqualität.

SEQUENCE
Pendant and Surface-Mounted LED Luminaire
LED-Pendel- und Anbauleuchte

Manufacturer
Zumtobel Lighting GmbH,
Dornbirn, Austria

In-house design
Zumtobel Lighting GmbH

Web
www.zumtobel.com

reddot award 2015
best of the best

Flexible adjustability

In computer age, the ways of working together keep changing constantly. The often quickly changing types of collaboration, whether in project teams, at conventional workstations or in open spaces, require lighting solutions that are flexibly adaptable. The design concept of the Sequence pendant and surface-mounted LED luminaire is based on a comprehensive study conducted in collaboration with the Fraunhofer IAO. The study has shown that users prefer individually controllable lighting solutions for working. The implementation of the luminaire meets these requirements of flexibility in terms of both form and function. It impresses with a minimalist, sleek housing that measures only 25 mm and thus acquires a visually unobtrusive appearance that blends in well with almost any interior. Embodying an elegant concept, the luminaire offers separately controlled LED modules for precise and flexible solutions. As such, it represents an integrative platform that merges the design possibilities offered by LED technology, lens technology and electronics into optimal harmony. To achieve outstanding lighting quality in the workplace, the luminaire is equipped with the innovative, high-performance advancedOptics lens technology. Featuring a design that perfectly matches form and technology, the Sequence luminaire emerges as a fascinating modular light source. It offers a high degree of adaptability and thus aestheticises both home interiors and workplaces alike.

Flexible Vielfalt

Im Zeitalter des Computers verändern sich die Formen der Zusammenarbeit. Der oftmals fließende Wechsel zwischen der Arbeit in Projektteams, am klassischen Arbeitsplatz oder im Open Space erfordert flexibel anpassbare Beleuchtungslösungen. Das Gestaltungskonzept der LED-Pendel- und Anbauleuchte Sequence basiert auf einer umfangreichen Studie in Zusammenarbeit mit dem Fraunhofer IAO. Sie führte zu dem Ergebnis, dass die Nutzer für ihre Tätigkeiten individuell steuerbare Beleuchtungslösungen bevorzugen. Als funktionale wie formale Umsetzung der Anforderungen ist diese Leuchte flexibel einsetzbar. Sie beeindruckt mit einem minimalistischen, nur 25 mm messenden und damit sehr flach gestalteten Leuchtenkörper, der sich im Raum visuell zurücknimmt. Dieses elegant anmutende Konzept bietet zudem den variablen Einsatz getrennt steuerbarer LED-Module. Als solches ist es eine integrative Plattform, welche die Möglichkeiten der LED-Technologie, der Linsentechnologie sowie der Elektronik in ein optimales Verhältnis bringt. Für eine hervorragend angepasste Lichtqualität am Arbeitsplatz ist die Leuchte ausgestattet mit der innovativen, leistungsfähigen advancedOptics-Linsentechnologie. In einer Form und Technologie perfekt vereinenden Gestaltung entstand mit Sequence eine faszinierende modulare Lichtquelle. Sie bietet ein hohes Maß an Variabilität und ästhetisiert sowohl die Wohnumgebung als auch den Arbeitsplatz.

Designer portrait
See page 56
Siehe Seite 56

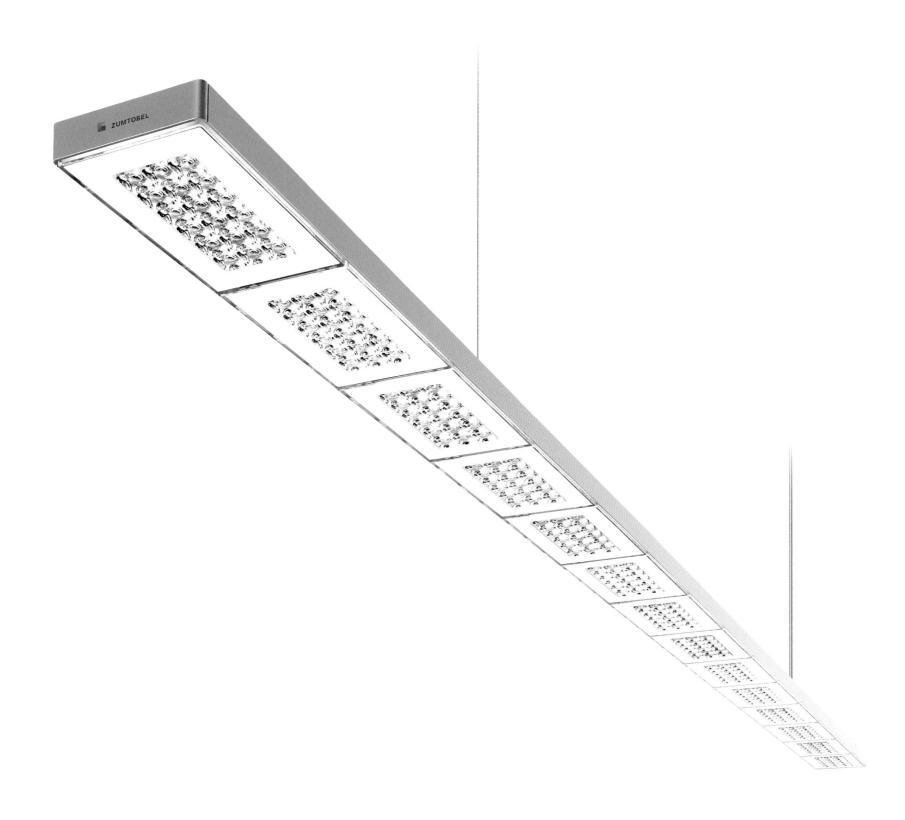

AXON
LED Pendant Luminaire
LED-Pendelleuchte

Manufacturer
Zumtobel Lighting GmbH,
Dornbirn, Austria
In-house design
Web
www.zumtobel.com

Like a slim light line, this LED pendant luminaire with a profile of 38 x 38 mm blends in with contemporary office architecture. The combination of high-performance LEDs and advancedOptics technology ensures efficient light distribution and glare control. In addition, a careful balance between indirect and direct light creates a pleasant atmosphere in the workplace. To suit individual user needs, Axon is available in different designs and warm white or neutral white colour temperature.

Wie eine schmale Lichtlinie fügt sich die LED-Pendelleuchte mit ihrem Querschnitt von 38 x 38 mm in die zeitgenössische Büroarchitektur ein. Aufgrund der Kombination von leistungsstarken LEDs mit einer advancedOptics-Linsentechnologie wird das Licht effektiv gelenkt und gleichzeitig entblendet. Zudem sorgt eine ausgewogene Abstimmung von indirekter und direkter Beleuchtung für eine angenehme Atmosphäre am Arbeitsplatz. Zur individuellen Anpassung an die Nutzerbedürfnisse ist Axon in unterschiedlichen Ausführungen sowie optional in warmweißer und neutralweißer Farbtemperatur verfügbar.

Statement by the jury
The Axon LED pendant luminaire achieves a high degree of lighting comfort. Its delicate colour palette harmoniously adapts to any ceiling.

Begründung der Jury
Ein hohes Maß an Beleuchtungskomfort erreicht die LED-Pendelleuchte Axon. Ihre dezente Farbgebung passt sich der Raumdecke harmonisch an.

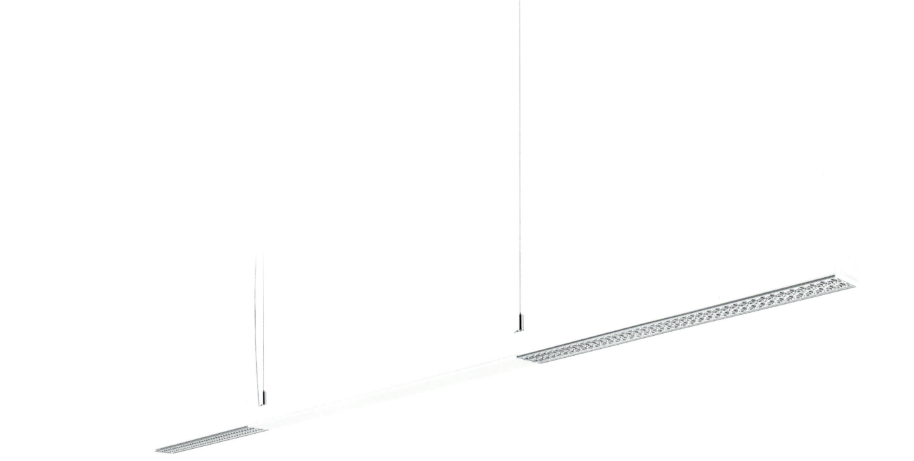

Oviso
Pendant Luminaire with Sensor
Pendelleuchte mit Sensorik

Manufacturer
Ribag Licht AG, Safenwil, Switzerland
In-house design
Daniel Kübler
Web
www.ribag.com

Oviso is an innovative pendant luminaire with OLED technology. Its area light consists of thin, organic layers. The ultra-slim 7 mm thin body produces a glare-free light without generating heat. The light spectrum comes close to that of sunlight. The high Colour Rendering Index (CRI) of 95 creates a pleasant ambience. The light control of the energy-efficient pendant is intuitive: thanks to integrated sensors, the brightness can be dimmed with a simple movement of the hand without the need for direct contact.

Statement by the jury
This pendant luminaire owes its convincing functionality to an innovative technology: an appealing and sustainable product solution.

Oviso ist eine innovative Pendelleuchte mit OLED-Technologie. Ihre Flächenlichtquelle besteht aus dünnen, organischen Schichten. Im nur 7 mm hohen Leuchtenkörper entsteht ein blendfreies Licht ohne Hitzeerzeugung. Das Lichtspektrum kommt dem des Sonnenlichts sehr nah. Eine hohe Farbwiedergabe CRI (Colour Rendering Index) von 95 sorgt für eine angenehme Raumatmosphäre. Die Lichtsteuerung der energiesparsamen Leuchte erfolgt mittels eingebauter Sensorik intuitiv: Durch eine einfache Handbewegung wird die Helligkeit berührungslos gedimmt.

Begründung der Jury
Ihre überzeugende Funktionalität verdankt diese Pendelleuchte dem Einsatz einer innovativen Technologie. Eine ansprechende und zugleich nachhaltige Produktlösung.

TASK
LED Pendant Luminaire
LED-Pendelleuchte

Manufacturer
XAL GmbH, Graz, Austria
In-house design
Web
www.xal.com

The efficient Task LED pendant luminaire was developed for work spaces and, with its well-balanced design, blends perfectly into any interior architecture. With a structural height of only 15 mm, Task demonstrates subtle aesthetic appeal. Glare suppression prisms integrated into the cover guarantee suitability for use with computer screens. The suspension cords can be adjusted continuously to any height without the use of tools. Both the direct and indirect light distribution ensure a pleasant ceiling lighting effect.

Statement by the jury
Task offers a high ease of use. Minimalist lines underline the filigree appeal of this LED pendant luminaire.

Die effiziente LED-Pendelleuchte Task wurde für den Arbeitsbereich konzipiert und passt sich mit ihrer ausgewogenen Formensprache der Raumarchitektur flexibel an. Mit einer Aufbauhöhe von nur 15 mm zeigt Task eine dezente Ästhetik. In ihre Abdeckung integrierte Entblendungsprismen stellen die Bildschirmtauglichkeit sicher. Die werkzeuglos höhenverstellbare Seilabhängung ermöglicht eine stufenlose Anpassung. Für eine angenehme Deckenaufhellung sorgt eine sowohl direkte als auch indirekte Lichtverteilung.

Begründung der Jury
Task bietet einen hohen Beleuchtungskomfort. Eine minimalistische Linienführung unterstreicht die filigrane Anmutung dieser LED-Pendelleuchte.

Ello
LED Luminaire
LED-Leuchte

Manufacturer
Wever & Ducré BVBA, Roeselare, Belgium
Design
Serge & Robert Cornelissen BVBA, Kortrijk (Marke), Belgium
Web
www.weverducre.com
www.sergecornelissen.com

The Ello LED luminaire can be either suspended from or mounted flush with the ceiling. Its slim body follows a purist design approach, featuring harmonious lines. The high light output of its LED technology allows for the comfortable illumination of a room. Thanks to a plug-in system, Ello luminaires are quick and easy to combine. A long service life and a good price-performance ratio complement the product concept.

Statement by the jury
Well-balanced proportions and a slim silhouette lend this powerful LED luminaire a distinctive overall appearance.

Die LED-Leuchte Ello kann sowohl abgehängt als auch deckenbündig montiert werden. Ihr schlanker Korpus folgt einem puristischen Gestaltungsansatz und zeigt eine ausgewogene Linienführung. Die hohe Lichtleistung der verwendeten LED-Technologie ermöglicht eine komfortable Ausleuchtung von Räumen. Mithilfe eines Stecksystems lassen sich Ello-Leuchten schnell und einfach kombinieren. Eine lange Lebensdauer sowie ein gutes Preis-Leistungsverhältnis ergänzen das Produktkonzept.

Begründung der Jury
Ausgewogene Proportionen und eine schmale Silhouette verleihen dieser leistungsstarken LED-Leuchte ein charakteristisches Gesamtbild.

Less is More
LED Pendant Luminaire
LED-Pendelleuchte

Manufacturer
RZB Rudolf Zimmermann, Bamberg GmbH,
Bamberg, Germany
In-house design
Helmut Heinrich
Web
www.rzb.de

Equipped with LED technology and featuring a minimalist design, the Less is More pendant luminaire serves as efficient general and accent lighting. It can be adapted to different room situations in terms of size and features. Users can select from various lumen packages and light colours. They can also determine the light distribution using a micro grid or either a microprismatic or an opal diffuser. The power supply running through the suspension cable underlines the delicate appeal of the luminaire.

Ausgestattet mit einer LED-Technologie dient die minimalistisch konzipierte Pendelleuchte Less is More als effiziente Allgemein- oder Akzentbeleuchtung. Sie lässt sich in puncto Größe und Ausstattung an die jeweilige Raumsituation anpassen. Optional stehen unterschiedliche Lumen-Pakete und Lichtfarben zur Auswahl, zudem lässt sich via Mikroraster sowie mikroprismatischer oder opaler Abdeckung die gewünschte Lichtverteilung bestimmen. Die Stromzufuhr erfolgt dezent über die Abhängeseile, was die grazile Anmutung des Aluminium-Korpus betont.

Statement by the jury
A convincing functionality in combination with a visual lightness characterises the Less is More LED pendant luminaire.

Begründung der Jury
Eine überzeugende Funktionalität in Kombination mit einer visuellen Leichtigkeit zeichnet die LED-Pendelleuchte Less is More aus.

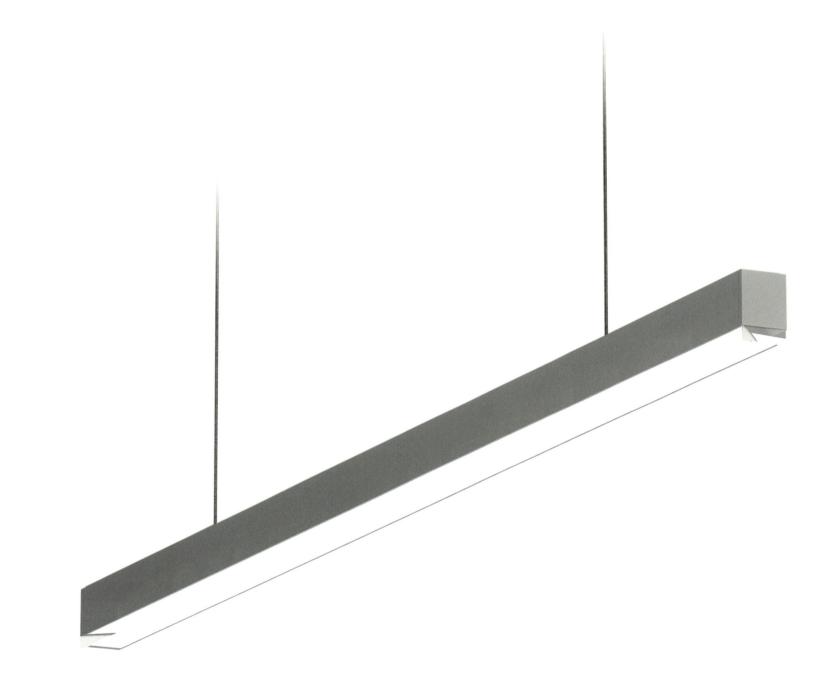

Sant
LED Pendant Luminaire
LED-Pendelleuchte

Manufacturer
Halla, a.s., Prague, Czech Republic
Design
RO-Architecten (Rob van Beek),
Soest, Netherlands
Web
www.halla.eu
www.ro-architecten.nl

Sant is a linear LED pendant luminaire with an opal diffuser, the light parameters of which are particularly well-balanced. It can be installed either as an individual light or as a continuous lighting system. The light intensity is dimmable, with the pendant luminaire providing both direct and indirect lighting. It also allows users to set the warmth of the light from warm-white (2,700 kelvins) to cold-white (6,500 kelvins). The aluminium body is available in white or black.

Sant ist eine lineare LED-Pendelleuchte mit Opaldiffusor, deren Lichtparameter besonders ausgewogen sind. Sie kann optional als Einzelleuchte oder als fortlaufendes Beleuchtungssystem installiert werden. Die Lichtintensität ist dimmbar, wobei die Pendelleuchte sowohl eine direkte wie indirekte Beleuchtung bietet. Zudem besteht die Möglichkeit, die Lichttemperatur bedarfsgerecht von Warmweiß (2.700 Kelvin) bis Kaltweiß (6.500 Kelvin) einzustellen. Der Korpus aus Aluminium ist in Weiß oder Schwarz erhältlich.

Statement by the jury
With its discreet, functional design, this energy-efficient pendant luminaire harmoniously blends into any interior.

Begründung der Jury
Mittels ihrer dezenten, funktionalen Gestaltung fügt sich diese energie-effiziente Pendelleuchte harmonisch in jedes Ambiente ein.

equilibra BALANS
LED Pendant Luminaire
LED-Hängeleuchte

Manufacturer
Aquaform Incorporated Sp. z o.o.,
Czernichów, Kraków County, Poland
Design
Piotr Jagiellowicz Design (Piotr Jagiellowicz),
Kraków, Poland
Web
www.aquaform.pl
www.jagiellowiczdesign.pl

The luminaires of the equilibra Balans product line introduce an innovative system that allows for an asymmetrical mounting with only one suspension cable. The main element of the patented system is made of carbon fibre composites. The design makes it possible to balance the lamp with only one suspension element, which is also connected to the power supply. The indirect light from the hidden LED modules is directed to a white reflector to achieve even light distribution.

In die Beleuchtungskörper der Produktreihe equilibra Balans wurde ein innovatives System integriert, welches eine unsymmetrische Aufhängung an nur einem Kabel ermöglicht. Der Hauptteil des patentierten Systems wird aus Verbundwerkstoffen aus Kohlenstofffasern hergestellt. Dank dieser Konstruktion ist die Ausbalancierung an nur einer mit dem Netzteil verbundenen Halterung möglich. Das indirekte Licht aus den versteckt platzierten LED-Modulen spiegelt sich in einem weißen Reflektor, der eine gleichmäßige Lichtstreuung erzielt.

Statement by the jury
The innovative suspension of these LED pendant luminaires is surprising and communicates pure elegance.

Begründung der Jury
Die innovative Halterung dieser LED-Hängeleuchte überrascht den Betrachter und vermittelt eine puristische Eleganz.

U–Line
LED Pendant Luminaire
LED-Pendelleuchte

Manufacturer
Ikizler Lighting, Istanbul, Turkey
Design
Omlet Istanbul Design Office,
Istanbul, Turkey
Web
www.ikizlerlighting.com
www.omletistanbul.com

U-line is an energy-efficient LED pendant luminaire, the product name alluding to its characteristic contours. The design concept combines a decorative and elegant appearance with modern lighting technology. Thanks to its modular design, the dimensions and connections can be adjusted to the requirements of different environments, for example by mounting the two lighting modules at different angles. The housing is recyclable while the mercury-free LED luminaire is designed for a long lifetime of 50,000 hours.

U-Line ist eine energieeffiziente LED-Pendelleuchte, deren Produktname auf ihre charakteristische Kontur anspielt. Das Gestaltungskonzept kombiniert eine dekorative, elegante Anmutung mit einer zeitgemäßen Beleuchtungstechnik. Dank ihres modularen Aufbaus können die Abmessungen und Verbindungen an die jeweiligen Anforderungen angepasst werden: So lassen sich die beiden Lichtmodule beispielsweise in unterschiedlichen Winkeln montieren. Das Gehäuse aus Aluminium ist recycelbar, während die quecksilberfreie LED-Leuchte auf eine Lebensdauer von 50.000 Stunden ausgelegt ist.

Statement by the jury
The three-dimensional interplay between two luminaires lends this lamp a distinctive overall appearance.

Begründung der Jury
Das dreidimensionale Zusammenspiel zweier Leuchtkörper erreicht bei dieser LED-Pendelleuchte ein eigenständiges Gesamtbild.

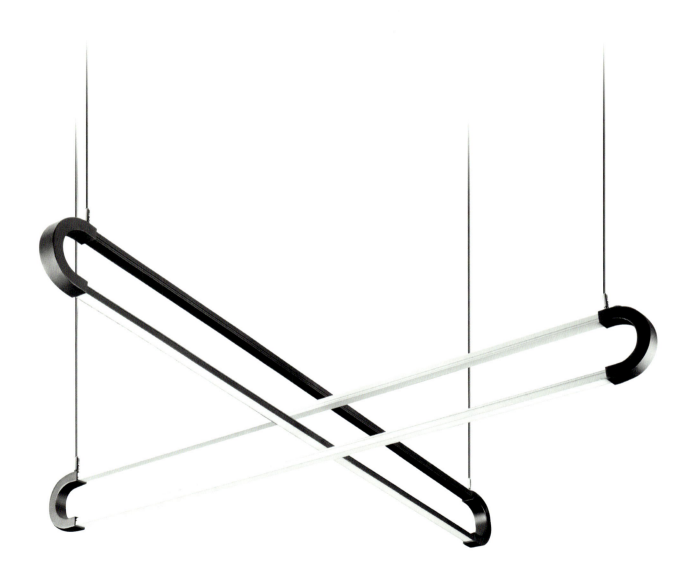

Arboreus
Pendant Luminaire
Hängeleuchte

Manufacturer
Herrmann Systems GmbH,
Albluxx Lichtmanufaktur,
Bisingen, Germany
In-house design
Volker Herrmann
Web
www.albluxx.com

The Arboreus pendant luminaire follows a minimalist design concept. The rectangular wood frame houses a hidden light source that only becomes visible in the form of an illuminated panel when the luminaire is switched on. The light distribution of 50 per cent direct and 50 per cent indirect light creates a pleasant ambience. Arboreus uses low-voltage technology in incorporation with a suspension system. Made of German Oak, the pendant luminaire combines high-grade materials with cutting edge LED technology. When the high-performance luminaire is switched off, it looks like a frame floating in the air.

Die Hängeleuchte Arboreus folgt einem minimalistischen Gestaltungskonzept. Der rechteckige Holzrahmen beherbergt eine versteckte Lichtquelle, die erst beim Einschalten in Form einer illuminierten Scheibe sichtbar wird. Die zu je 50 Prozent direkte und indirekte Lichtausbreitung erzeugt eine angenehme Raumatmosphäre. Die Arboreus setzt Niedervolt ein und integriert diese Technik in die Seilabhängung. Hergestellt aus Eichenholz, kombiniert die Hängeleuchte hochwertige Materialien mit einer innovativen LED-Technik. Sobald die lichtstarke Leuchte ausgeschaltet wird, entsteht der faszinierende Eindruck eines scheinbar frei schwebenden Holzrahmens.

Statement by the jury
Arboreus fascinates with its apparently invisible lighting technology while the wide frame construction conveys a distinctive appeal.

Begründung der Jury
Arboreus fasziniert aufgrund ihrer vermeintlich unsichtbaren Beleuchtungstechnik, während die breite Rahmenkonstruktion eine prägnante Anmutung vermittelt.

XOOTUBE 38
LED Pendant Luminaire
LED-Pendelleuchte

Manufacturer
LED Linear GmbH,
Neukirchen-Vluyn, Germany
In-house design
Sören Bleul
Web
www.led-linear.com

Xootube 38 is a linear design LED pendant luminaire with a diameter of 38 mm, combining an industrial style with elegant aesthetics. Through a defined pre-curvature of the luminaire, a length of almost two meters with a straight form can be created. It allows an individual choice between 11 end caps of different colour tones. The luminaire provides up to 2,203 lm/m and ensures homogenous light at a beam angle of almost 360 degrees. Its reduced and iconic form factor creates smart accents in lighting design, which is the reason why it is ideal for use at workplaces or in residential areas.

Statement by the jury
The Xootube 38 pendant luminaire conveys a fine aesthetic balance while the coloured end caps create attractive accents.

Xootube 38 ist eine lineare LED-Pendelleuchte mit einem Durchmesser von 38 mm. Sie kombiniert einen industriell anmutenden Stil mit einer eleganten Ästhetik. Durch eine definierte Krümmung der Leuchte kann eine Länge von fast zwei Metern mit einem linearen Formverlauf geschaffen werden. Eine individuelle Auswahl zwischen Endkappen in elf unterschiedlichen Farbnuancen ist möglich. Die Leuchte bietet bei bis zu 2.203 lm/m ein homogenes Lichtbild sowie einen Abstrahlwinkel von circa 360 Grad. Ihr reduzierter und ikonischer Formfaktor schafft smarte Akzente in der Lichtgestaltung und ist daher für den Einsatz an Arbeitsplätzen oder in Wohnbereichen sehr gut geeignet.

Begründung der Jury
Eine ästhetische Ausgewogenheit vermittelt die LED-Pendelleuchte Xootube 38, wobei ihre farbigen Endkappen reizvolle Akzente setzen.

Igloo
Pendant Lighting System
Hängeleuchten-System

Manufacturer
FontanaArte, Corsico (Milano), Italy
Design
Studio Klass, Milan, Italy
Web
www.fontanaarte.com
www.studioklass.com

Igloo is an innovative, self-supporting pendant lighting system for private living spaces and contract installations. The modular system with electromechanical connections, angles and spacers allows users to connect up to 100 modules to just one power supply. The brightness of the LED lights is adjustable through the use of a standard dimmer. The double shell allows for both a vertical and horizontal installation of the modules, while modules without LEDs can also be included.

Statement by the jury
The representative appearance of the Igloo pendant lighting system impresses with a high degree of flexibility and functionality.

Igloo ist ein innovatives, selbsttragendes Hängeleuchten-System für den Wohn- und Objektbereich. Als modulares System ermöglicht es aufgrund elektromechanischer Verbindungen, Winkel und Abstandshaltern die Verbindung von bis zu 100 Modulen mit einer einzigen Stromversorgung. Die Helligkeit der LED-Lampen ist per Dimmer einstellbar. Der Doppelaufbau lässt optional eine senkrechte oder waagerechte Installation der Module zu, zudem können auch Module ohne LEDs eingefügt werden.

Begründung der Jury
Das repräsentativ wirkende Hängeleuchten-System Igloo überzeugt durch seinen hohen Grad an Flexibilität und Funktionalität.

Infinito
LED Pendant Luminaire
LED-Hängeleuchte

Manufacturer
Qisda Corporation, Taipei, Taiwan
Design
BenQ Lifestyle Design Center, Taipei, Taiwan
Web
www.qisdesign.com
www.benq.com

Infinito is a multi-functional LED pendant luminaire that not only serves to illuminate the room but also creates atmospherical coloured light effects. Inspired by the symbol for infinity, these two functions are formally separated via an elegant twist: the big loop provides a warm, white light for reading while the coloured light flexibly designs the ambience of the room. Integrated sensors detect the movement of hands to control, among other things, the brightness and colour of the light.

Statement by the jury
With its characteristic lines and multi-functionality, Infinito creates a convincingly distinctive ambience in any room.

Infinito ist eine multifunktionale LED-Hängeleuchte, die sowohl der Raumausleuchtung dient als auch atmosphärische Farblicht-Effekte bietet. Inspiriert durch das Symbol für Unendlichkeit trennt sie ihre beiden Funktionsbereiche formal mittels einer eleganten Drehung: Die große Schleife liefert warmes, weißes Licht zum Lesen, während das Farblicht die flexible Gestaltung der Raumatmosphäre ermöglicht. Integrierte Sensoren erfassen Handbewegungen, wodurch unter anderem die Helligkeit und Farbe des Lichtes gesteuert werden.

Begründung der Jury
Mittels einer charakteristischen Linienführung und funktionaler Vielfalt ermöglicht Infinito eine überzeugend eigenständige Raumwirkung.

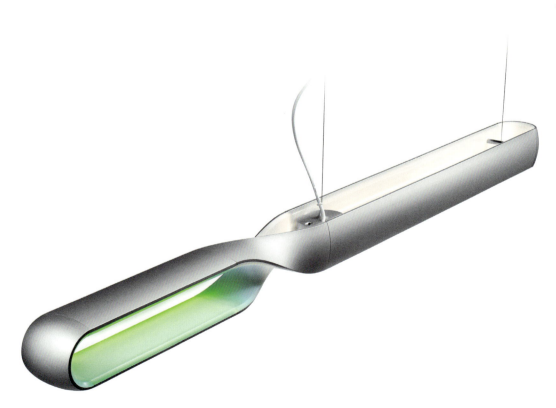

Nari
LED Luminaire
LED-Leuchte

Manufacturer
Innoluce, Architectural Lighting Design Company,
Bucheon, Gyeonggi, South Korea
In-house design
Eunjin Choi
Web
www.inno-luce.co.kr

The design concept of the Nari LED luminaire was inspired by nature. Its petal-like reflectors are made from high-quality reflective sheets and achieve a clear light quality. The individual reflectors seamlessly merge with each other. Groups of four LEDs, which are reminiscent of the stamen of a flower, are arranged to allow even distribution of light. With an energy-efficiency of 120 lm/W, Nari can be used flexibly in various environments as a surface-mounted light, pendant light, or running light.

Das Gestaltungskonzept der LED-Leuchte Nari wurde von der Natur inspiriert. Ihre blütenähnlichen Reflektoren bestehen aus einer hochwertigen Reflexionsfolie und erreichen eine klare Lichtqualität. Die einzelnen Reflektoren gehen nahtlos ineinander über. Jeweils vier LEDs, die an Pollenstängel erinnern, sind derart angeordnet, dass eine gleichmäßige Verteilung des Lichts ermöglicht wird. Mit einem energieeffizienten Wirkungsgrad von 120 lm/W kann Nari flexibel in unterschiedlichen Umgebungen als Aufbauleuchte, Pendelleuchte oder Lauflicht genutzt werden.

Statement by the jury
Thanks to the combination of organic looking details and geometrical contours, Nari achieves a highly distinctive overall appearance.

Begründung der Jury
Aufgrund einer Kombination organisch anmutender Details mit geometrischen Konturen erreicht Nori ein unverwechselbares Gesamtbild.

The Z
LED Desk Lamp
LED-Schreibtischleuchte

Manufacturer
Wow Energy Saving Lighting Co. Ltd.,
New Taipei City, Taiwan
In-house design
Web
www.lianglight.com

The Z desk lamp is an innovative light source, which redefines the relationship between humans and light. Different from conventional mechanical arm systems, the Z lamp allows for an unusual interaction with the light source: with its rotatable arm, the Z lamp provides the possibility to intuitively adjust the light precisely. Carefully balanced proportions create an elegant form language, which turns the lamp into a piece of art on the desk.

Statement by the jury
A formally attractive overall appearance characterises this LED desk lamp. Infinite inclination angles enable the lamp to adjust to individual lighting demands.

Die Z-Schreibtischleuchte ist eine innovative Lichtquelle, welche die Beziehung zwischen Mensch und Licht neu definieren möchte. Anders als herkömmliche Armsysteme ermöglicht die Z-Leuchte eine ungewöhnliche Interaktion mit der Lichtquelle: Mit ihrem drehbaren Arm bietet sie eine intuitive Möglichkeit zur exakten Lichteinstellung. Aufgrund sorgfältig abgestimmter Proportionen entsteht eine elegante Formensprache, welche die Schreibtischleuchte artifiziell erscheinen lässt.

Begründung der Jury
Ein formal ansprechendes Gesamtbild zeichnet diese LED-Schreibtischleuchte aus. Stufenlose Neigungswinkel passen sich Lichtbedürfnissen individuell an.

Cubert
Lamp, Power Outlet, Charging Device
Leuchte, Steckdose, Ladegerät

Manufacturer
Colebrook Bosson Saunders,
London, Great Britain
In-house design
Web
www.colebrookbossonsaunders.com
Honourable Mention

The vision for Cubert was to design a multifunctional desktop product that combines a light, two mains sockets, and two smart USB power outlets with the intention of supporting technology such as laptops, tablets and mobile phones. A small form factor of 80 x 80 mm, Cubert is versatile and is proving popular within commercial office space and also in hotels, both in reception or plaza areas, and as a bedside light or personal charge device. Cubert effectively brings power and light to the user.

Statement by the jury
Thanks to its sophisticated multifunctionality, Cubert impresses as a light source and a power outlet as well as a charging device.

Cubert folgt dem Gestaltungsziel eines multifunktionalen Desktop-Produkts, welches ein Licht, zwei Steckdosen und zwei USB-Steckdosen kombiniert und für Laptops, Tablet-PCs und Mobiltelefone kompatibel ist. Mit seinem kleinen Korpus von nur 80 x 80 mm ist Cubert vielseitig und eignet sich für Büros und Hotels, sowohl an der Rezeption oder im Empfangsbereich, als Nachtlicht oder als persönliche Ladevorrichtung. Cubert dient den Nutzern als effektive Licht- und Energiequelle.

Begründung der Jury
Aufgrund seiner durchdachten Multifunktionalität überzeugt Cubert zugleich als Licht- und Stromquelle sowie als Auflade- gerät.

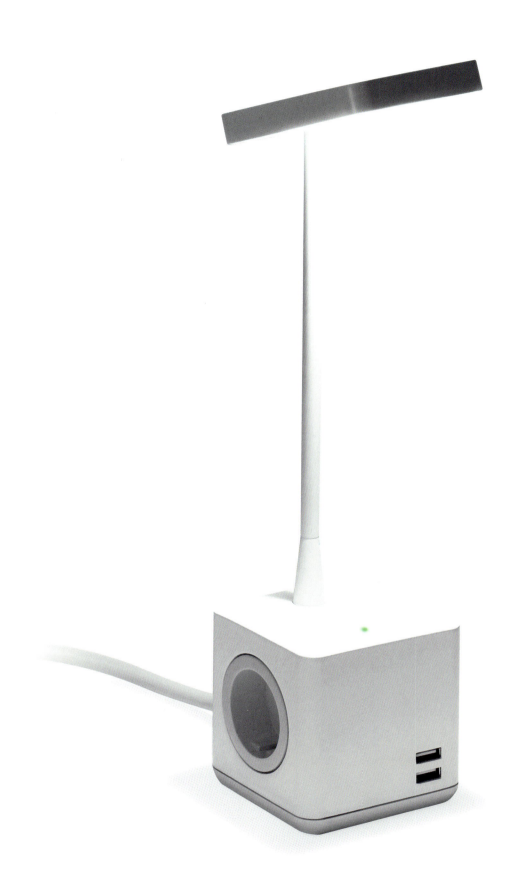

Compasso
LED Desk Lamp
LED-Schreibtischleuchte

Manufacturer
Qisda Corporation, Taipei, Taiwan
Design
BenQ Lifestyle Design Center,
Taipei, Taiwan
Web
www.qisdesign.com
www.benq.com

The Compasso desk lamp allows for a variety of adjustments and is dimmable. Its filigree lamp arm connects a sliding track with a highly flexible silicon hoop. The lighting angle can be adjusted by positioning the lamp arm back and forth, or by directing the lamp head up and down. The brightness can be set either by a touch-dimmer or by rotating the swivel base 360 degrees. The LED fibre optic technology is both energy efficient and bright.

Statement by the jury
The high functionality of this LED desk lamp allows for an even lighting that meets the demands of any user while conveying a multifaceted overall appearance.

Die Compasso Schreibtischleuchte kann vielseitig verstellt und gedimmt werden. Ihr filigraner Leuchtenarm verbindet eine Gleitschiene mit einer hochflexiblen Silikonschlaufe. Somit lässt sich der Beleuchtungswinkel durch das Positionieren des Leuchtenarms nach vorne und hinten oder des Leuchtenkopfes nach oben und unten anpassen. Die Helligkeit kann optional per Touch-Dimmer oder mittels einer 360-Grad-Drehung am Drehfuß eingestellt werden. Die LED-Lichtleiter-Technologie ist zugleich energieeffizient und lichtstark.

Begründung der Jury
Die hohe Funktionalität der LED-Schreibtischleuchte ermöglicht eine bedarfsgerechte Ausleuchtung, wobei Compasso ein facettenreiches Gesamtbild vermittelt.

Hatha
LED Table Lamp
LED-Tischleuchte

Manufacturer
Qisda Corporation, Taipei, Taiwan
Design
BenQ Lifestyle Design Center,
Taipei, Taiwan
Web
www.qisdesign.com
www.benq.com

Inspired by Hatha yoga, Hatha is an LED table lamp that conveys the flexibility of the human body. A zinc alloy metal base ensures stability while the rubber body of the lamp freely bends at different angles. This allows the lamp to flexibly adjust to individual lighting requirements. High power LEDs, with lens design, deliver sufficient brightness for reading and reach a high light efficiency. Hatha is available in several different colours.

Statement by the jury
Hatha uses a design that is convincing both formally and functionally, making this LED table lamp a true eye-catcher.

Hatha ist eine LED-Tischleuchte, die in Anlehnung an Hatha-Yoga die Beweglichkeit eines menschlichen Körpers darstellt. Ein Metallfuß aus einer Zinklegierung verschafft der Leuchte Stabilität, während der Gummikörper in flexible Neigungswinkel verbogen werden kann. So passt sich die Leuchte individuellen Beleuchtungsbedürfnissen variabel an. Leistungsstarke LEDs mit dem Design einer Linse bieten eine ausreichende Helligkeit zum Lesen und erzielen eine hohe Lichtausbeute. Hatha ist in unterschiedlichen Farben erhältlich.

Begründung der Jury
Hatha ist ein formal und funktional überzeugender Entwurf, der diese LED-Tischleuchte zum Blickfang macht.

Luctra
LED Workplace Lamp
LED-Arbeitsplatzleuchte

Manufacturer
Durable Hunke & Jochheim GmbH & Co. KG,
Iserlohn, Germany
Design
yellow design | yellow circle (Guenter Horntrich),
Cologne, Germany
Web
www.luctra.eu
www.yellowdesign.com

The Luctra product range comprises innovative LED workplace lamps, the light of which individually adapts to the diurnal rhythm of the user. The Vitacore electronic system was especially designed to enable intuitive control of the colour and intensity of light via its touch screen panel or app. The lamps simulate the natural cycle of daylight, which fosters a sense of wellbeing and boosts concentration. Uniform illumination, effective light output, low energy consumption and high durability are further evidence for the quality of the lamps.

Die Produktserie Luctra umfasst innovative LED-Arbeitsplatzleuchten, deren Lichtleistung sich individuell dem Tagesrhythmus des Nutzers anpassen lässt. Die eigens entwickelte Vitacore-Elektronik ermöglicht eine intuitive Steuerung von Lichtfarbe und -intensität per Touchscreen oder App. Die Leuchten simulieren dadurch den natürlichen Tageslichtverlauf, wodurch Wohlbefinden und Konzentration gefördert werden. Eine gleichmäßige Ausleuchtung, eine effektive Lichtausbeute, ein geringer Energiebedarf und eine sehr lange Lebensdauer sind weitere Qualitätsmerkmale.

Statement by the jury
The Luctra lamp series demonstrates both a formally and functionally convincing design. It provides a high degree of usability.

Begründung der Jury
Eine gleichsam formal wie funktional überzeugende Gestaltung zeigt die Leuchtenserie Luctra. Sie bietet ein hohes Maß an Bedienkomfort.

Lumio
Lamp
Leuchte

Manufacturer
Lumio, San Francisco, USA
In-house design
Max Gunawan
Web
www.hellolumio.com

When closed, Lumio resembles a book, the cover of which features an aesthetic wood grain effect. However, when Lumio is opened, it unfolds into a sculptural light. The lumina re's body is reminiscent of the fanning out of book pages. Since the luminaire opens up to any conceivable degree, it can be flexibly positioned on surfaces or edges. Being equipped with a rechargeable battery, the luminaire can be used without cable connection. The portable light provides powerful lighting regardless of location.

Im geschlossenen Zustand sieht Lumio aus wie ein Buch, dessen Einband eine ästhetische Holzmaserung zeigt. Sobald Lumio allerdings aufgeklappt wird, entfaltet sich eine skulptural anmutende Leuchte. Der Leuchtenkörper erinnert den Betrachter an aufgefächerte Buchseiten. Da sich die Leuchte in beliebigen Winkeln öffnen lässt, kann sie flexibel auf Flächen oder Kanten positioniert werden. Die Stromversorgung über einen Akku sorgt dafür, dass sie kabellos eingesetzt werden kann. Die tragbare Leuchte bietet eine leistungsstarke und ortsunabhängige Beleuchtung.

Statement by the jury
Lumio follows an original design idea, which had been implemented in a formally and functionally convincing manner. The lamp allows for intuitive operation.

Begründung der Jury
Lumio folgt einer originellen Gestaltungsidee, die formal und funktional überzeugend umgesetzt wurde. Die Leuchte ermöglicht eine intuitive Handhabung.

Lichtglas
Nick Veasey Edition:
The Alpha
Lamp
Leuchte

Manufacturer
GLIFS GmbH, Heitersheim, Germany
In-house design
Andreas Schall, Philipp Fischer
Web
www.glifs.de
Honourable Mention

The Alpha combines rustic timber, illuminated glass and eye-catching X-ray artwork to create an inventive table lamp. The base is made of upcycled wood from dismantled half-timbered houses in the Black Forest. The timeworn surfaces and historic tool marks of the untreated wood make each luminaire unique. The glass panel, embedded in the base, shows an X-ray image of a three-dimensional table lamp while a concealed light source illuminates the image.

Statement by the jury
The appealing implementation of this artful design idea achieves an elegant impression.

The Alpha vereint rustikales Holz, leuchtendes Glas und eine aufmerksamkeitsstarke Röntgenkunst zu einer originellen Tischleuchte. Den Sockel bildet upcycletes Holz von zurückgebauten Fachwerkhäusern aus dem Schwarzwald. Das unbehandelte Holz mit seinen Zeichen der Zeit, historischen Bearbeitungsspuren und Markierungen macht jede Leuchte zu einem Unikat. Die in den Sockel eingelassene Glasscheibe zeigt ein Röntgenbild von einer dreidimensionalen Tischleuchte, eine verborgene Lichtquelle lässt es erstrahlen.

Begründung der Jury
Die ansprechende Umsetzung dieser artifiziellen Gestaltungsidee erreicht eine elegante Anmutung.

Tap
Table Lamp
Tischleuchte

Manufacturer
Klasik M d.o.o., Zagreb, Croatia
Design
Havranek+Lugonja (Filip Havranek, Kristina Lugonja), Zagreb, Croatia
Web
www.havraneklugonja.com
Honourable Mention

Manufactured from a square aluminium plate, the Tap table lamp looks like a lighting object and is reminiscent of a folded sheet of paper. The bending of the aluminium plate creates two connected elements, one of them serving as a base while the other one serves as an illuminating surface. Furthermore, the aluminium constitutes the heat sink for the LED light source, which is placed on its edge. The structure is covered by synthetic paper, which pleasantly diffuses the light.

Statement by the jury
The artful production of this table lamp leads to a homogenous and aesthetically pleasing overall appearance.

Gefertigt aus einer quadratischen Aluminiumplatte, wirkt die Tischleuchte Tap wie ein Lichtobjekt und erinnert an ein gefaltetes Blatt Papier. Durch Biegung der Alu-Platte entstehen zwei zusammenhängende Teile, wobei das eine als Basis für den Ständer fungiert und das andere als beleuchtende Fläche dient. Das Aluminium bildet zudem den Kühlkörper für die am Rand platzierte LED-Lichtquelle. Bedeckt wird die Konstruktion mit einem synthetischen Papier, was die Lichtintensität angenehm abschwächt.

Begründung der Jury
Die kunstvolle Fertigung dieser Tischleuchte führt zu einem homogenen und ästhetisch ansprechenden Gesamtbild.

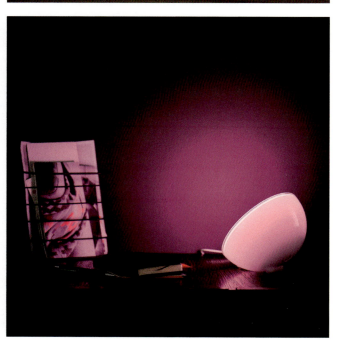

Hue Go
Lamp
Leuchte

Manufacturer
Royal Philips, Einchoven, Netherlands
In-house design
Web
www.philips.com

Hue Go is a portable lamp that can be positioned, controlled and used in multiple ways. It is suitable for direct and indirect lighting, adding particular highlights to the ambience of any room. Hue Go can be controlled directly or via a smart device, offering a wide selection of light effects such as cozy candlelight or enchanted forest. The thick-walled spherical body expresses solidity and material richness while its translucent surface gives the feel of pure, natural light.

Statement by the jury
Hue Go is an innovative lamp that allows the creative use of light. Its compact dimensions support individual use.

Als tragbare Leuchte kann Hue Go flexibel positioniert, angesteuert und eingesetzt werden. Sie eignet sich zur direkten oder indirekten Beleuchtung und setzt gezielt Akzente. Hue Go ist unmittelbar oder über intelligente Kommunikationsgeräte bedienbar und bietet die Auswahl zwischen diversen Lichteffekten wie gemütliches Kerzenlicht oder Zauberwald-Raumlicht. Der dickwandige, kugelförmige Leuchtenkörper drückt Stabilität und materielle Fülle aus, während seine transparente Oberfläche den Eindruck von purem Licht vermittelt.

Begründung der Jury
Hue Go ist eine innovative Leuchte, die den kreativen Umgang mit Licht ermöglicht. Ihr handliches Format unterstützt den individuellen Einsatz.

Meteorite
Luminaire
Leuchte

Manufacturer
Artemide S.p.A., Pregnana Milanese, Italy
Design
Pio & Tito Toso, Treviso, Italy
Web
www.artemide.com
www.pioetitotoso.com

Meteorite was developed with the aim of finding a new expressive language for luminaires made of blown glass. The idea for Meteorite derived from the fascination for Venetian glass with its uneven, seemingly broken surface. The specially developed glass-blowing and grinding processes are characterised by a double-layer artistic glass diffuser and a frost-covered moulding technique. The depth of the light created in the illuminated object impresses the beholder. This luminaire is part of a series of lights consisting of table, suspension, wall and ceiling versions.

Statement by the jury
The extraordinary materials of these luminaires achieve an entirely self-contained style. Their visually fascinating glass surface is eye-catching.

Meteorite entstand aus dem Bestreben nach einer neuen Formensprache für Leuchten aus mundgeblasenem Glas. Fasziniert von venezianischem Glas mit einer unebenen Oberfläche, das wie zerbrochen wirkt, entstand die Idee für Meteorite. Die speziell entwickelte Fertigung der Glasbläserei und der Schleifprozess zeichnen sich durch eine doppelschichtige Kunstglasabdeckung und Formtechnik aus. Die Lichttiefe, die innerhalb des beleuchteten Objekts erzeugt wird, beeindruckt den Betrachter. Diese Tischleuchte ist Teil einer Leuchtenserie, bestehend aus Tisch-, Pendel-, Wand- und Deckenversionen.

Begründung der Jury
Eine gänzlich eigenständige Anmutung erzielt die außergewöhnliche Materialität dieser Leuchten. Ihre visuell reizvolle Glasoberfläche ist ein Blickfang.

Hue Beyond
Lamp
Leuchte

Manufacturer
Royal Philips, Eindhoven, Netherlands
In-house design
Web
www.philips.com

Hue Beyond is an innovative lighting series with a dual light source that can be controlled through the Internet. Users can select and schedule different lighting modes via smart devices. The network lamp provides a versatile lighting experience combining its light quality with multiple colour tones and colour gradients. The satin glass of the lower illuminant captures and diffuses the main bulk of the light while the aluminium body on top creates direct highlights.

Statement by the jury
Through the expression of an original design idea, this lighting series offers versatile lighting options. Hue Beyond can be controlled from anywhere.

Hue Beyond ist eine innovative Leuchten-Serie mit Doppellichtquelle, die sich über das Internet steuern lässt. Via intelligenter Kommunikationsgeräte können unterschiedliche Beleuchtungsmodi ausgewählt und terminiert werden. Die vernetzte Leuchte bietet ein vielseitiges Lichterlebnis und kombiniert ihre Lichtqualität mit diversen Farbnuancen und Farbverläufen. Das satinierte Glas des unteren Leuchtschirms bündelt den Großteil des Lichts und verbreitet es, während der oben aufgesetzte Aluminiumkörper für direkte Lichtakzente sorgt.

Begründung der Jury
Als Ausdruck einer originellen Gestaltungsidee bietet Hue Beyond vielseitige Beleuchtungsmöglichkeiten und ist dabei ortsunabhängig steuerbar.

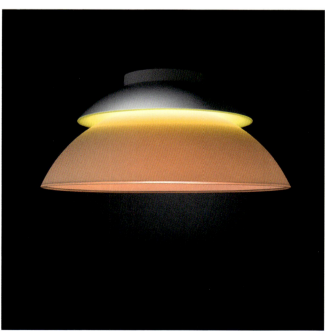

Luce Verde Slim
LED Wall and Pendant Luminaire
LED-Wand- und Pendelleuchte

Manufacturer
Sattler GmbH, Göppingen, Germany
In-house design
Web
www.sattler-lighting.com

Luce Verde Slim is an LED luminaire which integrates dried moss. It is available either as a pendant or wall-mounted luminaire. With the structure of its surface, the luscious green and the natural feel of the moss, the delicate LED ring luminaire conveys an atmosphere of well-being and simultaneously improves the acoustics of the room. The moss does not require any care and has an appealing tactile feel. The circular aluminium profile with integrated acrylic glass provides direct and homogenous lighting.

Statement by the jury
This design concept is both visually and haptically impressive. The moss-covered LED ring luminaire evokes pleasant emotions.

Luce Verde Slim ist eine LED-Leuchte mit integriertem Trockenmoos. Sie ist optional als Pendel- oder Wandleuchte erhältlich. Mit ihrer Oberflächenstruktur, dem satten Grün und der natürlichen Anmutung der Moose vermittelt die filigrane Ringleuchte ein Wohlfühlambiente und verbessert zudem die Raumakustik. Das pflegefreie Moosmaterial ist beständig und haptisch reizvoll. Das ringförmige Aluminiumprofil mit eingesetztem Acrylglas unterstützt eine direkte und homogene Beleuchtung.

Begründung der Jury
Dieses Gestaltungskonzept überzeugt sowohl visuell als auch haptisch. Die moosbegrünte LED-Ringleuchte erweckt angenehme Emotionen.

Forte
LED Wall Lamp
LED-Wandleuchte

Manufacturer
Grossmann Leuchten GmbH & Co. KG, Ense, Germany
In-house design
Stephanie Byrdus
Web
www.grossmann-leuchten.de

The Forte wall lamp conveys a minimalist elegance. Placed horizontally above a mirror or a picture, it emits a naturally coloured, glare-free light over its entire length. The stringent geometry of the lamp follows this function and is repeated on the frontal light edge, as a fine detail. The chrome plated aluminium profile is used to cool the interior, high-efficiency LED modules and harmoniously blends in with its environment. Forte is available in two lengths.

Statement by the jury
The delicate lines of this wall lamp capture the zeitgeist. In addition, the functionality impresses at a high level.

Die Wandleuchte Forte vermittelt eine minimalistische Eleganz. Horizontal über einem Spiegel oder einem Bild platziert, spendet sie über ihre gesamte Länge ein farbnatürliches und blendfreies Licht. Die stringente Geometrie der Leuchte folgt dieser Funktion und wiederholt sich als feines Detail in der frontalen Lichtkante. Das verchromte Aluminiumprofil dient der Kühlung der innen liegenden, leistungsstarken LED-Module und passt sich harmonisch seiner Umgebung an. Forte ist in zwei Längen erhältlich.

Begründung der Jury
Die filigrane Linienführung dieser Wandleuchte trifft den Zeitgeist. Zudem überzeugt die Funktionalität auf hohem Niveau.

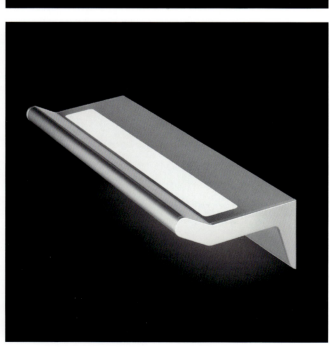

Sanesca
LED Wall Lamp
LED-Wandleuchte

Manufacturer
Trilux GmbH & Co. KG, Arnsberg, Germany
Design
Deck5 Dix Frankowski Noster Steber Industrialdesigner Partnerschaft, Essen, Germany
Web
www.trilux.com
www.deck5.de

Specifically designed for the health care sector, the Sanesca LED wall lamp blends harmoniously into patient rooms. It combines an indirect room light with a direct, glare-free reading light. The modular design enables the integration of sockets, data modules, nurse call buttons and further functional equipment. A long service life, long maintenance cycles as well as high reliability make the lamp a cost-efficient lighting solution. Thanks to its design, Sanesca is also easy to clean.

Statement by the jury
This wall lamp for the health care sector offers a high practical value. Its timeless elegance underlines the usefulness of its variable light output.

Gezielt konzipiert für einen Einsatz im Gesundheitswesen fügt sich die LED-Wandleuchte Sanesca harmonisch in Patientenzimmern ein. Sie verbindet ein indirektes Raumlicht mit einem direkten, blendfreien Leselicht. Ihr modularer Aufbau bietet die Integrationsmöglichkeit von Steckdosen, Daten-Modulen, Schwesternruf und weiterem funktionalen Equipment. Eine lange Lebensdauer und Wartungszyklen sowie Zuverlässigkeit machen sie zu einer kosteneffizienten Beleuchtungslösung. Aufgrund ihrer Konstruktion ist Sanesca leicht zu reinigen.

Begründung der Jury
Einen hohen Gebrauchswert bietet diese Wandleuchte für den Gesundheitsbereich. Ihre zeitlose Eleganz unterstreicht die variable Lichtleistung.

RS PRO LED Q1
Sensor LED Light
LED-Sensor-Leuchte

Manufacturer
Steinel GmbH,
Herzebrock-Clarholz, Germany
Design
Eckstein Design (Stefan Eckstein),
Munich, Germany
Web
www.steinel.de
www.eckstein-design.com

With its flat, square silhouette, the RS PRO LED Q1 presents itself as an indoor sensor LED light for demanding lighting situations. Despite the integrated sensor system, the luminaire features an installation height of only 6 cm. The ceiling and wall-mounted luminaire has a coverage angle of 360 degrees and can be linked with other luminaires. Equipped with a 27.5 watts LED light system, it yields a luminous flux of 1,960 lumens at 70 lm/W.

Statement by the jury
This sensor LED light stands out with a functionality that is thought-through to the last detail. Its form language lends it a timeless appeal.

Mit ihrer flachen, quadratischen Silhouette präsentiert sich die RS PRO LED Q1 als LED-Sensor-Innenraumleuchte für ansprechende Lichtsituationen. Trotz integrierter Sensorik hat die Leuchte eine Aufbauhöhe von nur knapp 6 cm. Die Decken- und Wandleuchte hat einen Erfassungswinkel von 360 Grad und ist mit weiteren Leuchten vernetzbar. Bestückt mit einem 27,5 Watt LED-Lichtsystem sorgt sie für einen Lichtstrom von 1.960 Lumen bei 70 lm/W.

Begründung der Jury
Eine bis ins Detail durchdachte Funktionalität zeichnet diese LED-Sensor-Leuchte aus. Ihre Formensprache verleiht ihr eine zeitlose Ästhetik.

Julia
LED Lamp
LED-Leuchte

Manufacturer
Luxuni GmbH, Leer, Germany
In-house design
Design
Phenix Lighting (Xiamen) Co., Ltd.
(Jean Chen), Xiamen, China
Web
www.luxuni.com
www.phenixlighting.com

Julia is a dimmable lamp with a strikingly flat, curvaceous shape. Although it is a pure surface-mounted luminaire, it is only 7.2 mm thick at the edge and 25 mm in the centre. Nevertheless, the LED light module provides a convincingly even light distribution. Consisting of only two parts, the luminaire is made of high-quality materials and easy to install. Julia is available in four sizes with diameters ranging from 300 to 600 mm and a power range from 15 to 30 watts.

Statement by the jury
This remarkably flat LED lamp achieves homogenous illumination. Its elegant contour discretely blends into the interior of any room.

Julia ist eine dimmbare Leuchte mit einer auffallend flachen, geschwungenen Bauform. Obwohl es sich um eine reine Aufbauleuchte handelt, ist sie am Rand nur 7,2 mm und in der Mitte 25 mm stark. Trotzdem bietet das LED-Lichtmodul eine überzeugend gleichmäßige Lichtverteilung. Da die aus hochwertigen Materialien gefertigte Leuchte nur aus zwei Teilen besteht, ist ihre Montage einfach. Julia ist in vier Abmessungen mit einem Durchmesser von 300 bis 600 mm sowie einer Leistung von 15 bis 30 Watt erhältlich.

Begründung der Jury
Ein homogenes Lichtbild erreicht diese bemerkenswert flach konstruierte LED-Leuchte. Ihre elegante Kontur fügt sich dezent ins Raumambiente ein.

INTRO
LED Lighting System
LED-Beleuchtungssystem

Manufacturer
Zumtobel Lighting GmbH, Dornbirn, Austria
Design
Sottsass Associati, Milan, Italy
Web
www.zumtobel.com
www.sottsass.it

By integrating different types of spotlights, the modular Intro lighting system offers an efficient and customisable lighting solution for different shop zones, including shop windows, display surfaces and shelves. The development of the liteCarve reflector technology enables a balanced rectangular light distribution. Starting from a point light source, the reflector directs the light indirectly and accurately, ensuring precise illumination of vertical surfaces, shelves and displays.

Durch die Integration unterschiedlicher Strahler-Typen bietet das modulare Beleuchtungssystem Intro eine effiziente und individualisierbare Beleuchtungslösung für Shop-Bereiche wie Schaufenster, Aktionsflächen oder Regale. Mit der Entwicklung der liteCarve-Reflektortechnologie gelingt eine ausgewogene, rechtwinklige Lichtverteilung. Von einer Punktlichtquelle ausgehend, leitet der Reflektor das Licht indirekt und zielgerichtet und ermöglicht das Ausleuchten vertikaler Flächen, Regale und Displays.

Statement by the jury
A design concept targeting different lighting requirements in retail lends Intro a high degree of functionality.

Begründung der Jury
Ein auf die Beleuchtungsanforderungen im Einzelhandel abgestimmtes Gestaltungskonzept verleiht Intro ein hohes Maß an Funktionalität.

Baldur, Skadi
LED Light Line, Spotlight
LED-Lichtband, Spot

Manufacturer
Nordeon GmbH, Springe, Germany
Design
Van Berlo, Eindhoven, Netherlands
Web
www.nordeon.com

The Baldur modular LED light line system offers a high light quality at reduced energy consumption. A subtle design with minimised dimensions and a small cross-sectional area characterises the luminaire's body. It is suitable for both general and accent lighting, harmoniously blending into any room design concept. The LED drivers for the optional Skadi spotlights are concealed inside the rail. Thanks to a consistent design, Skadi and Baldur constitute a single formal unit.

Statement by the jury
The Baldur LED light line system impresses as a formally and functionally sophisticated product solution, which can be complemented with the Skadi spotlights in a comfortable and easy way.

Das modulare LED-Lichtbandsystem Baldur bietet eine hohe Lichtqualität bei gleichzeitiger Energieeinsparung. Den Leuchtenkorpus mit seinen minimierten Abmessungen und seiner geringen Querschnittsfläche prägt eine dezente Formensprache. Er eignet sich für eine Allgemein- und Akzentbeleuchtung und fügt sich harmonisch in jedes Raumkonzept ein. Die LED-Treiber für den optional einsetzbaren Spot Skadi bleiben in der Tragschiene verborgen. Mittels einer einheitlichen Gestaltung bilden Skadi und Baldur eine formale Einheit.

Begründung der Jury
Als formal und funktional durchdachte Produktlösung überzeugt das LED-Lichtbandsystem Baldur, das sich komfortabel einfach mit den Spots Skadi ergänzen lässt.

Shopstar
Spotlight
Strahler

Manufacturer
Molto Luce GmbH, Wels, Austria
In-house design
Web
www.moltoluce.com

The energy-efficient Shopstar track spotlight meets the high demands of shop lighting with its powerful lighting technology. The spotlight is 350 degrees pivotable and 90 degrees tiltable for precise targeting of objects. Its cylindrical head is made of die-cast aluminium and available in white, black and matt silver. Shopstar features an innovative heat management system based on passive cooling.

Statement by the jury
Shopstar blends a remarkably functional quality with a timeless minimalist design. A consistent overall concept.

Der energieeffiziente Stromschienenstrahler Shopstar erfüllt mit seiner leistungsstarken Lichttechnik die gehobenen Ansprüche einer Ladenbeleuchtung. Der Strahler lässt sich um 350 Grad drehen sowie um 90 Grad schwenken und ermöglicht dadurch eine flexible Ausrichtung auf das Zielobjekt. Sein zylindrischer Kopf ist aus druckgegossenem Aluminium gefertigt und in den Farben Weiß, Schwarz und Mattsilber erhältlich. Shopstar verfügt über ein innovatives Wärmemanagement, welches auf passiver Kühlung basiert.

Begründung der Jury
Eine bemerkenswert funktionale Qualität vereint Shopstar mit einer zeitlos minimalistischen Linienführung. Ein stimmiges Gesamtbild.

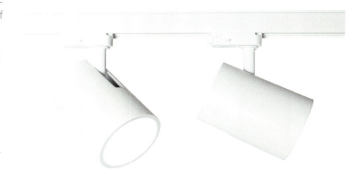

Clockwork
Spot-Lighting System
Wandstrahler-System

Manufacturer
Royal Philips, Eindhoven, Netherlands
In-house design
Web
www.philips.com

Clockwork is an affordable spot-lighting system designed to meet the requirements of private homes. Its patented click!FIX system makes installation fast and easy. Swiveling the spotlights allows the light beams to be flexibly directed while the brightness can be dimmed. Thanks to hidden connection elements, the construction made of aluminium die-cast parts yields a purist appearance. The subtle colour combinations of the powder coating lend the luminaire a distinctive look.

Statement by the jury
Clockwork presents a successful and attractive reinterpretation of a classic spotlight rail. It also convinces with functionally sophisticated details.

Clockwork ist ein preisgünstiges Wandstrahler-System, das den Anforderungen in Privathaushalten gerecht wird. Sein patentiertes click!FIX-System vereinfacht den Einbau. Durch Drehen der Strahler kann der Lichtstrahl flexibel ausgerichtet werden, zudem ist die Lichtstärke dimmbar. Aufgrund versteckter Anschlussteile erreicht die Konstruktion aus pulverbeschichteten Aluminium-Druckgussteilen eine puristische Anmutung. Die dezenten Farbkombinationen verleihen der Leuchte zusätzlich Charakter.

Begründung der Jury
Auf ansprechende Weise gelingt Clockwork die Neuinterpretation einer klassischen Strahler-Schiene. Zudem überzeugen die funktional ausgereiften Details.

Swap
Recessed LED Downlight
LED-Einbau-Downlight

Manufacturer
Arkoslight, Ribarroja del Turia, Spain
In-house design
Rubén Saldaña Acle
Web
www.arkoslight.com

The Swap recessed LED downlight features a minimalist design. Its recessed reflector provides a high degree of visual comfort. With a bezel that is reduced to minimal dimensions, Swap appears to have no frame at all and thus blends discreetly with the architecture. The downlight is available in white, gold and red, as well as in three different sizes and a dimmable version. It comes with an optional ring accessory to extend the bezel and replace old light fittings without having to alter or repaint the ceiling.

Eine minimalistische Formensprache zeigt das LED-Einbau-Downlight Swap. Sein zurückgesetzter Reflektor sorgt für einen hohen Sichtkomfort. Dank seinem auf ein Kleinstmaß reduzierten Zierring erweckt Swap den Eindruck einer rahmenlosen Leuchte, die sich dezent in die Architektur integriert. Das Downlight wird in Weiß, Gold und Rot angeboten, es ist in drei Größen sowie in einer dimmbaren Version erhältlich. Im Lieferumfang ist ein optionaler Zusatzring inbegriffen, der den Zierring vergrößert, um alte Deckenleuchten ohne Handwerksarbeiten oder erneutes Streichen austauschen zu können.

Statement by the jury
Whether in subtle white or as one of the colour versions, Swap allows for an appealing and energy-efficient lighting of interiors.

Begründung der Jury
Ob dezent in Weiß oder in einer Farbausführung, Swap ermöglicht eine ansprechende und energieeffiziente Raumausleuchtung.

Ergetic
LED Recessed Luminaire
LED-Einbauleuchte

Manufacturer
Intra lighting, Miren, Slovenia
Design
Serge Cornelissen BVBA,
Kortrijk (Marke), Belgium
Web
www.intra-lighting.com
www.sergecornelissen.com

The Ergetic LED recessed luminaire was conceived as a versatile lighting solution for a variety of commercial architectural spaces such as offices but also for health care and wellness facilities. The luminaire features three modes: the general light mode with white LEDs provides a homogenous and well-balanced light quality while the indirect colour-changing RGB mode creates different ambient atmospheres. The centre piece of the luminaire integrates an LED spotlight for accent lighting.

Statement by the jury
Focussed on different lighting requirements, this LED recessed luminaire presents the successful implementation of versatile functionality.

Die LED-Einbauleuchte Ergetic wurde als vielseitig einsetzbare Beleuchtungslösung für verschiedene Objektbereiche wie Büros, aber auch für Räumlichkeiten im Gesundheitswesen, Wellness etc. konzipiert. Die Leuchte umfasst drei Funktionsmodi: Der allgemeine Lichtmodus mit weißen LEDs bietet eine gleichmäßige und ausgewogene Lichtqualität, während die indirekte RGB-Farbwechselkomponente unterschiedliche Atmosphären kreiert. Im Mittelpunkt der Leuchte wurde zudem ein LED-Strahler integriert, der für eine akzentuierte Beleuchtung sorgt.

Begründung der Jury
Fokussiert auf unterschiedliche Beleuchtungsbedürfnisse gelingt mit dieser LED-Einbauleuchte die Umsetzung einer variablen Funktionalität.

Quantum
LED Downlight

Manufacturer
Linea Light Group, Treviso, Italy
In-house design
Web
www.linealight.com

Quantum is a downlight with a bell-shaped housing. However, its flange forms visual unity with a suspended ceiling. The heat sink is made of die-cast aluminium while the casing is coated aluminium with an either embossed or golden finish. The contours of the housing contribute to directing the light outward and limit dispersion. Quantum uses an innovative filter system that maximises the clarity of the light beam and the visual comfort. The spotlight is available in a tilting and a rotatable version as well as in four models of different light intensity.

Statement by the jury
The Quantum LED downlight harmoniously blends into any ceiling and impresses with a high degree of lighting comfort.

Quantum ist ein Downlight, dessen Gehäuse eine glockenförmige Kontur zeigt. Sein Flansch bildet hingegen eine visuelle Einheit mit der abgehängten Decke. Der Kühlkörper ist aus Aluminiumspritzguss und das Gehäuse aus lackiertem Aluminium mit gaufrierter oder vergoldeter Oberfläche gefertigt. Die Linienführung des Gehäuses trägt dazu bei, die Lichtabstrahlung nach außen zu lenken und Verluste einzuschränken. Quantum nutzt ein innovatives Filter-System, das die Reinheit des Lichtstrahls und den Sehkomfort maximiert. Der Strahler ist in einer schwenkbaren oder drehbaren Version und in vier unterschiedlich lichtstarken Modellen erhältlich.

Begründung der Jury
Harmonisch fügt sich das LED-Downlight Quantum in die Zimmerdecke ein und überzeugt durch ein hohes Maß an Beleuchtungskomfort.

OLED One
OLED Under-Cabinet Light Fixture
OLED-Unterbauleuchte

Manufacturer
Hera GmbH & Co. KG, Enger, Germany
Design
Studio Ambrozus
(Marco Vorberg, Stefan Ambrozus),
Cologne, Germany
Web
www.hera-online.de
www.studioambrozus.de

OLED One is a delicate under-cabinet light fixture, designed for use in kitchens and living rooms. The use of organic LEDs allows for an extremely flat body with a thin LED plate of only approximately 1 mm. The lighting panel virtually floats in mid-air below the wall cabinet and, with a visible edge of only 2.5 mm, looks like a thin sheet of paper. The OLED illuminant creates a pleasant atmospheric light and evenly illuminates work surfaces. The lighting panel also swivels.

Statement by the jury
A high degree of innovation characterises this comfortable under-cabinet light fixture. Its extremely flat body surprises the beholder.

OLED One ist eine filigrane Unterbauleuchte, konzipiert für Küche und Wohnbereich. Der Einsatz organischer Leuchtdioden ermöglicht die Gestaltung eines sehr flachen Korpus, wobei die flächige LED-Platine nur circa 1 mm dick ist. Der Leuchtschirm scheint unter dem Hängeschrank zu schweben und wirkt mit seiner 2,5 mm hohen Sichtkante wie ein Blatt Papier. Das OLED-Leuchtmittel erzeugt ein angenehm atmosphärisches Licht und leuchtet die Arbeitsfläche homogen aus. Der Leuchtschirm ist schwenkbar.

Begründung der Jury
Ein hoher Innovationsgrad zeichnet diese komfortable Unterbauleuchte aus, ihr sehr flacher Korpus überrascht den Betrachter.

PowerBalance gen2
Office Luminaire
Büroleuchte

Manufacturer
Royal Philips, Eindhoven, Netherlands
In-house design
Web
www.philips.com

The PowerBalance gen2 LED office luminaire was designed following a modular principle in order to provide users with versatile lighting solutions for different office environments. It offers sophisticated and energy-efficient interior illumination suitable to replace T5 fluorescent tubes. PowerBalance gen2 can easily be integrated into ceilings with either exposed or concealed T-bars, as well as in plaster ceilings and modular grid ceilings. The LED light is also available in a surface-mounted version, and in different shapes and sizes.

Statement by the jury
With impressive functionality, the PowerBalance gen2 flexibly adapts to different interior lighting requirements.

Die LED-Büroleuchte PowerBalance gen2 wurde in Modulbauweise konzipiert, um verschiedene Lösungen für die jeweilige Büroumgebung zu ermöglichen. Sie bietet eine hochwertige Innenraum-Beleuchtung und eignet sich für den energieeffizienten Ersatz von T5-Leuchtstoffröhren. Die PowerBalance gen2 kann unkompliziert in Decken mit freiliegenden oder verdeckten T-Profilen sowie in Putzdecken und Bandrasterdecken eingebaut werden. Die LED-Leuchte ist auch in einer Aufbauversion und in unterschiedlichen Formen und Größen verfügbar.

Begründung der Jury
Mit einer überzeugenden Funktionalität lässt sich die PowerBalance gen2 variabel an unterschiedliche Raumlicht-Anforderungen anpassen.

GreenUp LED Highbay
Industrial Luminaire
Industrieleuchte

Manufacturer
Royal Philips, Eindhoven, Netherlands
In-house design
Web
www.philips.com

GreenUp LED Highbay is a compact industrial luminaire for the Chinese market. A robust and energy efficient solution for industrial warehouses, it is powered by modern LED technology and controlled by motion and daylight sensors. It is made of sheet metal for a cost efficient manufacturing process while its minimal volume and low weight allow for economical transport. The luminaire, which is quick and easy to install and operate, offers high light efficiency and intelligent control.

Statement by the jury
This industrial luminaire presents a successful implementation of a purely functionally oriented design concept.

GreenUp LED Highbay ist eine kompakte Industrieleuchte für den chinesischen Markt. Die robuste, energieeffiziente Lösung für Industriehallen beruht auf einer zeitgemäßen LED-Technologie und ist mit einem Bewegungs- und Tageslichtsensor ausgestattet. Für den kostengünstigen Herstellungsprozess wird Feinblech verwendet. Das minimale Volumen und geringe Gewicht ermöglichen einen ökonomischen Transport. Die Leuchte bietet eine gute Lichtausbeute und eine intelligente Steuerung, die leicht einzubauen und zu handhaben ist.

Begründung der Jury
Bei dieser Industrieleuchte gelingt die stringente Umsetzung eines rein funktional ausgerichteten Gestaltungskonzepts.

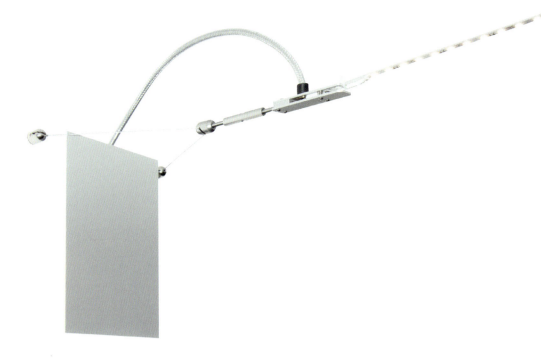

STRAPLED Advanced
LED Lighting System
LED-Beleuchtungssystem

Manufacturer
KIRRON light components GmbH & Co. KG,
Korntal-Münchingen, Germany
In-house design
Ronny Kirschner
Web
www.kirronlightcomponents.com

Strapled Advanced is an LED lighting system that looks like a light beam crossing the room. This effect is based on a transparent plastic composite band with attached linear LEDs. In order to appear like a light beam, the band needs to be strongly tightened, which requires mechanical fixtures. The transition from round to elongated shapes is achieved by reducing the radii and increasing the elements towards the band. A reduction of the thread's visible part can be achieved by the tolerance of the two fixtures to each other. The electrical connection is integrated in a longitudinal direction between the fixtures.

Statement by the jury
In this LED lighting system, a functionally sophisticated and unobtrusive design puts the light effect centre stage.

Strapled Advanced ist ein LED-Beleuchtungssystem, das wie ein durch den Raum „fliegender" Lichtstrahl wirkt. Grundlage des Effekts ist ein transparentes Komposit-Band, auf dem lineare LEDs befestigt sind. Das Band muss stark gespannt sein, um wie ein Lichtstrahl zu wirken. Dazu sind mechanische Anschlusselemente nötig. Der Übergang von runden zu länglichen Formen wird durch eine Reduzierung der Radien und länger werdende Elemente bis zum Spannband gelöst. Eine Reduzierung des sichtbaren Gewindeteils wird durch das Spiel der beiden Spannvorrichtungen zueinander erreicht. Die elektrische Verbindung ist in Längsrichtung zwischen den Spannelementen integriert und abgedeckt.

Begründung der Jury
Eine funktional durchdachte und dezente Gestaltung rückt bei diesem LED-Beleuchtungssystem gekonnt den Lichteffekt in den Vordergrund.

Fylo
Integrated LED Profile
Integrierbares LED-Profil

Manufacturer
Linea Light Group, Treviso, Italy
In-house design
Design
Studio Mathesis2012, Treviso, Italy
Web
www.linealight.com

Available for interior and exterior environments, Fylo is an easy-to-install LED profile. Its housing is customisable with different colour coatings and houses a linear circuit of LEDs to illuminate coves or frames. Thanks to the lightness and versatility of expanded polystyrene (EPS), the installation of Fylo is cost-effective. The integrated LED profile provides all the comfort of indirect lighting and achieves an impressive aesthetic ambience. The outdoor version can either be installed in the external wall insulation or in the masonry.

Statement by the jury
This integrated LED profile stands out through a high practical value and delivers a light that makes a distinctive architectural statement.

Erhältlich in einer In- und Outdoor-Version ist Fylo ein einfach zu installierendes LED-Profil. Sein Gehäuse ist in unterschiedlichen Farbbeschichtungen individualisierbar und beherbergt einen linearen LED-Schaltkreis zur Ausleuchtung von Nischen oder Rahmen. Aufgrund der Leichtigkeit und Vielseitigkeit von expandiertem Polystyrol (EPS) ermöglicht Fylo eine kostengünstige Installation. Das integrierbare LED-Profil bietet allen Komfort einer indirekten Beleuchtung und erzielt eine eindrucksvolle Raumästhetik. In seiner Outdoor-Version kann es entweder in der Außenwand-Dämmung oder im Mauerwerk installiert werden.

Begründung der Jury
Dieses integrierbare LED-Profil zeichnet ein hoher Gebrauchswert aus. Es liefert ein Raumlicht, das architektonische Akzente setzt.

HYDRALUX
LED Linear Luminaire
LED-Linearleuchte

Manufacturer
LED Linear GmbH,
Neukirchen-Vluyn, Germany
In-house design
Michael Kramer
Web
www.led-linear.com

Hydralux is a line of LED lights that is flexible in use. The luminaire can be cut to the desired length on site and easily be installed by adhesive bonding. It is field attachable from 62.5 mm to a maximum of 7.5 metres with only one electrical supply. The filigree linear luminaire in IP67 is available with up to 1,637 lm/m. The white base profile with lateral fins and reflective surface ensures a high light output. Hydralux lights are waterproof, dustproof and UV-protected by a polyurethane encapsulation. The well thought-through concept and the small cross-section, with a width of 12.5 mm and a height of 3.5 mm, allow for a flexible use in different areas such as shops and retail lighting, or applications in furniture and shelves.

Statement by the jury
The Hydralux LED linear luminaire offers a high practical value. It creates a distinctive ambience whilst capturing the zeitgeist with their style.

Hydralux ist eine flexibel einsetzbare LED-Linearleuchte. Sie kann direkt vor Ort auf die gewünschte Länge gekürzt und dann einfach durch Aufkleben montiert werden. Mit nur einer elektrischen Einspeisung ist sie von 62,5 mm bis maximal 7,5 m frei konfektionierbar. Die filigran anmutende Linearleuchte in IP67 ist mit bis zu 1.637 lm/m erhältlich. Ihr weißes Grundprofil mit seitlichen Stegen und einer reflektierenden Oberfläche begünstigt eine hohe Lichtausbeute. Hydralux ist durch einen Polyurethan-Verguss wasser-, staub- und UV-geschützt. Das durchdachte Konzept und der geringe Querschnitt von 12,5 mm Breite x 3,5 mm Höhe ermöglichen einen flexiblen Einsatz, so z.B. im Bereich Shop- und Retail-Beleuchtung oder Applikationen in Möbeln und Regalen.

Begründung der Jury
Einen hohen Gebrauchswert bietet die LED-Linearleuchte Hydralux. Ihre charakteristische Raumwirkung trifft den Zeitgeist.

VarioLED Flex AMOR
LED Light Line
LED-Lichtlinie

Manufacturer
LED Linear GmbH,
Neukirchen-Vluyn, Germany
In-house design
Michael Kramer, Michael Titze,
Sören Bleul
Web
www.led-linear.com

VarioLED Flex Amor is a miniaturised, opal encapsulated, flexible LED design light line. It is the uniformity of the light in combination with a very small cross section of only 5 mm width x 13 mm height that characterises Amor. It is horizontally bendable with a minimum radius of 3 cm. It also offers easy installation, high resistance against salt water and UV-light, and the connection of individual luminaires with a nearly seamless light-to-light transition. The LED light line is suitable for a variety of applications in the areas of drywall and furniture construction, decorative accent lighting, and exterior façades. The installation of the luminaire does not require any substructure.

Statement by the jury
High light output and innovative mounting options characterise this LED light line as a versatile product solution.

Die VarioLED Flex Amor ist eine miniaturisierte, opal vergossene und flexible LED-Design-Lichtlinie. Die Gleichmäßigkeit des Lichts in Kombination mit einem sehr kleinen Querschnitt von 5 mm Breite x 13 mm Höhe zeichnet Amor aus. Sie ist horizontal mit einem minimalen Radius von 3 cm biegsam. Gleichzeitig bietet sie eine einfache Installation, eine hohe Beständigkeit gegenüber Salzwasser und UV-Licht sowie die Verbindung einzelner Leuchten mit einem fast nahtlosen Leuchte-zu-Leuchte-Übergang. Die LED-Lichtlinie eignet sich für eine Vielzahl an Applikationen in den Bereichen Trockenbau, Möbelbau, dekorative Akzentbeleuchtung und Außenfassade. Dabei ist für den Einbau der Leuchte keine Unterkonstruktion notwendig.

Begründung der Jury
Eine hohe Lichtleistung und innovative Montagemöglichkeiten zeichnen die LED-Lichtlinie als flexibel einsetzbare Produktlösung aus.

Baldachin
Ceiling Connection
Deckenanschluss

Manufacturer
Georg Bechter Licht,
Langenegg, Austria
In-house design
Georg Bechter
Web
www.georgbechterlicht.at

As an integrated ceiling connection for standard pendant lights, the Baldachin series is a sophisticated solution. It is perceived as a deformation of the ceiling and not recognisable as a separate unit. Baldachin is available with the outlet facing the inside or the outside. The prefabricated plaster module is fitted into the ceiling; the edges are leveled out and painted together with the ceiling. It integrates the technical requirements for mounting a bayonet joint. The power cord is connected to the luminaire and the entire luminaire, including the bayonet pin, is locked onto the module.

Als flächenbündig integrierter Deckenanschluss für handelsübliche Hängeleuchten ist die Baldachin-Serie eine raffinierte Lösung. Der Anschluss wird als Verformung der Decke wahrgenommen und ist als Einzelelement nicht erkennbar. Baldachin ist mit einem nach innen oder außen gestülpten Anschluss erhältlich. Das vorgefertigte Gipsmodul wird in die abgehängte Decke eingespachtelt und mitgestrichen. Es integriert die technischen Erfordernisse zum Aufhängen eines Bajonettverschlusses. Die Elektrozuleitung wird an die Leuchte angeschlossen und die gesamte Leuchte inklusive des Bajonettbolzens am Gipsmodul arretiert.

Statement by the jury
The playful approach to the perception of the product characterises this original design idea – a functionally and formally extraordinary design.

Begründung der Jury
Der spielerische Umgang mit der Wahrnehmung zeichnet diese originelle Gestaltungsidee aus – ein funktional und formal außergewöhnlicher Entwurf.

OrbiLED
LED Light Series
LED Leuchtmittel-Serie

Manufacturer
Econlux GmbH, Cologne, Germany
In-house design
Daniel Muessli
Web
www.econlux.de
Honourable Mention

The OrbiLED light series is equipped with an innovative LED technology, the COD (Chip On Device). Thanks to this technology, the LED does not need a housing, a plate or a heat sink. The cooling takes place exclusively by convection. Nevertheless, the contours of the LED light bulb comply with industry standards. OrbiLED is particularly light and, thanks to a 360 degree beam angle, achieves a homogenous lighting. The light bulbs are available in clear and frosted versions, both with a light output of 800 lumens.

Statement by the jury
As an innovative product solution, the OrbiLED light series provides a high degree of lighting comfort.

Die Leuchtmittel-Serie OrbiLED ist mit einer innovativen LED-Technik ausgestattet, der COD (Chip On Device). Daher kommt die COD LED ohne Gehäuse, Platine und Kühlkörper aus. Die Kühlung erfolgt ausschließlich über Wärmeübertragung. Dennoch entsprechen die Konturen der LED-Lampe den genormten Standards. Die OrbiLED ist besonders leicht und erreicht dank eines 360-Grad-Ausstrahlwinkels ein homogenes Lichtbild. Die Lampen sind in einer klaren und einer milchigen Variante, jeweils mit einer Leistung von 800 Lumen, erhältlich.

Begründung der Jury
Als innovative Produktlösung bietet die Leuchtmittel-Serie OrbiLED ein hohes Maß an Beleuchtungskomfort.

epoFlorium
LED Light Bulb
LED-Glühbirne

Manufacturer
Epoch Chemtronics Corp., Epoch Lighting, Chupai City, Hsinchu Hsien, Taiwan
In-house design
epoFlorium Design Team, Epoch Lighting
Web
www.lighting-epoch.co.uk/en

Inspired by a blooming lily, epoFlorium features a streamlined contour with a flat lamp head. Its efficient LED and light guide plate technology fills the room with direct light that provides highly pleasant visual comfort. The E27 lamp cap fits into standard lamp holders. The LED light bulb is available in a variety of eye-catching colours, intensifying the light effect of the flower's contour. epoFlorium is an eco-friendly and creative product solution.

Statement by the jury
The epoFlorium LED light bulb follows an emotionally appealing design concept which is convincing formally as well as functionally.

Inspiriert von einer aufblühenden Lilie zeigt epoFlorium eine stromlinienförmige Kontur mit einem flachen Lampenkopf. Ihre effiziente LED- und Lichtleiterplatten-Technologie taucht den Raum in ein direktes Licht, das als angenehm empfunden wird. Der Sockel E27 passt in gängige Leuchten-Gewinde. Die LED-Lampe ist in einer Vielzahl von aufmerksamkeitsstarken Farben erhältlich, was den Lichteffekt der Blumenkontur visuell verstärkt. epoFlorium ist eine umweltfreundliche und kreative Produktlösung.

Begründung der Jury
Die LED-Glühbirne epoFlorium folgt einer emotional ansprechenden Gestaltungsidee, die formal wie funktional überzeugt.

Misfit Bolt
LED Light Bulb
LED-Glühbirne

Manufacturer
Misfit, Burlingame (California), USA
In-house design
Diana Chang, Timothy Golnik
Web
www.misfit.com

Misfit Bolt is a wireless light bulb that allows users to personalise their home lighting with numerous light effects, such as a sunrise simulation, and can be controlled via mobile devices. At over 800 lumens, Bolt replaces a standard incandescent bulb while using less than a quarter of the energy. The aesthetic overall appearance is dominated by an anodised aluminium body which encases the heat sink. The twisting fins direct heat away from the electronic components and cradle a glass diffusor that achieves a homogenous LED light with no hot-spots.

Misfit Bolt ist eine drahtlose LED-Lampe, die eine individuelle Raumbeleuchtung mittels zahlreicher Lichteffekte wie z.B. einer Sonnenaufgangssimulation ermöglicht und sich über Mobilgeräte steuern lässt. Mit über 800 Lumen ersetzt Bolt herkömmliche Glühbirnen und verbraucht dabei weniger als ein Viertel der Energie. Das ästhetische Gesamtbild wird durch ein eloxiertes Aluminiumgehäuse dominiert, das als Kühlkörper fungiert. Seine gewundenen Kühlrippen leiten die Wärme von den elektronischen Bauteilen ab und fassen den Glasschirm ein, der ein homogenes LED-Licht ohne störende Lichtpunkte erzeugt.

Statement by the jury
Misfit Bolt convinces as an innovative and durable product solution. The LED light bulb features an autonomous design language and provides a high ease-of-use.

Begründung der Jury
Als innovative und langlebige Produktlösung überzeugt Misfit Bolt. Die LED-Glühbirne zeigt eine eigenständige Formensprache und bietet einen hohen Bedienkomfort.

Linno Crystal
LED Lamp
LED-Lampe

Manufacturer
Linno, Seoul, South Korea
In-house design
Web
www.linno.com

The Linno Crystal LED lamp with its acrylic diffuser lens features a unique aesthetic. An especially eye-catching feature is the colour contrast between lens and socket. The user-friendly device allows the setting of brightness, colour temperature and up to 16 million colours. It is either controlled by a remote control or, after downloading the Linno App, by another smart communication device. The Unified Control System allows the central control of several lamps.

Statement by the jury
This distinctively designed LED lamp achieves a high level of recognition. Its functionality offers impressive ease of use.

Die LED-Lampe Linno Crystal zeigt mit ihrer Streulinse aus Acryl eine eigenständige Ästhetik. Auffällig ist insbesondere der Farbkontrast zwischen Sockel und Linse. Benutzerfreundlich können bei dieser Lampe Helligkeit, Farbtemperatur und bis zu 16 Millionen Farben eingestellt werden. Die Steuerung erfolgt mithilfe einer Fernbedienung oder wird, nach dem Herunterladen der Linno-App, über intelligente Kommunikationsgeräte kontrolliert. Das Unified Control System ermöglicht die zentrale Bedienung mehrerer Lampen.

Begründung der Jury
Einen hohen Wiedererkennungseffekt erreicht diese markant gestaltete LED-Lampe. Ihre Funktionalität bietet einen beeindruckenden Bedienkomfort.

Nanoleaf One
LED Light Bulb
LED-Glühbirne

Manufacturer
Nanoleaf, Hong Kong
In-house design
Web
www.nanoleaf.me

The Nanoleaf One LED light bulb follows an independent design concept, which aims to redefine current energy efficiency standards. The circuit board housing features the geometry of a dodecahedron and guarantees optimal cooling, allowing Nanoleaf One to do entirely without an aluminium heat sink. In addition, the durable light bulb achieves a light output of 120 lumens per watt, thanks to a special power supply and a performance-oriented LED housing. It is produced in a resource-friendly way.

Statement by the jury
This innovative LED light bulb impresses with a characteristic design as well as a sophisticated functionality.

Die LED-Lampe Nanoleaf One folgt einem eigenständigen Gestaltungskonzept, welches die gängigen Energieeffizienz-Standards neu definieren möchte. Ihr Platinen-Gehäuse zeigt die Geometrie eines Dodekaeders und gewährleistet eine sehr gute Kühlung, wodurch Nanoleaf One gänzlich auf einen Aluminium-Kühlkörper verzichten kann. Darüber hinaus erreicht die langlebige Lampe mittels einer speziellen Stromversorgung und ihres leistungsgerechten LED-Gehäuses eine Lichtleistung von 120 Lumen pro Watt. Sie wird ressourcenschonend produziert.

Begründung der Jury
Diese innovative LED-Glühbirne beeindruckt mit einer charakteristischen Linienführung sowie ihrer ausgereiften Funktionalität.

Nanoleaf Bloom
LED Light Bulb
LED-Glühbirne

Manufacturer
Nanoleaf, Hong Kong
In-house design
Web
www.nanoleaf.me

Nanoleaf Bloom is an energy-efficient LED light bulb that is dimmable via a traditional light switch. As a result, there is no need for the installation of an external dimmer. The desired brightness can be conveniently and continuously adjusted by the switch and can be flexibly adapted to individual lighting requirements. The self-contained design language features austere geometrical contours and combines them with the colour accents of the circuit-board structure.

Statement by the jury
An eye-catching colour palette as well as an unconventional bulb profile underline the convincing ease of use of this light bulb.

Nanoleaf Bloom ist eine energieeffiziente LED-Lampe, die sich mit jedem herkömmlichen Lichtschalter dimmen lässt. Die komfortable Bedienung per Lichtschalter macht den Einbau eines externen Dimmers überflüssig. Die gewünschte Lichtleistung ist über den Schalter stufenlos regelbar und passt sich den individuellen Beleuchtungsbedürfnissen flexibel an. Die eigenständige Formensprache nutzt eine streng geometrische Kontur und verbindet diese mit einer farblich akzentuierten Platinen-Struktur.

Begründung der Jury
Eine aufmerksamkeitsstarke Farbwahl sowie eine unkonventionelle Lampenkontur unterstreichen den überzeugenden Bedienkomfort dieser Glühbirne.

AruMAZE! E27
LED Lamp
LED-Lampe

Manufacturer
GIXIA Group Co.,
Jhubei City, Hsinchu County, Taiwan
In-house design
Web
www.gixia-group.com

Made entirely of plastics, the AruMAZE! E27 is an innovation in the LED lighting segment. Instead of metal glow filaments, it uses thermally conductive plastics and electrically conductive glue to connect the light with the power supply, resulting in a simplified assembly. The carefully designed LGP micro-structure of the lamp shade offers brilliant lighting effects. The seamless transition of the screw base and the reflector unites both elements into a single visual entity.

Statement by the jury
This plastic LED lamp impresses as an innovative product solution. AruMAZE! E27 demonstrates a discrete form language.

Komplett aus Kunststoff gefertigt, stellt AruMAZE! E27 eine Innovation im LED-Lampen-Segment dar. Anstelle von Metall-Glühwendeln werden thermisch leitfähige Kunststoffe und elektrisch leitende Klebstoffe zur Stromnetzverbindung eingesetzt – das sorgt für eine einfache Montage. Die sorgfältig entwickelte LGP-Mikrostruktur des Lampenschirms ermöglicht eine brillante Beleuchtungswirkung. Der Übergang zwischen Schraubsockel und Reflektor ist nahtlos und vereint beides zu einer visuellen Einheit.

Begründung der Jury
Als innovative Produktlösung überzeugt diese LED-Lampe aus Kunststoff. AruMAZE! E27 zeigt dabei eine eigenständige Formensprache.

Trick
LED Wall and
Ceiling Luminaire
LED-Wand- und Deckenleuchte

Manufacturer
iGuzzini illuminazicne,
Recanati (Macerata), Italy
Design
Skira d.o.o. (Dean Skira), Pula, Croatia
Web
www.iguzzini.com
www.skira.hr

Trick offers a variety of light effects such as a light blade, a wall washer and a radial light distribution. The functional concept of this wall and ceiling light is based on a toroidal lens with a microprismatic surface for optimised heat dispersion. The LED lights allow for a patented 180-degree light effect. Trick comprises a die-cast aluminium base as well as a screen with a weather resistant coating.

Statement by the jury
This wall and ceiling luminaire of discreet appearance convinces with its impressive light effects – a functionally and formally sophisticated product solution.

Mit Trick können verschiedene Lichteffekte wie eine Lichtlanze, Wall Washer oder eine radiale Lichtverteilung erzielt werden. Das Funktionskonzept dieser Wand- und Deckenleuchte basiert auf dem Einsatz einer torischen Linse mit mikroprismierter Oberfläche, was wiederum die Wärmeableitung optimiert. Die LED-Lichtquellen ermöglichen einen patentierten 180-Grad-Lichteffekt. Trick umfasst sowohl einen Montagesockel aus Aluminiumguss als auch einen Leuchtenschirm, der witterungsbeständig lackiert ist.

Begründung der Jury
Die dezent anmutende Wand- und Deckenleuchte überrascht mit beeindruckenden Lichteffekten – eine funktional und formal ausgereifte Produktlösung.

Playbulb Adapter

Manufacturer
Shenzhen Baojia Battery Technology Co., Ltd.,
Shenzhen, China
In-house design
Stanley Yeung
www.mipow.com

The Playbulb Adapter allows users to conveniently control light bulbs via smartphone. The adapter is simply screwed into any E26 or E27 conventional light bulb socket. Via a Bluetooth 4.0 platform, any number of light bulbs can be switched on and off via an iOS or Android app. In addition, a timer can be programmed and the brightness can be controlled. Optionally, all light bulbs can be set to energy-saving mode if required.

Mit dem Playbulb Adapter können Glühbirnen komfortabel per Smartphone gesteuert und programmiert werden. Dazu wird der Adapter einfach vor den E26- oder E27-Sockel herkömmlicher Glühlampen geschraubt. Über eine Bluetooth 4.0 Plattform lassen sich dann beliebig viele Glühbirnen per iOS- oder Android-App ein- und ausschalten. Darüber hinaus ist der Zeitraum der Beleuchtung wählbar oder die Lichtleistung dimmbar. Optional können die Glühlampen bei Bedarf auch in einen Energiesparmodus versetzt werden.

Statement by the jury
The stringent realisation of an innovative design idea allows for the high ease-of-use of this light bulb adapter.

Begründung der Jury
Die stringente Umsetzung einer innovativen Gestaltungsidee ermöglicht bei diesem Glühlampen-Adapter einen hohen Bedienkomfort.

Birdie
Lamp
Leuchte

Manufacturer
Royal Philips, Eindhoven, Netherlands
In-house design
Web
www.philips.com

The Birdie feeding light is designed to give just the right amount of light during night-feedings: bright enough to check on the baby, but gentle enough to ensure the baby can return to sleep. Pressing the top part turns on the lamp without the hassle of finding a switch and the light changes from a nightlight to a gentle glow. The nightlight also changes its colour in accordance with the room temperature, glowing red for too hot, green for too cold and yellow for the right temperature.

Statement by the jury
The Birdie feeding light owes its emotionally appealing appearance to a formally and functionally conclusive design concept.

Das Stilllicht Birdie bietet eine variable Lichtmenge für das nächtliche Stillen: hell genug, um nach dem Baby zu sehen, aber sanft genug, damit das Baby wieder einschlafen kann. Durch Drücken des Oberteils wird die Leuchte ohne Suche nach einem Schalter eingeschaltet. Sie wechselt dann vom Nachtlicht zu einem dezenten Leuchten. Das Nachtlicht wechselt zudem seine Farbe abhängig von der Raumtemperatur, es scheint Rot bei zu hoher, Grün bei zu niedriger und Gelb bei der richtigen Temperatur.

Begründung der Jury
Sein emotional ansprechendes Gesamtbild verdankt das Stilllicht Birdie einem formal und funktional schlüssigen Gestaltungskonzept.

Disney SoftPal
Silicone 3D Luminaire
Silikon-3D-Leuchte

Manufacturer
Royal Philips, Eindhoven, Netherlands
In-house design
Web
www.philips.com

The Disney SoftPal range comprises innovative, multi-coloured silicone 3D luminaires for use by children at night. With an appealing soft and flexible touch function, the iconic Disney characters are pleasant to hold and easy for little fingers to grip. The silicone acts as a diffuser, providing a soft light that is well suited as a portable guide light or to reading a book. The silicone adjusts quickly to body temperature, making the luminaires cuddly companions to fall asleep with.

Statement by the jury
Thanks to a child-oriented design, these silicone 3D luminaires convince as versatile and friendly lighting options in children's rooms.

Das Disney-SoftPal-Sortiment umfasst innovative, mehrfarbige Silikon-3D-Leuchten, die als Nachtlichter für Kinder dienen. Die ikonischen Comicfiguren mit einer sanften und flexiblen Touch-Funktion sind angenehm anzufassen und für kleine Finger leicht zu greifen. Das Silikon dient als Diffusor für ein sanftes Licht, welches als tragbares Orientierungslicht oder zum Lesen geeignet ist. Das Silikon passt sich schnell der Körpertemperatur an, wodurch sich die Leuchten als kuschelige Begleiter zum Einschlafen eignen.

Begründung der Jury
Dank einer kindgerechten Gestaltung überzeugen diese Silikon-3D-Leuchten als vielseitige und sympathische Beleuchtungsmöglichkeit in Kinderzimmern.

Recharge
LED Book Light
LED-Buchleuchte

Manufacturer
Mighty Bright, Santa Barbara (California), USA
In-house design
Armand van Oord
Web
www.mightybright.com

The Recharge LED book light combines a clear design language with innovative LED technology. Made of silicone, the dual flex neck provides both stability and flexibility. Its two long-life LEDs are integrated into the light head, casting a bright white light onto any reading material. The durable plastic clip firmly affixes to books and magazines. The slim, rechargeable battery is encompassed within the light base and easily charged via a micro USB cable. It provides power for up to 16 hours.

Die LED-Buchleuchte Recharge vereint eine klare Formensprache mit einer innovativen LED-Technik. Aus Silikon gefertigt, bietet der Dual-Flex-Hals sowohl Stabilität als auch Flexibilität. Die beiden langlebigen LEDs im Leuchtenkopf richten ein helles weißes Licht auf die Lektüre. Komfortabel einfach lässt sich der robuste Kunststoffclip an Büchern und Zeitschriften befestigen. Der schlanke, integrierte Akku kann über ein Micro-USB-Kabel aufgeladen werden und bietet eine Leistung von bis zu 16 Stunden.

Statement by the jury
The stringent and colourful manifestation of a user-oriented design idea characterises the Recharge LED book light.

Begründung der Jury
Die stringente und farbenfrohe Umsetzung einer anwendungsorientierten Gestaltungsidee zeichnet die LED-Buchleuchte Recharge aus.

LED LENSER F1 White
LED Torch
LED-Taschenlampe

Manufacturer
Zweibrüder Optoelectronics GmbH & Co.KG,
Solingen, Germany
In-house design
Andre Kunzendorf, Tobias Schleder
Web
www.ledlenser.com
Honourable Mention

The LED Lenser F1 White torch is remarkably compact, powerful and only 88 mm long. It features electronic regulation with a temperature sensor as well as a replaceable tactical ring, which guarantees water resistance level IPX8. The torch stands firmly on its tail cap with integrated roll protection. The battery-powered luminous flux of 500 lumens allows for ideal light quality.

Statement by the jury
Accommodated in a compact white housing, the LED Lenser F1 White offers a surprisingly high light output.

Die Taschenlampe LED Lenser F1 White ist mit ihrer Länge von nur 88 mm auffallend kompakt und leistungsstark. Sie bietet eine elektronische Regelung mit Temperatursensor sowie einen austauschbaren Tacticalring, der eine Wasserdichtigkeit nach IPX8 sicherstellt. Die Taschenlampe lässt sich auf ihrer Endkappe aufstellen und verbleibt aufgrund des integrierten Rollschutzes stabil in dieser Position. Die batteriebetriebene Lichtstärke von 500 Lumen ermöglicht ein optimiertes Lichtbild.

Begründung der Jury
Mit ihrem kompakt konstruierten, weißen Gehäuse bietet die LED Lenser F1 White eine überraschend hohe Lichtleistung.

LED LENSER M3R
LED Torch
LED-Taschenlampe

Manufacturer
Zweibrüder Optoelectronics GmbH & Co.KG,
Solingen, Germany
In-house design
Andre Kunzendorf, Tobias Schleder
Web
www.ledlenser.com

The LED Lenser M3R is a focusable and remarkably small torch with an external charging station. Including batteries, the torch weighs only 45 grams, yet still reaches a light performance of more than 200 lumens. Depending on the light mode, battery life is between two and six hours. The M3R is equipped with a patented system that, among other features, includes a one-hand focusing mechanism. Its Smart Light Technology offers three lighting modes (power, low power, and stroboscope).

Statement by the jury
This LED torch convinces as a highly compact as well as environmentally friendly product. It provides users with a high ease of use.

Die fokussierbare LED Lenser M3R ist eine auffallend kleine Taschenlampe mit einer externen Ladestation. Inklusive Akku-Batterien wiegt die Taschenlampe nur 45 Gramm und kommt dennoch auf eine Lichtleistung von über 200 Lumen. Die Leuchtdauer beträgt je nach Lichtmodus zwischen zwei und sechs Stunden. Die M3R ist mit einem patentierten System, welches unter anderem eine Einhandfokussierung beinhaltet, ausgestattet. Ihre Smart-Light-Technology bietet die drei Leuchtmodi Power, Low Power und Stroboskop.

Begründung der Jury
Als sehr kompakte und zugleich umweltfreundliche Produktlösung überzeugt diese LED-Taschenlampe. Sie bietet ihrem Nutzer einen hohen Bedienkomfort.

3 Watt LED High Optics Light 3AAA
Torch
Taschenlampe

Manufacturer
VARTA Consumer Batteries GmbH & Co.
KGaA, Ellwangen, Germany
In-house design
Web
www.varta-consumer.com

The 3 Watt LED High Optics Light 3AAA flexibly adapts to any situation. With its adjustable head, the torch gives users full control of the light beam. The high-performance LEDs can be dimmed or set to strobe mode as required. In addition, they offer a continuous and precise focus with a high light output of 200 lumens and a light range of up to 141 metres. Powered by three AAA alkaline batteries, the torch features an operating time of 26 hours.

Statement by the jury
Sophisticated technical features and rugged workmanship make this torch a high-quality product solution.

Die 3 Watt LED High Optics Light 3AAA passt sich jeder Situation flexibel an. Mit ihrem beweglichen Leuchtenkopf ermöglicht die Taschenlampe die volle Kontrolle über den Lichtkegel. Die Hochleistungs-LEDs lassen sich je nach Anforderung dimmen oder in einen Stroboskop-Modus setzen. Zudem bieten diese LEDs eine stufenlose und präzise Fokussierung mit einer hohen Lichtausbeute von 200 Lumen. Die Leuchtweite beträgt bis zu 141 Meter. Die mit drei AAA Alkaline Batterien betriebene Taschenlampe hat eine Laufzeit von 26 Stunden.

Begründung der Jury
Technische Raffinessen und eine robuste Verarbeitung machen diese Taschenlampe zu einer qualitativ hochwertigen Produktlösung.

LED LENSER XEO19R
Focusable LED Headlamp
Fokussierbare LED-Stirnlampe

Manufacturer
Zweibrüder Optoelectronics GmbH & Co.KG,
Solingen, Germany
In-house design
Andre Kunzendorf, Tobias Schleder
Web
www.ledlenser.com

The comfortable LED Lenser XEO19R headlamp offers up to 2,000 lumens with a range of 300 metres. The passive cooling system automatically adjusts the lamp and protects it against overheating. Two independently focusable reflector lenses offer comprehensive illumination at both close and far range. In addition, the two dimmable LEDs can be controlled individually, allowing different modes and brightness settings. The versatile universal lamp can be used as a head- or chestlamp, bike lamp, or torch.

Die komfortable Stirnlampe LED Lenser XEO19R bietet bis zu 2.000 Lumen bei einer Reichweite von 300 Metern. Das passive Kühlsystem regelt die Lampe automatisch und schützt diese vor Überhitzung. Die zwei getrennt fokussierbaren Reflektorlinsen bieten nah wie fern eine umfassende Ausleuchtung. Zudem lassen sich die beiden dimmbaren LEDs einzeln ansteuern, sodass unterschiedliche Modi und Helligkeitsstufen möglich sind. Die flexible Universalleuchte kann als Stirn-, Fahrrad-, Helm-, Brust- oder Taschenlampe verwendet werden.

Statement by the jury
Multiple technical features mark this LED headlamp as a multifunctional and convenient product solution.

Begründung der Jury
Eine Vielzahl an technischen Details zeichnet die LED-Stirnlampe als multifunktionale und komfortabel zu handhabende Produktlösung aus.

Bars	Bars
Cafés	Briefkästen
Carpets	Cafés
Cultural and	Fenster
public amenities	Fliesen
Cushion material	Geländer
Doors	Glas
Floor tiles	Griffsysteme
Glass	Holz
Handle systems	Inneneinrichtungen
Interior fittings	Kulturelle und
Keys	öffentliche Einrichtungen
Natural materials	Natürliche Materialien
Object furniture	Objektmöbel
Parquet flooring	Parkett
Partition systems	Polstermaterialien
Postboxes	Restaurants
Railings	Schalter
Residential buildings	Schlüssel
Restaurants	Shops
Security systems	Sicherheitssysteme
Shops	Sonnenschutz
Sun protection	Tapeten
Switches	Teppiche
Wallpapers	Trennwandsysteme
Windows	Türen
Wood	Wohnhäuser

Interior design
Interior Design

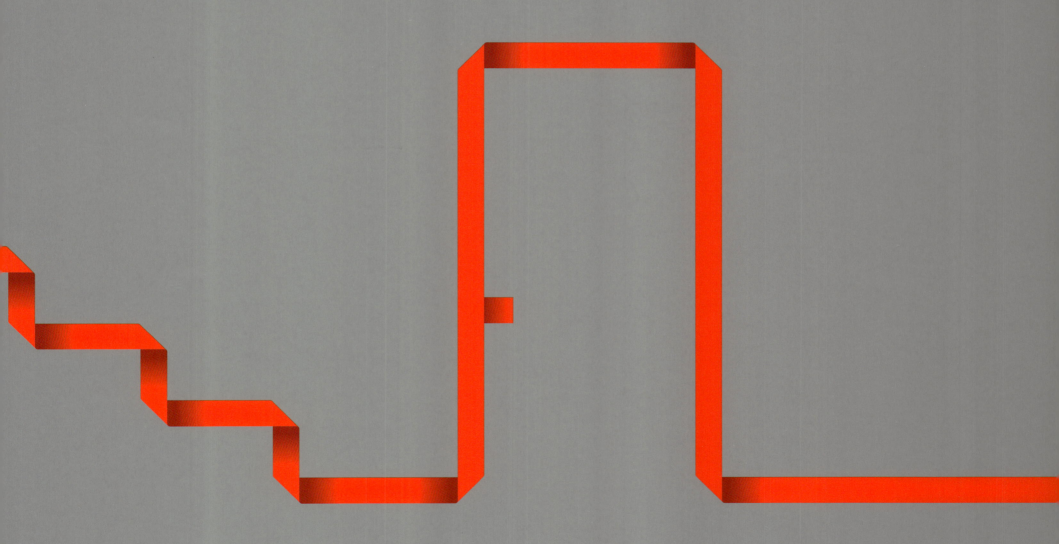

Bitter Bamboo Room
Working Space
Arbeitsraum

Manufacturer
Deve Build, Shenzhen, China

In-house design
Yu Feng

Web
www.devebuild.com

Pure originality

Round the globe, the world of work is changing and is subject to ever new demands and scenarios. As designer Yu Feng said of the basic concept for the Bitter Bamboo Room: "From the beginning, I wanted to design a place which does not look like anywhere else, trying to avoid the characteristics of space as much as possible." It was intended as a place devoid of the typical features of a workplace in order to create an environment that allows one to focus on the essentials. The guiding principles for this interior focused on providing a humane and emotional experience. The aim was to make the space as simple and natural as possible, and original enough to speak for itself. This was achieved through the use of clear lines and reduced fittings, which nonetheless fulfil all the needs of the person working in it. To replicate the concept of originality visually, only natural materials were used. A low-cost, simple type of bamboo was chosen as the main material for the construction, as it is at its most expressive in this form. In line with the maxim of simplicity, only basic craftsmanship skills were used to produce the architectural structure of the Bitter Bamboo Room. The overall impression is one of authenticity and its naturalness is captivating. Bitter Bamboo Room's stringent interior design offers a seminal new alternative to the conventional workspace.

Pure Ursprünglichkeit

Global ist die Arbeitswelt im Wandel begriffen und stetig neuen Anforderungen und Szenarien unterworfen. „Von Anfang an wollte ich einen Ort schaffen, der keinem anderen gleicht, und typische Raumcharakteristika so weit wie möglich vermeiden", beschreibt der Designer Yu Feng das Grundkonzept des Bitter Bamboo Room. Ohne die Merkmale einer Arbeitsstätte soll dieser ein Ambiente schaffen, in dem man sich wieder auf das Wesentliche besinnen kann. Der humane Aspekt und das emotionale Erleben waren zentrale Leitgedanken dieses Interieurs. Es sollte möglichst einfach und natürlich sein und in seiner Originalität für sich selbst stehen können. Umgesetzt wird dies durch klare Linien und eine insgesamt reduzierte Einrichtung, die dennoch alle Bedürfnisse der hier arbeitenden Menschen erfüllt. Um den Gedanken der Ursprünglichkeit eindringlich zu visualisieren, kamen ausschließlich natürliche Materialien zum Einsatz. Dabei wurde für die Gestaltung vor allem Bambus in einer kostengünstigen und einfachen Ausführung gewählt, da er in dieser Form die stärkste Ausdruckskraft besitzt. Entsprechend der Maxime der Einfachheit kamen für die architektonische Strukturierung des Bitter Bamboo Room zudem nur einfache Handwerkstechniken zum Einsatz. Er wirkt dadurch überaus authentisch und verführt die Sinne durch seine Natürlichkeit. Mit einem sehr stringent umgesetzten Interior Design bietet der Bitter Bamboo Room wegweisend neue Alternativen zu herkömmlichen Arbeitsräumen.

Statement by the jury

The subtle staging of simplicity has resulted in a particularly compelling workspace. The design of Bitter Bamboo Room brings to life the underlying concept of creating a new type of work environment that differs from conventional structures. All the elements blend harmoniously and the structural influence of bamboo is given free rein. This interior exemplifies the idea of "back to basics" in a fascinating way.

Begründung der Jury

Die feinsinnige Inszenierung des Einfachen ließ hier einen Arbeitsraum von besonderer Anziehungskraft entstehen. Die Gestaltung des Bitter Bamboo Room verwirklicht eindrucksvoll das zugrunde liegende Konzept, um eine neue Art von Arbeitsatmosphäre abseits gängiger Strukturen zu kreieren. Überaus harmonisch fügen sich alle Elemente zusammen und das Material Bambus entfaltet seine prägende Kraft. Auf faszinierende Weise visualisiert dieses Interieur eine Rückbesinnung auf das Wesentliche.

Designer portrait
See page 58
Siehe Seite 58

Gu Yuan – Specious Feeling
Commercial Space
Gewerbefläche

Client
Fudiyuan Estate Company, Xinjiang, China
Design
Deve Build, Shenzhen, Guangdong, China
Web
www.devebuild.com
Honourable Mention

Crossing the narrow, zigzag-shaped bridge enables the area to be discovered from varying angles. Modern methods were used to lay traditional floor tiles. The purely white walls appear soft like paper, whereas the wood effect of the bamboo poles adds a sense of well-being. Natural light illuminates the premises in a way that changes with the time of day in various ways and spreads a changing but thoroughly harmonious ambience.

Statement by the jury
The design of Gu Yuan evokes elements of a traditional Chinese garden by creating an abstract effect at first and then reinterpreting it in a contemporary way.

Der Gang über die schmale Brücke, die im Zickzack verläuft, ermöglicht es, den Raum aus verschiedenen Blickwinkeln zu entdecken. Auf dem Boden wurden nach moderner Methode traditionelle Fliesen verlegt. Die reinen weißen Wände wirken sanft wie Papier, während die Holzoptik der Bambusstäbe für ein angenehmes Wohlgefühl sorgt. Natürliche Lichtquellen erhellen die Räumlichkeiten in Abhängigkeit von der Tageszeit auf verschiedene Weise und schaffen so ein veränderliches, aber spürbar harmonisches Ambiente.

Begründung der Jury
Die Gestaltung von Gu Yuan zitiert Elemente eines traditionellen chinesischen Gartens, indem sie sie zunächst abstrahiert und dann auf zeitgemäße Weise neu interpretiert.

Restaurant RAW, Taipei

Client
Restaurant RAW, Hasmore Ltd.,
Taipei, Taiwan
Design
Weijenberg Pte. Ltd., Singapore
Web
www.raw.com.tw
www.weijenberg.co

Leaving the turbulent streets of Taipei, visitors step across a "wooden path" to enter a restaurant that exudes a sense of complete calm and rest. Upon entering the lounge area, visitors are greeted by wave-like and organic structures that present a fitting contrast to the bustle of the city. In the main dining area, the gentle ambiance of the interior arrangement continues with specially designed tables for two offering ample privacy. For those who wish to hide away completely in the peace and quiet, the restaurant even has a semi private dining area towards the rear.

Statement by the jury
The organic forms and gently curvated wooden elements of the interior architecture of the Raw restaurant convey a calm, relaxed atmosphere while the lighting above each table creates an accent which focuses on the cuisine.

Von den belebten Straßen Taipeis gelangt der Besucher über einen „hölzernen Pfad" in ein Restaurant, das völlige Ruhe ausstrahlt. Er kommt in einen Loungebereich, der mit seinen wellenartigen und organischen Strukturen einen gelungenen Ausgleich zum Lärm der Stadt schafft. Im Hauptspeiseraum setzt sich die sanfte Gestaltung des Interieurs in eigens entworfenen Esstischen fort, die den Gästen mit zwei Sitzmöglichkeiten ausreichend Privatsphäre bieten. Wer sich komplett zurückziehen möchte, findet in einem hinteren, halbprivaten Speiseraum des Restaurants Ruhe.

Begründung der Jury
Die Innenarchitektur des Raw Restaurants vermittelt mit organisch geformten, sanft geschwungenen Holzelementen eine ruhige, entspannte Atmosphäre, während die Akzentbeleuchtung über jedem Tisch die Speisen in den Mittelpunkt rückt.

Yakiniku Tsuruichi
Restaurant

Design
Golucci International Design
(Hsuheng Lee), Beijing, China
Web
www.golucci.com

Strip-type partition boards made of wood laths, which are widely seen in Japanese architecture, were used in the construction of this restaurant. For its design, the owners were inspired by a picture from the renowned works of Utawaga Hiroshige: "Sudden Shower Over Shin-Ohasi Bridge and Atake", which captures the moment when a summer rain shower suddenly changes the weather and surprises pedestrians. By presenting the rain as thick stripes, not only the Japanese painter's technique was adopted but also a unique atmosphere was created.

Statement by the jury
The traditional and, at the same time, modern appearance of this restaurant is characterised by geometrically arranged, warm timber elements.

Für den Bau des Restaurants wurden, wie in der japanischen Architektur verbreitet, Trennwände aus einzelnen Latten verwendet. Die Inhaber ließen sich bei der Gestaltung außerdem von den bekannten Kunstwerken Utawaga Hiroshiges inspirieren: In seinem Bild „Sudden Shower Over Shin-Ohashi Bridge and Atake" fängt er den Moment eines Sommerregens ein, der mit seinem plötzlichen Umschwung des Wetters die Passanten überrascht. Durch die Darstellung von Regen in Form dicker Streifen wurde neben der Technik des japanischen Malers auch eine einzigartige Atmosphäre übernommen.

Begründung der Jury
Die traditionelle und gleichzeitig sehr moderne Anmutung dieses Restaurants wird von geometrisch angeordneten, warmen Holzelementen geprägt.

Songshan Culture and
Creative Park Café & Souvenir Shop

Design
Akuma Design (Christine Pang, Tim Chou),
Taipei, Taiwan
Web
www.akumagroup.com

The cigarette factory was built during Taiwan's industrialisation period and still retains the charm of simplicity the early Japanese modernist style imparts. It has an open interior with high ceilings and symmetrical columns and the large windows ensure it is filled with natural light. The building has been put under monumental protection in order to preserve its historic value and has not been remodelled, so that minor flaws are visible.

Die Zigarettenfabrik wurde zu Zeiten der Industrialisierung Taiwans erbaut und besticht noch heute mit der Einfachheit des frühmodernistischen japanischen Stils. Die hohen, offenen Räumlichkeiten sind mit symmetrischen Säulen ausgestattet und werden dank großer Fenster durch natürliches Licht ausreichend beleuchtet. Um den historischen Wert beizubehalten, blieb das unter Denkmalschutz stehende Gebäude unberührt und ist daher von kleinen sichtbaren Makeln gekennzeichnet.

Statement by the jury
By the use of concept that puts modularity and mobility centre stage, this conversion of a building protected as historic monument has been highly successful. The thoughtful integration of modern elements has preserved its original purist and open character.

Begründung der Jury
Mit einem Konzept, das auf Modularität und Mobilität setzt, ist hier die Umnutzung denkmalgeschützter Bausubstanz gut gelungen. Die rücksichtsvolle Integration moderner Elemente erhält den ursprünglichen puristischen und offenen Charakter des Gebäudes.

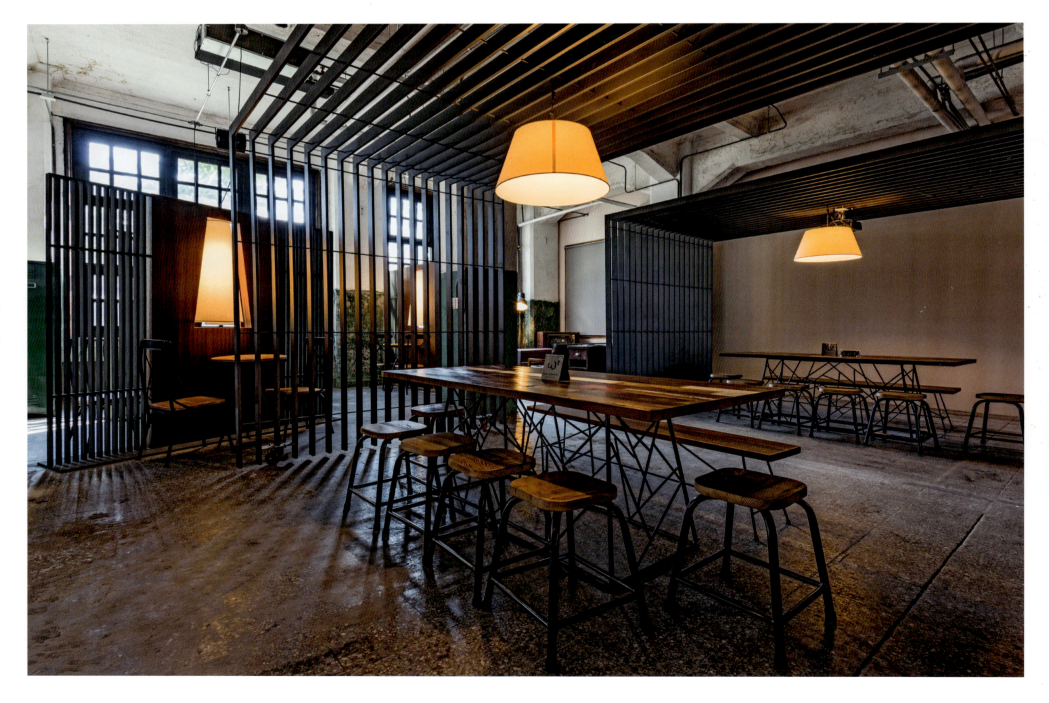

Lin Mao Sen – New. Release. Sweetness.
Traditional Tea Shop
Traditioneller Teeshop

Design
Ahead Design (Onion Yang),
Taipei, Taiwan
Web
www.aheadesign.com

The traditional tea ceremony is reflected in every aspect of the realisation of this store. By choosing an appropriate interior, an ambience was created which strengthens the authentic way of presenting the products. Whereas the unusual structure of the ceiling ensures a dynamic interplay of the individual elements and the generous use of wood creates a feeling of warmth.

Die Tradition der Teezeremonie spiegelt sich in der gesamten Umsetzung des Ladens wider. Indem das Interieur entsprechend gewählt wurde, entsteht ein Ambiente, das die authentische Präsentation der Produkte unterstützt. Die außergewöhnliche Deckenstruktur sorgt zugleich für ein dynamisches Zusammenspiel der einzelnen Elemente, die durch die großzügige Verwendung des Materials Holz Wärme verbreiten.

Statement by the jury
An atmosphere of tranquillity and contemplation has been created in this tea store through a clear symmetry, an artistic design of the ceiling, warm lighting and the generous use of wood.

Begründung der Jury
Eine Atmosphäre der Ruhe und Kontemplation entsteht in diesem Teegeschäft durch eine klare Symmetrie, kunstvolle Deckengestaltung, warmes Licht und den großflächigen Einsatz von Holz.

Reversal
Showroom

Client
Ne. Sense, Taipei, Taiwan
Design
Yun-Yih Interior Design, Taipei, Taiwan
Web
www.ne-sense.com
www.yundyih.com.tw
Honourable Mention

High concrete walls and a hidden entrance make it almost impossible to find the showroom. Only the horizontal windows high up the walls give a vague indication that there might be something inside to be discovered. The combination of grey facades with red bricks and an extravagant LED illumination showcase the products on display in an artful way. These are to be found placed on partial mezzanines which allow the visitor's view to spiral upwards towards the ceiling.

Statement by the jury
Reversal draws attention to itself with its experimental design concept. It breaks with the traditional idea of a showroom and thus showcases products in a new and interesting manner.

Hohe Wände aus Beton und ein versteckter Eingang bewirken, dass der Showroom von außen kaum einzusehen ist. Lediglich die horizontalen Fenster im oberen Teil des Gebäudes lassen vage erkennen, dass es im Inneren etwas zu entdecken gibt. Die Kombination markanter grauer Fassaden mit rotem Backstein und eine extravagante LED-Beleuchtung setzen die Produkte raffiniert in Szene. Diese finden unter anderem auf Wandvorsprüngen Platz, die den Blick des Besuchers spiralförmig in Richtung Decke leiten.

Begründung der Jury
Reversal macht mit einem experimentellen Gestaltungskonzept auf sich aufmerksam, das gelernte Muster für Showrooms aufbricht und die Waren so auf eine neue, interessante Weise in Szene setzt.

Dolby Cinema
Cinema
Kino

Client
Dolby Laboratories, Inc.,
San Francisco, USA
Design
Dolby Laboratories, Inc.
(Donald Burlock, Vince Voron),
San Francisco, USA
Eight Inc.
(Mark Little, Wilhelm Oehl),
San Francisco, USA
Web
www.dolby.com
www.eightinc.com

The concept of this futuristic cinema is focused wholly on the expectations of the visitors. As visitors cross the threshold of the Dolby Cinema signature entrance, they are guided along a curved hallway immersed in full-height projected imagery and sound, creating a sensory transition and separation from the bright theatre lobby and into the auditorium. The resulting integration of inspired design and the latest technology is a cinema where the stories you love come alive before your eyes through Dolby sight and sound technologies in ways you never thought possible.

Das Konzept dieses futuristischen Kinos konzentriert sich voll und ganz auf die Erwartungshaltung des Publikums. Sobald die Besucher die Schwelle zum Dolby Cinema überschreiten, tauchen sie in Bildprojektionen und in eine Geräuschkulisse ein, die in einem kurvenförmigen Gang einen sensorischen Übergang und gleichzeitig eine Trennung zur hellen Lobby und dem Vorführungssaal bilden. Durch die sich ergebende Kombination von beseelter Gestaltung und modernster Technologie entsteht schließlich ein Kino, in dem Dolby Licht- und Tontechnik auf außergewöhnliche Weise beliebte Geschichten zum Leben erweckt.

Coastal Cinema
Cinema
Kino

Client
Coastal Group, Wuxi, China
Design
One Plus Partnership Limited
(Ajax Law, Virginia Lung), Hong Kong
Web
www.onepluspartnership.com

Inspiration for this cinema's design was the interplay between water and land and it is thus characterised by a particularly dynamic impact. The oval-shaped lobby that greets visitors has a floor in rectangular tiles reminiscent of a stony coast. Powder-coated metal rods stretching from floor to ceiling simulate the rising and falling movement of waves. The maritime look is a leitmotif through the whole design concept and makes the cinema an experience of a very special kind.

Statement by the jury
The sophisticated concept of the interior architecture of Coastal Cinema astonishes. The theme of the coast is ever present and, through skilful abstraction, the architecture conveys an impression of elegance.

Der Entwurf dieses Kinos greift das Zusammenspiel von Wasser und Land auf und zeichnet sich somit durch eine besonders dynamische Wirkung aus. Die Besucher werden in einer ellipsenförmigen Lobby empfangen, die mit ihren rechteckigen Fliesen an eine steinige Küste erinnert. Pulverbeschichtete Metallstäbe reichen von der Decke zum Boden und greifen so die Bewegung ansteigender und fallender Wellen auf. Die maritim inspirierte Gestaltung zieht sich durch das gesamte Konzept und macht das Kino zu einem Erlebnis ganz besonderer Art.

Begründung der Jury
Das Coastal Cinema begeistert mit einem anspruchsvollen innenarchitektonischen Konzept, bei dem das Thema Küste allgegenwärtig ist und durch gekonnte Abstraktion eine moderne und elegante Anmutung erhält.

Hankou City Plaza International Cinema
Cinema
Kino

Client
Hubei Insun Cinema Film Co., Ltd.,
Wuhan, China
Design
One Plus Partnership Limited
(Ajax Law, Virginia Lung), Hong Kong
Web
www.onepluspartnership.com

From the moment visitors enter this cinema they feel themselves to be the centre of attention. They are welcomed by powerful spotlights just like at a shooting and straight away feel as if they are part of a movie themselves. Lazy tongs of varying lengths hang like chandeliers from the ceiling and large LCD screens show trailers of the films currently running. The combination of dark and metallic materials stresses the technical aspect of the cinema, thus awaking the interest of particularly the male visitors.

Statement by the jury
Simple means were used in the design of the Hankou City Plaza International Cinema in order to create a setting which enables visitors to enter the world of cinema even before the actual film starts.

Gleich nach dem Eintritt in das Kino bekommt der Besucher vollste Aufmerksamkeit: Er wird von hellen Scheinwerfern begrüßt, wie man sie von Drehorten kennt, und fühlt sich so selbst als Teil eines Films. Greifzangen in unterschiedlichen Länge hängen wie Kronleuchter von der Decke, große LCD-Bildschirme zeigen laufend Trailer der aktuellen Filme. Die Kombination dunkler und metallener Materialien unterstreicht den technischen Aspekt des Kinos und weckt damit vor allem das Interesse der männlichen Besucher.

Begründung der Jury
Bei der Gestaltung des Hankou City Plaza International Cinema wurde mit einfachen Mitteln eine Szenerie kreiert, die die Besucher schon vor Beginn des eigentlichen Films in die Welt des Kinos eintauchen lässt.

Ganna Studio, Taipei
Residential Building and Office
Wohngebäude und Büro

Design
Ganna Design, Interior Design Company,
Taipei, Taiwan
Web
www.ganna-design.com
Honourable Mention

The origin of both the living area and office of the Ganna Studio is the architecture of ancient Taipei. In order to maintain historic elements, the building structure of the complex was left untouched. Every area displays multifunctionality, which makes a perfect balance between the work and private spheres possible. The living room doubles as a showroom and business meetings may be held around the dining table. If required, the living and working areas can be separated by integrated panels to ensure that quietness is maintained.

Sowohl der Wohnbereich als auch das Büro des Ganna Studios sind auf die Architektur des antiken Taipei zurückzuführen. Um die historischen Elemente beibehalten zu können, ließ man die Baustruktur des Komplexes unverändert. Jeder Bereich zeichnet sich durch eine Multifunktionalität aus, die die richtige Balance zwischen Arbeit und Privatleben ermöglicht. Das Wohnzimmer dient zugleich als Showroom, während am Esstisch geschäftliche Meetings abgehalten werden können. Bei Bedarf kann durch fünf integrierte Paneele der Wohn- vom Arbeitsbereich abgetrennt werden, sodass entsprechend Ruhe gewährleistet ist.

Original

Residential Building
Wohngebäude

Client
Jerry Shyu, Hsinchu, Taiwan
Design
DINGRUI Design Studio, Taoyuan, Taiwan
Web
www.ding-rui.com
Honourable Mention

The functions of the various rooms are so well-coordinated that a clear, smooth transition exists between them. Not only the construction material, but also the furniture and other fixtures were chosen with nature in mind. Through its elegance, every element of the studio blends harmoniously into the overall effect and underlines the sense of nature.

Die Funktionen der Räume sind so aufeinander abgestimmt, dass ein klarer, fließender Übergang entsteht. Nicht nur die Baumaterialien, sondern auch das Mobiliar und weitere Einrichtungsgegenstände weisen Anklänge an die Natur auf. Jedes Element des Studios fügt sich mit seiner Eleganz in das Gesamtbild ein und unterstreicht so die sehr natürliche Wirkung.

Interface Microsfera
Carpet Tiles
Teppichfliesen

Manufacturer
Interface,
Scherpenzeel, Netherlands

In-house design
Interface

Design
Interface European Design Team,
Shelf, Great Britain

Web
www.interface.com

reddot award 2015
best of the best

Pleasant ambiance

Carpet tiles are an attractive product for interior design, as they enable architects to create a range of defined spaces throughout any organisation. Combining an aesthetic and functional quality, with an environmentally friendly production process, Microsfera is a ground-breaking range. Made entirely of a newly developed yarn and manufactured using an innovative modular fusion technology, the carpet tile has a remarkably low carbon footprint with emissions of only 3 kg carbon dioxide per square metre. The low VOC emissions also guarantee a healthy indoor air quality in rooms using this product. Based on a solid design concept, Microsfera showcases sophisticated, linear design variations, while the 100 x 25 cm Skinny Plank tile enables material saving throughout the installation process. Equally, the colour palette comprises a range of harmonious, balanced shades; industrial-style greys, trendy denim-inspired blues and raw copper hues, allow for countless compositions, as well as striking effects and transitions. This new range, Microsfera, enables architects and designers to have the freedom to create individual and inspiring spaces that are environmentally sound.

Angenehme Atmosphäre

Teppichfliesen sind für Architekten ein attraktives Produkt der Raumgestaltung, da sie mit diesen in beliebiger Organisation verschiedene, klar definierte Bereiche schaffen können. Bei der wegweisenden Microsfera-Kollektion verbindet sich eine hohe ästhetische und funktionale Qualität mit umweltfreundlichen Herstellungsverfahren. Die Teppichfliesen bestehen zu 100 Prozent aus einem neu entwickelten Garn und bei ihrer Produktion kommt eine innovative modulare Verschmelzungstechnologie zum Einsatz. Dies führt zu einer erstaunlich guten Umweltbilanz, da im Produktionsprozess lediglich 3 kg Kohlendioxid pro Quadratmeter an Emissionen entstehen. Dank sehr niedriger VOC-Werte gewährleisten diese Teppichfliesen ein gesundes Raumklima in den Räumen, in denen sie verlegt werden. Auf der Grundlage eines durchdachten Konzepts bietet Microsfera zudem raffinierte lineare Entwurfsvariationen. Das Skinny Plank-Format in den Maßen 100 x 25 cm erlaubt eine materialsparende Anordnung. Die Farbpalette besteht aus sorgfältig aufeinander abgestimmten Farbtönen. Grautöne mit industriellem Charakter, trendige Denim-inspirierte Blautöne oder natürliche Kupfertöne ermöglichen unzählige Kompositionen sowie auffällige Effekte und Übergänge. Die neue Microsfera-Kollektion bietet Architekten und Designern damit die Freiheit, Innenräume individuell zu gestalten und ein anregendes Ambiente zu schaffen, das gleichzeitig umweltgerecht ist.

Statement by the jury

With their environmental soundness, the Microsfera carpet tiles set new standards. They impress with an extraordinarily low carbon dioxide footprint and extremely low VOC emissions. Thanks to a newly developed yarn and an innovative production process, only 3 kg of carbon dioxide per square metre are emitted during the production. Their variety of available patterns allows for a creative interior design together with a very healthy indoor climate.

Begründung der Jury

Mit ihrer Umweltfreundlichkeit setzen die Teppichfliesen Microsfera neue Standards. Sie beeindrucken mit einer außerordentlich guten Kohlendioxidbilanz und äußerst niedrigen VOC-Werten. Dank eines neu entwickelten Garns und des innovativen Herstellungsverfahrens entstehen bei der Fertigung der Teppichfliesen nur 3 kg Kohlendioxid pro Quadratmeter an Emissionen. Ihre Mustervielfalt ermöglicht eine kreative Innenraumgestaltung im Einklang mit einem sehr gesunden Raumklima.

Designer portrait
See page 60
Siehe Seite 60

Dickson Rugs
In & Out Woven Vinyl Rugs
In & Out Vinyl-Teppiche

Manufacturer
Dickson Constant, Wasquehal, France
In-house design
Web
www.dickson-constant.com

Conceived for extreme wear, these rugs can be used both inside as well as outside and withstand all weathers. They combine robust vinyl with the aesthetics and textures of traditional floor coverings and thus benefit in their functionality from the use of innovative fibers. Even the sun's extreme UV radiation cannot damage the pigments used so that they retain their brightness for a long time. These rugs are available in two types of finishes, Sunbrella fabric or Dickson woven fabric.

Konzipiert für starke Belastung können diese Teppiche sowohl drinnen als auch draußen verwendet werden und dort jeder Witterung standhalten. Hier wurde robustes Vinyl mit der Ästhetik und Textur traditioneller Teppichböden vereint und dabei durch den Einsatz neuartiger Fasern noch in seinen Gebrauchseigenschaften verbessert. Auch extreme Sonneneinstrahlung kann den verwendeten Pigmenten nicht schaden, sodass ihr Glanz über lange Zeit erhalten bleibt. Die Teppiche sind in den zwei Ausführungen Sunbrella und Dickson erhältlich.

Breaking Form
Carpet Tile
Teppichfliese

Manufacturer
Mohawk Group, Calhoun (Georgia), USA
Design
Massive Design Sp. z o.o (Mac Stopa),
Warsaw, Poland
Web
www.mohawkgroup.com
www.massivedesign.pl

Together with industrial designer and architect Mac Stopa, Mohawk Group has developed a carpet tile whose vivid patterns and geometric shapes are eye-catching. Available in 12 and 36 inch size, its playful colours energise any environment and it skilfully adapts to it. Once laid, the effect of these carpet tiles changes from being two- to three-dimensional, giving users the impression to be on a different level.

Statement by the jury
Vibrant colours, flowing lines and geometric patterns define the eye-catching, high-contrast appearance of these carpet tiles.

In Zusammenarbeit mit Industriedesigner und Architekt Mac Stopa hat die Mohawk Group eine Teppichfliese entwickelt, die sowohl durch lebhafte Muster als auch geometrische Strenge überzeugt. Erhältlich in 12 und 36 Zoll versprüht sie in verspielten Farben in jeder Umgebung Energie und passt sich mit ihrer Gestaltung gekonnt an. Auf dem Boden verlegt, verändert sich ihre Wirkung vom Zwei- zum Dreidimensionalen und löst so beim Endverbraucher das Gefühl aus, sich auf einer anderen Ebene zu befinden.

Begründung der Jury
Leuchtende Farben, fließende Linien und geometrische Muster prägen das auffällige und kontrastreiche Erscheinungsbild dieser Teppichfliesen.

Flat 01 – 04
Textile Hard Flooring
Textilhartbelag

Manufacturer
ANKER Gebr. Schoeller GmbH & Co. KG,
Düren, Germany
In-house design
Web
www.anker.eu

The Flat hard textile flooring is characterised in its surface structure by its purist design. Thanks to a high-end weaving technique and sophisticated construction, this flooring is available in four different finishes that show extreme durability using little material. In spite of their flat shape they appear three-dimensional.

Statement by the jury
The use of modern weaving techniques has led to the creation of a thin, textile floor covering that is purist in appearance, but nonetheless extremely hard-wearing.

Der Textilhartbelag Flat ist in seinem Flächenbild durch eine puristische Gestaltung gekennzeichnet. Dank High-End-Webtechnik und ausgefeilter Konstruktion ist er in vier unterschiedlichen Oberflächen erhältlich, die bei geringem Materialeinsatz eine sehr hohe Strapazierfähigkeit aufweisen. Trotz der flachen Ausführung ergibt sich eine dreidimensionale Wirkung.

Begründung der Jury
Durch den Einsatz moderner Webtechnik ist mit Flat ein textiler, flacher Bodenbelag mit puristischer Anmutung und hoher Widerstandsfähigkeit entstanden.

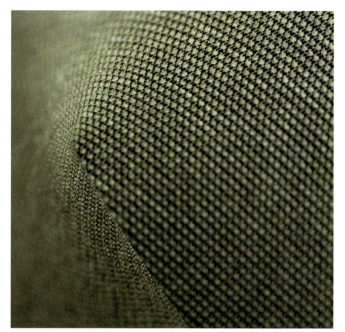

FINETT FEINWERK
himmel und erde / buntes treiben
Needled Carpet
Nadelvlies

Manufacturer
Findeisen GmbH, Ettlingen, Germany
In-house design
Web
www.nadelvlies.de
Honourable Mention

"himmel und erde" and "buntes treiben" are two products that harmoniously complement one another and are made from fine nonwoven material in a range of colours. While himmel und erde, German for heaven and earth, is available in natural, soft tones reminiscent of nature, buntes treiben, colourful hustle and bustle, as its name implies in German, sticks to its promise and lightens any room with strong shades. They are pleasant to walk on, even barefoot, thanks to their soft feel and can be used as floor coverings in various situations and locations, from educational institutions to hotels or shops.

Statement by the jury
This nonwoven fine fibre carpet in natural tones appeals thanks to its pleasantly soft, felt-like feel.

Die beiden sich harmonisch ergänzenden Produkte „himmel und erde" und „buntes treiben" bestehen aus feinem Nadelvlies in verschiedenen Farben: Während himmel und erde in Anlehnung an die Natur in dezenten Ausführungen erhältlich ist, hält buntes treiben, was es verspricht, und akzentuiert den Raum mit kräftigen Tönen. Die Haptik erlaubt selbst barfuß ein angenehmes Gehen über den Bodenbelag, der in Bildungseinrichtungen und Hotels bis hin zu Ladengeschäften verlegt werden kann.

Begründung der Jury
Diese Nadelvliesböden in dezenten Naturtönen beziehungsweise kräftigen Farben überzeugen mit einer angenehm weichen, filzartigen Haptik.

Flush-Mount Skirting Board
Wandbündige Sockelleiste

Manufacturer
Admonter,
STIA Holzindustrie GmbH,
Admont, Austria

In-house design
Martin A. Dolkowski

Web
www.admonter.at

reddot award 2015
best of the best

Harmonious transition
Skirting boards mounted to the wall are generally intended to form a visual transition between wall and floor. The skirting board mounted flush with the wall by Admonter now presents itself as a fascinating new solution: as a self-sufficient design element it embodies a flush connection between two planes as it creates a seamless transition from floor to wall. The result is a room environment characterised by an overall harmonious, purist appearance. The system consists of a basic profile, preinstalled during the construction phase, and the skirting board that completes the system during interior finishes. A particularly innovative aspect is presented by the integrated height-gauge, which offers many functional benefits for the screed installation process. It not only makes the work for screed-installers substantially easier and more comfortable, it also minimises the risk of errors. Technical details such as fine adjustment for the compensation of height tolerances as well as the possibility of reducing impact sound transmission further enhance the comfort aspect of this skirting board. Oriented toward sustainability, the boards are made of PEFC-certified solid wood, whereas the screed gauge is composed of wood-based, compostable bioplastics. In addition, the boards can also be used to create an extraordinary ambiance when installed with the optional energy-efficient LED lighting system. In this configuration, the aestheticising quality of the boards in almost any interior space is further enhanced by impressive lighting effects.

Harmonisch überbrückt
An der Wand angebrachte Sockelleisten bilden in der Regel den visuellen Übergang zwischen Fußbodenbelag und Wand. Die wandbündigen Sockelleisten von Admonter stellen hier eine faszinierend neue Lösung dar: Als eigenständiges Designelement verbinden sie flächenbündig zwei Ebenen miteinander und schaffen dabei einen nahtlosen Übergang vom Fußboden zur Wand. Es entsteht ein harmonisches, puristisch anmutendes Gesamtbild im Raum. Das System besteht aus einem bereits in der Rohbauphase montierten Grundprofil sowie einer Sockelleiste, die nach der Fertigstellung des Innenausbaus eingesetzt wird. Ein besonders innovativer Aspekt ist dabei vor allem eine integrierte Höhenlehre. Für den Innenausbau bietet das viele funktionale Vorteile, denn die Arbeit des Estrichlegers wird dadurch wesentlich vereinfacht und das Fehlerrisiko beim Verlegen verringert. Technische Details wie eine Feinjustierung zum Ausgleich von Höhentoleranzen sowie die Möglichkeit der Trittschallentkopplung erhöhen den Komfort dieser Sockelleisten. Orientiert an Nachhaltigkeit, werden sie aus PEFC-zertifiziertem Massivholz gefertigt, für die Estrichlehre kommt ein holzbasierter, kompostierbarer Biokunststoff zum Einsatz. Eine außergewöhnliche Atmosphäre erzeugt diese Sockelleiste in der optionalen Ausstattung mit einer energieeffizienten LED-Beleuchtung. Ihre den Raum harmonisierende Ästhetik wird dabei mit beeindruckenden Lichteffekten akzentuiert.

Statement by the jury
The flush-mount skirting boards by Admonter are self-sufficient design elements that connect walls and floors in a brilliant transitional manner. An architectural detail has thus emerged as a new interpretation with an aesthetic appeal of its own. The optional LED lighting fixture is fascinating as it creates an impressive ambiance in any given interior. The innovative concept is brought to perfection by an integrated, highly functional height gauge that makes installation an easy task.

Begründung der Jury
Die wandbündigen Sockelleisten von Admonter verbinden als eigenständiges Designelement auf brillante Weise Wand und Fußboden miteinander. Ein architektonisches Detail erhält dadurch neu interpretiert eine ganz eigene Ästhetik. Faszinierend ist die optionale Ausstattung mit einer LED-Beleuchtung, welche ein eindrucksvolles Ambiente im Raum schafft. Das innovative Konzept wird perfektioniert durch eine integrierte, sehr funktionale Höhenlehre, welche die Montage erheblich vereinfacht.

Designer portrait
See page 62
Siehe Seite 62

Desire from The Miracles Collection
Three-Layer Wooden Floors
Dreischicht-Holzböden

Manufacturer
Baltic Wood, Jaslo, Poland
In-house design
Web
www.balticwood.pl

This wooden floor with brushed surface finish has been treated with a white, environmentally-friendly oil. The floorboards has micro bevels at all four sides and consist of three layers in total. Whereas the wearing surface is made of oak word, the middle layer, comprised of transverse spruce slats crosswise, enhances elasticity. The bottom layer made of solid spruce veneer stabilises the entire structure of the floor, which thus can be laid without requiring any special tools or adhesive.

Statement by the jury
Desire is outstanding due to its beautiful, vibrant surface finish and its well thought-out structure.

Dieser Holzboden mit gebürstetem Oberflächenfinish wurde mit weißem, umweltfreundlichem Öl behandelt. Die Dielenbretter sind an allen vier Seiten mit Mikrofasen versehen und bestehen aus insgesamt drei Schichten. Während die Nutzschicht aus Eichenholz gefertigt ist, erhöht die mittlere mit über Kreuz gelegten Fichtenleisten die Flexibilität. Die untere Schicht aus durchgehendem Fichtenfurnier stabilisiert schließlich die gesamte Struktur des Bodens, dessen Verlegung ohne Spezialwerkzeug oder Klebstoff erfolgt.

Begründung der Jury
Desire zeichnet sich durch eine schöne, lebendige Oberflächenstruktur und einen durchdachten Aufbau aus.

Grigo bog oak flooring
Engineered Flooring
Fertigparkett

Manufacturer
UAB Grigo, Kaunas, Lithuania
In-house design
Vytautas Grigas
Web
www.grigostudio.com

The rare wood of the bog oak and its wide range of natural colours were the inspiration for this parquet flooring and these solid floorboards. Fossilised wood, which is characterised by its high durability, has spent centuries buried in damp areas, submerged under water in moors or in bogs. In this case, it has been imitated using fresh oak that has been artificially aged. The top layer of the floor is shaded grey or black throughout its complete thickness, whereas the natural transition of these colour shades is another property of the bog oak. A finishing with oil or beeswax gives the surface an additional silky shine.

Statement by the jury
The innovative and speedy transformation of natural oak into bog oak produces a parquet flooring that has all the benefits of bog oak, but is readily and easily available.

Das seltene Holz der Mooreiche dient dem Parkett und den Massivholzdielen als Vorbild und inspiriert diese mit einer breiten Palette an natürlichen Farben. Das fossile Holz, das über Jahrhunderte in Feuchtgebieten, unter Wasser in Mooren und Sümpfen gelegen hat und sich durch seine hohe Widerstandsfähigkeit auszeichnet, wird hierfür mittels einer speziellen Technologie imitiert. Die Nutzschicht ist über ihre ganze Dicke hinweg grau oder schwarz, wobei der natürliche Übergang der Schattierungen eine weitere Eigenschaft der Mooreiche ist. Die Veredelung mit Öl oder Bienenwachs verleiht der Oberfläche einen zusätzlichen seidigen Glanz.

Begründung der Jury
Durch die innovative und schnelle Umwandlung normaler Eiche in Mooreiche wird hier ein Parkett realisiert, das die Vorzüge der Mooreiche bietet und gleichzeitig leicht verfügbar ist.

RELAZZO style
with NATURAL surface
WPC Deckingsystem
WPC-Terrassensystem

Manufacturer
REHAU GmbH, Guntramsdorf, Austria
In-house design
Productmanagement WPC
Web
www.rehau.com/relazzo

Benefits such as exceptional colour fastness, very few maintenance efforts and the natural wood appearance make this patio decking material stand out. By developing a surface with a wood grain effect surface, created by deep brushing, an authentic visual appearance of the material is achieved. Relazzo highlights this with the natural-looking trend colours Terra, Tasso and Pino.

Vorteile wie außergewöhnliche Farbechtheit, ein sehr geringer Pflegeaufwand sowie eine natürliche Holzoptik machen die Besonderheit dieser Terrassendielen aus. Mit der Entwicklung einer intensiv geprägten Oberfläche, die durch eine tiefe Bürstung entsteht, wird eine authentische visuelle Erscheinung des Materials erzielt. Relazzo betont diese mit den drei ebenfalls naturnahen Trendfarben Terra, Tasso und Pino.

Statement by the jury
This patio decking system impresses with its durable, but at the same time natural surface texture, whose wooden look is achieved with a special brushing technique.

Begründung der Jury
Dieser Terrassenbelag überzeugt mit einer widerstandsfähigen und, dank besonderer Bürstung, zugleich natürlich wirkenden Oberflächenstruktur in Holzoptik.

Neolith Calacatta
Architectural Surface
Architekturoberfläche

Manufacturer
TheSize Surfaces S.L,
Almassora (Castellón), Spain
In-house design
Carlos García
Web
www.neolith.com

This product offers a marble-like design with bold dramatic veining that closely resembles nature. It has the feel of the shimmering stone and is characterised by its resistance to scratch, heat and frost. Neolith Calacatta is suitable for floors, walls, façades, kitchen or bathroom surfaces. It is available in sheets measuring 3.2 x 1.5 metres and is 6 or 12 mm thick.

Statement by the jury
Neolith Calacatta combines the aesthetics of Carrara marble with the functional properties of sintered surface. It opens up a wide spectrum of design possibilities for architects and designers.

Dieses Produkt ist mit Adern versehen, die seine Oberfläche natürlichem Marmor ähnlich erscheinen lassen. Es fühlt sich wie das schimmernde Gestein an und zeichnet sich zudem durch Kratzfestigkeit sowie Hitze- und Frostbeständigkeit aus. Geeignet ist Neolith Calacatta für Böden, Wände, Fassaden oder auch Flächen in Küche oder Bad und kommt dort mit einer Größe von 3,2 x 1,5 Metern bei 6 oder 12 mm Stärke zum Einsatz.

Begründung der Jury
Neolith Calacatta verbindet die Ästhetik des Carrara-Marmors mit den funktionalen Eigenschaften einer gesinterten Oberfläche. Es eröffnet Architekten und Designern so neue Möglichkeiten bei der Gestaltung.

Chromatic Collection
Wall and Floor Tiles
Wand- und Bodenfliesen

Manufacturer
Revigrés, Águeda, Portugal
In-house design
Web
www.revigres.com

These full body porcelain wall and floor tiles are available in 40 colours, 11 sizes and in four different surface finishes. In addition to timeless primary colours, the Chromatic Collection also comes in a range of pure and saturated shades. A diversity of colours that can either bring harmony or create a contrast to the surroundings in their respective field of application. The aesthetic qualities of the wall/floor tiles make them a reliable solution for interior and exterior, public and residential areas.

Diese Porzellan-Wand- und Bodenfliesen sind in 40 Farben, elf Größen und vier verschiedenen Oberflächenausführungen erhältlich. Zusätzlich zu den zeitlosen Grundfarben beinhaltet die Chromatic-Kollektion auch eine Reihe von reinen und satten Farbtönen. Diese Vielfalt an Farben macht es möglich, in der Anwendung sowohl Harmonie als auch Kontraste zu der Umgebung zu schaffen. Dank ihrer ästhetischen Eigenschaften bieten diese Wand- und Bodenfliesen eine verlässliche Lösung für Interieur und Exterieur im öffentlichen oder im privaten Bereich.

Statement by the jury
With many different colours, sizes and surface finishes, these ceramic floor tiles open up a wide range of creative possibilities for architects.

Begründung der Jury
Mit vielen verschiedenen Farben, Formaten und Oberflächenstrukturen bieten diese Bodenfliesen Architekten eine große Bandbreite an Gestaltungsmöglichkeiten.

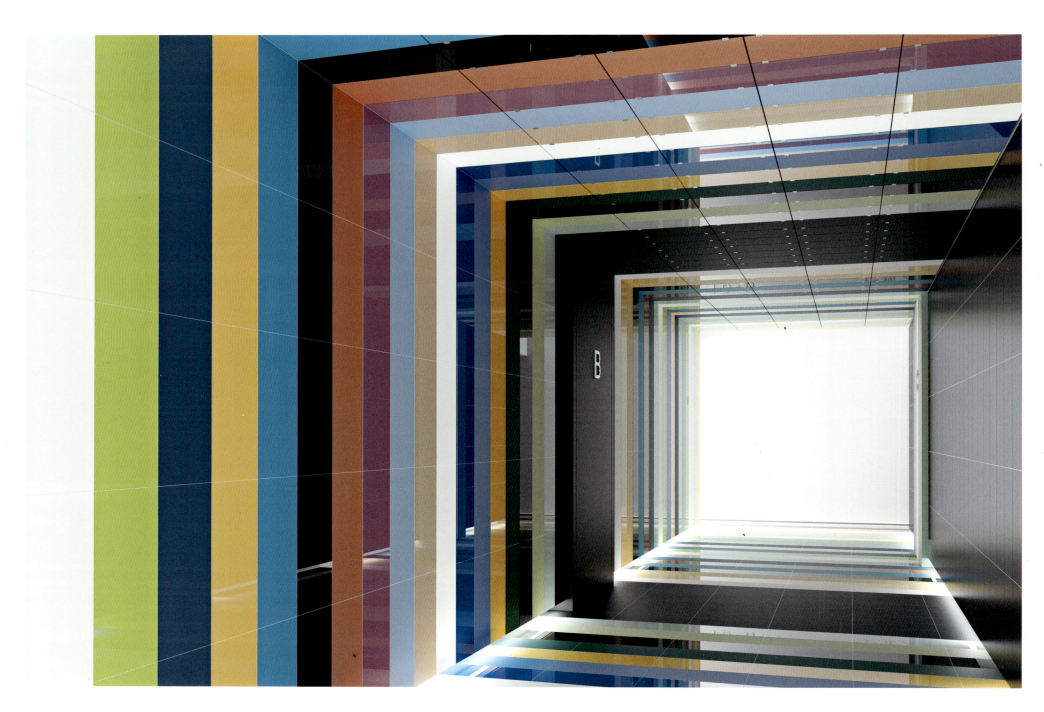

New Wood
Floor Tiles
Bodenfliesen

Manufacturer
Seranit Granit Seramik Sanayi Ve Tic AS,
Istanbul, Turkey
In-house design
Web
www.seranit.com.tr

Two contrasting properties come together in the New Wood floor tile and
endow it with a unique appeal. A modern
interpretation of the natural appearance
of wood and a glossy surface has been
made possible through novel production
techniques and these tiles therefore create a warm atmosphere in any room in
which they are used. The collection is easy
to clean and hence very hygienic.

Statement by the jury
New Wood is a thoroughly appealing
floor covering combining the straightforward surface qualities of tiles with
the warm appearance of wood.

In der Bodenfliese New Wood treffen zwei
kontrastierende Eigenschaften aufeinander
und zeichnen sie so mit einer einzigartigen
Wirkung aus: Die natürliche Holzoptik und
eine glänzende Oberfläche ermöglichen
die technische Umsetzung des Produkts
nach modernen Kriterien und erzeugen in
allen Räumen eine warme Atmosphäre.
Die Kollektion ist sehr einfach zu reinigen
und damit auch besonders hygienisch.

Begründung der Jury
New Wood überzeugt als Bodenbelag, der
die unkomplizierten Oberflächeneigen-
schaften von Fliesen mit einer warmen
Holzoptik kombiniert.

Duostone
Outdoor Tiles
Außenfliesen

Manufacturer
Seranit Granit Seramik Sanayi Ve Tic AS,
Istanbul, Turkey
In-house design
Web
www.seranit.com.tr

These ceramic tiles prevent the formation of algae and moss whether they
are used on roof terraces, balconies,
patios or courtyards. This makes cleaning them significantly easier so that with
Duostone lasting stains caused by red
wine, coffee or water are finally a thing
of the past. The tiles can be cleaned in
exactly the same way as indoor floor
tiles and their colour will not fade even
under strong sunlight.

Statement by the jury
Duostone combines a high-quality
ceramic surface with stain-resistant
properties and thus offers a trouble-
free solution for outdoor tiling.

Auf der Dachterrasse, dem Balkon, der
Veranda oder auch im Innenhof verhindert
diese Keramikfliese die Bildung von Algen
und Moos. Damit wird auch die Reinigung
maßgeblich erleichtert, sodass dauerhaf-
te Flecken wie von Rotwein, Kaffee oder
Wasser mit der Duostone nicht mehr zu
befürchten sind. Sie wird ebenso gereinigt
wie Bodenfliesen für den Innenbereich
und bleicht auch bei starker Sonneneinstrahlung nicht aus.

Begründung der Jury
Duostone verbindet eine hochwertige
Keramikoberfläche mit schmutzabweisen-
den Eigenschaften und überzeugt so als
unkomplizierte Fliese für den Außenbereich.

CODE 519/Filigrane
Wall Tiles
Wandfliesen

Manufacturer
Seranit Granit Seramik Sanayi Ve Tic AS,
Istanbul, Turkey
In-house design
Web
www.seranit.com.tr

The elegant wall tile Filigrane attracts attention with a combination of traditional values and modern artisanship. In the style of Mediterranean crochet work and given its final shape through the use of natural wood materials and textures, a technologically advanced tile has emerged whose impressive historical elements enrich everyday life.

Statement by the jury
This wall tile appeals with its textured, aesthetic appearance that harmoniously links tradition and modernism.

Die elegante Wandfliese Filigrane spricht den Nutzer mit ihrer Kombination aus traditionellen Werten und moderner Handwerkskunst an. In Anlehnung an mediterrane Häkelarbeiten entstand durch den Einsatz natürlicher Holzmaterialien und Texturen ein Produkt, das neben seiner technologischen Ausgereiftheit auch mit eindrucksvollen historischen Elementen zu einer Bereicherung im Alltag wird.

Begründung der Jury
Diese Wandfliese gefällt mit ihrem strukturierten, ästhetischen Erscheinungsbild, das Tradition und Moderne harmonisch verbindet.

Terra Nova
Porcelain Floor Tiles
Porzellan-Bodenfliesen

Manufacturer
VitrA Karo San. ve Tic. A.S,
Istanbul, Turkey
In-house design
Web
www.vitra.com.tr

With colours such as Vanilla, Nero, Tobacco and Mocha, the floor tiles of this collection create a warm, inviting atmosphere. They are available in four different sizes, can be used outdoors and inside and have practical anti-slip properties as well as a great light refraction and a tremendous bending strength. That makes them just as suitable for use in living areas as in high traffic zones. An alternative decorative version is available in two sizes and eight different patterns.

Statement by the jury
It is Terra Nova's great aesthetic qualities in skilful combination with practical features such as their anti-slip properties and flexural strength that make these floor tiles so appealing.

Die Grundfliesen dieser Kollektion schaffen mit Farben wie Vanille, Nero, Tabak und Mokka eine warme und gemütliche Atmosphäre. Sie sind in vier alternativen Größen erhältlich, im Innen- oder Außenbereich einsetzbar, mit Rutschhemmung ausgestattet und zeichnen sich durch hohe Lichtbrechung und außergewöhnliche Biegefestigkeit aus. Die Fliesen eignen sich neben der Nutzung im Wohnbereich ebenso für Verkehrszonen. Eine Dekoralternative ist in zwei Größen und acht verschiedenen Mustern erhältlich.

Begründung der Jury
Terra Nova besticht durch eine hohe ästhetische Qualität und verbindet diese gekonnt mit praktischen Eigenschaften wie Rutsch- und Biegefestigkeit.

TOSA
System-Dry-Bricks
System-Trockenmauerstein

Manufacturer
Diephaus Betonwerk GmbH,
Vechta, Germany
In-house design
Web
www.diephaus.de

The concrete material emphasises the natural appearance of stone and also promotes its robust and low-maintenance character. Large format and smooth visible surfaces as well as slightly chamfered edges and lively patterns accentuate the harmonious appearance of these bricks and, together with nuanced colour shades such as quartzite and shell limestone, form visual highlights. Thanks to their light weight, it is easy to mount these hollow bricks in a height of up to two metres, the top row can be planted in an individual way as desired.

Der Werkstoff Beton lässt den Stein natürlich wirken und verleiht ihm zudem seinen robusten und pflegeleichten Charakter. Großformatige, glatte Sichtflächen sowie leicht gefaste Kanten und lebhafte Maserungen unterstreichen die harmonische Wirkung und setzen mit nuancierten Farbtönen wie Quarzit und Muschelkalk optische Akzente. Die hohlen Steine sind dank ihrer Leichtigkeit problemlos bis zu einer Höhe von zwei Metern zu montieren, die oberste Reihe kann individuell begrünt werden.

Statement by the jury
Thanks to its high-quality design and exceptionally aesthetic appeal, this system-dry wall brick offers flexibility and a high degree of creative freedom when it comes to building walls.

Begründung der Jury
Dieser System-Trockenmauerstein bietet dank seiner hochwertigen Ausführung und besonderen ästhetischen Qualität Flexibilität und gestalterische Freiheit beim Mauerbau.

SGG Master-Soft
Patterned Glass
Ornamentglas

Manufacturer
Saint-Gobain Innovative
Materials sp. z o.o.
Oddział Glass w Dąbrowie
Górniczej, Poland

In-house design
Saint-Gobain Glass

Web
www.saint-gobain-glass.pl

reddot award 2015
best of the best

Material poetry

Newly developed materials have always stimulated the imagination of architects and designers. They offer exciting new possibilities that often also lead to new approaches in thinking and creating. SGG Master-Soft is such a material, as its distinctive properties expand the way glass can be used. This patterned glass possesses an extraordinarily high quality. The surface itself virtually touches and challenges the perception of the viewer as it features an impressive, 1 mm deep textured engraving. The texture is obtained by an unusual casting process followed by rolling the glass between two cylinders that emboss a motif into the glass. The result is a glass surface that creates the impression as if it was delicately making waves. When looking from a distance, the entire glass surface seems to be moving like a textile curtain. Due to the soft visual feel of the pattern, this glass is particularly suitable in bigger applications such as shower screens or partitions. The glass possesses the ability to both provide a high degree of intimacy and let through enough light while at the same time sufficiently brightening a given space. Therefore, the SGG Master-Soft patterned glass lends itself for use both on the inside and outside of buildings. The glass fascinates with a deeply enticing poetic appeal that redefines a familiar material in a pioneering new approach.

Die Poesie des Materials

Neu entwickelte Materialien beflügeln stets auch die Entwürfe der Architekten und Designer. Es bieten sich spannende, neue Chancen, die oftmals zu veränderten Denkweisen führen. SGG Master-Soft ist ein solches Material, da es durch seine besonderen Eigenschaften die Einsatzmöglichkeiten von Glas erweitert. Dieses Ornamentglas besitzt eine außerordentlich hohe Glasqualität. Seine Oberfläche fordert auf faszinierende Weise die Wahrnehmung heraus, da sie mit einem eindrucksvollen, 1 mm tief geprägten Relief gestaltet ist. Die Struktur entsteht durch ein außergewöhnliches Gießverfahren und ein danach erfolgendes Walzen zwischen zwei Zylindern, auf die ein Motiv aufgebracht ist. Das Ergebnis ist eine Glasfläche, die den Eindruck erweckt, als schlage sie Wellen. Aus größerer Entfernung betrachtet, scheint sich die gesamte Glasfläche wie ein aufgehängter Vorhang zu bewegen. Mit seiner sehr sanft anmutenden Ornamentik eignet sich dieses Glas insbesondere für größere Flächen wie etwa bei Duschkabinen oder Trennwänden. Dabei besitzt es die Eigenschaft, die nötige Privatsphäre zu schaffen und zugleich genügend Licht hindurch zu lassen, um den Raum ausreichend zu beleuchten. Daher ist das Ornamentglas SGG Master-Soft vielfältig innerhalb und außerhalb von Gebäuden einsetzbar. Es begeistert dabei mit seiner sehr eindringlichen Poesie, die ein bekanntes Material zukunftsweisend neu definiert.

Statement by the jury

An impressive design achievement has emerged here as a patterned glass of excellent quality. The way viewers can discern ever changing wave forms and movements in the glass surface is fascinating. The distinctive structure exudes a poetic appeal that enriches the environment and the lives of people around it. This patterned glass can be installed in both indoor and outdoor environments – as it challenges and inspires the creativity of architects and planners alike.

Begründung der Jury

Eine beeindruckende Designleistung ließ hier ein Ornamentglas von exzellenter Qualität entstehen. Faszinierend ist, wie der Betrachter immer wieder neue Wellenformen und Bewegungsmuster in der Glasfläche wahrnimmt. Die besondere Struktur strahlt eine poetische Anmutung aus, die ihre Umgebung und das Leben der Menschen in dieser bereichert. Dieses Ornamentglas kann im Innen- wie im Außenbereich vielfältig eingesetzt werden – die Kreativität von Architekten und Planern gleichermaßen wird dabei herausgefordert und inspiriert.

Designer portrait
See page 64
Siehe Seite 64

MOTION FACET
Glass Mosaic
Glasmosaik

Manufacturer
Thai Ceramic, Saraburi, Thailand
In-house design
Wiriya Wattanayon
Web
www.cotto.co.th

The dynamic aesthetics created by this glass mosaic is based on the interplay of glittering crystals with the variations of light and shade caused by the facets. Viewers see a new pattern every time they change the viewing angle – an optical illusion. Thus, when viewed like this, the glass mosaic interacts with its surroundings, highlighting the effect of choppiness.

Statement by the jury
The modern, captivating design of these three-dimensional mosaic tiles is brought to life by the interplay of light and shade allowing it to constantly appear new and different.

Die dynamische Ästhetik der Glasmosaik-fliesen basiert auf dem Zusammenspiel zwischen funkelnden Kristallen und der Licht- und Schattenwirkung der Facetten. Der Betrachter sieht je nach Sichtwinkel ein anderes Muster – eine optische Täuschung. Somit entsteht beim Anblick des Glas-mosaiks eine gewisse Interaktion mit der Umgebung, die den Effekt der Bewegtheit zusätzlich unterstreicht.

Begründung der Jury
Das moderne, reizvolle Design dieser dreidimensionalen Mosaikfliesen erwacht durch Licht und Schatten gewissermaßen zum Leben und wirkt so immer wieder neu und anders.

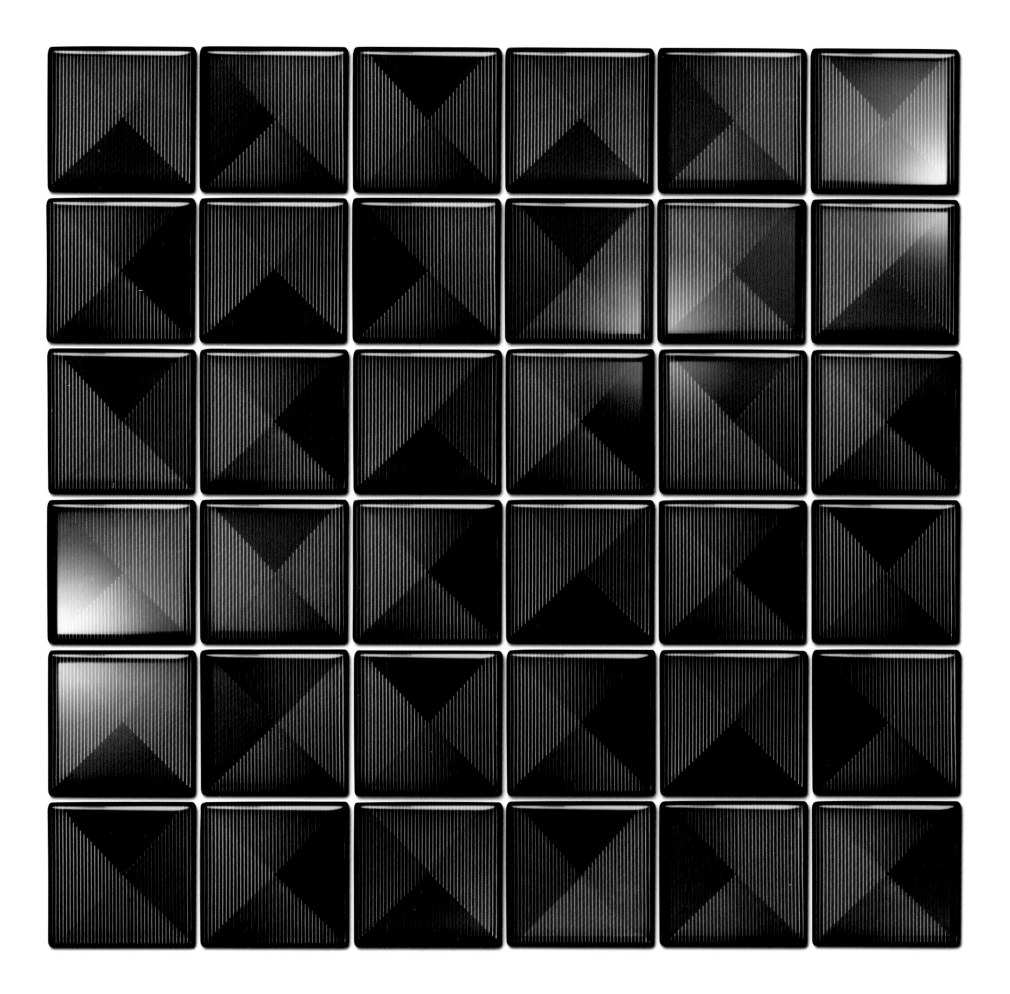

Hybrid Collection
Glass with 3D Pattern for Interior Design
Glas mit 3D-Muster für Innenarchitektur

Manufacturer
Casali, Cesenatico (Emilia-Romagna), Italy
Design
Massive Design Sp. z o.o (Mac Stopa),
Warsaw, Poland
Web
www.casali.net/en-us
www.massivedesign.pl

This novel coating concept presents colour and shade effects on glass in the form of artistic, parametrically designed circles, triangles and many other geometrical shapes. The skilful combination of glass and paint thereby gives the product its three-dimensional appearance. Available in different colours, this design was conceived for large surfaces such as partitions, doors and all types of furniture.

Dieses neuartige Konzept der Lackierung arbeitet mit Farb- und Schatteneffekten auf Glas in künstlerisch-parametrisch gestalteten Kreisen, Dreiecken und vielen weiteren Formen. Die gekonnte Kombination der beiden Materialien Glas und Lack erzeugt dabei die dreidimensionale Wirkung des Produktes. Verfügbar in verschiedenen Farben, wurde der Entwurf für große Flächen wie zum Beispiel Trennwände, Türen und verschiedenartiges Mobiliar konzipiert.

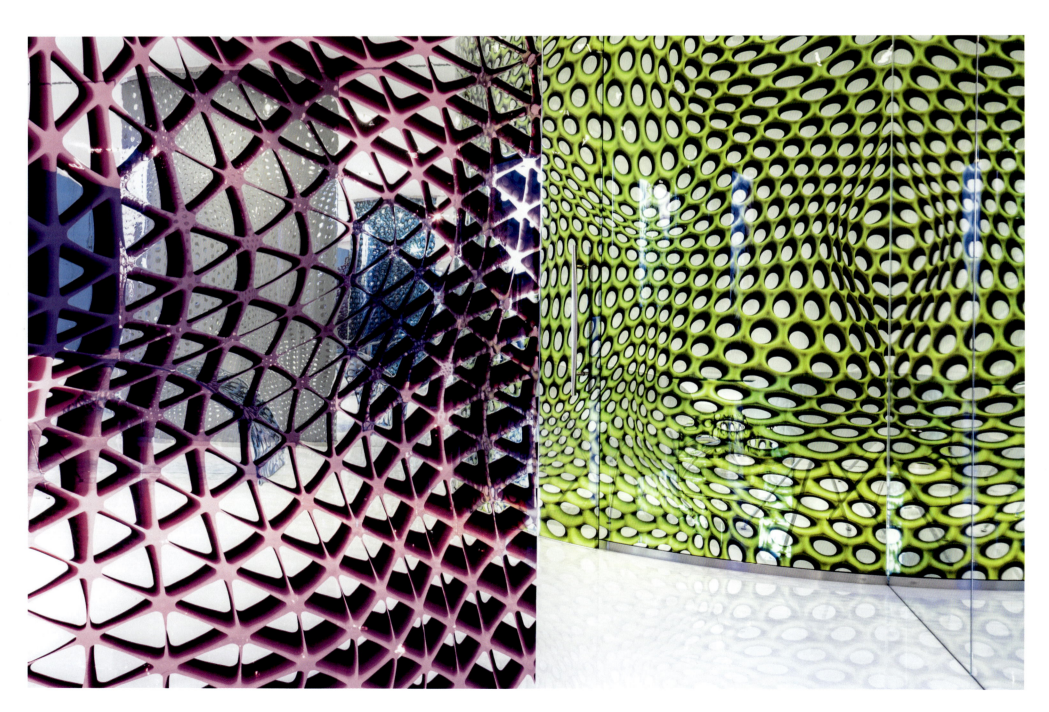

XSpark
Effect for Automotive Coatings
Effekt für Fahrzeugserienlack

Manufacturer
BASF Coatings Division, Münster, Germany
In-house design
Mark Gutjahr, Florina Trost
Web
www.basf-coatings.com

This colour effect for OEM car paints has an extremely strong sparkle and therefore a more intense appearance than other pigments of this type. It is made up of extremely small glass particles that reflect light more precisely and create a noticeable sparkle. The glass particles are applied in a single process step with the coloured paint in a thin layer. The result is a thin, homogenous coating of paint, which sparkles and thereby creates a depth effect.

Dieser Farbeffekt für Fahrzeugserienlacke glitzert außergewöhnlich stark und hat somit auch eine intensivere Wirkung als andere Pigmente seiner Art. Die präzise Reflexion des Lichts durch kleine Glaspartikel, die in einem Arbeitsschritt zusammen mit dem farbgebenden Lack und in einer dünnen Schicht aufgetragen werden, erzeugt ein ausgeprägtes Funkeln. Dies wird außerdem durch eine Tiefe im Lack verstärkt, die in der homogenen Gestaltung der Oberfläche begründet liegt.

Statement by the jury
XSpark gives car paints a distinctive sense of depth thanks to an elegant, subtle sparkling effect.

Begründung der Jury
XSpark verleiht der Autolackierung durch einen eleganten, dezenten Glitzereffekt eine markante Tiefenwirkung.

Swarovski Active Panel

Back-lit Glas Panel covered with Crystal

Hinterleuchtetes Glaspaneel mit Kristallen

Manufacturer
D. Swarovski KG, Wattens, Austria
In-house design
Simon Grubinger
Web
www.swarovski.com

Precisely polished crystals make up the surface of this backlit glass panel. Thanks to an open structure these can also be clearly felt. Approximately 200,000 crystals are glued to every square metre of a glass substrate and are ultimately backlit using special LEDs. The boards are modular and can be combined freely, enabling the creation of a wide variety of colours and shapes. The corresponding Swarovski Active Panel casing is custom-made and can be individually fitted.

Statement by the jury
The stunning design and modular concept of these glass panels offers exciting decorative possibilities for the world of interior design.

Präzise geschliffene Kristalle bilden die Oberfläche dieses hinterleuchteten Glaspaneels und sind aufgrund der offenen Struktur auch haptisch wahrnehmbar. Pro Quadratmeter sind 200.000 Kristalle auf einem Glasträger verklebt und werden schließlich mit speziellen LEDs hinterleuchtet. Die Platinen sind modular kombinierbar, was eine hohe Vielfalt an Farben und Formen erlaubt. Das zugehörige Gehäuse des Swarovski Active Panel wird projektspezifisch angepasst und dementsprechend gefertigt.

Begründung der Jury
Mit ihrem reizvollen Design und einer modularen Bauweise ermöglichen diese Glaspaneele spannende Dekorationsmöglichkeiten im Bereich des Interior Designs.

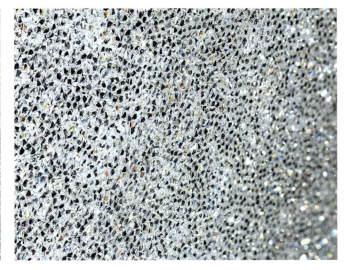

Carp
Decorative Material
Dekomaterial

Manufacturer
Jessica Owusu Boakye,
Hamburg, Germany
In-house design

Carp is a material based on natural components. It is made entirely from citrus fruits and is therefore 100 per cent biodegradable. The surface is characterised by a marbled finish, which is translucent or opaque depending on its thickness. The influence of heat, air or cold changes the colours, depending on its exposure to light Carp either appears matt or glossy. The attractive marbling as well as the elasticity and durability of the material round off its overall impression.

Statement by the jury
This decorative material made entirely of citrus fruit is environmentally friendly and appealing thanks to its distinctive marbling.

Carp ist ein aus natürlichen Baustoffen bestehendes Material, dessen Masse ausschließlich aus Zitrusfrüchten hergestellt wird und somit zu 100 Prozent biologisch abbaubar ist. Die Oberfläche ist durch eine marmorierte Struktur charakterisiert, die je nach Stärke transluzent oder blickdicht ist. Durch den Einsatz von Wärme, Luft oder Kälte verändert sich die Farbe, unter Lichteinstrahlung tritt Carp matt oder glänzend in Erscheinung. Sowohl eine attraktive Maserung als auch die hohe Flexibilität und Belastbarkeit des Materials runden den Gesamteindruck ab.

Begründung der Jury
Dieses aus Zitrusfrüchten gefertigte Dekormaterial ist umweltfreundlich und begeistert zugleich mit seiner prägnanten Maserung.

Organoider Bildabsorber
Acoustic Panel
Akustikpanee

Manufacturer
Organoid Technologies GmbH, Fliess, Austria
In-house design
Klemens Baier
Web
www.organoids.at

The organoid surfaces used for this sound absorber work according to the Helmholtz resonance principle. Depending on the nature of the slits and holes in the support, any acoustic performance can be created using the corresponding damping. The ready-made, 70 mm-thick acoustic elements come in any size required and have an uninterrupted, folded edge. In addition to their natural fragrance, authenticity is one of the key characteristics of these sound absorbers.

Statement by the jury
Thanks to the fragrance of these surfaces made of natural materials such as alpine hay or wood, these acoustic panels not only absorb sound, but also appeal to all senses.

Die für diesen Schallabsorber verwendeten organoiden Oberflächen wirken nach dem Helmholtz-Prinzip des akustischen Resonators. Je nach Schlitzung und Bohrung des Trägermaterials lässt sich mit der entsprechenden Dämpfung jede auditive Leistung realisieren. Die fertig konfektionierten, 70 mm starken Akustikelemente sind in beliebiger Größe mit einer gefalteten Kante versehen, die unterbrechungsfrei verläuft. Neben ihrem natürlichen Duft zählt Authentizität zu den wichtigsten Merkmalen der Schallabsorber.

Begründung der Jury
Mit ihren duftenden Oberflächen aus natürlichen Materialien wie Almgras oder Holz absorbieren diese Paneele den Schall und sprechen zugleich alle Sinne an.

zeyko Metal-X^2
Structured Surface for Kitchen Furniture
Strukturoberfläche für Küchenmöbel

Manufacturer
Zeyko Möbelwerk GmbH & Co. KG,
Mönchweiler, Germany
In-house design
Zeyko Design Team
Web
www.zeyko.de

To manufacture this structured surface for kitchen furniture, naturally sourced metal is milled, liquefied by admixtures and subsequently applied by hand. The resulting texture produces a surface that appears soft and wafer-like, standing out due its particular hand-crafted characteristics. The polished metal surfaces contrast with the matt indentations and shimmer depending on the viewing angle. This creates an unexpected sense of depth.

Statement by the jury
Innovative processing of material has been combined with craftsmanship to create zeyko Metal-X^2, resulting in surface textures that are both visually enticing as also pleasing to the touch.

Für die Verarbeitung dieser Strukturoberfläche für Küchen wird aus der Natur gewonnenes Metall gemahlen, durch Zugaben verflüssigt und anschließend von Hand aufgetragen. Die daraus entstehende Struktur erzeugt eine weich wirkende und waffelartige, gespachtelte Fläche, die sich durch ihren besonderen handwerklichen Charakter auszeichnet. Die polierten Flächen des Metalls stehen in Kontrast zu den matten Vertiefungen und changieren je nach Blickwinkel, was eine unerwartete Tiefe schafft.

Begründung der Jury
Innovative Materialverarbeitung und Handwerkskunst sind bei zeyko Metal-X^2 so kombiniert, dass neue, visuell ebenso wie haptisch reizvolle Oberflächenstrukturen entstehen.

skai® Solaris EN
Upholstery Synthetics
Polsterbezugsstoff

Manufacturer
Konrad Hornschuch AG,
Weißbach, Germany
In-house design
Web
www.hornschuch.com

The three-dimensional effect created by the honeycomb cube is the result of an optical illusion that is confusing and fascinating at the same time. Similar to a game of deception, the impression conveyed by the surface of the polyester upholstery fabric changes depending on the viewing angle and the degree of exposure to light. Variously oriented hexagon clusters form rhombuses. Three rhombuses each form the fourth side of a cube in front of the observer's eyes. This effect is intensified by a special lacquering on the hexagons.

Statement by the jury
This upholstery fabric is captivating thanks to its exciting effects which create an exceptional optical effect and also give the fabric an interesting feel.

Die dreidimensionale Wirkung des Wabenwürfels entsteht durch eine optische Täuschung, die verwirrend und faszinierend zugleich scheint. Einem Vexierspiel gleich ändert sich je nach Betrachtungswinkel und Lichteinfall auch der Eindruck der Oberfläche aus Polsterbezugsstoff. Unterschiedlich ausgerichtete Hexagon-Cluster formieren sich zu Rhomben. Je drei Rhomben fügen sich vor dem Auge des Betrachters zu den drei Seiten eines Würfels. Verstärkt wird diese Wirkung durch eine spezielle Lackierung der Hexagone.

Begründung der Jury
Dieser Polsterbezugsstoff besticht mit spannenden Effekten, die eine außergewöhnliche optische Wirkung erzielen und dem Material zudem eine interessante Haptik verleihen.

DecoSOUL
Individual Decorative Surface
Individuelle Dekoroberfläche

Manufacturer
Dräxlmaier Group, Vilsbiburg, Germany
In-house design
Isabella Schmiedel
Web
www.draexlmaier.com

This modern composite material is manufactured by stacking its basic components in layers and bonding them together to create a solid block. Thin sheets are then sliced from the end face of the block. If transparent material is included in the layers, additional backlighting is possible. Regardless whether on the basis of leather, metal or natural fibres – the composite material is highly flexible and therefore opens up a manifold range of applications.

Statement by the jury
The novel use of raw material makes DecoSOUL an innovative composite material, which has a wide range of applications and is a successful example of upcycling.

Für die Herstellung dieses modernen Verbundwerkstoffs wird das Ausgangsmaterial in Lagen aufeinandergeschichtet und verklebt. So entsteht ein Block, von dessen Stirnseite dünne Schichten abgeschnitten werden. Wird transparentes Material in die Schichtung mit aufgenommen, ist eine zusätzliche Hinterleuchtung möglich. Ob auf Basis von Leder, Metall oder Naturfaser – der Werkstoff ist sehr flexibel und daher auch vielfältig in seinen Einsatzmöglichkeiten.

Begründung der Jury
Durch eine neue Art der Verarbeitung von Werkstoffen entstand mit DecoSOUL ein innovativer und vielseitig einsetzbarer Verbundwerkstoff, der ein gelungenes Beispiel für Upcycling darstellt.

Post-it® Dry Erase Surface
Office Productivity Tool
Büro-Arbeitsoberfläche

Manufacturer
3M, St. Paul (Minnesota), USA
In-house design
Web
www.3m.com/design

Post-it Dry Erase Surface makes it possible to record collaborative ideas immediately during planned or spontaneous meetings so that there is no risk for them to be forgotten. Attached to walls, glass surfaces, steel installations or wood, this media device enables collaborative communication and hence flexible cooperation at any time. Once rolled up, an in-house adhesive technology makes it possible to attach it to various surfaces without additional tools.

Statement by the jury
Post-It Dry Erase Surface is a captivatingly simple solution to turn any smooth surface into a temporary white board.

Auf geplanten oder spontanen Meetings können mithilfe der Post-it Dry Erase Surface Ideen unmittelbar festgehalten werden und geraten somit nicht in Vergessenheit. An Mauern, Glasflächen, Stahlmobiliar oder Holz befestigt, bietet dieses Medium stets eine Möglichkeit zu gemeinsamer Kommunikation und dadurch einer flexiblen Zusammenarbeit. Dank eigens entwickelter Klebetechnik kann das Material nach dem Aufrollen ohne zusätzliches Werkzeug auf variablen Untergründen angebracht werden.

Begründung der Jury
Post-it Dry Erase Surface ist eine bestechend einfache Möglichkeit, aus jeder glatten Oberfläche ein temporäres White Board zu machen.

Anyway Pivoting Room Divider
Pivoting Interior Doors
Pivottüren

Manufacturer
Anyway Doors, Indoor Collection NV,
Massenhoven, Belgium
In-house design
Web
www.anywaydoors.be

These elegant room dividers have hinges integrated into the door frames that allow them to open through 360 degrees. A mechanical, self-closing pivot system with 90-degree positioning makes opening in both directions easy, despite a weight of 150 kg. As the bolts and plugs only require holes to a depth of 35 mm, the dividers may be used with any type of floor structure, even if it contains underfloor heating.

Der elegante Raumteiler wird mit verdeckten, in den Türrahmen eingebauten Aufhängungen befestigt und ist um 360 Grad schwenkbar. Ein mechanisches, selbstschließendes Drehgelenksystem mit einer 90-Grad-Positionierung erleichtert das Öffnen in beide Richtungen des bis zu 150 kg schweren Raumteilers. Da die Dübel und Bolzen eine Montagetiefe von lediglich 35 mm erfordern, ist die Anwendung unabhängig von der Bodenstruktur und auch bei Vorhandensein einer Fußbodenheizung unproblematisch.

Statement by the jury
The Anyway Pivot Doors are elegant, easy to use room dividers and doors in one. Their sophisticated construction allows a wide variety of uses.

Begründung der Jury
Die Pivot-Tür Anyway ist eleganter, leicht wirkender Raumteiler und Tür in einem und eröffnet mit ihrer raffinierten Konstruktion vielfältige Einsatzmöglichkeiten.

MUTO
Multifunctional Manual Sliding Door System
Multifunktionales, manuelles
Schiebetürsystem

Manufacturer
Dorma Deutschland GmbH,
Ennepetal, Germany
In-house design
Bernhard Heitz
Web
www.dorma.com/innovation
Honourable Mention

The manual sliding door system MUTO
with self-closing action is both easy
to operate and to install. The elegant
concealed lock can be controlled by
remote control or from a wall switch.
The status indicator enables constant
monitoring of the door, while the proven
DORMOTION damping mechanism gently
cushions the door panel when closing.
All relevant functions can be set from
the front thanks to a removable front
cover, making the installation particularly easy.

Statement by the jury
The clearly thought-through functionality of this sliding door makes it easy to
handle, both during installation and in
daily use.

Das manuelle Schiebetürsystem
MUTO mit Self-Closing ist sowohl in der
Handhabung als auch in der Montage
kinderleicht. Die elegante integrierte Ver-
riegelung kann bequem per Fernbedienung
oder Wandschalter betätigt werden. Der
Statusindikator ermöglicht die kontinu-
ierliche Überwachung der Tür, während
die DORMOTION-Dämpfungseinheit das
Türblatt beim Schließen sanft stoppt.
Alle relevanten Funktionen können dank
abnehmbarer Frontklappe von vorne eingestellt werden, was die Montage besonders
einfach macht.

Begründung der Jury
Die durchdachte Funktionalität dieses
Schiebetürsystems erlaubt eine unkom-
plizierte Handhabung sowohl bei der
Montage als auch im täglichen Gebrauch.

Centor Integrated Doors
Folding Doors
Flügeltüren

Manufacturer
Centor Holdings Pty Ltd,
Eagle Farm, Queensland, Australia
In-house design
Web
www.centor.com

The integration of screens, shades and hardware ensures that these folding doors can be used flexibly, at different times of the day. While the screens keep insects at bay, the shades filter light and provide privacy. The screens and shades are built-in, and retract completely into the doorframe, while the hinges and locks are concealed, so there is nothing to hinder the view. This practical system is available with a wood or aluminium interior and a thermally insulated aluminium exterior.

Die Integration von Fliegengitter, Rollos und Beschlägen stellt sicher, dass diese Flügeltüren zu unterschiedlichen Tageszeiten flexibel verwendet werden können. Das Fliegengitter hält Insekten fern, die Rollos filtern das Sonnenlicht und schützen die Privatsphäre. Beide sind in den Türrahmen eingebaut und verschwinden gänzlich, während die Beschläge und das Türschloss komplett in den schlanken Flügelrahmen versteckt sind und so nicht von der Sicht nach draußen ablenken. Dieses praktische System ist wahlweise mit einer Holz- oder Aluminiumverkleidung innen und einer wärmegedämmten Aluminiumverkleidung außen erhältlich.

Statement by the jury
This range of doors leaves a convincing impression thanks to its innovative multi-functionality. As its features are built into the structure, it also has a clean purist overall appearance.

Begründung der Jury
Diese Türenkollektion überzeugt mit einer innovativen Multifunktionalität, gleichzeitig zeigt sie dank der integrierten Bauweise ein klares, puristisches Gesamtbild.

Nevos GLASS
Frameless, Automatic Front Door in Glass Design
Rahmenlose, automatische Haustür in Glasoptik

Manufacturer
Josko Fenster und Türen GmbH,
Kopfing, Austria
In-house design
Johann Ecker
Web
www.josko.at
Honourable Mention

The Nevos GLASS front door has a surface of enamelled glass that can be either tinted black or given a high-gloss metallic coating as desired. Aside from the LED illumination, which accentuates the door's silhouette, the absence of a frame, handles or fittings underlines its purist appearance. The transparent side panel melds with the opaque leaf to create a continuous front.

Statement by the jury
Through the absence of a frame, handles and fittings, this automatic front door takes on an avant-garde and purist appearance.

Die Oberfläche der Nevos GLASS besteht aus emailliertem Glas, das auf Wunsch schwarz eingefärbt oder mit einer hoch-reflektierenden Metallic-Beschichtung versehen werden kann. Neben einer akzentuierten Beleuchtung der Silhouette durch LEDs unterstreicht der Verzicht auf Rahmen, Griffe und Beschläge die puristi-sche Wirkung der Tür. Das lichtdurchlässige Seitenteil verschmilzt optisch mit dem blickdichten Flügel, sodass eine durchgän-gige Front entsteht.

Begründung der Jury
Durch den Verzicht auf Rahmen, Griff und Beschläge entsteht das avantgardistische, puristische Erscheinungsbild dieser auto-matischen Haustür.

KeraTür – Space 63701
Front Door
Haustür

Manufacturer
KeraTür GmbH & Co. KG,
Raesfeld, Germany
In-house design
Julian Kemming
Web
www.keratuer.de

This entrance door is 220 mm thick and its exterior is completely covered with glass, thus allowing it to blend in harmo-niously with the house entrance. On the side is a functional channel in black ano-dised aluminium which contains the bell and a door-opening module together with the letterbox. The inside of the door is made of fine woods, such as varnished German wild oak, a feature which under-lines its exclusive character.

Statement by the jury
The Kera door has an extremely purist appearance which is accentuated by the clear separation between the door and its functional area.

Die Stärke dieser Haustür beträgt 220 mm. Ihre verglaste Außenseite bedeckt den Blendrahmen an der Seite und fügt sich so ohne störende Unterbrechung in den Hauseingang ein. Der seitlich angebrachte Funktionskanal ist mit schwarz eloxiertem Aluminium verkleidet und enthält neben einer Klingel und Modulen zum Öffnen der Tür einen Briefkasten. Die Innenseite aus edlen Holzarten, wie zum Beispiel deut-scher Wildeiche lasiert, unterstreicht den exquisiten Charakter dieser Tür.

Begründung der Jury
Die KeraTür überzeugt mit einem extrem puristischen Erscheinungsbild, das außen durch die klare Trennung zwischen Tür und Funktionsbereich noch verstärkt wird.

WC Door Knob „INSAFE"
Handle for Public WC Doors
Beschlag für öffentliche
WC-Türen

Manufacturer
Schäfer Trennwandsysteme GmbH,
Horhausen, Germany
In-house design
Design
Barski Design GmbH
(Olaf Barski, Claas Wellhausen),
Frankfurt/Main, Germany
Web
www.schaefer-tws.de
www.barskidesign.com

Used for public and semi-public toilets, this door handle offers a range of functions, such as the red-white indicator which can be seen, via two large openings, both from inside and outside the cubicle. This ensures that users can always tell whether the door is properly locked. It can be locked by turning the ergonomic wheel whose high-quality, anodised aluminium guarantees long durability.

Statement by the jury
A clean appearance that clearly indicates whether the cubicle is open or locked, an ergonomic design and a high degree of self-explanatory use characterise this door knob.

Angebracht in öffentlichen und halböffentlichen Toiletten verfügt der Türbeschlag über verschiedene Funktionen. So ist etwa die Rot-Weiß-Anzeige über zwei große Sichtfenster und sowohl auf der Außen- als auch auf der Innenseite der Kabine erkennbar. Dies garantiert, dass auch der Nutzer stets sehen kann, ob die Tür richtig verriegelt ist. Sie lässt sich durch Drehen des ergonomischen Rades verschließen, das sich dank hochwertigem Aluminium mit eloxierter Oberfläche durch eine lange Lebensdauer auszeichnet.

Begründung der Jury
Ein klares Erscheinungsbild, das deutlich kommuniziert, ob die Kabine frei oder besetzt ist, eine ergonomische Gestaltung und eine hohe Selbsterklärungsqualität zeichnen diesen Türbeschlag aus.

ILLUNOX
Illuminated Handrail and Balustrade System
Beleuchtetes Handlauf- und
Geländer-System

Manufacturer
Illunox, Lable of Lagusski,
Wijchen, Netherlands
In-house design
Jan-Willem Wessels
Web
www.illunox.com

The special feature of this handrail and balustrade system is its LED illumination. The diameter and shape of the stainless-steel elements as well as the colour of the illumination may be chosen according to individual taste before the system is installed. It is suitable for use indoors or out. The wall brackets and baluster adapters can be moved into the desired position by sliding them along the profile and can be fixed without the need for additional welding, so that the flow of light remains uninterrupted.

Statement by the jury
The energy-saving LED light strip integrated into the ILLUNOX handrail offers attractive, indirect lighting and increased safety.

Die Besonderheit dieses Geländer- und Brüstungssystems liegt in einer Beleuchtung durch LEDs. Durchmesser und Form des Edelstahls sowie Stärke und Farbe der Beleuchtung können individuell gewählt werden, bevor das System schließlich drinnen oder draußen zum Einsatz kommt. Die Handlaufträger und die Geländerbefestigungen werden über das Profil auf die gewünschte Position geschoben und ohne zusätzliche Schweißarbeiten befestigt, sodass der Lichtlauf nicht unterbrochen wird.

Begründung der Jury
Mit einer in den Handlauf integrierten, energiesparenden LED-Lichtleiste bietet ILLUNOX sowohl eine angenehme indirekte Beleuchtung als auch mehr Sicherheit.

albo® fenix letter box
Design Letter Box
Design-Briefkasten

Manufacturer
Argent Alu NV, Kruishoutem, Belgium
In-house design
Web
www.argentalu.com

The flat, vertical flap of this aluminium letterbox opens inwards which makes it easy to use. If desired, the plate inside the letter box can be removed in order to double the volume. The floor has drainage holes and a special profiling to prevent condensation from damaging the post. Furthermore, an anti-theft device is built into the flap opening for increased security.

Statement by the jury
The extremely minimalist, strictly geometric design of the exterior surface and its well-balanced proportions make this free-standing letter box resemble a sculpture. The vertical, inwardly opening letter flap is an attractive detail.

Die flache, senkrechte Klappe des Aluminiumbriefkastens lässt sich nach innen drehen und ermöglicht so eine einfache Handhabung. Auf Wunsch kann eine Platte im Inneren herausgenommen werden, sodass das Volumen sich verdoppelt. Damit die Post nicht durch Kondensation beschädigt wird, ist der Boden mit Entwässerungsbohrungen und einer speziellen Profilierung versehen. Ferner sorgt der Einbau einer Diebstahlsicherung an der Öffnungsklappe für ein hohes Maß an Sicherheit.

Begründung der Jury
Mit seiner äußerst minimalistischen, streng geometrischen Oberflächengestaltung und ausgewogenen Proportionen wirkt dieser freistehende Briefkasten wie eine Skulptur. Ein schönes Detail ist die senkrechte, nach innen öffnende Klappe.

W80
Partition System
Trennwandsystem

Manufacturer
Tecno S.p.A., Milan, Italy
Design
Centro Progetti Tecno, Daniele Del Missier,
Elliot Engineering & Consulting,
Pordenone, Italy
Web
www.tecnospa.com

A state of the art partition, W80, with its new patented system, satisfies all possible structural, acoustic, equipment and flexibility needs; customizable to suit any typology of space. W80 is a partition system with a total thickness of 80 mm available in single and double glazed versions which can integrate blinds, LED lights and pocket sliding doors. Multiple finishing solutions for the panels combined with a sophisticated snap system make W80 a very versatile product suitable for any type of project.

Statement by the jury
This modern room divider system impresses with its high versatility and well thought-out details such as the discreetly integrated construction elements that contribute to the overall purist appearance.

Als eine hochmoderne Trennwand erfüllt W80 mit ihrem neuen, patentgeschützten System alle möglichen baulichen, akustischen, Ausrüstungs- und Anpassungsanforderungen. Außerdem lässt sie sich beliebig der Raumstruktur anpassen. Es handelt sich um ein Trennwandsystem mit einer Gesamtdicke von 80 mm, das sowohl als einfach- oder doppelverglaste Variante verfügbar ist. Jalousien, LED-Beleuchtung und Schiebetüren sind einfach zu integrieren. Aufgrund der verschiedenen Oberflächenbehandlungen für die Panele und einer ausgeklügelten Schnappverriegelung ist W80 ein vielseitiges Produkt, das sich für jede Art von Projekt eignet.

Begründung der Jury
Dieses moderne Trennwandsystem beeindruckt mit großer Vielseitigkeit und durchdachten Detaillösungen wie unauffällig integrierten Konstruktionselementen, die zu einem puristischen Gesamteindruck beitragen.

VARIA
Planter and Partition
Pflanzgefäß und Raumteiler

Manufacturer
Degardo GmbH, Bad Oeynhausen, Germany
In-house design
Andreas Arndt
Design
Volker Hundertmark Design AB
(Volker Hundertmark), Vittsjö, Sweden
Web
www.degardo.de
www.vhform.com

The Varia XL plant holder offers many uses. Its tubular form, which is angled in various directions, makes it possible to use the holders together, so that they can be positioned to create an island of flowers, which can take the shape of petals or of a bunch of tubes or be set up as an opaque room divider. Manufactured in a single piece using a complex moulding process, Varia can appear with a slim or wide silhouette depending on requirements. The UV- and weather resistant container comes in a range of colours such as concrete or terracotta and is also available in a translucent version with warm white lighting.

Das XL Pflanzgefäß Varia kann vielfältig eingesetzt werden: Die gegensätzlich geknickte Röhrenform ermöglicht es, dass mehrere Gefäße interagieren und etwa als Blumeninsel in einer Blüten- oder Röhrenform sowie als blickdichter Raumteiler aufgestellt werden können – je nach Ausrichtung erscheinen sie dann mit einer schlanken oder breiten Silhouette. Im Rotationsschmelzgussverfahren wird Varia in einem Stück gefertigt. Das UV- und witterungsbeständige Produkt ist in unterschiedlichen Farbnuancen wie Beton oder Terrakotta und ebenso in einer transluzenten Version mit warmweißer Beleuchtung erhältlich.

Statement by the jury
Thanks to its asymmetric, tubular shape, Varia has an interesting appeal when standing on its own as well as in groups or when arranged as a room divider, thus becomes an aesthetic plant holder that attracts attention.

Begründung der Jury
Varia wirkt dank seiner asymmetrischen Röhrenform interessant als Solitär – in Gruppen oder als Raumteiler arrangiert wird das Pflanzgefäß zum ästhetischen Blickfang.

li-lith
Commercial Chair
Objektstuhl

Manufacturer
rosconi GmbH, Kippenheim, Germany
Design
Eichinger Offices (Gregor Eichinger),
Vienna, Austria
Web
www.rosconi.de
www.eichingeroffices.com
Honourable Mention

li-lith was originally created by designer Gregor Eichinger for Viennese restaurants so that ladies' handbags could be stored comfortably and safely. The wooden chair with feminine touches, the name of which symbolises women's independence, is characterised by its lightness and its delicate, curved armrest. The seat and back of the chair are made of moulded plywood which contrasts with the sturdy beech or oak frame.

Dieses Möbel von Designer Gregor Eichinger wurde ursprünglich für die Wiener Gastronomie entwickelt, damit die Handtaschen weiblicher Gäste bequem und sicher verwahrt werden können. Der Holzstuhl mit femininen Zügen, dessen Produktname symbolisch für die Selbstständigkeit der Frau steht, zeichnet sich durch seine Leichtigkeit und die filigran geschwungene Armlehne aus. Sitz und Rücken des Stuhls bestehen aus Formsperrholz und bilden einen Kontrast zum stabilen Buchen- bzw. Eichenholzgestell.

Statement by the jury
li-lith's exceptionally broad seat surface offers space for handbags and its high armrest protects them from theft.

Begründung der Jury
Mit einer breiteren Sitzfläche als üblich bietet li-lith Platz für Handtaschen und schützt sie mit einer elegant hochgezogenen Seitenlehne vor Diebstahl.

SmartCare
Security System
Sicherheitssystem

Manufacturer
Haier Group, Qingdao, China
Design
Haier Innovation Design Center,
Jiang Qi,Chen Yichuan, Xue Lu,
Xia Xu, Chen Yiwen, Zhao Hang,
Qingdao, China
Web
www.haier.com

Aside from integrating a multifunctional sensor, a water sensor and a sensor for windows and doors, SmartCare also has a socket and a gateway that automatically adapt. Every component reacts to movement and is controlled wirelessly via an app. The equipment is made from a material with a matt surface and has a slightly curved shape, which makes it unobtrusive and suitable for any home environment.

SmartCare umfasst neben einem Multifunktions-, Wasser-, Fenster- und Türsensor eine Steckdose und ein passendes Gateway. Jede dieser Komponenten kann kabellos und gezielt über eine App gesteuert werden und reagiert außerdem auf Bewegung. Das verwendete Material weist eine matte Oberfläche auf und ist leicht gewölbt, wodurch sich die Geräte diskret in jedes Wohnumfeld integrieren.

Statement by the jury
Throughout, the design of SmartCare is clean and of a high quality. It links together all the various elements of the range and uses discreet LED lighting to accentuate its technical qualities.

Begründung der Jury
Eine durchgehend klare, hochwertige Gestaltung verbindet die Elemente der SmartCare-Reihe miteinander, während dezente LED-Beleuchtung den technologischen Anspruch der Geräte betont.

Myfox Security System
Security System
Sicherheitssystem

Manufacturer
Myfox, Labege, France
Design
Quaglio Simonelli (Manuella Simonelli),
Paris, France
Web
www.getmyfox.com
www.quagliosimonelli.com

This security system is fitted with a
110 dB siren to protect the home from
burglars. As it does not need a cable, it
is quick and easy to install. The Myfox
home alarm can be controlled from a
smartphone via an intuitive app and is
deactivated using an intelligent key fob.
An emergency battery and an automatic
WiFi back up, linked anonymously to
systems in the neighbourhood, ensure
that the system also works during a
power cut.

Statement by the jury
The Myfox Security System appeals because of its unobtrusive, minimalist appearance, which allows it to fit elegantly
into domestic environments.

Dieses Sicherheitssystem ist mit einer
110 Dezibel lauten Sirene ausgestattet
und schützt so vor Einbruchsversuchen.
Da es ohne Kabel auskommt, lässt es
sich schnell und einfach installieren.
Myfox Home Alarm wird über eine intuitive App für das Smartphone gesteuert
und mit dem Schlüsselanhänger wieder
deaktiviert. Eine Notstrombatterie und
ein automatisches WiFi-Backup, über das
das Gerät sich anonym mit Systemen
der Nachbarschaft verbindet, sichern
dessen Funktionsfähigkeit auch bei einem
Stromausfall.

Begründung der Jury
Das Myfox Security System gefällt mit
einer zurückhaltenden, minimalistischen
Gestaltung, durch die es sich elegant und
unauffällig in die häusliche Umgebung
integrieren lässt.

LEO Smartkey
Automated Bunch of Keys
Vollautomatischer
Schlüsselbund

Manufacturer
Keyos GmbH, Potsdam, Germany
In-house design
Web
www.leo-smartkey.de

This fully automatic key box with illuminated buttons can store up to six keys.
Each one is given a number and can
be called up on demand. The keys are
copied or adapted so that LEO Smartkey
may be used without the need to alter
the doors involved. Accessories such
as a coloured cover or USB flash drive
round off the optical and functional
aspects of the product.

Statement by the jury
LEO makes an everyday action easier
for people with or without disabilities
and has a convincing, compact and
functional form.

Der vollautomatische Schlüsselbund mit
beleuchteten Tasten erlaubt die Aufbewahrung von bis zu sechs Schlüsseln, die
jeweils einer Zahl zugeordnet sind und
nach Bedarf ausgefahren werden können.
Die eigenen Schlüssel werden kopiert oder
umgearbeitet, sodass der Gebrauch des
LEO Smartkey ohne Umbaumaßnahmen
an den Türen möglich ist. Zubehör wie ein
buntes Gehäuse oder ein USB-Stick ist frei
wählbar und rundet das Produkt in Optik
und Nutzen ab.

Begründung der Jury
LEO erleichtert Menschen mit oder ohne
Einschränkungen einen alltäglichen
Handgriff und überzeugt auch mit einer
kompakten, funktionalen Gestaltung.

schlüsselbrett.ch
Multifunctional Key Rack
Multifunktionales
Schlüsselbrett

Manufacturer
schlüsselbrett.ch, Widnau, Switzerland
In-house design
Karin Sieber-Graf
Web
www.schluesselbrett.ch
Honourable Mention

Thanks to this simple key rack users will
always be able to find their keys as they
can easily be inserted into the felt. Optional extras on the design ledge such as
rings, buttons or test-tube holders offer
space for glasses, flowers or jewellery.
The combination of aluminium and replaceable designer felt, which is made of
pure wool and is available in 20 different colours, offers a range of different
value-added design options.

Statement by the jury
This key rack is an aesthetically pleasing,
locally produced and sustainable solution for the storage of keys.

Dank dieses schlichten Schlüsselbretts
behält sein Nutzer stets die Übersicht über
seine Schlüssel, welche sich ganz einfach
in den Filz stecken lassen. Durch optionales
Zubehör wie Ringe, Knöpfe oder Reagenz-
glashalter finden auch Brillen, Blumen oder
Schmuck an dieser Designleiste einen Platz.
Die Kombination aus Aluminium und aus-
wechselbarem Designfilz aus reiner Schur-
wolle, in 20 verschiedenen Farben, erlaubt
eine Gestaltungsvielfalt mit Mehrwert.

Begründung der Jury
Dieses Schlüsselbrett stellt eine schlichte,
ästhetische und regional nachhaltig pro-
duzierte Möglichkeit der Schlüsselaufbe-
wahrung dar.

Enter your destination floor

1 2 3
4 5 6
7 8 9
★ 0 -

HSP-A15
Lift Operating Panel
Fahrstuhl-Bedienfeld

Manufacturer
Mitsubishi Electric Corporation,
Tokyo, Japan
In-house design
Eunjin Choi
Web
www.mitsubishielectric.co.jp

Thanks to a new destination-oriented
system, this updated lift-operating panel
ensures that users reach the desired
floor the quickest way possible and thus
more efficiently. It is set at an ergonomically engineered angle, which improves
handling and makes it significantly easier
to use. The high-quality, robust aluminium frame, which holds the upper and
lower horizontal edges, ensures that the
operating panel suits any surroundings.

Das modernisierte Bedienfeld für Fahr-
stühle leitet den Benutzer zu dem
Fahrstuhl, der ihn am schnellsten in das
gewünschte Stockwerk bringt. Die Ausführung in einem ergonomischen Winkel
erhöht den Komfort und erleichtert so
die Bedienung maßgeblich. Mit einem
hochwertigen und robusten Rahmen aus
Aluminium, der nicht das gesamte Feld
umfasst, sondern sich auf die obere und
untere Waagerechte beschränkt, passt
es sich jeder Umgebung an.

Statement by the jury
The HSP-A15 operating panel is particularly user-friendly thanks to its underlying operating concept, which increases
the efficiency of lift systems.

Begründung der Jury
Das Bedienfeld HSP-A15 überzeugt durch
seine hohe Benutzerfreundlichkeit und das
zugrundeliegende Funktionskonzept, das
die Effizienz von Fahrstuhlanlagen erhöht.

kamereon
KNX Switch
KNX-Schalter

Manufacturer
dakanimo GmbH, Hamburg, Germany
In-house design
Sven Bär
Web
www.dakanimo.com
Honourable Mention

With its purist design and "piano lacquer-black" and "snow peak-white" colours, this switch gives sensory, acoustic and visual feedback via vibration, clicking sounds and the illuminated contours of five capacitive surfaces. The contours are highlighted with a number of different colours provided by RGB LED lights. Moreover, the switch can be customised to suit user habits by offering a free choice of programming. The operation of this hand-made switch is intuitive using all the senses.

Statement by the jury
The switch interacts with the user via numerous methods of feedback and is therefore intuitive to use.

In den Farben Klavierlackschwarz oder Schneekoppenweiß gibt der puristisch gestaltete Schalter durch Vibration, Klickgeräusche und die beleuchteten Konturen der fünf kapazitiven Flächen ein sensorisches, akustisches und optisches Feedback. Die Konturen sind mit RGB-LEDs in frei wählbaren Farben hinterlegt – auch die Programmierung lässt sich individuell auf die Gewohnheiten des Anwenders abstimmen. Die Bedienung des handgefertigten Schalters erfolgt intuitiv.

Begründung der Jury
Durch seine vielfältigen Arten des Feedbacks interagiert der Schalter mit dem Nutzer und ist so intuitiv bedienbar.

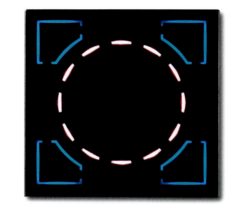

Panno
Lighting Control Panel
Lichtsteuerungs-Panel

Manufacturer
Enno Electronics Co., Ltd.,
London, England
In-house design
Web
www.enno.com

Panno is a stand-alone, wall-mounted lighting controller. Its innovative hardware-software-system is conveniently operated via a multi-touch user interface. The geometrically divided key panel is made of high gloss polished aluminium and creates a visual contrast to the OLED display. Panno can be used to control both lights and electronic roller shutters, connecting up to ten electric circuits. Integrated infrared sensors detect motion in the dark and light up the room automatically. Pre-programmed lighting scenes are activated at the touch of a button.

Statement by the jury
A high degree of user comfort is emphasised by an elegant overall appearance. The materials used visually accentuate its high quality standards.

Panno ist ein eigenständiges Lichtsteuerungsgerät, das an der Wand montiert wird. Sein innovatives Hardware-Software-System lässt sich komfortabel per Multitouch-Benutzeroberfläche bedienen. Das geometrisch unterteilte Tastenfeld besteht aus hochglanzpoliertem Aluminium und kontrastiert optisch mit dem OLED-Bildschirm. Panno dient sowohl der Steuerung von Leuchten als auch von elektronischen Rollladen. Es lassen sich bis zu zehn Stromkreise mit ihm verbinden. Integrierte Infrarotsensoren reagieren im Dunkeln auf Bewegungen im Raum und erleuchten diesen automatisch. Vorprogrammierte Lichtstimmungen lassen sich per Tastendruck aktivieren.

Begründung der Jury
Ein elegantes Gesamtbild betont den gehobenen Bedienkomfort von Panno. Die verwendeten Materialien bringen den hohen Qualitätsanspruch zum Ausdruck.

Anti-surge Power Strip
Power Strip
Steckdosenleiste

Manufacturer
Bull Electric Co., Ltd., Cixi, China
Design
Shenzhen ARTOP Design Co., Ltd.,
Shenzhen, China
Web
www.gongniu.cn
www.artopcn.com
Honourable Mention

The sockets in this power strip are set at an angle to ensure maximum use of space, an additionally integrated cable tie helps to avoid the tangling up of cables. The anti-surge system solves the problem of unstable power fluctuations that could damage electrical equipment. As high voltage caused by such fluctuations also poses a risk for people, this product offers users an increased level of protection in the office or at home.

Statement by the jury
This power strip combines clearly thought-through design with intelligent technology.

Die Löcher der Steckdosenleiste sind schräg angeordnet, sodass der Platz optimal genutzt werden kann, zudem verhindert ein integrierter Kabelbinder lästiges Verheddern der Kabel. Der Überspannungsschutz steuert Leistungsschwankungen entgegen, die den angeschlossenen elektrischen Geräten schaden könnten. Da hohe Spannungen auch für Menschen eine Gefahr darstellen, bietet das Produkt dem Nutzer im Büro oder zu Hause ein erhöhtes Maß an Sicherheit.

Begründung der Jury
Diese Steckdosenleiste kombiniert eine durchdachte Gestaltung mit intelligenter Technologie.

BORK AA581
Double Adapter
Doppelsteckdose

Manufacturer
Bork-Import LLC, Moscow, Russia
Design
Cube Design China, Ltd., Shenzhen, China
Web
http://bork.com

What makes the BORK AA581 different from standard double sockets is the addition of two USB ports that are identified by a hidden symbol which only illuminates when charging begins. The double adapter has a stark appearance and merely comprises a master switch in an elegant and ergonomic design. The practical appliance is simple to use and pleasantly soft to the touch.

Statement by the jury
Despite little elaborate detailing, BORK AA581 is a product whose minimalist appearance is a pleasure to look at.

BORK AA581 bietet neben den normalen Steckeranschlüssen eine Erweiterung um zwei USB-Ports. Diese zeichnen sich durch ein zunächst verstecktes Symbol aus, das erst mit Beginn des Aufladens aufleuchtet. Die Doppelsteckdose ist äußerlich schlicht gehalten und umfasst lediglich einen eleganten und ergonomisch gestalteten Hauptschalter. Das praktische Gerät ist einfach in der Bedienung und hat eine angenehm glatte Oberfläche.

Begründung der Jury
Diese Doppelsteckdose überzeugt mit einer minimalistischen Gestaltung und einer erweiterten Funktionalität dank zweier USB-Ports.

Push
Shop Fitting System
Ladenbausystem

Manufacturer
Visplay International GmbH,
Weil am Rhein, Germany
Design
Volker Otto, Düren, Germany
Web
www.visplay.com

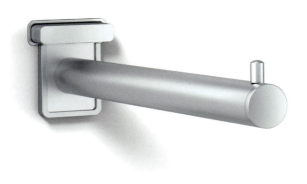

Push is a single point support system with sophisticated technology: with just one click, the push-button adapters for merchandise supports can either pop-out or be flush to the wall panel, as required. Depending on the version, holders are either virtually invisible or appear as design elements, and can even become part of the Corporate Design. The Push shop fitting system is available in 43 mm or 28 mm and can therefore be used for large and small ready-to-wear accessories, from lingerie and beachwear to shoes.

Statement by the jury
Thanks to the carefully thought through functionality of this purist merchandise carrier, only the push of a button is needed to make it pop out when needed and be hidden unobtrusively in the wall panel when not in use.

Push ist ein Einpunkt-Tragsystem mit raffinierter Technik: Mit einem Klick lassen sich die Hülsen für die Warenträger ganz nach gewünschter Nutzung hervorzaubern oder im Wandpaneel versenken. Je nach Ausführung sind die Hülsen fast unsichtbar, wirken als Designelement oder werden zum Bestandteil des Corporate Designs. Das Tragsystem Push ist in den Größen 43 mm und 28 mm erhältlich und kann somit für Großkonfektion sowie Accessoires, Lingerie und Bademoden bis hin zu Schuhen verwendet werden.

Begründung der Jury
Dank einer durchdachten Funktionalität kommen diese puristischen Warenträger auf Knopfdruck hervor, wenn sie gebraucht werden, und verbergen sich bei Nichtgebrauch dezent im Wandpaneel.

Create Cabinet
Clothing Cabinet
Kleiderspind

Manufacturer
AJ Produkter AB, Halmstad, Sweden
In-house design
Cecilia Stööp
Web
www.ajprodukter.se

These clothes locker sets out to fulfil user requirements and is suitable for staff rooms, changing rooms and other public spaces. His form encourages creative interior design and will result in a clearly laid out and hence safe space. Depending on the respective height and whether the doors are chosen to be flat or sharp-edged, his appearance is versatile and changes due to the construction, which includes ventilation and an observation window made of synthetic material.

Dieser Kleiderspind ist auf die Bedürfnisse des Benutzers ausgerichtet und passt sich in Personalräumen, Umkleidekabinen oder anderen öffentlichen Bereichen dem Umfeld an. Seine Form begünstigt eine kreative und übersichtliche Raumgestaltung, die dadurch hohe Sicherheit gewährleistet. Je nach Höhe und ob mit flacher oder erhabener Tür gewählt, erzeugt das Objekt – ausgestattet mit Belüftung und Sichtfenster aus Kunststoff – ganz unterschiedliche optische Wirkungen.

Statement by the jury
With its modular construction and numerous customisation options, Create provides great scope for interior design.

Begründung der Jury
Mit einer modularen Bauweise und vielfältigen Individualisierungsmöglichkeiten eröffnet Create Innenarchitekten großen Gestaltungsspielraum.

WECO 2C
Window System
Fenstersystem

Manufacturer
Weco Windows S.L., Majadahonda, Spain
In-house design
Iciar de las Casas, Rosario Chao Foriscot
Web
www.wecowindows.com

The sash in this wood window incorporates glass and hardware in a single assembly, without a frame, maximizing the window's open area. The double or triple-glazed sash is mounted on the interior face of the wood frame in the wall. The windows are available in fixed, tilt-and-turn and sliding options, and lengths up to six meters. The triple gasketing offers optimal thermal and acoustic performance.

Die Scheibe und die Beschläge werden bei diesem Holzfenster direkt in den Schiebeflügel ohne einen getrennten Rahmen integriert, sodass die offene Fläche des Fensters maximiert wird. Die Scheibe ist doppelt oder dreifach verglast und wird direkt auf der Innenseite des äußeren Rahmens befestigt. Mit einer Länge bis zu sechs Metern ist WECO 2C als festes, Schwenk- oder Kippfenster erhältlich und bietet durch eine dreifache Abstufung des Rahmens eine optimale thermische und akustische Isolierung.

Statement by the jury
WECO 2C impresses with its large-scale and elegant appearance made possible by the use of new technologies and high-quality construction elements.

Begründung der Jury
WECO 2C imponiert mit einem großflächigen, eleganten Erscheinungsbild, das durch den Einsatz neuer Technologien und hochwertiger Bauteile ermöglicht wurde.

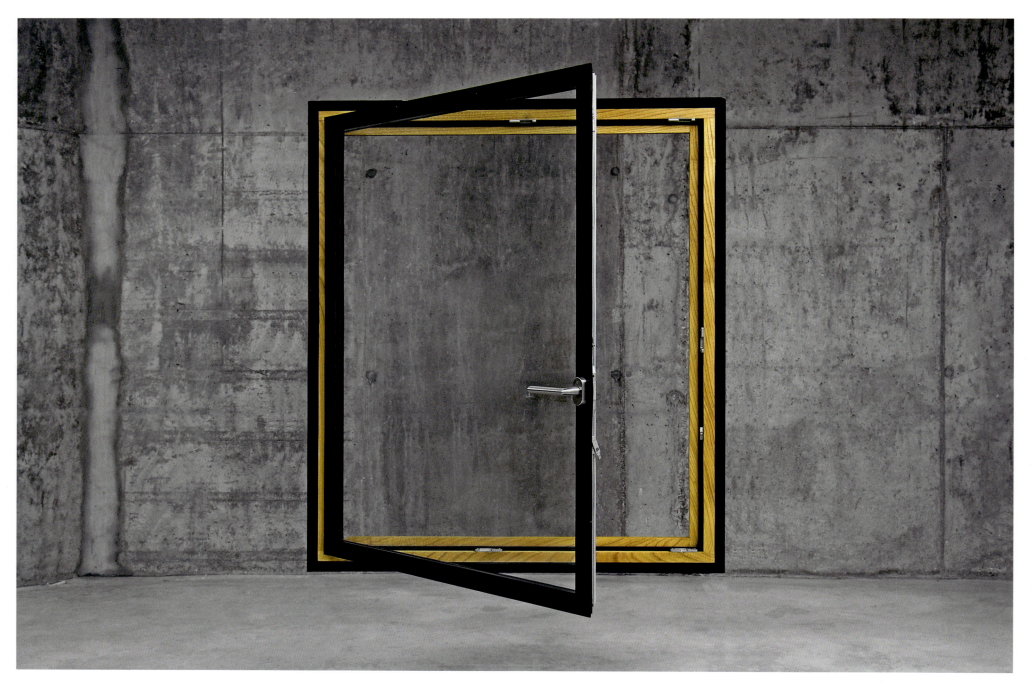

markilux MX-1 / MX-1 compact
Patio Awning System
Terrassenmarkisen-System

Manufacturer
markilux, Schmitz-Werke GmbH + Co. KG,
Emsdetten, Germany
Design
kramerDesign (Prof. Andreas Kramer),
Wildeshausen, Germany
Web
www.markilux.com
www.kramer-produkt-design.de

The awning's cassette, including the fixed canopy section, extends 62 cm from the wall and protects windows and façades from rain and, on sunny days, affords enough shadow to protect the terrace. When completely extended, the system measures 700 x 437 cm and a tilting mechanism gives a pitch of 5 to 25 degrees. An integrated gutter ensures that excess water can easily run off. An aluminium screen at the front comes in a wide choice of colours and any model can be equipped with dimmable LED lighting.

Die Kassette inklusive Vorbau mit einer Bautiefe von 62 cm schützt Fenster und Fassaden vor Regen und spendet an sonnigen Tagen ausreichend Schatten. Komplett über der Terrasse ausgefahren misst das System 700 x 437 cm und kann über ein Kippgelenk um 5 bis 25 Grad geneigt werden. Durch die integrierte Regenrinne läuft überschüssiges Wasser problemlos ab. Die vordere Sichtblende aus Aluminium ist in zahlreichen Farben erhältlich und ist in allen Ausführungen mit einer dimmbaren LED-Beleuchtung ergänzt.

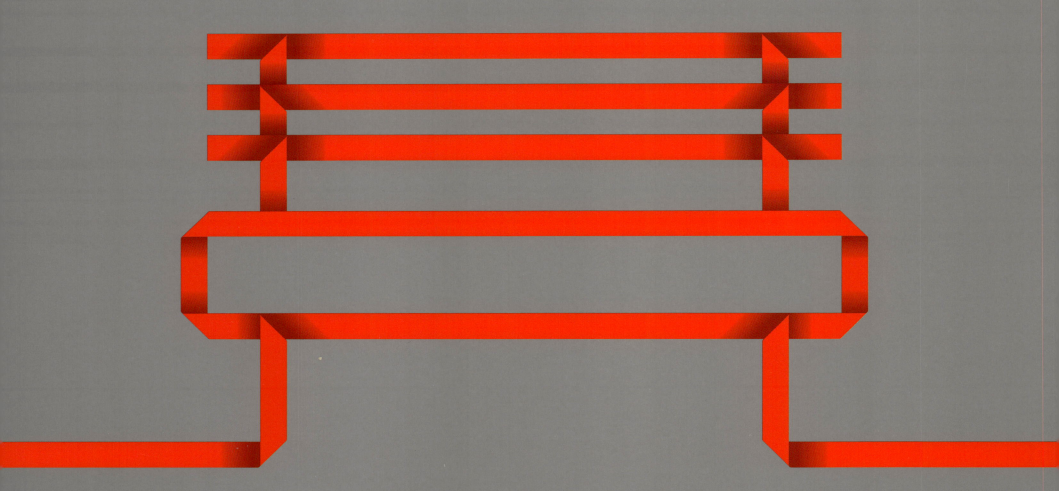

Urban design and public spaces
Urban Design und öffentlicher Raum

DropBucket
Waste Bin
Abfalleimer

Manufacturer
DS Smith Packaging Denmark A/S,
Taastrup, Denmark

Design
DropBucket ApS
(Heiða Gunnarsdóttir Nolsøe,
Marie Stampe Berggreen),
Frederiksberg, Denmark

Web
www.dropbucket.dk

reddot award 2015
best of the best

Thoroughly well thought out

The growing number of parties and public events brings with the problem of mountains of rubbish produced in a short period of time. It is vital to have sufficient rubbish containers available. The problem is compounded by the fact that the volume of waste is often difficult to calculate in advance. The concept for the DropBucket waste bin perfectly addresses such situations and offers a thoroughly well thought-out solution even for unanticipated shortages. The functional waste bin is based on a simple, folded cardboard structure. It is simple to transform into an attractive, pyramid-shaped container that easily fits into its surroundings. Assembly is intuitive and large numbers of bins can be put up within a short period of time. The DropBucket is made of robust corrugated cardboard, consisting entirely of recycled material. A water-resistant coating ensures that the cardboard does not get soggy with moisture. Impressively, the concept for this waste bin is based on closed-circuit processes – once the bin is full, it can be used as a cost-effective, disposable container. It is easy to compress and dispose of with the rubbish. That protects the environment and reduces greenhouse gas emissions. Its successful, functional and aesthetic design makes the innovative DropBucket waste bin an impressive product.

Rundum durchdacht

Die wachsende Anzahl von Partys und öffentlichen Veranstaltunger wird von dem Problem begleitet, dass in kurzer Zeit sehr viel Müll anfällt. Dafür müssen ausreichend Behälter bereitstehen und oftmals ist die Müllmenge vorab kaum zu kalkulieren. Das Konzept des Abfalleimers DropBucket ist perfekt auf derartige Situationen abgestimmt und bietet auch für unvorhergesehene Engpässe eine rundum durchdachte Lösung. Dem funktionalen Abfalleimer liegt eine einfache, zusammengefaltete Kartonstruktur zugrunde. Mittels weniger Handgriffe verwandelt sich diese in einen gefälligen pyramidenförmigen Behälter, der sich gut in die Umgebung einfügt. Der Aufbau geschieht selbsterklärend, und in kurzer Zeit können große Stückzahlen aufgestellt werden. Gefertigt wird DropBucket aus einer stabilen Wellpappe, die wiederum aus Recyclingmaterial besteht. Damit die Pappe bei Feuchtigkeit nicht durchweicht, ist sie zusätzlich mit einer wasserbeständigen Schicht versehen. Das Konzept dieses Abfalleimers folgt auf beeindruckende Weise in sich geschlossenen Kreisläufen: Ist er gefüllt, kann er als kostengünstiger Wegwerfbehälter genutzt werden. Leicht lässt er sich zusammenpressen und mit dem Müll entsorgen. Das schützt die Umwelt und senkt die Emission von Treibhausgasen. Der innovative Abfalleimer DropBucket überzeugt daher durch seine funktional wie ästhetisch gelungene Gestaltung.

Statement by the jury

The DropBucket waste bin offers a very functional and coherent solution for straightforward waste disposal at events. Both concept and design are impressive. In hardly any time at all, folded cardboard changes into a practical, aesthetically appealing waste bin that communicates its purpose at first sight. Once it is no longer needed, it can be disposed of easily along with the rubbish.

Begründung der Jury

Der Abfalleimer DropBucket bietet eine sehr funktionale und in sich schlüssige Lösung für eine unkomplizierte Müllentsorgung auf Veranstaltungen. Konzept und Gestaltung sind beeindruckend. In kürzester Zeit verwandelt sich ein zusammengefalteter Karton in einen praktikablen, ästhetisch ansprechenden Abfalleimer, der seine Bestimmung auf den ersten Blick kommuniziert. Und wenn er nicht mehr gebraucht wird, kann er leicht mitsamt dem Müll entsorgt werden.

Designer portrait
See page 66
Siehe Seite 66

IPOMEA simplex
Innovative Sunshade for Events
Innovativer Eventschirm

Manufacturer
Ipomea GmbH, Munich, Germany
Design
Studio MSB UG (haftungsbeschränkt),
Schorndorf, Germany
Web
www.ipomea.com
www.studio-msb.de

The system consists of a basic framework which, together with the corresponding fabric can be fitted together into a parasol within five minutes. This can be enhanced with light and sound modules, creating an individual atmosphere. The high-tech materials are characterised by their light weight as well as great stability. The parasol can be combined with further devices of its kind to a whole complex and after use, can be stored in a compact carrying bag.

Das System besteht aus einem Grundgestell, das sich mit dem zugehörigen Stoff innerhalb von fünf Minuten zu einem Schirm zusammenbauen lässt. Dieser kann mit Licht- und Soundmodulen ergänzt werden, die eine individuelle Atmosphäre schaffen. Die verwendeten Hightech-Materialien zeichnen sich neben ihrem geringen Gewicht durch eine hohe Stabilität aus. Der Schirm kann mit weiteren Exemplaren zu einer großen Überdachung kombiniert und nach Gebrauch in einer kompakten Tragetasche verstaut werden.

Statement by the jury
When folded up, this light-weight parasol for events captivates by its handy format and, when opened up, by its filigree design which provides at the same time functional advantages such as a central water drainage.

Begründung der Jury
Dieser leichte Eventschirm besticht zusammengeklappt durch sein handliches Format und aufgespannt durch seine filigrane Konstruktion, die zugleich funktionale Vorteile wie eine zentrale Wasserableitung mit sich bringt.

Gigseat Open-Air Tribune |
Gigseat Outdoor Event Seating
Seats to be connected for Outdoor Events
Flexible Tribüne für Ereignisse im Freien

Manufacturer
Gigtrigger AS, Bodø, Norway
Design
EGGS Design AS
(Carl André Nørstebø, Carl-Gustaf Lundholm),
Trondheim, Norway
Minoko Design AS (Audun Sneve), Levanger, Norway
Web
www.gigseat.com
www.eggsdesign.no

You can also find this product in
Dieses Produkt finden Sie auch in
Doing
Page 195
Seite 195

This seating system offers comfort for the audience at outdoor events such as festivals and sports events. Gigseat can be used on both sides, according to the terrain gradient. A triangular and removable plug in the back prevents the seat from sliding and keeps the user's bottom dry. On the adjacent side, a triangular hole in the seat provides a drink holder. Gigseat has sufficient grip on grass, stone, sand or snow. Seats may be joined one to the other via a simple and strong connector. This lightweight seat is made of 100 per cent recyclable, robust thermoplastic and is easy to stack and store. The Gigseat Outdoor Event Seating is available in many colours.

Das Sitzsystem bietet bei Freilandveranstaltungen wie Festivals und sportlichen Großveranstaltungen Komfort für das Publikum. Der Gigseat kann beidseitig, je nach Gefälle des Terrains, benutzt werden. Ein dreieckiger, entfernbarer Stöpsel verhindert, dass der Sitz rutscht, und hält das Gesäß des Benutzers trocken. Auf der gegenüberliegenden Seite bietet ein weiteres dreieckiges Loch im Sitz die Möglichkeit, Trinkgefäße abzustellen. Der Gigseat hat genügend Halt auf Gras, Stein, Sand und Schnee und kann mithilfe einer einfachen, aber robusten Steckverbindung mit weiteren Sitzen gekoppelt werden. Er ist extrem leicht, ist aus 100-prozentig recycelbarem Thermoplast gefertigt und einfach stapel- und verstaubar. Der Gigseat Open-Air Tribune ist in vielen Farben erhältlich.

Statement by the jury
Gigseat constitutes an effective solution as uncomplicated, versatile seat with well-conceived functionality.

Begründung der Jury
Gigseat überzeugt als unkomplizierter, vielseitig einsetzbarer Sitz mit durchdachter Funktionalität.

Versio Juno
Urban Furniture
Stadtmobiliar

Manufacturer
Gemeinnützige Westeifel Werke der
Lebenshilfen Bitburg, Daun und Prüm GmbH,
Gerolstein, Germany
Design
Büro Wehberg (Max Wehberg),
Hamburg, Germany
Web
www.freiraumausstattung.de
www.buero-wehberg.de

As the formal and technical features of the seating bench focus on what is necessary, it is particularly suitable for heavily public places. Discreet and reduced to essentials in its design, the bench is characterised by its durability as well as its easy cleaning and maintenance properties. The 12 cm wide concrete side element is cast in light or dark grey; the seat and backrest of the bench are constructed of untreated FSC-certified hardwood.

Statement by the jury
The particular charm of Versio Juno consists of its formal and functional reduction to essentials, making it suitable for heavily frequented public places.

Da sich die formalen und technischen Anforderungen der Sitzbank auf das Notwendige konzentrieren, eignet sie sich besonders für stark beanspruchte Nutzungsbereiche. Zurückhaltend und reduziert gestaltet zeichnet sie sich durch ihre Langlebigkeit sowie eine einfache Pflege und Wartung aus. Der 12 cm starke Betonfuß wird in den Farben Hell- oder Dunkelgrau gegossen, Sitzfläche und Lehne der Bank bestehen aus unbehandeltem, FSC-zertifiziertem Hartholz.

Begründung der Jury
Der besondere Charme von Versio Juno besteht in ihrer formalen wie funktionalen Reduktion auf das Wesentliche, wodurch sie sich gut für stark frequentierte öffentliche Bereiche eignet.

Campus levis
modular system
Urban Furniture
Stadtmobiliar

Manufacturer
Gemeinnützige Westeifel Werke der
Lebenshilfen Bitburg, Daun und Prüm GmbH,
Gerolstein, Germany
Design
Büro Wehberg (Max Wehberg),
Hamburg, Germany
Web
www.freiraumausstattung.de
www.buero-wehberg.de

This series was designed especially for public and semi-public premises such as universities, schools or company grounds. Various seating options for taking a rest offer flexible use which includes the construction of a coherent communication landscape, thanks to variable extensions sideways and lengthways. All functions are interchangeable, which allows for individual adaptability of the diverse sitting, standing and lying possibilities at the chosen place of installation.

Statement by the jury
With its clear, geometrical use of forms as well as seating and working surfaces made of hard wood, Campus levis radiates peace. The modularity of the system facilitates flexible integration in various environments.

Konzipiert wurde diese Serie speziell für öffentliche und halböffentliche Bereiche wie Universitäten, Schulen oder auch Firmengelände. Unterschiedliche Sitzmöglichkeiten zum Verweilen bieten eine flexible Nutzung, zu der dank variabler Längen- und Tiefenreihung auch die Konstruktion einer zusammenhängenden Kommunikationslandschaft zählt. Alle Funktionen sind untereinander kombinierbar, was die Anpassung der abwechslungsreichen Sitz-, Steh- und Liegemöglichkeiten an den jeweiligen Ort erlaubt.

Begründung der Jury
Mit einer klaren, geometrischen Formensprache sowie Sitz- und Arbeitsflächen aus Hartholz strahlt Campus levis Ruhe aus. Die Modularität des Systems ermöglicht eine flexible Einbindung in unterschiedliche Umgebungen.

duplus
Modular Bench System
Modulares Sitzbanksystem

Manufacturer
L. Michow & Sohn GmbH, Hamburg, Germany
Design
bbz landschaftsarchitekten, Berlin, Germany
Web
www.michow.com
www.bbz.la

This modular bench system for outdoors has a double-sided alignment of seating surfaces, so that the reverse side can also be used. Due to its medium height backrest the bench features a high degree of sitting comfort and creates a special type of spatial zoning without disturbing the view. A further highlight is its foot and the supporting construction made of zinc galvanised and powder-coated steel, crafted to the form of a sword and secured underground. Both the seat and the backrest are made of solid FSC-certified hardwood.

Die Sitzflächen dieses modularen Systems für den Außenbereich sind nach zwei Seiten ausgerichtet, somit können auch die Stirnseiten genutzt werden. Aufgrund ihrer mittelhohen Lehnen bietet die Bank einen hohen Sitzkomfort und kreiert eine räumliche Zonierung, ohne dass der Blick beeinträchtigt wird. Eine Besonderheit liegt in ihrem Fuß und der Unterkonstruktion aus feuerverzinktem und pulverbeschichtetem Stahl, gefertigt in Form eines Schwertes und unterirdisch fundamentiert. Sowohl Auflage als auch Lehne sind aus massivem, FSC-zertifiziertem Hartholz gefertigt.

Sirius
Playground Equipment
Spielplatzgerät

Manufacturer
smb Seilspielgeräte GmbH Berlin in Hoppegarten,
Hoppegarten, Germany
In-house design
Web
www.smb-seilspielgeraete.de

The large seat membrane of this playground equipment was installed at the highest point of the area net, so that the goal of the climbing experience serves simultaneously as communication platform with a good view. Inside the construction the user can almost stand upright, depending on the respective body height, and move around freely without needing to leave it. Popular games such as tag played within the area net require a higher degree of skills than just on the ground, before players finally can take a rest in one of the three nets specifically designed for lying.

Die große Sitzmembran dieser Spielgerätebau-Konstruktion wurde an den höchsten Punkt des Raumnetzes gelegt, sodass das Ziel der Kletterpartie gleichzeitig als Kommunikationsplattform mit guter Aussicht dient. Im Inneren der Konstruktion kann der Benutzer abhängig von seiner Größe fast aufrecht stehen und sich frei bewegen, ohne sie verlassen zu müssen. Beliebte Spiele wie Fangen stellen im Raumnetz höhere Anforderungen an die Geschicklichkeit als zu ebener Erde, bevor man sich anschließend in einem der drei eingebauten Liegenetze erholen kann.

Statement by the jury
Sirius playfully promotes children's coordination skills and dexterity. At the same time, the playground equipment impresses with a well-conceived construction which is easily assembled.

Begründung der Jury
Sirius fördert spielerisch die Koordinationsfähigkeit und Geschicklichkeit von Kindern. Zugleich überzeugt das Spielgerät mit einer durchdachten, leicht aufzubauenden Konstruktion.

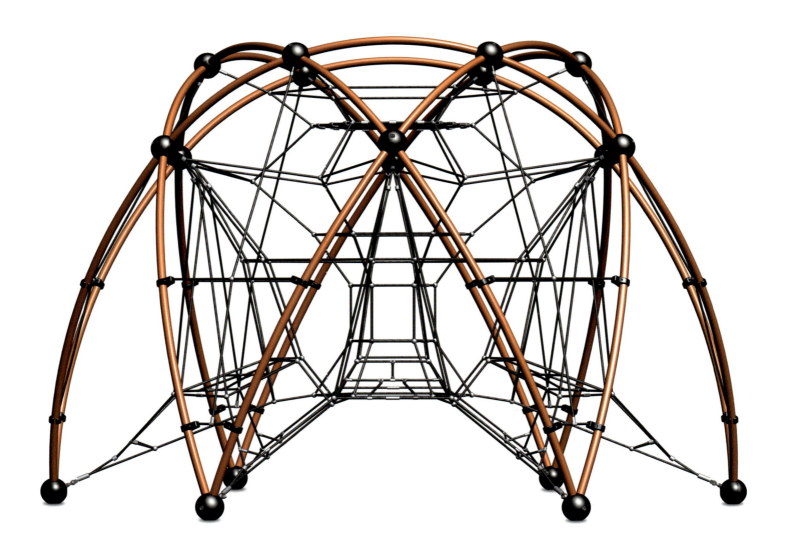

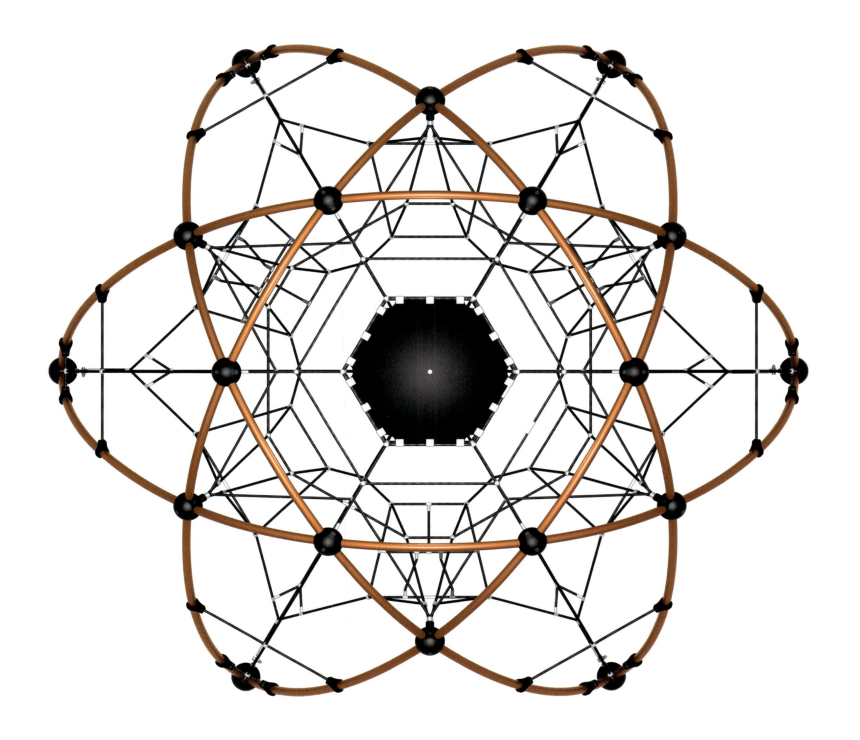

Gemini
Double-Slit Grates
Doppelschlitzroste

Manufacturer
Richard Brink GmbH & Co. KG,
Schloß Holte-Stukenbrock,
Germany

In-house design
Markus Brink, Stefan Brink

Web
www.richard-brink.de

reddot award 2015
best of the best

Pure surface

Drains are an important part of architecture at ground level. As they have to do their job reliably when it is raining or when cleaning is in progress, they play a key role. Here, the Gemini double-slit grates for drains offer an elegant and versatile solution. They are based on a design with a universal panel cover. This makes it possible to keep the visible drainage technology of these innovative double-slit grates to a minimum, giving them a very minimalist appearance. They are suitable for use with every surface material and for every type of slab thickness. Thanks to the grates' puristic design, only the two slits with their gleaming stainless steel frames are visible and create an elegant contrast to the surface material. This allows the double-slit grates to fit in with surrounding paving slabs discretely and stylishly, whether it is in court-yards, patios or on balconies. Their level elegance is combined with innovative functionality. In contrast to conventional slit drains, they can be removed in their entirety so that drains can be cleaned quickly and easily. This is of particular benefit in public areas as it results in cost-effective low maintenance. With its puristic appearance, the Gemini double-slit grates open up new possibilities for the planning of exterior areas. Their functionality sets new standards.

Pure Fläche

Entwässerungsrinnen sind ein wichtiger Bestandteil der Architektur am Boden. Da sie ihre Aufgabe bei Regen oder Reinigungsarbeiten zuverlässig erfüllen müssen, spielen sie eine zentrale Rolle. Die Doppelschlitzroste für Entwässerungsrinnen Gemini stellen hier eine elegante und vielseitige Lösung dar. Sie basieren auf der Gestaltung mit einer universell belegbaren Platten-schale. Die Form dieser innovativen Doppelschlitzroste konnte daher radikal reduziert werden, weshalb sie eine sehr minimalistische Anmutung haben. Sie eignen sich für jede Art von Bodenbelag und jede gewünschte Plattenstärke. Dank der puristischen Gestaltung bilden dabei lediglich die beiden Schlitze mit ihren glänzenden Edelstahleinfassungen einen eleganten Kontrast zum Bodenbelag. Die Doppelschlitzroste fügen sich daher stilvoll und dezent in das umgebende Plattenmaterial von Höfen, Terrassen oder Balkonen ein. Ihre plane Eleganz geht einher mit einer innovativen Funktionali-tät: Anders als bei üblichen Schlitzrinnen lassen sie sich komplett entnehmen und ermöglichen so eine schnelle und unproblematische Reinigung der Rinne. Dies hat gerade im öffentlichen Bereich den Vorteil einer kosten-günstigen und wenig arbeitsintensiven Wartung. Mit ihrer puristischen Anmutung eröffnen die Doppel-schlitzroste Gemini neue Horizonte für die Planung von Außenflächen – ihre Funktionalität setzt dabei neue Standards.

Statement by the jury

What is so appealing about Gemini is the grates' abso-lutely consistent moulding. Visually kept to a minimum, these double-slit grates for drains effortlessly fit into the surrounding paving material of courtyards, bal-conies or patios thanks to their discreet appearance. As they can be removed completely, they are very functional and easy to clean. They offer planners and architects versatile new options to create accentedly extensive and exceptionally elegant exterior design.

Begründung der Jury

Bei Gemini begeistert die überaus konsistente Ausfor-mung. Visuell auf ein Minimum reduziert, fügen sich diese Doppelschlitzroste für Entwässerungsrinnen mit ihrer unaufdringlichen Ästhetik gekonnt in das umgebende Plattenmaterial von Höfen, Balkonen oder Terrassen ein. Da sie sich vollständig entnehmen lassen, sind sie sehr funktional und pflegeleicht. Planern und Architekten bieten sich dadurch vielseitige neue Mög-lichkeiten einer betont flächigen und äußerst eleganten Außengestaltung.

Designer portrait
See page 68
Siehe Seite 68

ACO Drainlock
Linear U–Profile Grating
Grating
Abdeckrost

Manufacturer
ACO Severin Ahlmann GmbH & Co. KG,
Büdelsdorf, Germany
In-house design
Jaroslav Řička, Marco Wandkowski
Web
www.aco.com

The continuous profile of the filling bars is U-shaped and, together with the accurate slot sizes, provides a balanced visual effect, which emphasises once more the durability of the stainless steel material used. A non-slip structure with raised supporting bars assures sure-footedness. Since the slots with an integrated heelguard feature are only 8 mm wide, even shoes with narrow heels cannot get stuck.

Statement by the jury
This clearly designed, functionally well-considered stainless steel grating emphasises elegant features as visible element of a drainage system in public areas.

Das durchgehende Profil der Füllstäbe verläuft in U-Form und sorgt gemeinsam mit den exakten Schlitzdimensionen für eine ausgewogene Optik, sodass der langlebige Werkstoff Edelstahl gut zur Geltung kommt. Eine rutschhemmende Struktur mit emporragenden Tragstäben sorgt für eine gute Trittsicherheit. Dank einer Schlitzweite von nur 8 mm können im integrierten Absatzschutz auch Schuhe mit schmalen Absätzen nicht steckenbleiben.

Begründung der Jury
Dieses klar gestaltete, funktional durchdachte Edelstahl-Abdeckgitter setzt als sichtbares Element eines Entwässerungssystems elegante Akzente im öffentlichen Raum.

Paddle
Drinking Faucet
Wasserhahn

Manufacturer
MIZSEI MFG CO., LTD., Yamagata City,
Gifu Prefecture, Japan
In-house design
Hiroaki Negishi, Tohru Hayakawa
Design
Keisuke Funahashi Design
(Keisuke Funahashi), Inuyama, Japan
Web
www.mizsei.co.jp
www.keisukefunahashi.com

The elegant drinking water tap with automatic shut-off function is intended for universal use in public facilities and parks. Due to an integrated, light spring it is easy to use also for senior citizens, children and people with disabilities. Water pressure supports the shut-off function and the typical T-shape also helps the simplified use. The large, wing-shaped handles on both sides can be used by left-handers as well as right-handers and can be pulled, pushed or turned.

Statement by the jury
Paddle impresses with a convenient operating concept and an organically curved, T-shaped design.

Der elegante Trinkwasserhahn mit automatischer Schließfunktion ist für den vielfältigen Einsatz in öffentlichen Einrichtungen und Parks gedacht. Aufgrund einer eingebauten, leichten Feder ist er auch von Senioren, Kindern oder Menschen mit Behinderung problemlos zu bedienen. Der Wasserdruck unterstützt das Abstellen, auch die typische T-Form trägt zu einer vereinfachten Nutzung bei. Die großen, flügelartig geformten Griffe auf beiden Seiten können von Links- und Rechtshändern bedient und gezogen, gedrückt oder gedreht werden.

Begründung der Jury
Paddle beeindruckt mit einem komfortablen Bedienkonzept und einer organisch geschwungenen, T-förmigen Gestaltung.

KONE KT100
Sensor Barrier
Sensorschleuse

Manufacturer
KONE Corporation, Hyvinkää, Finland
In-house design
Web
www.kone.com

With its slim overall appearance, the sensor barrier fits seamlessly into any building and offers safe pedestrian traffic flow management. Due to its half-height design, the user is immediately very clearly guided around and, thanks to an inwardly tilted front surface, receives the assurance of walking in the right direction. An illuminated transition inspires optimal barrier freedom and imparts a precise sense of space.

Mit einem schlanken Gesamtbild fügt die Sensorschleuse sich nahtlos in jedes Gebäude ein und bietet dort ein sicheres Personenflussmanagement. Durch eine halbhohe Gestaltung wird der Benutzer gleich sehr deutlich gelenkt, und die nach innen geneigten Stirnflächen vermitteln ihm das Gefühl, die Schleuse in der richtigen Richtung zu betreten. Ein gut ausgeleuchteter Durchgang sorgt für optimierte Barrierefreiheit und schafft ein präzises Raumgefühl.

Statement by the jury
With its light and seemingly transparent design, simple yet elegant signalling elements and a pleasant illumination, this barrier ensures relaxed usage.

Begründung der Jury
Mit einer leicht und transparent wirkenden Gestaltung, schlicht-eleganten Signalisierungselementen und einer angenehmen Ausleuchtung gewährleistet diese Schleuse eine entspannte Benutzung.

Schüco Parametric System
System Façade
Systemfassade

Manufacturer
Schüco International KG, Bielefeld, Germany
In-house design
Web
www.schueco.com
Honourable Mention

The Schüco Parametric System enables the design of geometrically free and three-dimensional building façades. The continuously flexible components can be individually created and thus allow efficient planning: The system offers various options of usage, from sun radiation and shading adjustment via optimisation of daylight admission to solar energy generation. Due to the consistent process chain from concept to construction, the production of elements not supplied as standard is also feasible.

Das Schüco Parametric System ermöglicht die Gestaltung geometrisch freier und dreidimensionaler Gebäudefassaden. Die durchgängig flexiblen Bauteile können individuell entworfen werden und erlauben so eine effiziente Planung: Von der Anpassung von Sonneneinstrahlung und Verschattung über die Optimierung des Tageslichteinfalls bis hin zur solaren Energiegewinnung bietet das System verschiedene Möglichkeiten der Nutzung. Durch die lückenlose Prozesskette vom Entwurf bis zur Fertigung ist auch die Herstellung nichtserieller Elemente realisierbar.

Statement by the jury
The flexible components of this façade system offer architects a great deal of design freedom and contribute to the optimisation of building efficiency with various functional elements.

Begründung der Jury
Die flexiblen Bauteile dieser Systemfassade bieten Architekten große gestalterische Freiheit und tragen mit verschiedenen Funktionselementen dazu bei, die Gebäudeeffizienz zu optimieren.

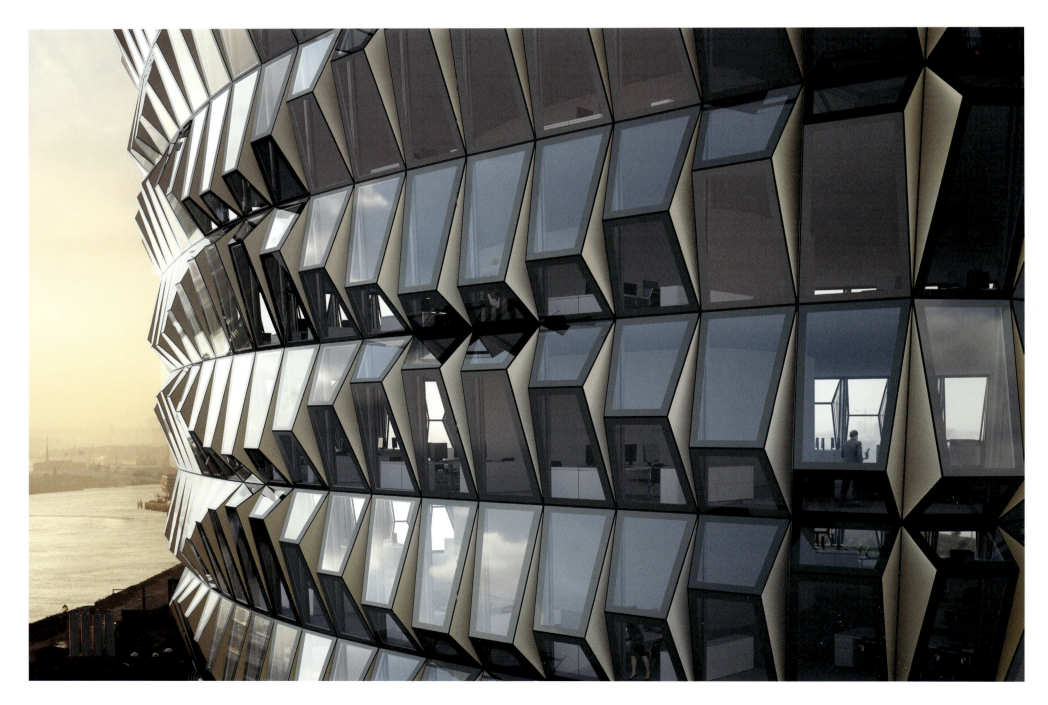

essertop Karat
Flat Glass Skylight
Flachdachfenster

Manufacturer
Eternit Flachdach GmbH, Neuss, Germany
Design
Ingo Fitzel Atelier für Design, Berlin, Germany
Web
www.eternit-flachdach.de

A V-form with distinctive sides inclined towards the outside enlarges the lighting and ventilation surfaces by 125 per cent compared to conventional constructions. Shading, which is dimmed via a wall-mounted control switch, is integrated in the thermal insulation glass. Also the ventilation can be conveniently operated by remote control, smartphone or tablet. With clear separation, vertically mounted connections and a flat construction give the window the appearance of floating over the rooftop and blend in functionally with their surroundings.

Eine V-Form mit markant nach außen geneigten Seiten vergrößert die Licht- und Lüftungsflächen im Vergleich zu konventionellen Konstruktionen um 125 Prozent. In die Wärmedämmverglasung ist ein Sonnenschutz integriert, der über einen Wandtaster gedimmt wird. Auch die Lüftung kann bequem über eine Fernbedienung, das Smartphone oder ein Tablet gesteuert werden. Mit klarer Trennung vertikal ausgeführte Anschlüsse und eine flache Konstruktion lassen das Fenster visuell über der Dachfläche schweben und passen sich auch in den Funktionen ihrer Umgebung an.

Statement by the jury
essertop Karat stands out due a linear, accentuated design, behind which a complex functionality is concealed.

Begründung der Jury
essertop Karat beeindruckt mit einer geradlinigen, akzentuierten Gestaltung, hinter der sich eine komplexe Funktionalität verbirgt.

The jury 2015
International orientation and objectivity
Internationalität und Objektivität

The jurors of the Red Dot Award: Product Design
All members of the Red Dot Award: Product Design jury are appointed on the basis of independence and impartiality. They are independent designers, academics in design faculties, representatives of international design institutions, and design journalists.

The jury is international in its composition, which changes every year. These conditions assure a maximum of objectivity. The members of this year's jury are presented in alphabetical order on the following pages.

Die Juroren des Red Dot Award: Product Design
In die Jury des Red Dot Award: Product Design wird als Mitglied nur berufen, wer völlig unabhängig und unparteiisch ist. Dies sind selbstständig arbeitende Designer, Hochschullehrer der Designfakultäten, Repräsentanten internationaler Designinstitutionen und Designfachjournalisten.

Die Jury ist international besetzt und wechselt in jedem Jahr ihre Zusammensetzung. Unter diesen Voraussetzungen ist ein Höchstmaß an Objektivität gewährleistet. Auf den folgenden Seiten werden die Jurymitglieder des diesjährigen Wettbewerbs in alphabetischer Reihenfolge vorgestellt.

01

Prof.
Werner Aisslinger
Germany
Deutschland

The works of the designer Werner Aisslinger cover the spectrum of experimental and artistic approaches, including industrial design and architecture. He makes use of the latest technologies and has helped introduce new materials and techniques to the world of product design. His works are part of prestigious, permanent collections of international museums such as the Museum of Modern Art and the Metropolitan Museum of Art in New York, the Fonds National d'Art Contemporain in Paris, the Victoria & Albert Museum London, the Neue Sammlung Museum in Munich, and the Vitra Design Museum in Weil, Germany. In 2013, Werner Aisslinger opened his first solo show called "Home of the future" in the Berlin museum Haus am Waldsee. Besides numerous international awards he received the prestigious A&W Designer of the Year Award in 2014. Werner Aisslinger lives in Berlin and works for companies such as Vitra, Foscarini, Haier, Canon, 25hours Hotel Bikini and Moroso, for whom he has developed the "Hemp chair" together with BASF, which is the first ever biocomposite monobloc chair.

Die Arbeiten des Designers Werner Aisslinger umfassen das Spektrum experimenteller und künstlerischer Ansätze samt Industriedesign und Architektur. Er bedient sich modernster Technologien und hat dazu beigetragen, neue Materialien und Techniken in die Welt des Produktdesigns einzuführen. Seine Arbeiten sind Bestandteil bedeutender Museumssammlungen, darunter Museum of Modern Art und Metropolitan Museum of Art in New York, Fonds National d'Art Contemporain in Paris, Victoria & Albert Museum in London, Die Neue Sammlung in München und Vitra Design Museum in Weil. 2013 eröffnete Werner Aisslinger seine erste Solo-Show „Home of the future" im Berliner Museum „Haus am Waldsee". Neben zahlreichen internationalen Auszeichnungen wurde er 2014 als „A&W-Designer des Jahres" ausgezeichnet. Der in Berlin lebende Designer arbeitet u. a. für Unternehmen wie Vitra, Foscarini, Haier, Canon, 25hours Hotel Bikini und Moroso, für das er zusammen mit BASF den „Hemp chair", den weltweit ersten biologisch zusammengesetzten Monoblock-Stuhl, entwickelt hat.

01 Upcycling & tuning
The textile cover for Porsche 928
from the 1970s as a colourful
reminder of the possible further uses
of old cars. Upcycling and tuning are
considered as a tool for prolonging
the life cycles of products. Exhibition
at Kölnischer Kunstverein 2014.
Textilhülle für den Porsche 928 aus den
1970ern als farbenfrohe Erinnerung
an die weitere mögliche Nutzung alter
Fahrzeuge. Upcycling und Tuning lassen
sich als Werkzeuge verstehen, um
den Lebenszyklus eines Produktes zu
verlängern. Ausstellung im Kölnischen
Kunstverein 2014.

02 Luminaire
Osram OLED, concept study 2014
Osram OLED, Konzeptstudie 2014

02

"Design is one of the most
important profffessions of our time."
„Design ist eine der wichtigsten
Professionen unserer Zeit."

What is, in your opinion, the significance of design quality in the industries you evaluated?
The design quality of the established brands is consistently high with a tendency towards a greater focus on product design, which is overall a positive trend.

How important is design quality in the global market?
Design quality is the ultimate distinguishing criterion. Particularly in product groups in which technical evolution creates a level playing field, design becomes the only relevant distinguishing criterion for products.

Do you see a dominating trend in this year's designs?
Innovative materials and new manufacturing methods are almost the norm in ambitious product developments. What is remarkable this year, however, is the sophistication with which these topics are combined with product design.

Wie würden Sie den Stellenwert von Designqualität in den von Ihnen beurteilten Branchen einschätzen?
Die Designqualität der etablierten Marken ist gleichbleibend hoch, mit Tendenz zu einem größeren Fokus auf das Produktdesign – alles in allem ein erfreulicher Trend.

Wie wichtig ist Designqualität im globalen Markt?
Designqualität ist das Unterscheidungskriterium schlechthin – gerade in Produktgruppen, in denen sich die technische Evolution nivelliert, wird Design zum einzigen relevanten Unterscheidungsfaktor für Produkte.

Sehen Sie eine Entwicklung im Design, die sich in diesem Jahr durchsetzt?
Innovative Materialien und neue Herstellungsmethoden sind bei ambitionierten Produktentwicklungen fast schon Standard – erstaunlich in diesem Jahr ist jedoch die „Sophistication", diese Themen mit dem Produktdesign zu verknüpfen.

01

Manuel
Alvarez Fuentes
Mexico
Mexiko

Manuel Alvarez Fuentes studied industrial design at the Universidad Iberoamericana, Mexico City, where he later served as Director of the Design Department. In 1975, he also received a Master of Design from the Royal College of Art, London. He has over 40 years of experience in the fields of product design, furniture and interior design, packaging design, signage and visual communications. From 1992 to 2012, he was Director of Diseño Corporativo, a product design consultancy. Currently, he is head of innovation at Alfher Porcewol in Mexico City. He has acted as consultant for numerous companies and also as a board member of various designers' associations, including as a member of the Icsid Board of Directors, as Vice President of the National Chamber of Industry of Mexico, Querétaro, and as Director of the Innovation and Design Award, Querétaro. Furthermore, Alvarez Fuentes has been senior tutor in Industrial Design at Tecnológico de Monterrey Campus Querétaro since 2009. In 2012 and 2013, he was president of the jury of Premio Quorum, the most prestigious design competition in Mexico.

Manuel Alvarez Fuentes studierte Industriedesign an der Universidad Iberoamericana, Mexiko-Stadt, wo er später die Leitung des Fachbereichs Design übernahm. 1975 erhielt er zudem einen Master of Design vom Royal College of Art, London. Er hat über 40 Jahre Erfahrung in Produkt- und Möbelgestaltung, Interior Design, Verpackungsdesign, Leitsystemen und visueller Kommunikation. Von 1992 bis 2012 war er Direktor von Diseño Corporativo, einem Beratungsunternehmen für Produktgestaltung. Heute ist Alvarez Fuentes Head of Innovation bei Alfher Porcewol in Mexiko-Stadt. Er war als Berater für zahlreiche Unternehmen sowie als Vorstandsmitglied in verschiedenen Designverbänden tätig, z. B. im Icsid-Vorstand, als Vizepräsident der mexikanischen Industrie- und Handelskammer im Bundesstaat Querétaro und als Direktor des Innovation and Design Award, Querétaro. Seit 2009 ist Alvarez Fuentes Senior-Dozent für Industriedesign am Tecnológico de Monterrey Campus Querétaro. 2012 und 2013 war er Präsident der Jury des Premio Quorum, des angesehensten Designwettbewerbs in Mexiko.

01 ELICA MAFdi 1
Kitchen hood, limited edition,
ceiling mount, island type.
The design of the hood is based on
Elica's patented, highly effective
Evolution ventilation system.
The steel carcase finish in porcelain
enamel (Peltre) is by Alfher Porcewol.
Küchenhaube, limitierte Auflage,
Deckenmontage, Insel-Typ.
Die Gestaltung der Haube basiert auf
dem patentierten, hocheffizienten
Ventilationssystem Elica Evolution.
Das Finish aus Porzellan-Emaille
(Peltre) ist von Alfher Porcewol.

02 ELICA MAFdi 1
Kitchen hoods, limited edition,
wall mounted.
The hood includes illumination
offered also in the Twin Evolution
system. Peltre finish colours:
orange, blue and black.
Küchenhauben, limitierte Auflage,
Wandmontage.
Die Haube bietet die gleiche Be-
leuchtung wie das „Twin Evolution"-
System. Farben des Peltre-Finishs:
Orange, Blau und Schwarz.

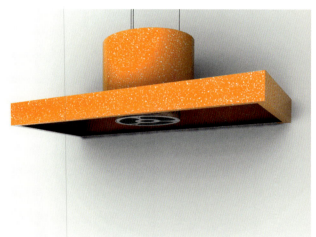

02

"My work philosophy is based on
the idea that design is not only
responsible for continuous innovation. Its main focus should be
the users, their happiness and
satisfaction."

„Meine Arbeitsphilosophie basiert
auf der Idee, dass Gestaltung
nicht nur für laufende Innovation
verantwortlich ist. Das Hauptaugenmerk sollte vielmehr auf
den Benutzern, ihrem Glück und
ihrer Zufriedenheit liegen."

What are your sources of inspiration?
Living with a very talented and creative Mexican
architect, my father Augusto Alvarez, was a source
of inspiration at the beginning and later in my life.
Subsequently, the most reliable sources came from
a thorough observation of people.

**What do you take special notice of when you are
assessing a product as a jury member?**
I search for intelligent, simple, meaningful and
purposeful products; in many respects I try to
differentiate originality "per se" from true innovation.
I disregard those products that are only good-looking
but do not contribute to a better material world.

**What significance does winning an award in
a design competition have for the designer?**
It should mean a triggering motivation for his or her
creative work, to do better every time.

Was sind Ihre Inspirationsquellen?
Das Zusammenleben mit einem sehr talentierten und
kreativen mexikanischen Architekten, meinem Vater
Augusto Alvarez, war eine Inspirationsquelle für mich,
sowohl zu Beginn als auch später in meinem Leben.
Danach ergaben sich die zuverlässigsten Inspirations-
quellen aus der sorgfältigen Beobachtung von
Menschen.

**Worauf achten Sie besonders, wenn Sie ein
Produkt als Juror bewerten?**
Ich bin auf der Suche nach intelligenten, einfachen,
bedeutungsvollen und zweckmäßigen Produkten.
In vielerlei Hinsicht versuche ich, zwischen Originalität
per se und wahrer Innovation zu unterscheiden.
Ich ignoriere diejenigen Produkte, die lediglich gut
aussehen und keinen Beitrag zu einer besseren
materiellen Welt leisten.

**Welche Bedeutung hat die Auszeichnung in einem
Designwettbewerb für den Designer?**
Sie sollte einen Motivationsschub für die kreative
Arbeit des Designers darstellen, einen Ansporn, jedes
Mal etwas besser zu machen.

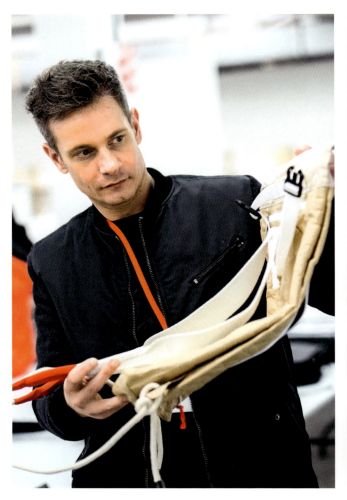

David Andersen
Denmark
Dänemark

David Andersen, born in 1978, graduated from the Glasgow School of Art and the Copenhagen Academy of Fashion and Design in 2003. In 2004, he was awarded "Best Costume Designer" and received the award "Wedding Gown of the Year" from the Royal Court Theatre in Denmark. He develops designs for ready-to-wear clothes, shoes, perfume, underwear and home wear and emerged as a fashion designer working as chief designer at Dreams by Isabell Kristensen as well as designing couture for artists and dance competitions under his own name. He debuted his collection "David Andersen" in 2007 and apart from Europe it also conquered markets in Japan and the US. In 2010 and 2011, the Danish Fashion Award nominated him in the category "Design Talent of the Year", and in 2010 as well as 2012, David Andersen received a grant from the National Art Foundation. He regularly shows his sustainable designs, which are worn by members of the Royal Family, politicians and celebrities, at couture exhibitions around the world. Furthermore, he is a guest lecturer at TEKO, Scandinavia's largest design and management college.

David Andersen, 1978 geboren, graduierte 2003 an der Glasgow School of Art und an der Copenhagen Academy of Fashion and Design. 2004 wurde er als „Best Costume Designer" sowie für das „Hochzeitskleid des Jahres" vom Royal Court Theatre in Dänemark prämiert. Er entwirft Konfektionskleidung, Schuhe, Parfüm, Unterwäsche und Heimtextilien und wurde als Modedesigner bekannt, als er als Chefdesigner bei Dreams von Isabell Kristensen sowie unter seinem Namen für Künstler und Tanzwettbewerbe arbeitete. Seine Kollektion „David Andersen", erschienen ab 2007, eroberte neben Europa Märkte in Japan und den USA. 2010 und 2011 nominierte ihn der Danish Fashion Award in der Kategorie „Design Talent of the Year" und 2010 sowie 2012 erhielt er ein Stipendium der National Art Foundation. David Andersen präsentiert seine nachhaltigen Entwürfe, die von Mitgliedern der Königsfamilie, Politikern oder Prominenten getragen werden, regelmäßig in Couture-Ausstellungen weltweit und ist zudem Gastdozent an der TEKO, der größten Hochschule für Design und Management Skandinaviens.

"My work philosophy is to be true to myself."

„Meine Arbeitsphilosophie besteht darin, mir selbst treu zu bleiben."

In your opinion, what will never be out of fashion?
Personality.

Which of the projects in your lifetime are you particular proud of?
The first dress I made for HRH Crown Princess Mary of Denmark was a cerise-coloured dress she wore at the royal wedding in Monaco. It was both thrilling and challenging. Meeting her privately was one thing, but knowing that this dress will be showcased in the press all around the world was just nerve-racking.

The Red Dot Award has been uncovering the best designs for 60 years now. Which product innovation would you like to see in the next 60 years, and why?
I hope that we as designers will own up to our responsibility for the environment through the production of our products and that the focus will shift from getting turnover to what actually happens on the journey there.

Was ist Ihrer Meinung nach nie unmodern?
Persönlichkeit.

Auf welche Ihrer bisherigen Projekte sind Sie besonders stolz?
Das erste Kleid, das ich für Ihre Königliche Hoheit Kronprinzessin Mary von Dänemark entwarf, war ein kirschrotes Kleid, das sie zur königlichen Hochzeit in Monaco trug. Dieses Projekt war sowohl aufregend als auch eine Herausforderung. Sie privat zu treffen, war eine Sache, aber zu wissen, dass dieses Kleid von der internationalen Presse weltweit zur Schau gestellt werden würde, war schlicht nervenaufreibend.

Der Red Dot Award ermittelt seit 60 Jahren die besten Gestaltungen. Welche Produktinnovation würden Sie sich für die nächsten 60 Jahre wünschen?
Ich hoffe, dass wir unserer Verantwortung als Designer für die Umwelt durch die Herstellung unserer Produkte gerecht werden und sich das Augenmerk weg von dem Ziel des Umsatzes hin zu dem verschiebt, was auf dem Weg dorthin tatsächlich passiert.

01

Prof. Martin Beeh
Germany
Deutschland

Professor Martin Beeh is a graduate in Industrial Design from the Darmstadt University of Applied Sciences in Germany and the ENSCI-Les Ateliers, Paris, and completed a postgraduate course in business administration. In 1995, he became design coordinator at Décathlon in Lille/France, in 1997 senior designer at Electrolux Industrial Design Center Nuremberg and Stockholm and furthermore became design manager at Electrolux Industrial Design Center Pordenone/Italy, in 2001. He is a laureate of several design awards as well as founder and director of the renowned student design competition "Electrolux Design Lab". In the year 2006 he became general manager of the German office of the material library Material ConneXion in Cologne. Three years later, he founded the design office beeh_innovation. He focuses on Industrial Design, Design Management, Design Thinking and Material Innovation. Martin Beeh lectured at the Folkwang University of the Arts in Essen and at the University of Applied Sciences Schwäbisch Gmünd. Since 2012, he is professor for design management at the University of Applied Sciences Ostwestfalen-Lippe in Lemgo/Germany.

Professor Martin Beeh absolvierte ein Studium in Industriedesign an der Fachhochschule Darmstadt und an der ENSCI-Les Ateliers, Paris, sowie ein Aufbaustudium der Betriebswirtschaft. 1995 wurde er Designkoordinator bei Décathlon in Lille/Frankreich, 1997 Senior Designer im Electrolux Industrial Design Center Nürnberg und Stockholm sowie 2001 Design Manager am Electrolux Industrial Design Center Pordenone/ Italien. Er ist Gewinner diverser Designpreise und gründete und leitete den renommierten Designwettbewerb für Studierende, das „Electrolux Design Lab". Im Jahr 2006 wurde er General Manager der deutschen Niederlassung der Materialbibliothek „Material ConneXion" in Köln. Drei Jahre später gründete Martin Beeh das Designbüro beeh_innovation. Seine Schwerpunkte liegen in den Bereichen Industriedesign, Design Management, Design Thinking und Materialinnovation. Martin Beeh hatte Lehraufträge an der Folkwang Universität der Künste in Essen und an der Hochschule für Gestaltung Schwäbisch Gmünd. Seit 2012 ist er Professor für Designmanagement an der Hochschule Ostwestfalen-Lippe in Lemgo.

02

"My work philosophy is to aim for the best for and together with the customer, because every task is a new adventure."

„Meine Arbeitsphilosophie ist es, dass ich das Beste für und mit dem Kunden erreichen will, denn jede Aufgabe ist ein neues Abenteuer."

What are the main challenges in a designer's everyday life?
The movement between inspiration and stringency, order and creative chaos. Those who only follow a routine dry out and stop growing.

What impressed you most during the Red Dot judging process?
The concentrated competence of jurors and our interaction, the professionalism of the Red Dot team and the courage of the large number of companies that take part in the global competition.

What do you take special notice of when you are assessing a product as a jury member?
That the product is suitable for daily use, its ergonomic quality, and that there are no unnecessary frills. I also check if the design matches the brand, and I look for the product that stands out in a positive way.

What significance does winning an award in a design competition have for the manufacturer?
The role design plays in a company is strengthened and it is "officially" recognised as a successful tool.

Was sind große Herausforderungen im Alltag eines Designers?
Die Bewegung zwischen Inspiration und Stringenz, Ordnung und schöpferischem Chaos. Wer nur Routine macht, trocknet aus und wächst nicht mehr.

Was hat Sie bei der Red Dot-Jurierung am meisten beeindruckt?
Die geballte Kompetenz der Juroren und unser Zusammenspiel, die Professionalität des Red Dot-Teams und der Mut der vielen Unternehmen, die sich dem globalen Wettbewerb stellen.

Worauf achten Sie besonders, wenn Sie ein Produkt als Juror bewerten?
Auf Alltagstauglichkeit, Ergonomie, gestalterische Ordnung und den Verzicht auf Schnickschnack. Ich schaue auch, ob die Gestaltung zur Marke passt, und suche das Produkt, das sich positiv abhebt.

Welche Bedeutung hat die Auszeichnung in einem Designwettbewerb für den Hersteller?
Die Position von Design im Unternehmen wird gestärkt und „offiziell" als erfolgreiches Instrument anerkannt.

01

Dr Luisa Bocchietto
Italy
Italien

Luisa Bocchietto, architect and designer, graduated from the Milan Polytechnic. She has worked as a freelancer undertaking projects for local development, building renovations and urban planning. As a visiting professor she teaches at universities and design schools, she takes part in design conferences and international juries, publishes articles and organises exhibitions on architecture and design. Over the years, her numerous projects aimed at supporting the spread of design quality. She was a member of the Italian Design Council at the Ministry for Cultural Heritage, the Polidesign Consortium of the Milan Polytechnic, the CIDIC Italo-Chinese Council for Design and Innovation and the CNAC National Anti-Counterfeiting Council at the Ministry of Economic Development. From 2008 until 2014 she was National President of the ADI (Association for Industrial Design). Currently, she is a member of the Icsid Board, and since January 2015 she has been Editorial Director of PLATFORM, a new magazine about Architecture and Design.

Luisa Bocchietto, Architektin und Designerin, graduierte am Polytechnikum Mailand. Sie arbeitet freiberuflich und führt Projekte für die lokale Entwicklung, Gebäudeumbauten und Stadtplanung durch. Als Gastprofessorin lehrt sie an Universitäten und Designschulen, sie nimmt an Designkonferenzen und internationalen Jurys teil, veröffentlicht Artikel und betreut Ausstellungen über Architektur und Design. Ihre zahlreichen Projekte über die Jahre hinweg verfolgten das Ziel, die Verbreitung von Designqualität zu unterstützen. Sie war Mitglied des italienischen Rates für Formgebung am Kulturministerium, der Polidesign-Vereinigung des Polytechnikums Mailand, der CIDIC, der italienisch-chinesischen Vereinigung für Design und Innovation, und der CNAC, der Nationalen Vereinigung gegen Fälschungen am Ministerium für wirtschaftliche Entwicklung. Von 2008 bis 2014 war sie nationale Präsidentin der ADI, des Verbandes für Industriedesign. Aktuell ist sie Gremiumsmitglied des Icsid und seit Januar 2015 Chefredakteurin von PLATFORM, einem neuen Magazin über Architektur und Design.

02

"My work philosophy is to simplify, initiate sensible projects, generate quality and share joy."

„Meine Arbeitsphilosophie besteht darin, Dinge und Vorgänge zu vereinfachen, sinnvolle Projekte anzustoßen, Qualität zu schaffen und Freude zu teilen."

You are the editorial director of the new magazine "PLATFORM". What fascinates you about this job?
It allows me to explore the design aspects that I am more passionate about while at the same time keeping in touch with all the extraordinary people I have met during my ADI presidency.

The Red Dot Award has been uncovering the best designs for 60 years now. Which product innovation would you like to see in the next 60 years?
Products that are meant to improve people's lives – not only in an aesthetic and functional sense, but also in a social and ethical one.

What is the advantage of commissioning an external designer compared to having an in-house design team?
Working with an internal technical team means optimising in a linear way, while engaging with external designers entails the opportunity of a "lateral deviation", which often leads to unexpected and great results.

Sie sind die Chefredakteurin des neuen Magazins „PLATFORM". Was fasziniert Sie an dieser Rolle?
Sie erlaubt mir, die Gestaltungsaspekte, die mich besonders begeistern, näher zu untersuchen und zugleich mit all den außergewöhnlichen Menschen, die ich während meiner ADI-Präsidentschaft kennengelernt habe, in Kontakt zu bleiben.

Der Red Dot Award ermittelt seit 60 Jahren die besten Gestaltungen. Welche Produktinnovation würden Sie sich für die nächsten 60 Jahre wünschen?
Produkte, die darauf abzielen, das Leben zu verbessern – nicht nur in einem ästhetischen und funktionalen Sinn, sondern auch in einem sozialen und ethischen.

Worin liegt der Vorteil, einen externen Designer zu beauftragen, im Vergleich zu einem Inhouse-Designteam?
Mit einem internen technischen Team zu arbeiten, bedeutet, Dinge auf lineare Weise zu optimieren, wohingegen die Zusammenarbeit mit externen Designern die Gelegenheit für eine „laterale Abweichung" bietet, die oft zu unerwarteten und großartigen Ergebnissen führt.

01

Gordon Bruce
USA
USA

Gordon Bruce is the owner of Gordon Bruce Design LLC and has been a design consultant for 40 years working with many multinational corporations in Europe, Asia and the USA. He has worked on a very wide range of products, interiors and vehicles – from aeroplanes to computers to medical equipment to furniture. From 1991 to 1994, Gordon Bruce was a consulting vice president for the Art Center College of Design's Kyoto programme and, from 1995 to 1999, chairman of Product Design for the Innovative Design Lab of Samsung (IDS) in Seoul, Korea. In 2003, he played a crucial role in helping to establish Porsche Design's North American office. For many years, he served as head design consultant for Lenovo's Innovative Design Center (IDC) in Beijing and he is presently working with Bühler in Switzerland and Huawei Technologies Co., Ltd. in China. Gordon Bruce is a visiting professor at several universities in the USA and in China and also acts as an author and design publicist. He recently received Art Center College of Design's "Lifetime Achievement Award".

Gordon Bruce ist Inhaber der Gordon Bruce Design LLC und seit mittlerweile 40 Jahren als Designberater für zahlreiche multinationale Unternehmen in Europa, Asien und den USA tätig. Er arbeitete bereits an einer Reihe von Produkten, Inneneinrichtungen und Fahrzeugen – von Flugzeugen über Computer bis hin zu medizinischem Equipment und Möbeln. Von 1991 bis 1994 war Gordon Bruce beratender Vizepräsident des Kioto-Programms am Art Center College of Design sowie von 1995 bis 1999 Vorsitzender für Produktdesign beim Innovative Design Lab of Samsung (IDS) in Seoul, Korea. Im Jahr 2003 war er wesentlich daran beteiligt, das Büro von Porsche Design in Nordamerika zu errichten. Über viele Jahre war er leitender Designberater für Lenovos Innovative Design Center (IDC) in Beijing. Aktuell arbeitet er für Bühler in der Schweiz und Huawei Technologies Co., Ltd. in China. Gordon Bruce ist Gastprofessor an zahlreichen Universitäten in den USA und in China und als Buchautor sowie Publizist tätig. Kürzlich erhielt er vom Art Center College of Design den Lifetime Achievement Award.

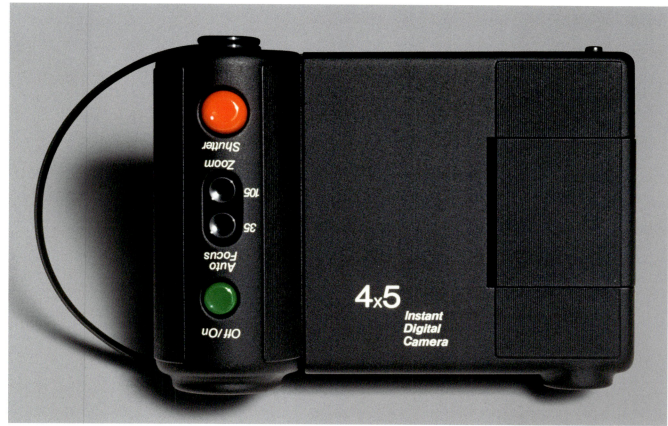

02

"My work philosophy is to maintain a sense of curiosity. Then, I conceptualise an idea from my inquisitiveness, test it and start the process over again."

„Meine Arbeitsphilosophie besteht darin, Neugierde zu bewahren. In einem zweiten Schritt konzipiere ich eine Idee auf Basis meiner Wissbegierde, teste sie und beginne den Prozess von vorne."

Which IT project that you were involved in are you most proud of?
I was one of three designers who helped design the first "Massively Parallel Processor", a super computer produced by the Thinking Machines Company. The unique design, using a series of black translucent boxes while accommodating many issues based on improved usability, resulted in a paradigm shift for large computers.

The Red Dot Award has been uncovering the best designs for 60 years now. Which product innovation would you like to see in the next 60 years, and why?
I would like to see a common theme in design move more and more towards Mother Nature's design.

What significance does winning an award in a design competition have for the designer?
Winning a design award from a very well-recognised design competition gives designers affirmation that they are making the right choices in their design decision processes.

Auf welches IT-Projekt, an dem Sie beteiligt waren, sind Sie besonders stolz?
Ich war einer von drei Designern, die dabei halfen, den ersten „Massively Parallel Processor", einen Super-Computer der Thinking Machines Company, zu gestalten. Das einzigartige Design, das aus einer Reihe lichtdurchlässiger Kästen besteht und gleichzeitig viele Aspekte verbesserter Benutzerfreundlichkeit beherbergt, begründete einen Paradigmenwechsel für große Computer.

Der Red Dot Award ermittelt seit 60 Jahren die besten Gestaltungen. Welche Produktinnovation würden Sie sich für die nächsten 60 Jahre wünschen, und warum?
Ich würde gerne einen einheitlichen Trend im Design-bereich sehen, der sich mehr und mehr zu einer Gestaltung im Stil von Mutter Natur hinbewegt.

Welche Bedeutung hat die Auszeichnung in einem Designwettbewerb für den Designer?
Eine Auszeichnung in einem allgemein anerkannten Designwettbewerb zu gewinnen, gibt Designern die Bestätigung, dass sie die richtigen Entscheidungen im Designprozess getroffen haben.

01

Rüdiger Bucher
Germany
Deutschland

Rüdiger Bucher, born in 1967, graduated in political science from the Philipps-Universität Marburg and subsequently completed the postgraduate study course "Interdisciplinary studies on France" in Freiburg, Germany. While still at school he wrote for daily newspapers and magazines, before joining publishing house Verlagsgruppe Ebner Ulm in 1995, where he was in charge of "Scriptum. Die Zeitschrift für Schreibkultur" (Scriptum. The magazine for writing culture) for five years. In 1999 he became Chief Editor of Chronos, the leading German-language special interest magazine for wrist watches, with the same publishing house. During his time as Chief Editor, since 2005, Chronos has positioned itself internationally with subsidiary magazines and licensed editions in China, Korea, Japan and Poland. At the same time, Rüdiger Bucher established a successful corporate publishing department for Chronos. Since 2014, he has been Editorial Director of the business area "Watches" at the Ebner publishing house and besides Chronos he has also been in charge of the sister magazines concerning watches and classic watches as well as the New York-based "WatchTime".

Rüdiger Bucher, geboren 1967, studierte Politikwissenschaft an der Philipps-Universität Marburg und schloss daran den Aufbaustudiengang „Interdisziplinäre Frankreich-Studien" in Freiburg an. Schon als Schüler schrieb er für verschiedene Tageszeitungen und Zeitschriften, bevor er 1995 zum Ebner Verlag Ulm kam und dort fünf Jahre lang „Scriptum. Die Zeitschrift für Schreibkultur" betreute. Im selben Verlag wurde er 1999 Redaktionsleiter von „Chronos", dem führenden deutschsprachigen Special-Interest-Magazin für Armbanduhren. Während seiner Amtszeit als Chefredakteur, seit 2005, hat sich Chronos mit Tochtermagazinen und Lizenzausgaben in China, Korea, Japan und Polen international aufgestellt. Gleichzeitig baute Rüdiger Bucher für Chronos einen erfolgreichen Corporate-Publishing-Bereich auf. Seit 2014 verantwortet er als Redaktionsdirektor des Ebner Verlags im Geschäftsbereich „Uhren" neben Chronos auch die Schwestermagazine „Uhren-Magazin", „Klassik Uhren" sowie die in New York beheimatete „WatchTime".

01 **Chronos Special Uhrendesign**
Published once a year each
September since 2013.
Erscheint seit 2013 einmal
jährlich im September.

02 Chronos is present around the
globe with issues of differing
themes plus special issues.
Mit verschiedenen Ausgaben
und Sonderheften ist Chronos
rund um den Globus vertreten.

02

"My work philosophy is to inform
our readers competently, while at
the same time entertaining and
occasionally surprising them with
something new and unexpected."
„Meine Arbeitsphilosophie ist es,
den Leser kompetent zu informieren,
ihn dabei zu unterhalten und ihn
gelegentlich durch Neues, Unerwar-
tetes zu überraschen."

**What fascinates you about your job as Chief
Editor of Chronos?**
It is a very multi-faceted position. We distribute our
information using different channels and at the same
time we are closely in touch with what's happening
in a highly interesting industry, which is characterised
by a fascinating tension between past and future.

When did you first think consciously about design?
As a child I already preferred using specific glasses
and a specific cutlery set. Apart from the look, it was
above all the well-designed haptic qualities that
appealed to me.

**The Red Dot Award has been uncovering the
best designs for 60 years now. Which product
innovations would you like to see in the next
60 years, and why?**
A technology that does away with cables and at the
same time is non-hazardous to health.

**What significance does winning an award in
a design competition have for the manufacturer?**
An objective quality seal increases credibility, generates attention and enhances its profile.

**Was fasziniert Sie an Ihrem Beruf als Chef-
redakteur von Chronos?**
Er ist sehr vielseitig: Wir verbreiten unsere Informationen auf unterschiedlichsten Kanälen und sind zugleich
nah dran an einer hochinteressanten Branche, die sich
in Technik und Design in einem faszinierenden Span-
nungsfeld zwischen Vergangenheit und Zukunft bewegt.

**Wann haben Sie das erste Mal bewusst über Design
nachgedacht?**
Schon als Kind benutzte ich bestimmte Gläser und
bestimmtes Besteck lieber als andere. Neben der Optik
sprach mich vor allem eine gelungene Haptik an.

**Der Red Dot Award ermittelt seit 60 Jahren die
besten Gestaltungen. Welche Produktinnovation
würden Sie sich für die nächsten 60 Jahre
wünschen, und warum?**
Eine Technik, die es erlaubt, auf Kabel zu verzichten,
und dabei gesundheitlich unbedenklich ist.

**Welche Bedeutung hat die Auszeichnung in einem
Designwettbewerb für den Hersteller?**
Ein objektives Gütesiegel erhöht die Glaubwürdigkeit,
schafft Aufmerksamkeit und schärft sein Profil.

01

Wen-Long Chen
Taiwan
Taiwan

Wen-Long Chen has been CEO of the Taiwan Design Center since 2013 and has been designated as "Taiwan's Top Boss in Design". He has accumulated over 25 years of practical experience in design management and product design. In 1988, he founded Nova Design with support from Chinfon Trading Group, and has since led and completed hundreds of design projects. During his time at Nova Design, Wen-Long Chen led the company to tremendous growth with visionary design thinking. In response to clients' needs, he utilised the "Design System Competitiveness" as the core value of development, established a KMO (Knowledge Management Officer) system to oversee six branch offices in Asia, North America and Europe, and invested millions of dollars in the most advanced facilities. Today, Nova Design has over 200 employees worldwide and has won over 100 international design awards since 2006. Wen-Long Chen continues to propel the development of Taiwan's design industry using "Design x Knowledge Management", striving to shape Taiwan into a powerful international design force.

Wen-Long Chen ist seit 2013 CEO des Taiwan Design Centers und ist zu „Taiwan's Top Boss in Design" ernannt worden. Er verfügt über mehr als 25 Jahre praktische Erfahrung im Designmanagement und Produktdesign. 1988 gründete er mit Unterstützung der Chinfon Trading Group Nova Design und hat seitdem Hunderte Designprojekte durchgeführt. Während dieser Zeit führte und leitete er Nova Design mit visionärer Denkweise zu enormem Erfolg. Als Antwort auf die Bedürfnisse der Kunden nutzte er „Design System Competitiveness" als zentralen Entwicklungswert, etablierte das System „KMO" (Knowledge Management Officer), um sechs Zweigbüros in Asien, Nordamerika und Europa zu betreuen, und investierte mehrere Millionen Dollar in die fortgeschrittensten Einrichtungen. Heute hat Nova Design mehr als 200 Angestellte weltweit und erhielt seit 2006 über 100 internationale Designauszeichnungen. Wen-Long Chen treibt weiterhin die Entwicklung der taiwanesischen Designindustrie voran, indem er „Design x Knowledge Management" nutzt und danach strebt, Taiwan zu einer weltweit starken Designmacht zu formen.

01 SYM Fighter 4V 150

02 **Transformed Crane**
Commissioned by SANY Heavy
Industry Co., Ltd. and exhibited
at "bauma China 2012", the
International Trade Fair for
Construction Machinery,
Building Material Machines,
Construction Vehicles and
Equipment.
Transformierter Kran
In Auftrag gegeben von SANY
Heavy Industry Co., Ltd. und
ausgestellt auf der „bauma
China 2012", der internationalen
Messe für Baumaschinen, Bau-
materialmaschinen, Konstruktion,
Baufahrzeuge und Baugeräte.

02

"My work philosophy is: do the
right thing right, which will
eventually lead you to the right
people and the right resources."

„Meine Arbeitsphilosophie lautet:
Mache die richtige Sache richtig,
denn das wird dich letztlich zu
den richtigen Menschen und den
richtigen Quellen führen."

**Which product area will Taiwan have most
success with in the next ten years?**
Our efficient design industry will achieve success as
a whole. There are both manufacturers with strong
technology expertise and high adaptivity as well as
a rich amount of creative talent in various fields.

**Which country do you consider to be a pioneer
in product design?**
I would still consider Germany the pioneer in product
design, due to the value they place on craftsmanship.
Even though nowadays most breakthrough innovation
comes from US-based start-ups, I believe in the end,
it is hard to outperform Germany, both in terms of
aesthetics and practicality.

**What significance does winning an award in
a design competition have for the designer?**
While winning the award is proof of their career
progress, the most beneficial effect comes from
designers having the means to position themselves
in the industry and determine their status.

In welchen Produktbereichen wird Taiwan in den
nächsten zehn Jahren den größten Erfolg haben?
Unsere effiziente Designindustrie wird als Ganzes
erfolgreich sein. Es gibt sowohl Hersteller mit großem
Technologiewissen und hoher Adaptionsfähigkeit als
auch eine Vielzahl an kreativen Talenten in den
verschiedensten Bereichen.

Welches Land halten Sie für einen Pionier im
Bereich Produktgestaltung?
Ich halte Deutschland nach wie vor für den Pionier
im Produktdesign, weil es großen Wert auf handwerk-
liches Können legt. Obwohl die meisten bahnbrechen-
den Innovationen heutzutage von Start-up-Unter-
nehmen aus den USA kommen, glaube ich, dass es
letztlich schwer ist, Deutschland zu übertreffen,
sowohl in puncto Ästhetik als auch in Praktikabilität.

Welche Bedeutung hat die Auszeichnung in einem
Designwettbewerb für den Designer?
Während die Auszeichnung eine Bestätigung des
Karrierefortschritts der Designer ist, besteht ihr hilf-
reichster Effekt darin, dass sie ihnen ermöglicht,
sich in der Branche zu positionieren und ihren eigenen
Status zu ermitteln.

01

Vivian Wai Kwan Cheng
Hong Kong
Hongkong

On leaving Hong Kong Design Institute after 19 years of educational service, Vivian Cheng founded "Vivian Design" in 2014 to provide consultancy services and promote her own art in jewellery and glass. She graduated with a BA in industrial design from the Hong Kong Polytechnic University and was awarded a special prize in the Young Designers of the Year Award hosted by the Federation of Hong Kong Industries in 1987, and the Governor's Award for Industry: Consumer Product Design in 1989, after joining Lambda Industrial Limited as the head of the Product Design team. In 1995 she finished her Master degree and joined the Vocational Training Council teaching product design and later became responsible for, among others, establishing an international network with design-related organisations and schools. Vivian Cheng was the International Liaison Manager at the Hong Kong Design Institute (HKDI) and member of the Chartered Society of Designers Hong Kong, member of the Board of Directors of the Hong Kong Design Centre (HKDC) and board member of the Icsid from 2013 to 2015. Furthermore, she has been a panel member for the government and various NGOs.

Nach 19 Jahren im Lehrbetrieb verließ Vivian Cheng 2014 das Hong Kong Design Institute und gründete „Vivian Design", um Beratungsdienste anzubieten und ihre eigene Schmuck- und Glaskunst weiterzuentwickeln. 1987 machte sie ihren BA in Industriedesign an der Hong Kong Polytechnic University. Im selben Jahr erhielt sie einen Sonderpreis im Wettbewerb „Young Designers of the Year", veranstaltet von der Federation of Hong Kong Industries, sowie 1989 den Governor's Award for Industry: Consumer Product Design, nachdem sie bei Lambda Industrial Limited als Leiterin des Produktdesign-Teams angefangen hatte. 1995 beendete sie ihren Master-Studiengang und wechselte zum Vocational Training Council, wo sie Produktdesign unterrichtete und später u. a. für den Aufbau eines internationalen Netzwerks mit Organisationen und Schulen im Designbereich verantwortlich war. Vivian Cheng war International Liaison Manager am Hong Kong Design Institute (HKDI), Mitglied der Chartered Society of Designers Hong Kong, Vorstandsmitglied des Hong Kong Design Centre (HKDC) und Gremiumsmitglied des Icsid. Außerdem war sie Mitglied verschiedener Bewertungsgremien der Regierung und vieler Nichtregierungsorganisationen.

02

"My work philosophy is: things
I do today will be a bit better than
yesterday's. I look into tomorrow,
while enjoying today, and articulate
what I experienced yesterday."
„Meine Arbeitsphilosophie ist:
Dinge, die ich heute mache, werden
ein bisschen besser als die gestrigen
sein. Ich schaue auf das Morgen,
genieße das Heute und spreche aus,
was ich gestern erfahren habe."

When did you first think consciously about design?
I always wanted to dress only in the way I like since
I was a small child. And I was told that I started to
draw on walls at the age of three. So I believe I was
born to appreciate art and design.

**What do you take special notice of when you are
assessing a product as a jury member?**
I pay special attention to whether the design combines well with the choice of materials and the
required technology, the craftsmanship, its positioning
in the market, and that the user's emotional needs
have been taken into account.

**What is the advantage of commissioning an
external designer compared to having an in-house
design team?**
An external designer can bring an outsider's view and
provide the opportunity for a second party to challenge the product's functions and the way it operates.
It might also help with the extension of product
functions as well as the application of new materials
and technology in different ways.

**Wann haben Sie das erste Mal bewusst über
Design nachgedacht?**
Schon als Kleinkind wollte ich nur die Kleidung tragen,
die mir gefiel. Und ich weiß von Erzählungen, dass ich
bereits im Alter von drei Jahren begann, Wände zu
bemalen. Daher glaube ich, dass ich dazu geboren bin,
Kunst und Design zu würdigen.

**Worauf achten Sie besonders, wenn Sie ein
Produkt als Jurorin bewerten?**
Ich achte besonders darauf, dass die Gestaltung gut
zur Wahl der Materialien und Technologien, der
Verarbeitung und der Marktpositionierung passt und
dass die emotionalen Bedürfnisse des Benutzers
berücksichtigt wurden.

**Worin liegt der Vorteil, einen externen Designer
zu beauftragen, im Vergleich zu einem Inhouse-
Designteam?**
Ein externer Designer kann eine Sicht von außen einbringen und Raum für eine zweite Seite schaffen, die
die Funktionen und die Handhabung der Produkte
hinterfragt. Zudem kann es dabei helfen, die Produkt-
funktionen zu erweitern und neue Materialien und
Technologien auf unterschiedliche Arten einzubeziehen.

01

Datuk Prof. Jimmy Choo OBE
Malaysia / Great Britain
Malaysia / Großbritannien

Datuk Professor Jimmy Choo OBE is descended from a family of Malaysian shoemakers and learned the craft from his father. He studied at Cordwainers College, which is today part of the London College of Fashion. After graduating in 1983, he founded his own couture label and opened a shoe shop in London's East End whose regular customers included the late Diana, Princess of Wales. In 1996, Choo launched his ready-to-wear line with Tom Yeardye. Choo sold his share in the business in November 2001 to Equinox Luxury Holdings Ltd, charging them with the ongoing use of the label on the luxury goods market, while he continued to run his couture line. Choo now spends his time promoting design education. He is an ambassador for footwear education at the London College of Fashion, a spokesperson for the British Council in their promotion of British Education to foreign students and also spends time working with the non-profit programme, Teach For Malaysia. In 2003, Jimmy Choo was honoured for his contribution to fashion by Queen Elizabeth II who appointed him "Officer of the Order of the British Empire".

Datuk Professor Jimmy Choo OBE, der einer malaysischen Schuhmacher-Familie entstammt und das Handwerk von seinem Vater lernte, studierte am Cordwainers College, heute Teil des London College of Fashion. Nach seinem Abschluss 1983 gründete er sein eigenes Couture-Label und eröffnete ein Schuhgeschäft im Londoner East End, zu dessen Stammkundschaft auch Lady Diana gehörte. 1996 führte Choo gemeinsam mit Tom Yeardye seine Konfektionslinie ein und verkaufte seine Anteile an dem Unternehmen im November 2001 an die Equinox Luxury Holdings Ltd. Diese beauftragte er damit, das Label auf dem Markt für Luxusgüter fortzuführen, während er sich weiter um seine Couture-Linie kümmerte. Heute fördert Jimmy Choo die Designlehre. Er ist Botschafter für Footwear Education am London College of Fashion sowie Sprecher des British Council für die Förderung der Ausbildung ausländischer Studenten in Großbritannien und arbeitet für das gemeinnützige Programm „Teach for Malaysia". Für seine Verdienste in der Mode verlieh ihm Königin Elisabeth II. 2003 den Titel „Officer of the Order of the British Empire".

01 Red Sandal
In modern curves, made with silk and a leather strap.
Rote Sandale
In modernen Kurven, hergestellt aus Seide und einem Lederband.

02 Light Brown Sandal
Made with traditional woven cotton from East Malaysia (Pua Kumbu) with leather strap.
Leichte braune Sandale
Hergestellt aus traditioneller, gewobener Baumwolle aus Ostmalaysia (Pua Kumbu) mit Lederband.

02

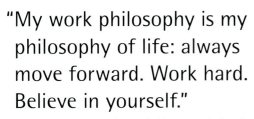

"My work philosophy is my philosophy of life: always move forward. Work hard. Believe in yourself."

„Meine Arbeitsphilosophie ist meine Lebensphilosophie: Gehe immer vorwärts. Arbeite hart. Glaube an dich selbst."

How does it feel to be a brand and a person of the same name?
I am very proud of what I have achieved with Jimmy Choo Couture. I no longer hold shares in Jimmy Choo London, but I am proud of co-founding a company that has become a global phenomenon.

Which country do you consider to be a pioneer in product design?
England has a rich history of producing leading design talent – and that continues to this day.

What impressed you most during the Red Dot judging process?
That professionals from around the world gather in one place, which gives us the opportunity to choose the most outstanding talent among the best.

What do you take special notice of when you are assessing a product as a jury member?
I always look for something that I haven't seen before – e.g. a new take on a well-known product. To catch my eye, the product also has to be of the highest quality, functional and durable.

Wie ist es, zugleich eine Marke und eine Person mit demselben Namen zu sein?
Ich bin sehr stolz auf das, was ich mit Jimmy Choo Couture erreicht habe. Ich besitze keine Anteile mehr an Jimmy Choo London, aber ich bin stolz darauf, ein Unternehmen mitgegründet zu haben, das sich zu einem globalen Phänomen entwickelt hat.

Welche Nation ist für Sie Vorreiter im Produktdesign?
England hat in der Vergangenheit eine Vielzahl führender Designtalente hervorgebracht – und das bis heute.

Was hat Sie bei der Red Dot-Jurierung am meisten beeindruckt?
Dass sich Profis aus aller Welt an einem Ort versammeln, was uns die Gelegenheit gibt, die hervorstechendsten Talente aus den Besten auszuwählen.

Worauf achten Sie besonders, wenn Sie ein Produkt als Juror bewerten?
Ich suche immer nach etwas, das ich vorher noch nicht gesehen habe – etwa eine neue Interpretation eines bekannten Produktes. Um meine Aufmerksamkeit zu erregen, muss es sowohl die höchsten Qualitätsstandards erfüllen als auch funktional und langlebig sein.

01

Vincent Créance
France
Frankreich

After graduating from the Ecole Supérieure de Design Industriel, Vincent Créance began his career in 1985 at the Plan Créatif Agency where he became design director and developed, among other things, numerous products for hi-tech and consumer markets, for France Télécom and RATP (Paris metro). In 1996 he joined Alcatel as Design Director for all phone activities on an international level. In 1999, he became Vice President Brand in charge of product design and user experience as well as all communications for the Mobile Phones BU. During the launch of the Franco-Chinese TCL and Alcatel Mobile Phones joint venture in 2004, Vincent Créance advanced to the position of Design and Corporate Communications Director. In 2006, he became President and CEO of MBD Design, one of the major design agencies in France, providing design solutions in transport design and product design. Créance is a member of the APCI (Agency for the Promotion of Industrial Creation), on the board of directors of ENSCI (National College of Industrial Creation), and a member of the Strategic Advisory Board for Strate College.

Vincent Créance begann seine Laufbahn nach seinem Abschluss an der Ecole Supérieure de Design Industriel 1985 bei Plan Créatif Agency. Hier stieg er 1990 zum Design Director auf und entwickelte u. a. zahlreiche Produkte für den Hightech- und Verbrauchermarkt, für die France Télécom oder die RATP (Pariser Metro). 1996 ging er als Design Director für sämtliche Telefonaktivitäten auf internationaler Ebene zu Alcatel und wurde 1999 Vice President Brand, zuständig für Produktdesign und User Experience sowie die gesamte Kommunikation für den Geschäftsbereich „Mobile Phones". Während des Zusammenschlusses des französisch-chinesischen TCL und Alcatel Mobile Phones 2004 avancierte Vincent Créance zum Design and Corporate Communications Director. 2006 wurde er Präsident und CEO von MBD Design, einer der wichtigsten Designagenturen in Frankreich, und entwickelte Designlösungen für Transport- und Produktdesign. Créance ist Mitglied von APCI (Agency for the Promotion of Industrial Creation), Vorstand des ENSCI (National College of Industrial Design) und Mitglied im wissenschaftlichen Beirat des Strate College.

02/03

"My work philosophy is: when risks appear, it becomes interesting!"

„Meine Arbeitsphilosophie lautet: Wenn Risiken auftauchen, wird es interessant!"

What are the main challenges in a designer's everyday life?
For a young one: to acquire experience in order to avoid big mistakes. For a senior one: to forget his or her experience in order to avoid big mistakes!

The Red Dot Award has been uncovering the best designs for 60 years now. Which product innovations would you like to see in the next 60 years, and why?
If everything we read about global warming etc. is true, it is obvious that product innovations have to find answers to these issues.

What significance does winning an award in a design competition have for the designer?
It's a recognition from his peers, always pleasant and helps to improve his image.

What significance does winning an award in a design competition have for the manufacturer?
It's a very good means to check the "non-rational" performance of their products. Because people also want pleasure, even from professional goods.

Was sind große Herausforderungen im Alltag eines Designers?
Für einen jungen Designer: Erfahrung sammeln, um große Fehler zu vermeiden. Für einen erfahrenen Designer: seine Erfahrung vergessen, um große Fehler zu vermeiden!

Der Red Dot Award ermittelt seit 60 Jahren die besten Gestaltungen. Welche Produktinnovationen würden Sie sich für die nächsten 60 Jahre wünschen, und warum?
Falls alles, was wir über die globale Erwärmung etc. lesen, wahr ist, ist offensichtlich, dass Produktinnovationen für diese Probleme Lösungen finden müssen.

Welche Bedeutung hat die Auszeichnung in einem Designwettbewerb für den Designer?
Sie ist eine Anerkennung seiner Partner, immer erfreulich und hilfreich, um sein Image zu verbessern.

Welche Bedeutung hat die Auszeichnung in einem Designwettbewerb für den Hersteller?
Es ist ein gutes Mittel, um die nicht-rationale Leistung seiner Produkte zu überprüfen. Denn Menschen wollen Wohlgefallen, selbst bei Fachprodukten.

01

Martin Darbyshire
Great Britain
Großbritannien

Martin Darbyshire is a founder and CEO of the internationally renowned design consultancy tangerine, working for clients such as Asiana, Azul, B/E Aerospace, Huawei, Nikon, The Royal Mint, Snoozebox.com and Virgin Australia. Before founding tangerine in 1989, he worked for Moggridge Associates and then in San Francisco at ID TWO (now IDEO). Martin Darbyshire is responsible for tangerine's commercial management, leading design projects and creating new business opportunities. Most notably, he led the multidisciplinary team that created both generations of the "Club World" business-class aircraft seating for British Airways – the world's first fully flat bed in business class which, since its launch in 2000, has remained the profit engine of the airline and transformed the industry. Besides, Martin Darbyshire has been a UK Trade and Investment ambassador for the UK Creative Industries sector, he is a recognised industry spokesperson, an advisor on design and innovation and was a board member of the Icsid.

Martin Darbyshire ist Gründer und CEO des international renommierten Designbüros tangerine, das für Kunden wie Asiana, Azul, B/E Aerospace, Huawei, Nikon, The Royal Mint, Snoozebox.com und Virgin Australia tätig ist. Bevor er tangerine 1989 gründete, arbeitete er für Moggridge Associates und danach in San Francisco bei ID TWO (heute IDEO). Martin Darbyshire verantwortet das kaufmännische Management von tangerine, wozu die Leitung von Designprojekten und die Entwicklung neuer Geschäftsmöglichkeiten gehört. Unter seiner Leitung stand insbesondere das multidisziplinäre Team, das beide Generationen der Business-Class-Sitze „Club World" für British Airways gestaltet hat – das weltweit erste komplett flache Bett in einer Business Class, das der Airline seit seiner Markteinführung im Jahr 2000 enorme Umsatzzahlen beschert und die Branche nachhaltig verändert hat. Darüber hinaus ist Martin Darbyshire für das Ministerium für Handel und Investition des Vereinigten Königreichs Botschafter für den Bereich der Kreativindustrie, ein anerkannter Sprecher der Branche sowie Berater für Design und Innovation und er war Gremiumsmitglied des Icsid.

01 tangerine's design work for Snoozebox.com created a premium hotel guest experience within a very small space (3.6m x 2m x 2m).
tangerines Gestaltungsentwurf für Snoozebox.com kreierte ein hochwertiges Hotelgasterlebnis auf sehr begrenzten Raum (3,6 m x 2 m x 2 m).

02 **British Airways' Club World**
In the second generation of Club World, by angling the pair of seats at two degrees to the centre line of the aircraft and making the arms drop, the bed was made 25 per cent wider within the same footprint.
In der zweiten Club-World-Generation wurde das Sitzpaar in einem Winkel von 2 Grad zur Mittellinie des Flugzeugs positioniert; die Armlehnen lassen sich herunterklappen. Dadurch wurde das Bett innerhalb der gleichen Grundfläche um 25 Prozent verbreitert.

02

"My work philosophy is about challenging preconceptions and creating breakthrough change, improving lives and generating wealth."
„Meine Arbeitsphilosophie besteht darin, vorgefasste Meinungen zu hinterfragen und bahnbrechende Veränderungen zu erzielen, die das Leben verbessern und Wohlstand erzeugen."

The Red Dot Award has been uncovering the best designs for 60 years now. Which product innovations would you like to see in the next 60 years, and why?
I would like to see more applications of modern technologies and a greater focus on using modern technological advances to develop cheaper medical products, for instance creating low-cost portable equipment for developing countries.

What do you take special notice of when you are assessing a product as a jury member?
When assessing a new car entry for instance, I look at the interior first, taking great interest in the detailing and finishes before looking at the exterior. For me it is important that both have been given equal importance and attention by the design team.

What significance does winning an award in a design competition have for the designer?
It is a good feeling to know that your peers have judged you worthy of it, because a Red Dot carries such prestige within the design and commercial community.

Der Red Dot Award ermittelt seit 60 Jahren die besten Gestaltungen. Welche Produktinnovationen würden Sie sich für die nächsten 60 Jahre wünschen, und warum?
Ich würde gerne mehr Anwendungen moderner Technologien und ein größeres Augenmerk auf moderne technologische Fortschritte sehen, um günstigere Medizinprodukte zu entwickeln, wie zum Beispiel bei der Herstellung von kostengünstigen, tragbaren Geräten für Entwicklungsländer.

Worauf achten Sie besonders, wenn Sie ein Produkt als Juror bewerten?
Wenn ich etwa ein Auto bewerte, schaue ich zuerst auf die Innenausstattung, wobei mich die Details und die Verarbeitung besonders interessieren. Erst dann inspiziere ich das Äußere. Für mich ist es wichtig, dass beide die gleiche Aufmerksamkeit vom Designteam erhalten haben.

Welche Bedeutung hat die Auszeichnung in einem Designwettbewerb für den Designer?
Es ist ein gutes Gefühl zu wissen, dass dich deine Fachkollegen als dessen würdig erachten, da ein Red Dot ein so großes Ansehen in der Design- und Geschäftswelt genießt.

01

Robin Edman
Sweden
Schweden

Robin Edman has been the chief executive of SVID, the Swedish Industrial Design Foundation, since 2001. After studying industrial design at Rhode Island School of Design he joined AB Electrolux Global Design in 1981 and parallel to this started his own design consultancy. In 1989, Robin Edman joined Electrolux North America as vice president of Industrial Design for Frigidaire and in 1997, moved back to Stockholm as vice president of Electrolux Global Design. Throughout his entire career he has worked towards integrating a better understanding of users, their needs and the importance of design in society at large. His engagement in design related activities is reflected in the numerous international jury appointments, speaking engagements, advisory council and board positions he has held. Robin Edman served on the board of Icsid from 2003 to 2007, the last term as treasurer. Since June 2015, he is the president of BEDA (Bureau of European Design Associations).

Robin Edman ist seit 2001 Firmenchef der SVID, der Swedish Industrial Design Foundation. Nach seinem Industriedesign-Studium an der Rhode Island School of Design kam er 1981 zu AB Electrolux Global Design. Zeitgleich startete er seine eigene Unternehmensberatung für Design. 1989 wechselte Edman zu Electrolux North America als Vizepräsident für Industrial Design für Frigidaire und kehrte 1997 als Vizepräsident von Electrolux Global Design nach Stockholm zurück. Während seiner gesamten Karriere hat er daran gearbeitet, ein besseres Verständnis für Nutzer zu entwickeln, für deren Bedürfnisse und die Wichtigkeit von Design in der Gesellschaft insgesamt. Sein Engagement in designbezogenen Aktivitäten weist sich durch zahlreiche Jurierungsberufungen aus sowie durch Rednerverpflichtungen und Positionen in Gremien sowie Beratungsausschüssen. Von 2003 bis 2007 war Robin Edman Mitglied im Vorstand des Icsid, in der letzten Amtsperiode als Schatzmeister. Seit Juni 2015 ist er Präsident von BEDA (Bureau of European Design Associations).

02

"My work philosophy is to create a better place to live in for as many people as possible ... and to have fun while doing so."

„Meine Arbeitsphilosophie besteht darin, für so viele Menschen wie möglich einen besseren Lebensraum zu schaffen ... und dabei Spaß zu haben."

Which product area will Sweden have most success with in the next ten years?
Assuming the definition of "product" follows the now common view of both goods and services, I would like to stress the enormous potential in the public sector. The areas ranging from the development of the health care sector to regional and local innovation will include an array of design-focused competences that will transform our future society.

The Red Dot Award has been uncovering the best designs for 60 years now. Which product innovations would you like to see in the next 60 years?
A combination of solutions that will secure the continuation of humankind on earth and decrease our footprint on earth.

What significance does winning an award in a design competition have for the manufacturer?
It means recognition and is a great marketing tool that gives a seal of credibility and drives internal innovation.

In welchem Produktbereich wird Schweden in den nächsten zehn Jahren den größten Erfolg haben?
Voraussetzend, dass sich die Definition des Begriffs „Produkt" auf dessen zurzeit gängige Ansicht als Güter und Dienstleistungen bezieht, möchte ich das enorme Potenzial im öffentlichen Sektor hervorheben. Die Bereiche von der Entwicklung der Gesundheitsversorgung bis hin zu regionalen und lokalen Innovationen werden ein breites Spektrum designbezogener Kompetenzen beinhalten, das unsere Gesellschaft zukünftig verändern wird.

Der Red Dot Award ermittelt seit 60 Jahren die besten Gestaltungen. Welche Produktinnovationen würden Sie sich für die nächsten 60 Jahre wünschen?
Eine Kombination aus Lösungen, die das Fortbestehen der Menschheit auf der Erde sichern und unseren ökologischen Fußabdruck auf der Erde verringern.

Welche Bedeutung hat die Auszeichnung in einem Designwettbewerb für den Hersteller?
Sie bedeutet Anerkennung und ist ein großartiges Marketingwerkzeug, das ein Siegel der Glaubwürdigkeit darstellt und die interne Motivation antreibt.

01

02

Hans Ehrich
Sweden
Schweden

Hans Ehrich, born 1942 in Helsinki, Finland, has lived and been educated in Finland, Sweden, Germany, Switzerland, Spain and Italy. From 1962 to 1967, he studied metalwork and industrial design at the University College of Arts, Crafts and Design (Konstfackskolan), Stockholm. In 1965, his studies as a designer took him to Turin and Milan. With Tom Ahlström, he co-founded and became a director of A&E Design AB in 1968, a company which he still heads as director. From 1982 to 2002, he was managing director of Interdesign AB, Stockholm. Hans Ehrich has designed for, among others, Alessi, Anza, ASEA, Cederroth, Colgate, Fagerhults, Jordan, RFSU, Siemens, Turn-O-Matic and Yamagiwa. His work has been exhibited at many international exhibitions and collections and he has received numerous awards.

Hans Ehrich, 1942 in Helsinki, Finnland, geboren, lebte und lernte in Finnland, Schweden, Deutschland, der Schweiz, Spanien und Italien. Von 1962 bis 1967 studierte er Metall- und Industriedesign am University College of Arts, Crafts and Design (Konstfackskolan) in Stockholm. 1965 führten ihn Studien als Designer nach Turin und Mailand. 1968 war er zusammen mit Tom Ahlström Gründungsdirektor von A&E Design AB, Stockholm, das er heute immer noch als Direktor leitet, und von 1982 bis 2002 war er geschäftsführender Direktor von Interdesign AB, Stockholm. Hans Ehrich gestaltete u. a. für Alessi, Anza, ASEA, Cederroth, Colgate, Fagerhults, Jordan, RFSU, Siemens, Turn-O-Matic und Yamagiwa. Seine Arbeiten sind in zahlreichen internationalen Ausstellungen und Sammlungen vertreten und er wurde vielfach ausgezeichnet.

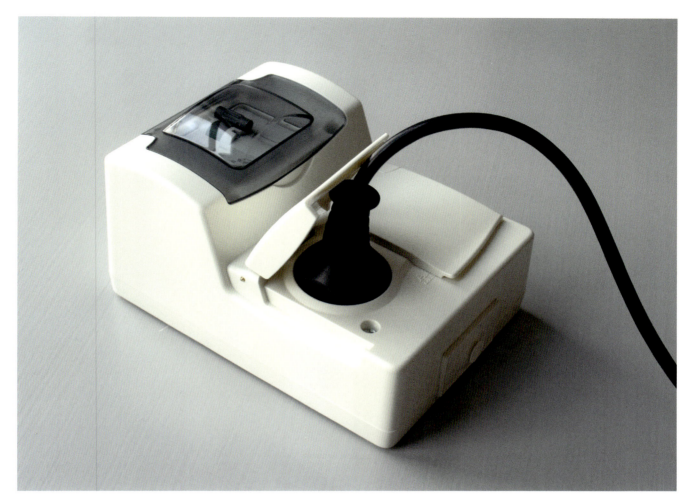

03

"My work philosophy has always been: create for tomorrow with a foothold on yesterday."
„Meine Arbeitsphilosophie war stets: Gestalte für morgen mit einem Halt in der Vergangenheit."

This year you have donated the complete collection from your 50 years of work to the Swedish National Museum. Which motto do you think would be most fitting for this collection?
"Handmade models and prototypes, approximately 400 objects, which represent half a century of successful Swedish product design."

Which of the projects in your lifetime are you particularly proud of?
I am particularly proud of the "Stockholm II" folding stool for museums, which is in use worldwide.

What impressed you most during the Red Dot judging process?
The high design standards, the good functionality and the appealing material quality of many of the submitted products.

What is the advantage of commissioning an external designer compared to having an in-house design team?
An external designer often has a broader spectrum of professional expertise with more influences than an in-house-designer.

Sie haben in diesem Jahr die gesamte Sammlung Ihrer 50 Berufsjahre dem schwedischen National-museum überlassen. Unter welches Motto möchten Sie diese Sammlung am liebsten stellen?
„Handgefertigte Modelle und Prototypen, etwa 400 Objekte, die ein halbes Jahrhundert erfolgreiches schwedisches Produktdesign repräsentieren."

Auf welches Projekt in Ihrem Leben sind Sie besonders stolz?
Auf den weltweit verbreiteten Museumsklapphocker „Stockholm II".

Was hat Sie bei der Red Dot-Jurierung am meisten beeindruckt?
Das hohe Gestaltungsniveau, die gute Funktionalität und die ansprechende Materialqualität einer Vielzahl der eingereichten Produkte.

Worin liegt der Vorteil, einen externen Designer zu beauftragen, im Vergleich zu einem Inhouse-Designteam?
Ein externer Designer verfügt meistens über eine breiter gefächerte Berufskompetenz mit mehr Einfalls-winkeln als ein Inhouse-Designer.

01

Joachim H. Faust
Germany
Deutschland

Joachim H. Faust, born in 1954, studied architecture at the Technical University of Berlin, the Technical University of Aachen, as well as at Texas A&M University (with Professor E. J. Romieniec), where he received his Master of Architecture in 1981. He worked as a concept designer in the design department of Skidmore, Owings & Merrill in Houston, Texas and as a project manager in the architectural firm Faust Consult GmbH in Mainz. From 1984 to 1986, he worked for KPF Kohn, Pedersen, Fox/Eggers Group in New York and as a project manager at the New York office of Skidmore, Owings & Merrill. In 1987, Joachim H. Faust took over the management of the HPP office in Frankfurt am Main. Since 1997, he has been managing partner of the HPP Hentrich-Petschnigg & Partner GmbH + Co. KG in Düsseldorf. He also writes articles and gives lectures on architecture and interior design.

Joachim H. Faust, 1954 geboren, studierte Architektur an der TU Berlin und der RWTH Aachen sowie – bei Professor E. J. Romieniec – an der Texas A&M University, wo er sein Studium 1981 mit dem Master of Architecture abschloss. Er war Entwurfsarchitekt im Design Department des Büros Skidmore, Owings & Merrill, Houston, Texas, sowie Projektleiter im Architekturbüro der Faust Consult GmbH in Mainz. Anschließend arbeitete er im Büro KPF Kohn, Pedersen, Fox/Eggers Group in New York und war Projektleiter im Büro Skidmore, Owings & Merrill in New York. 1987 übernahm Joachim H. Faust die Leitung des HPP-Büros in Frankfurt am Main und ist seit 1997 geschäftsführender Gesellschafter der HPP Hentrich-Petschnigg & Partner GmbH + Co. KG in Düsseldorf. Er ist zudem als Autor tätig und hält Vorträge zu Fachthemen der Architektur und Innenarchitektur.

02

"My work philosophy is:
to your own self be true."
„Meine Arbeitsphilosophie lautet:
Bleibe dir selbst treu."

What piece of advice has been useful in your youth or at the beginning of your professional career?
To face big things with composure, and small ones with close attention.

What will a single-family home look like in the year 2050?
Modular, prefabricated, interactive, adaptive, but hopefully also personal with human-centred standards and materials.

The Red Dot Award has been uncovering the best designs for 60 years now. Which product innovation would you like to see in the next 60 years, and why?
Intrinsic value is one of the highest goals in the design of a product or building. This means continuity in the best sense of a "classic".

What has impressed you most during the Red Dot judging process?
The wide range of products and the sensitivity for material conformity have increased every year. And with that my enthusiasm to take part.

Welcher Ratschlag hat Sie in Ihrer Jugend oder frühen Berufskarriere weitergebracht?
Großen Dingen begegnet man mit Gelassenheit, kleinen mit besonderer Aufmerksamkeit.

Wie wird ein Einfamilienhaus im Jahr 2050 aussehen?
Modular, vorfabriziert, interaktiv, adaptiv, aber hoffentlich auch privat mit menschlich angemessenem Maßstab und Material.

Der Red Dot Award ermittelt seit 60 Jahren die besten Gestaltungen. Welche Produktinnovation würden Sie sich für die nächsten 60 Jahre wünschen, und warum?
Werthaltigkeit ist eines der höchsten Ziele in der Gestaltung eines Produkts bzw. Bauwerks. Das bedeutet Kontinuität im besten Sinne des „Klassikers".

Was hat Sie bei der Red Dot-Jurierung am meisten beeindruckt?
Die Vielfalt der Produkte und die Sensibilität für Materialkonformität sind in jedem Jahr gestiegen. Damit natürlich auch meine Begeisterung, mitzuwirken.

01

Hideshi Hamaguchi
USA / Japan
USA / Japan

Hideshi Hamaguchi graduated with a Bachelor of Science in chemical engineering from Kyoto University. Starting his career with Panasonic in Japan, Hamaguchi later became director of the New Business Planning Group at Panasonic Electric Works, Ltd. and then executive vice president of Panasonic Electric Works Laboratory of America, Inc. In 1993, he developed Japan's first corporate Intranet and also led the concept development for the first USB flash drive. Hideshi Hamaguchi has over 15 years of experience in defining strategies and decision-making, as well as in concept development for various industries and businesses. As Executive Fellow at Ziba Design and CEO at monogoto, he is today considered a leading mind in creative concept and strategy development on both sides of the Pacific and is involved in almost every project this renowned business consultancy takes on. For clients such as FedEx, Polycom and M-System he has led the development of several award-winning products.

Hideshi Hamaguchi graduierte als Bachelor of Science in Chemical Engineering an der Kyoto University. Seine Karriere begann er bei Panasonic in Japan, wo er später zum Direktor der New Business Planning Group von Panasonic Electric Works, Ltd. und zum Executive Vice President von Panasonic Electric Works Laboratory of America, Inc. aufstieg. 1993 entwickelte er Japans erstes Firmen-Intranet und übernahm zudem die Leitung der Konzeptentwicklung des ersten USB-Laufwerks. Hideshi Hamaguchi verfügt über mehr als 15 Jahre Erfahrung in der Konzeptentwicklung sowie Strategie- und Entscheidungsfindung in unterschiedlichen Industrien und Unternehmen. Als Executive Fellow bei Ziba Design und CEO bei monogoto wird er heute als führender Kopf in der kreativen Konzept- und Strategieentwicklung auf beiden Seiten des Pazifiks angesehen und ist in nahezu jedes Projekt der renommierten Unternehmensberatung involviert. Für Kunden wie FedEx, Polycom und M-System leitete er etliche ausgezeichnete Projekte.

01 **Cintiq 24HD**
for Wacom, 2012
für Wacom, 2012

02 **FedEx World Service
Centre, 1999**
The concept that changed
the way FedEx understands
and treats its customers.
Das Konzept, das FedEx ein
besseres Verständnis für
seine Kunden und deren
Behandlung vermittelte.

02

"My work philosophy is:
all I need is less."

„Meine Arbeitsphilosophie lautet:
Alles, was ich brauche, ist weniger."

**What impressed you most during the Red Dot
judging process?**
This year still felt like the end of a transitional
period – a time to resolve some of the critical ten-
sions that have emerged out of the massive changes
of technology, consumer experience, and business
models in the past ten years.

**What challenges do you see for the future
in design?**
The challenge is finding the sweet spot between
what resonates with the consumer and what is true
to the brand.

**Do you see a correlation between the design
quality of a company's products and the economic
success of this company?**
I see a strong correlation between them. If a company
has a good design in all three phases of consumer
interaction – attract, engage and extend – it should
directly impact its success.

**Was hat Sie bei der Red Dot-Jurierung am meisten
beeindruckt?**
Dieses Jahr fühlte sich immer noch wie das Ende einer
Übergangzeit an – eine Zeit, um einen Teil der kriti-
schen Spannungen, die aus den massiven Veränderungen der Technologie, Konsumentenerfahrung und
Geschäftsmodelle während der letzten zehn Jahre
resultieren, aufzulösen.

**Welche zukünftigen Herausforderungen sehen
Sie im Designbereich?**
Die Herausforderung besteht darin, das richtige Verhältnis zwischen dem, was beim Konsumenten Anklang
findet, und dem, was der Wahrheit der Marke entspricht, zu finden.

**Sehen Sie einen Zusammenhang zwischen der
Designqualität, die sich in den Produkten eines
Unternehmens äußert, und dem wirtschaftlichen
Erfolg dieses Unternehmens?**
Ich sehe eine starke Korrelation zwischen beiden.
Wenn ein Unternehmen gute Gestaltung für alle drei
Phasen der Interaktion mit dem Konsumenten –
auffallen, einnehmen, ausbauen – bietet, dann sollte
sich das unmittelbar auf seinen Erfolg auswirken.

01

Prof. Renke He
China
China

Professor Renke He, born in 1958, studied civil engin-
eering and architecture at Hunan University in China.
From 1987 to 1988, he was a visiting scholar at the
Industrial Design Department of the Royal Danish
Academy of Fine Arts in Copenhagen and, from 1998
to 1999, at North Carolina State University's School of
Design. Renke He is dean and professor of the School
of Design at Hunan University and is also director of
the Chinese Industrial Design Education Committee.
Currently, he holds the position of vice chair of the
China Industrial Design Association.

Professor Renke He wurde 1958 geboren und studierte
an der Hunan University in China Bauingenieurwesen
und Architektur. Von 1987 bis 1988 war er als Gastprofessor für Industrial Design an der Royal Danish
Academy of Fine Arts in Kopenhagen tätig, und von
1998 bis 1999 hatte er eine Gastprofessur an der
School of Design der North Carolina State University
inne. Renke He ist Dekan und Professor an der Hunan
University, School of Design, sowie Direktor des
Chinese Industrial Design Education Committee. Er ist
derzeit zudem stellvertretender Vorsitzender der China
Industrial Design Association.

01/02 Fashion product designs inspired by the Dong nationality's traditional textile patterns, developed in the project New Channel Design & Social Innovation Summer Camp 2014, at the School of Design of Hunan University, China.
Modeprodukt-Designs inspiriert von den traditionellen Stoffmustern der Dong-Nationalität, entwickelt im Projekt „New Channel Design & Social Innovation Summer Camp 2014" an der School of Design der Hunan University, China.

02

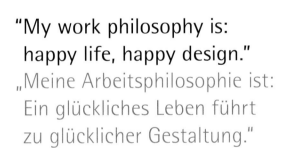

"My work philosophy is: happy life, happy design."
„Meine Arbeitsphilosophie ist: Ein glückliches Leben führt zu glücklicher Gestaltung."

Which country do you consider to be a pioneer in product design?
The USA, because companies like Apple combine technologies with service design, business model and interaction design in order to create brand new products for the global markets.

What are the main challenges in a designer's everyday life?
Finding a balance between business and social responsibility in design.

What significance does winning an award in a design competition have for the designer?
It is the ultimate recognition of a designer's professional skills and reputation in this competitive world.

What significance does winning an award in a design competition have for the manufacturer?
A design award is a wonderful ticket to success in the marketplace.

Welche Nation ist für Sie Vorreiter im Produktdesign?
Die USA. Unternehmen wie Apple vereinen Technologien mit Servicedesign, Geschäftsmodell und Interaction Design, um so brandneue Produkte für den globalen Markt zu entwerfen.

Was sind große Herausforderungen im Alltag eines Designers?
Das Gleichgewicht zwischen Geschäft und sozialer Verantwortung im Design zu finden.

Welche Bedeutung hat die Auszeichnung in einem Designwettbewerb für den Designer?
Sie ist die beste Bestätigung für die Fertigkeiten und den Ruf eines Designers in unserer wettbewerbsorientierten Welt.

Welche Bedeutung hat die Auszeichnung in einem Designwettbewerb für den Hersteller?
Eine Auszeichnung in einem Designwettbewerb ist eine wunderbare Fahrkarte zum Markterfolg.

01

Prof. Herman Hermsen
Netherlands
Niederlande

Professor Herman Hermsen, born in 1953 in Nijmegen, Netherlands, studied at the ArtEZ Institute of the Arts in Arnhem from 1974 to 1979. Following an assistant professorship, he began his career in teaching in 1985. Since 1979, he is an independent jewellery and product designer. Until 1990, he taught product design at the Utrecht School of the Arts (HKU), after which time he returned to Arnhem as lecturer at the Academy. Hermsen has been professor of product and jewellery design at the University of Applied Sciences in Düsseldorf since 1992. He gives guest lectures at universities and colleges throughout Europe, the United States and Japan, and began regularly organising specialist symposia in 1998. He has also served as juror for various competitions. Herman Hermsen has received numerous international awards for his work in product and jewellery design, which is shown worldwide in solo and group exhibitions and held in the collections of renowned museums, such as the Cooper-Hewitt Museum, New York; the Pinakothek der Moderne, Munich; and the Museum of Arts and Crafts, Kyoto.

Professor Herman Hermsen, 1953 in Nijmegen in den Niederlanden geboren, studierte von 1974 bis 1979 am ArtEZ Institute of the Arts in Arnheim und ging nach einer Assistenzzeit ab 1985 in die Lehre. Seit 1979 ist er unabhängiger Schmuck- und Produktdesigner. Bis 1990 unterrichtete er Produktdesign an der Utrecht School of the Arts (HKU) und kehrte anschließend nach Arnheim zurück, um an der dortigen Hochschule als Dozent zu arbeiten. Seit 1991 ist Hermsen Professor für Produkt- und Schmuckdesign an der Fachhochschule Düsseldorf; er hält Gastvorlesungen an Hochschulen in ganz Europa, den USA und Japan, organisiert seit 1988 regelmäßig Fachsymposien und ist Juror in verschiedenen Wettbewerbsgremien. Für seine Arbeiten im Produkt- und Schmuckdesign, die weltweit in Einzel- und Gruppenausstellungen präsentiert werden und sich in den Sammlungen großer renommierter Museen befinden – z. B. Cooper-Hewitt Museum, New York, Pinakothek der Moderne, München, und Museum of Arts and Crafts, Kyoto –, erhielt Herman Hermsen zahlreiche internationale Auszeichnungen.

02

"My work philosophy is: if 'less or more' is wanted, it should at least contribute to the poetry."

„Meine Arbeitsphilosophie ist: Wenn ‚weniger oder mehr' gewünscht wird, sollte es wenigstens zur Poesie beitragen."

What can laypeople learn, when they look at award-winning products in design museums?
They can learn that design is not a direct translation of emotions, but that a lot of thought has gone into which design language can aesthetically communicate the perception of the product's function, and how this perception has developed in different cultures and eras.

Which of the projects in your lifetime are you particularly proud of?
I have created designs which I believe provide a poetic answer; they communicate the essence of a concept, such as my lamp "Charis" for Classicon.

Which nation to you consider to be a pioneer in product design?
I have come to increasingly appreciate Scandinavian design, because the designers have developed a successful combination of design language, function, zeitgeist, innovation and high-quality manufacturing.

Was können Laien lernen, wenn sie in Design-museen hervorragend gestaltete Produkte betrachten?
Dass Gestaltung keine direkte Umsetzung von Emotionen ist, dass aber genau überlegt wurde, welche Formensprache die Sichtweise auf die Produktfunktion ästhetisch kommunizieren kann. Und wie sich diese Sichtweisen in den unterschiedlichen Kulturen und Epochen entwickelten.

Auf welches Projekt in Ihrem Leben sind Sie besonders stolz?
Ich habe natürlich Entwürfe, von denen ich denke, mit der gestalterischen Umsetzung eine poetische Antwort gefunden zu haben, die die Essenz des Konzeptes kommuniziert; z. B. meine Lampe „Charis" für Classicon.

Welche Nation ist für Sie Vorreiter im Produktdesign?
Immer mehr schätze ich das skandinavische Design, denn die Designer entwickelten ein gelungenes Zusammenspiel aus Formensprache, Funktion, Zeitgeist, Innovation und qualitativ sehr guter Herstellung.

01

Prof. Carlos Hinrichsen
Chile
Chile

Professor Carlos Hinrichsen, born in 1957, graduated as an industrial designer in Chile in 1982 and earned his master's degree in engineering in Japan in 1991. Currently, he is dean of the Faculty of Engineering and Business at the Gabriela Mistral University in Santiago. From 2007 to 2009 he was president of Icsid and currently serves as senator within the organisation. He has since been heading research projects in the areas of innovation, design and education, and in 2010 was honoured with the distinction "Commander of the Order of the Lion of Finland". From 1992 to 2010 he was director of the School of Design Duoc UC, Chile and from 2011 to 2014 director of the Duoc UC International Affairs. He has led initiatives that integrate trade, engineering, design, innovation, management and technology in Asia, Africa and Europe and is currently design director for the Latin American Region of Design Innovation. Since 2002, Carlos Hinrichsen has been an honorary member of the Chilean Association of Design. Furthermore, he has been giving lectures at various conferences and universities around the world.

Professor Carlos Hinrichsen, 1957 geboren, erlangte 1982 seinen Abschluss in Industriedesign in Chile und erhielt 1991 seinen Master der Ingenieurwissenschaft in Japan. Aktuell ist er Dekan der Fakultät „Engineering and Business" an der Universität Gabriela Mistral in Santiago. Von 2007 bis 2009 war er Icsid-Präsident und ist heute Senator innerhalb der Organisation. Seither leitet er Forschungsprojekte in den Bereichen Innovation, Design sowie Erziehung und wurde 2010 mit der Auszeichnung „Commander of the Order of the Lion of Finland" geehrt. Von 1992 bis 2010 war er Direktor der School of Design Duoc UC in Chile und von 2011 bis 2014 Direktor der Duoc UC International Affairs. In Asien, Afrika und Europa leitete Carlos Hinrichsen Initiativen, die Handel, Ingenieurwesen, Design, Innovation, Management und Technologie integrieren, und ist aktuell Designdirektor der Latin American Region of Design Innovation. Seit 2002 ist er Ehrenmitglied der Chilean Association of Design. Außerdem hält er Vorträge auf Konferenzen und in Hochschulen weltweit.

02

"My work philosophy is well
described by the following quote:
'If you can imagine it, you can
create it. If you can dream it, you
can become it.'"
„Meine Arbeitsphilosophie lässt sich
gut mit folgendem Zitat beschreiben:
‚Wenn du es dir vorstellen kannst,
kannst du es kreieren. Wenn du
davon träumen kannst, kannst du
es werden.'"

When did you first think consciously about design?
I realised as a child that design contributes to human
happiness, and I have seen this insight confirmed over
the years.

**What impressed you most during the Red Dot
judging process?**
This time I saw how products offer realisations of the
desires and dreams of the users, as well as those of
the designers and producers. In product categories,
design quality and innovation are playing a key role
in turning technological innovations into good and
useful solutions. For this the Red Dot Award is like
a mirror, always reflecting what is going on in the
current design industry and market. It reveals the
prevalent trends and enables us to foresee other
potential trends, all of which have opened an unpre-
cedented field of knowledge and expectations.

**Wann haben Sie das erste Mal bewusst über
Design nachgedacht?**
Als Kind habe ich erkannt, dass Design dazu beiträgt,
dass sich Menschen glücklich fühlen. Und diese
Erkenntnis hat sich über die Jahre hinweg bestätigt.

**Was hat Sie bei der Red Dot-Jurierung am meisten
beeindruckt?**
Dieses Mal habe ich gesehen, wie Produkte Umset-
zungen der Wünsche und Träume sowohl der
Nutzer als auch der Designer und Hersteller bieten.
In den Produktkategorien spielen Designqualität
und -innovation eine zentrale Rolle dabei, technische
Neuerungen in gute und nützliche Lösungen zu überführen. Daher ist der Red Dot Award wie ein Spiegel,
der wiedergibt, was momentan in der Designbranche
und auf dem Markt passiert. Er offenbart die vor-
herrschenden Trends und ermöglicht uns, andere po-
tenzielle Trends vorherzusehen, die alle zusammen
einen neuartigen Bereich an Wissen und Erwartungen
eröffnet haben.

01

Tapani Hyvönen
Finland
Finnland

Tapani Hyvönen graduated in 1974 as an industrial designer from the present Aalto University School of Arts, Design and Architecture. He founded the design agency "Destem Ltd." in 1976 and was co-founder of ED-Design Ltd. in 1990, one of Scandinavia's largest design agencies. He has served as CEO and president of both agencies until 2013. Since then, he has been a visiting professor at, among others, Guangdong University of Technology in Guangzhou and Donghua University in Shanghai, China. His many award-winning designs are part of the collections of the Design Museum Helsinki and the Cooper-Hewitt Museum, New York. Tapani Hyvönen was an advisory board member of the Design Leadership Programme at the University of Art and Design Helsinki from 1989 to 2000, and a board member of the Design Forum Finland from 1998 to 2002, as well as the Icsid from 1999 to 2003 and again from 2009 to 2013. He has been a jury member in many international design competitions, a member of the Finnish-Swedish Design Academy since 2003 and a board member of the Finnish Design Museum since 2011.

Tapani Hyvönen graduierte 1974 an der heutigen Aalto University School of Arts, Design and Architecture zum Industriedesigner. 1976 gründete er die Design-agentur „Destem Ltd." und war 1990 Mitbegründer der ED-Design Ltd., einer der größten Designagenturen Skandinaviens, die er beide bis 2013 als CEO und Präsident leitete. Seitdem lehrt er als Gastprofessor u. a. an der Guangdong University of Technology in Guangzhou und der Donghua University in Shanghai, China. Seine vielfach ausgezeichneten Arbeiten sind in den Sammlungen des Design Museum Helsinki und des Cooper-Hewitt Museum, New York, vertreten. Tapani Hyvönen war von 1989 bis 2000 in der Bera-tungskommission des Design Leadership Programme der University of Art and Design Helsinki, von 1998 bis 2002 Vorstandsmitglied des Design Forum Finland sowie von 1999 bis 2003 und von 2009 bis 2013 des Icsid. Er ist international als Juror tätig, seit 2003 Mitglied der Finnish-Swedish Design Academy und seit 2011 Vorstandsmitglied des Finnish Design Museum.

02

"My work philosophy is to be open to new ideas."

„Meine Arbeitsphilosophie ist es, für neue Ideen offen zu sein."

What advice did you find helpful in your younger years or in the early days of your career?
Be curious, look and study different aspects seriously and don't fall in love with the first idea that comes.

Are there any designers who are role models for you?
A designer who puts good design, function, aesthetics and structure in balance. Alvar Aalto's beautiful, simple and functional designs are something I appreciate.

The Red Dot Award has been uncovering the best designs for 60 years now. Which product innovation would you like to see in the next 60 years?
Technology will become invisible. We will have products similar to the ones from now but they will feature a new kind of intelligence.

What significance does winning an award in a design competition have for the manufacturer?
It challenges the company to invest in design.

Welcher Ratschlag hat Sie in Ihrer Jugend oder frühen Berufskarriere weitergebracht?
Neugierig zu sein, sich ernsthaft verschiedene Aspekte anzuschauen und zu studieren und sich nicht in die erstbeste Idee, die daherkommt, zu verlieben.

Gibt es Designer, die Ihnen als Vorbilder dienen?
Designer, die gute Gestaltung, Funktion, Ästhetik und Konstruktion in Einklang bringen. Alvar Aaltos schöne, schlichte und funktionale Gestaltungen schätze ich durchaus.

Der Red Dot Award ermittelt seit 60 Jahren die besten Gestaltungen. Welche Produktinnovation würden Sie sich für die nächsten 60 Jahre wünschen?
Technologie wird unsichtbar werden. Wir werden ähnliche Produkte wie die heutigen haben, aber sie werden eine neue Art von Intelligenz aufweisen.

Welche Bedeutung hat die Auszeichnung in einem Designwettbewerb für den Hersteller?
Es fordert das Unternehmen dazu heraus, in Gestaltung zu investieren.

01

Guto Indio da Costa
Brazil
Brasilien

Guto Indio da Costa, born in 1969 in Rio de Janeiro, studied product design and graduated from the Art Center College of Design in Switzerland in 1993. He is design director of Indio da Costa A.U.D.T., a consultancy based in Rio de Janeiro, which develops architectural, urban planning, design and transportation projects. It works with a multidisciplinary strategic-creative group of designers, architects and urban planners, supported by a variety of other specialists. Guto Indio da Costa is a member of the Design Council of the State of Rio de Janeiro, former Vice President of the Brazilian Design Association (Abedesign) and founder of CBDI (Brazilian Industrial Design Council). He has been active as a lecturer and a contributing writer to different design magazines and has been a jury member of many design competitions in Brazil and abroad.

Guto Indio da Costa, geboren 1969 in Rio de Janeiro, studierte Produktdesign und machte 1993 seinen Abschluss am Art Center College of Design in der Schweiz. Er ist Gestaltungsdirektor von Indio da Costa A.U.D.T., einem in Rio de Janeiro ansässigen Beratungsunternehmen, das Projekte in Architektur, Stadtplanung, Design- und Transportwesen entwickelt und mit einem multidisziplinären, strategisch-kreativen Team aus Designern, Architekten und Stadtplanern sowie mit der Unterstützung weiterer Spezialisten operiert. Guto Indio da Costa ist Mitglied des Design Councils des Bundesstaates Rio de Janeiro, ehemaliger Vize-Präsident der brasilianischen Designvereinigung (Abedesign) und Gründer des CBDI (Industrial Design Council Brasilien). Er ist als Lehrbeauftragter aktiv, schreibt für verschiedene Designmagazine und ist als Jurymitglied zahlreicher Designwettbewerbe in und außerhalb Brasiliens tätig.

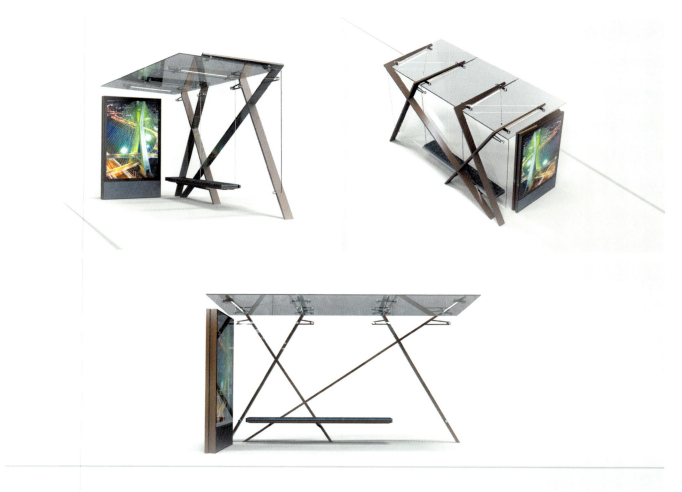

02

"My work philosophy focuses
on how to please and cleverly
surprise the user not only
through aesthetics but also
through functional and technical
innovations."
„Meine Arbeitsphilosophie konzentriert sich darauf, wie man Benutzer
erfreuen und geschickt überraschen
kann, nicht nur durch Ästhetik,
sondern auch durch funktionelle
und technische Innovationen."

Which product design area will Brazil have most success with in the next ten years?
Considering the world's urgent need for a more eco-friendly industrial production and that Brazil's vast and diversified natural resources could lead to the research and development of eco-friendly materials, Brazilian design has the opportunity to play a leading role in this new eco-production revolution.

What are your sources of inspiration?
People. Observing the way people behave, the way people work, live or enjoy life.

The Red Dot Award has been uncovering the best designs for 60 years now. Which product innovations would you like to see in the next 60 years, and why?
I would love to see innovations that lead to zero-footprint production, where waste is easily transformed into resources. Designers can play a leading role in this transformation.

In welchen Produktbereichen wird Brasilien in den nächsten zehn Jahren den größten Erfolg haben?
In Anbetracht dessen, dass die Welt ein dringendes Bedürfnis nach einer umweltfreundlicheren industriellen Herstellung hat und dass Brasiliens riesige Vorkommen an verschiedenen Rohstoffen zur Erforschung und Entwicklung umweltfreundlicher Materialien dienen können, hat brasilianisches Design die Gelegenheit, eine führende Rolle in dieser Revolution hin zu einer neuen, umweltfreundlichen Produktion zu spielen.

Was sind Ihre Inspirationsquellen?
Menschen. Zu beobachten, wie sie sich verhalten, wie sie arbeiten, leben oder das Leben genießen.

Der Red Dot Award ermittelt seit 60 Jahren die besten Gestaltungen. Welche Produktinnovationen würden Sie sich für die nächsten 60 Jahre wünschen, und warum?
Innovationen hin zu einer umweltneutralen Produktion, in der Abfälle einfach wieder in Rohstoffe umgewandelt werden. Designer können bei dieser Umwandlung eine führende Rolle spielen.

01

Prof. Cheng-Neng Kuan
Taiwan

Taiwan

In 1980, Professor Cheng-Neng Kuan earned a Master's degree in Industrial Design (MID) from the Pratt Institute in New York. He is currently a full professor and the vice president of Shih-Chien University, Taipei, Taiwan. With the aim of developing a more advanced design curriculum in Taiwan, he founded the Department of Industrial Design, in 1992. He served as department chair until 1999. Moreover, Professor Kuan founded the School of Design in 1997 and had served as the dean from 1997 to 2004 and as the founding director of the Graduate Institute of Industrial Design from 1998 to 2007. He had also held the position of the 16th chairman of the Board of China Industrial Designers Association (CIDA), Taiwan. His fields of expertise include design strategy and management as well as design theory and creation. Having published various books on design and over 180 research papers and articles, he is an active member of design juries in his home country and internationally. He is a consultant to major enterprises on product development and design strategy.

1980 erwarb Professor Cheng-Neng Kuan einen Master-Abschluss in Industriedesign (MID) am Pratt Institute in New York. Derzeit ist er ordentlicher Professor und Vizepräsident der Shih-Chien University in Taipeh, Taiwan. 1992 gründete er mit dem Ziel, einen erweiterten Designlehrplan zu entwickeln, das Department of Industrial Design in Taiwan. Bis 1999 war Professor Kuan Vorsitzender des Instituts. Darüber hinaus gründete er 1997 die School of Design, deren Dekan er von 1997 bis 2004 war. Von 1998 bis 2007 war er Gründungsdirektor des Graduate Institute of Industrial Design. Zudem war er der 16. Vorstandsvorsitzende der China Industrial Designers Association (CIDA) in Taiwan. Seine Fachgebiete umfassen Designstrategie, -management, -theorie und -kreation. Neben der Veröffentlichung verschiedener Bücher über Design und mehr als 180 Forschungsarbeiten und Artikel ist er aktives Mitglied von Designjurys in seiner Heimat sowie auf internationaler Ebene. Zudem ist er als Berater für Großunternehmen im Bereich Produktentwicklung und Designstrategie tätig.

02

"My work philosophy is to bridge the known and unknown, familiarity and strangeness and to make new and good things happen continuously."

„Meine Arbeitsphilosophie besteht darin, eine Brücke zwischen Bekanntem und Unbekanntem, Familiärem und Fremdem zu schlagen und ständig neue und gute Dinge zu verwirklichen."

What challenges will designers face in the future?
In response to the interaction of technology (Internet, IoT, Big Data, 3D printing), entrepreneurship (crowd-funding) and micro lifestyles (cross-culture, cross-age, cross-region), designers will not only have to discover new ways of thinking and working, but also face creative challenges from individuals without a design background.

What do you take special notice of when you are assessing a product as a jury member?
As a language, design has to integrate all criteria; however, what concerns me most in terms of the degree of integration is the expressive uniqueness and the exquisite qualities of a product.

What is the advantage of commissioning an external designer compared to having an in-house design team?
It can broaden the vision for the design with regards to creativity and through the differentiation of designs improves the discovery and interpretation of a brand's DNA from different perspectives.

Was sind Herausforderungen für Designer in der Zukunft?
Als Reaktion auf die Interaktion zwischen Technologie (Internet, Internet der Dinge, Big Data, 3D-Drucken), Unternehmergeist (Crowdfunding) und Mikro-Lebensstilen (ku tur-, alters- und regionenübergreifend) werden Designer nicht nur neue Arten des Denkens und Arbeitens entdecken müssen, sondern sich auch kreativen Herausforderungen von Individuen ohne Gestaltungshintergrund stellen müssen.

Worauf achten Sie besonders, wenn Sie ein Produkt als Juror bewerten?
Als eine Sprache muss Design alle Kriterien integrieren. Was mir jedoch am wichtigsten bei dem Grad der Integration ist, sind die Einzigartigkeit des Ausdrucks und die hervorragenden Qualitäten eines Produktes.

Worin liegt der Vorteil, einen externen Designer zu beauftragen, im Vergleich zu einem Inhouse-Designteam?
Es kann die Vision des Designs kreativ erweitern und verbessert durch Differenzierung der Entwürfe die Entdeckung und Interpretation der Marken-DNA aus verschiedenen Perspektiven.

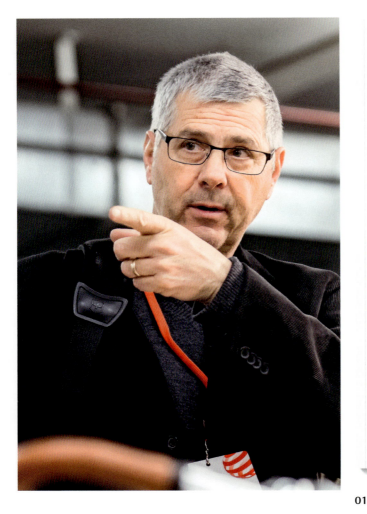

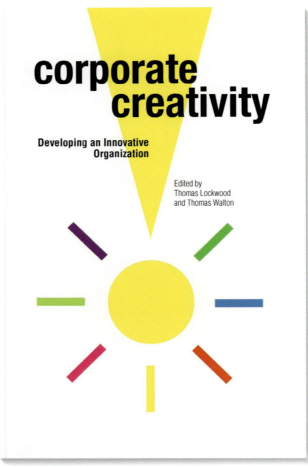

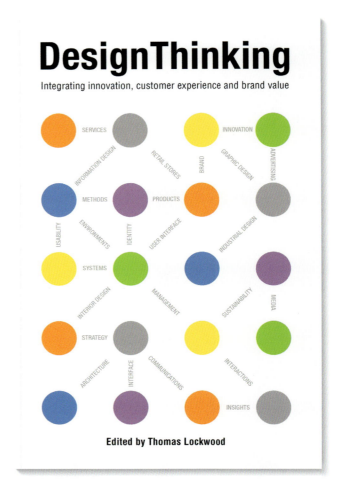

01

02

Dr Thomas Lockwood
USA
USA

Thomas Lockwood is the author of books on design thinking, design management, corporate creativity and design strategy. He has a PhD, works in design management and is regarded as a pioneer in the integration of design and innovation practices with business as we l as for founding international design and user-experience organisations. He was guest professor at Pratt Institute in New York City, is a Fellow at the Royal Society of the Arts in London and a frequent design award judge. Lockwood is a founding partner of Lockwood Resource, an international recruiting firm specialising in design leadership recruiting. From 2005 to 2011, he was president of DMI, the Design Management Institute, a non-profit research association in Boston. From 1995 to 2005, he was design director at Sun Microsystems and StorageTek, and creative director at several design and advertising firms for a number of years. In addition, he manages a blog about design leadership at lockwoodresource.com.

Thomas Lockwood ist Autor von Büchern zu den Themen Design Thinking, Designmanagement, Corporate Creativity und Designstrategie und als Doktor der Philosophie im Designmanagement tätig. Er gilt als Vordenker für die Integration der Design- und Innovationspraxis ins Geschäftsleben sowie für die Gründung internationaler Design- und User-Experience-Organisationen. Er war Gastprofessor am Pratt Institute in New York City, ist Mitglied der Royal Society of the Arts in London und regelmäßig Juror bei Designwettbewerben. Lockwood ist Gründerpartner von Lockwood Resource, einer internationalen Rekrutierungsfirma, die sich auf die Anwerbung von Führungspersonal für den Designbereich spezialisiert hat. Von 2005 bis 2011 war er Präsident des DMI (Design Management Institute), einer gemeinnützigen Forschungsorganisation in Boston. Von 1995 bis 2005 war er Designdirektor bei Sun Microsystems und StorageTek und über mehrere Jahre hinweg Kreativdirektor in verschiedenen Design- und Werbefirmen. Darüber hinaus führt er den Blog lockwoodresource.com über Mitarbeiterführung im Designsektor.

01 Corporate Creativity (2008) explores
    how to develop more creativity and
    innovation power at the individual,
    team and organisational levels.
    Corporate Creativity (2008) untersucht,
    wie sich Kreativität und Innovationskraft
    auf individueller, Team- und Organisati-
    onsebene steigern lassen.

02 Design Thinking (2010) presents best
    practice cases of design thinking
    methods applied to innovation, brand
    building, service design and customer
    experience.
    Design Thinking (2010) präsentiert
    Best-Practice-Fallstudien gestalterischer
    Denkansätze für Innovation, Marken-
    bildung, Service-Design und Kunden-
    erfahrung.

"My work philosophy is to learn, grow, and change. When we learn, we grow. When we grow, we change. I choose to embrace this process."

„Meine Arbeitsphilosophie ist zu lernen, zu wachsen und zu verän-dern. Wenn wir lernen, wachsen wir. Wenn wir wachsen, verändern wir uns. Ich habe entschieden, mir diesen Prozess zu eigen zu machen."

In your opinion, which decade brought the greatest progress in the design sector?
I think the years from 2006 to 2011, the Great Recession, were fabulous for design. Because many CEOs realised that recovery would not just mean going back to business as usual. These CEOs gradually realised that design is the connection between their innovation ideas and new customers.

The Red Dot Award has been uncovering the best designs for 60 years now. Which product innovations would you like to see in the next 60 years?
I would love to see innovations in food distribution, in energy development, in health, and in equality of wealth.

What impressed you most during the Red Dot judging process?
I love the process: to review all the entries by myself first, and then to review them again with the team of judges for discussions, which are very insightful.

Welches Jahrzehnt hat Ihrer Meinung nach den größten Fortschritt im Designbereich gebracht?
Ich glaube, die Jahre von 2006 bis 2011, die große Rezession, waren fabelhaft für Design. Vielen CEOs wurde bewusst, dass ein Aufschwung nicht einfach bedeuten würde, dass alles wie gehabt weitergeht. Diese CEOs haben nach und nach gemerkt, dass Design die Verbindung zwischen ihren Innovationsideen und neuen Kunden darstellt.

Der Red Dot Award ermittelt seit 60 Jahren die besten Gestaltungen. Welche Produktinnovationen würden Sie sich für die nächsten 60 Jahre wünschen?
Ich würde gerne mehr Innovationen in der Nahrungs-mittelverteilung, der Energieentwicklung, im Gesund-heitsbereich und im Angleichen des Wohlstands sehen.

Was hat Sie bei der Red Dot-Jurierung am meisten beeindruckt?
Ich liebe den Prozess: alle Einreichungen zuerst allein zu begutachten, um sie dann erneut zusammen mit dem Jurorenteam anzuschauen und sehr erkenntnis-reiche Diskussionen zu führen.

01

Lam Leslie Lu
Hong Kong
Hongkong

Lam Leslie Lu received a Master of Architecture from Yale University in Connecticut, USA in 1977, and was the recipient of the Monbusho Scholarship of the Japanese Ministry of Culture in 1983, where he conducted research in design and urban theory in Tokyo. He is currently the principal of the Hong Kong Design Institute and academic director of the Hong Kong Institute of Vocational Education. Prior to this, he was head of the Department of Architecture at the University of Hong Kong. Lam Leslie Lu has worked with, among others, Cesar Pelli and Associates, Hardy Holzman Pfeiffer and Associates, Kohn Pedersen Fox and Associates and Shinohara Kazuo on the design of the Centennial Hall of the Institute for Technology Tokyo. Moreover, he was visiting professor at Yale University and the Delft University of Technology as well as assistant lecturer for the Eero Saarinen Chair at Yale University. He also lectured and served as design critic at major international universities such as Columbia, Cambridge, Delft, Princeton, Yale, Shenzhen, Torgji, Tsinghua and the Chinese University Hong Kong.

Lam Leslie Lu erwarb 1977 einen Master of Architecture an der Yale University in Connecticut, USA, und war 1983 Monbusho-Stipendiat des japanischen Kulturministeriums, an dem er die Forschung in Design und Stadttheorie in Tokio leitete. Derzeit ist er Direktor des Hong Kong Design Institutes und akademischer Direktor des Hong Kong Institute of Vocational Education. Zuvor war er Leiter des Architektur-Instituts an der Universität Hongkong. Lam Leslie Lu hat u. a. mit Cesar Pelli and Associates, Hardy Holzman Pfeiffer and Associates, Kohn Pedersen Fox and Associates und Shinohara Kazuo am Design der Centennial Hall des Instituts für Technologie Tokio zusammengearbeitet, war Gastprofessor an der Yale University und der Technischen Universität Delft sowie Assistenz-Dozent für den Eero-Saarinen-Lehrstuhl in Yale. Er hielt zudem Vorträge und war Designkritiker an großen internationalen Universitäten wie Columbia, Cambridge, Delft, Princeton, Yale, Shenzhen, Tongji, Tsinghua und der chinesischen Universität Hongkong.

01 Centennial Hall of
the Tokyo Institute of
Technology
Die Centennial Hall
des Instituts für
Technologie Tokio

"My work philosophy is to advance knowledge and talent through calculated strategies and research while drawing in fresh thinking."
„Meine Arbeitsphilosophie besteht darin, Wissen und Talent durch wohl kalkulierte Strategien und Forschung voranzutreiben und gleichzeitig frische Denkansätze einzubeziehen."

Which of the projects in your lifetime are you particularly proud of?
I spent a period of time working with Shinohara Kazuo on the Centennial Hall building. The depth of thought and the level of experimentation in space, form and structure was the furthest I ever got in the design of a building. The interactive experience of designing as a duo set the standard for all my designs to follow.

What significance does winning an award in a design competition have for the designer?
The need to be recognised by one's peers is paramount and maybe more important than public acclaim. We need it not for our ego but for our soul.

What is the advantage of commissioning an external designer compared to having an in-house design team?
Change and new conceptions can only come about with the ebb and flow of talent, either through internal deployment or external injection.

Auf welches Ihrer bisherigen Projekte sind Sie besonders stolz?
Ich habe eine Zeit lang mit Shinohara Kazuo am Centennial-Hall-Gebäude zusammengearbeitet. Die Tiefe der Gedanken und das Niveau der Experimente in Bezug auf Raum, Form und Struktur waren die extremste Arbeit, die ich je bei der Gestaltung eines Gebäudes geleistet habe. Die interaktive Erfahrung des Gestaltens zu zweit hat Maßstäbe für alle meine nachfolgenden Entwürfe gesetzt.

Welche Bedeutung hat die Auszeichnung in einem Designwettbewerb für den Designer?
Die Notwendigkeit, von seinen Partnern anerkannt zu werden, ist entscheidend und womöglich wichtiger als öffentliche Anerkennung. Wir brauchen sie nicht für unser Ego, sondern für unsere Seele.

Worin liegt der Vorteil, einen externen Designer zu beauftragen, im Vergleich zu einem Inhouse-Designteam?
Veränderung und neue Ideen können nur mit der Ebbe und Flut an Talenten entstehen, entweder durch internen Personaleinsatz oder durch Zuführung von außen.

01

Wolfgang K. Meyer-Hayoz
Switzerland
Schweiz

Wolfgang K. Meyer-Hayoz studied mechanical engineering, visual communication and industrial design and graduated from the Stuttgart State Academy of Art and Design. The professors Klaus Lehmann, Kurt Weidemann and Max Bense have had formative influence on his design philosophy. In 1985, he founded the Meyer-Hayoz Design Engineering Group with offices in Winterthur/Switzerland and Constance/Germany. The design studio offers consultancy services for national as well as international companies in the five areas of design competence: design strategy, industrial design, user interface design, temporary architecture and communication design, and has received numerous international awards. From 1987 to 1993, Wolfgang K. Meyer-Hayoz was president of the Swiss Design Association (SDA). He is a member of the Association of German Industrial Designers (VDID), Swiss Marketing and the Swiss Management Society (SMG). Wolfgang K. Meyer-Hayoz also serves as juror on international design panels and supervises Change Management and Turnaround projects in the field of design strategy.

Wolfgang K. Meyer-Hayoz absolvierte Studien in Maschinenbau, Visueller Kommunikation sowie Industrial Design mit Abschluss an der Staatlichen Akademie der Bildenden Künste in Stuttgart. Seine Gestaltungsphilosophie prägten die Professoren Klaus Lehmann, Kurt Weidemann und Max Bense. 1985 gründete er die Meyer-Hayoz Design Engineering Group mit Büros in Winterthur/Schweiz und Konstanz/Deutschland. Das Designstudio bietet Beratungsdienste für nationale wie internationale Unternehmen in den fünf Designkompetenzen Designstrategie, Industrial Design, User Interface Design, temporäre Architektur und Kommunikationsdesign und wurde bereits vielfach ausgezeichnet. Von 1987 bis 1993 war Wolfgang K. Meyer-Hayoz Präsident der Swiss Design Association (SDA); er ist Mitglied im Verband Deutscher Industrie Designer (VDID), von Swiss Marketing und der Schweizerischen Management Gesellschaft (SMG). Wolfgang K. Meyer-Hayoz engagiert sich auch als Juror internationaler Designgremien und moderiert Change-Management- und Turnaround-Projekte im designstrategischen Bereich.

02

"My work philosophy is to see
technological innovations and
societal changes as opportunities
for the development of better
products."

„Meine Arbeitsphilosophie lautet,
die technologischen Innovationen
und gesellschaftlichen Veränderungen als Chancen für die Entwicklung
besserer Produkte zu begreifen."

**You have been a Red Dot juror for many years.
What significance does being appointed to the jury
have for you?**
I see it is an expression of the appreciation for my
work as a designer, and I appreciate the exchange
with international colleagues from all areas of design.

**What advice has been helpful in your youth or at
the beginning of your professional career?**
The advice to always be open to new developments,
be they technical or societal, and to fathom what
they mean for new design approaches.

**What do you take special notice of when you are
assessing a product as a juror?**
In the field of medical technology products in
particular, the aspects of safety, clarity of function,
manufacturing processes, hygienic properties and
stability of value are important criteria.

**What significance does winning an award in
a design competition have for the designer?**
A competition gives designers the opportunity to
compete internationally and thus get to know their
skills better.

**Sie sind langjähriger Red Dot-Juror. Welchen Wert
hat eine Berufung in diese Jury für Sie?**
Sie ist für mich Ausdruck der Wertschätzung meiner
gestalterischen Arbeit, und ich selbst schätze den
Austausch mit den internationalen Kollegen aus allen
Sparten des Designs.

**Welche Ratschläge haben Sie in Ihrer Jugend oder
frühen Berufskarriere weitergebracht?**
Immer offen zu sein für neue Entwicklungen, seien es
technische oder gesellschaftliche, und auszuloten, was
dies für neue Designansätze bedeutet.

**Worauf achten Sie besonders, wenn Sie ein
Produkt als Juror bewerten?**
Speziell in den Produktsortimenten der Medizintechnik
sind für mich die Aspekte Sicherheit, Sinnfälligkeit,
Herstellungsprozesse, Hygieneeigenschaften und
Wertbeständigkeit wichtige Kriterien.

**Welche Bedeutung hat die Auszeichnung in einem
Designwettbewerb für den Designer?**
Ein Wettbewerb gibt Designern die Gelegenheit, sich
international zu messen und so ihre Qualifikation
besser kennenzulernen.

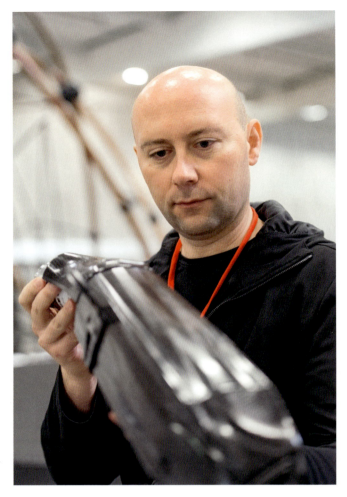

01

Prof. Jure Miklavc
Slovenia
Slowenien

Professor Jure Miklavc graduated in industrial design from the Academy of Fine Arts in Ljubljana, Slovenia, and has nearly 20 years of experience in the field of design. Miklavc started his career working as a freelance designer, before founding his own design consultancy, Studio Miklavc. Studio Miklavc works in the fields of product design, visual communications and brand development and is a consultancy for a variety of clients from the industries of light design, electronic goods, user interfaces, transport design and medical equipment. Sports equipment designed by the studio has gained worldwide recognition. From 2013 onwards, the team has been working for the prestigious Italian motorbike manufacturer Bimota. Designs by Studio Miklavc have received many international awards and have been displayed in numerous exhibitions. Jure Miklavc has been involved in design education since 2005 and is currently a lecturer and head of industrial design at the Academy of Fine Arts and Design in Ljubljana.

Professor Jure Miklavc machte seinen Abschluss in Industrial Design an der Academy of Fine Arts and Design in Ljubljana, Slowenien, und verfügt über nahezu 20 Jahre Erfahrung im Designbereich. Er arbeitete zunächst als freiberuflicher Designer, bevor er sein eigenes Design-Beratungsunternehmen „Studio Miklavc" gründete. Studio Miklavc ist in den Bereichen Produktdesign, visuelle Kommunikation und Markenentwicklung sowie in der Beratung zahlreicher Kunden der Branchen Lichtdesign, elektronische Güter, Benutzeroberflächen, Transport-Design und medizinisches Equipment tätig. Die von dem Studio gestalteten Sportausrüstungen erfahren weltweit Anerkennung. Seit 2013 arbeitet das Team für den angesehenen italienischen Motorradhersteller Bimota. Studio Miklavc erhielt bereits zahlreiche Auszeichnungen sowie Präsentationen in Ausstellungen. Seit 2005 ist Jure Miklavc in der Designlehre tätig und aktuell Dozent und Head of Industrial Design an der Academy of Fine Arts and Design in Ljubljana.

01 **Alpina ESK Pro**
Racing boots for cross-country skiing with a third of global market share and countless podium finishes for top athletes at the highest level of racing.
Racingschuhe für den Skilanglauf, die ein Drittel des Weltmarktes beherrschen und unzählige Podiumsgewinne für Topathleten in der höchsten Racingkategorie vorweisen können.

02 **Helmet for Bimota SA**
It is part of the redesign of the identity and strategy, new communication materials, fair stand design and merchandise products.
Helm für Bimota SA
Er ist Teil der Neugestaltung der Identität und Strategie, der neuen Kommunikationsmaterialien, Messestandgestaltung und Handelswaren.

02

"My work philosophy is: quality over quantity."
„Meine Arbeitsphilosophie lautet: Qualität über Quantität."

You are involved in design education. What fascinates you about this job?
The exchange between the practical knowledge that I gain in my studio and the theoretical knowledge that is based on the Academy. Mixing those influences enriches both parts of my professional life.

When did you first think consciously about design?
When I was in primary school, I spent all the time enhancing, modifying and redesigning my toys. This sounds like just playing, but I added electric motors, new metal parts or just enhanced their finish.

What impressed you most during the Red Dot judging process?
All jury members took the task very seriously and displayed a high level of professionalism. It is a privilege and remarkably easy to work with so many renowned experts. The work is quite intense and focused, but enjoyable.

Sie sind in der Designlehre tätig. Was fasziniert Sie an dieser Aufgabe?
Der Austausch zwischen praktischem Wissen, das ich in meinem Designstudio erwerbe, und dem theoretischen Wissen, das auf der Akademie vermittelt wird. Diese Einflüsse zu vermischen, bereichert beide Bereiche meines beruflichen Lebens.

Wann haben Sie das erste Mal bewusst über Design nachgedacht?
Als ich in der Grundschule war, verbrachte ich meine gesamte Zeit damit, mein Spielzeug zu verbessern, zu verändern und umzugestalten. Das klingt nach bloßem Spielen, aber ich baute auch elektrische Motoren ein, neue Teile aus Metall oder veränderte einfach die Lackierung.

Was hat Sie bei der Red Dot-Jurierung am meisten beeindruckt?
Alle Mitglieder der Jury nahmen die Aufgabe sehr ernst und zeigten sich sehr professionell. Es ist ein Privileg und bemerkenswert einfach, mit so vielen berühmten Experten zusammenzuarbeiten. Die Arbeit ist ziemlich intensiv und konzentriert, aber unterhaltsam.

01

Prof. Ron A. Nabarro
Israel

Israel

Professor Ron A. Nabarro is an industrial designer, strategist, entrepreneur, researcher and educator. He has been a professional designer since 1970 and has designed more than 750 products to date in a wide range of industries. He has played a leading role in the emergence of age-friendly design and age-friendly design education. From 1992 to 2009, he was a professor of industrial design at the Technion Israel Institute of Technology, where he founded and was the head of the graduate programme in advanced design studies and design management. Currently, Nabarro teaches design management and design thinking at DeTao Masters Academy in Shanghai, China. From 1999 to 2003, he was an executive board member of Icsid and now acts as a regional advisor. He is a frequent keynote speaker at conferences, has presented TEDx events, has lectured and led design workshops in over 20 countries and consulted to a wide variety of organisations. Furthermore, he is co-founder and CEO of Senior-touch Ltd. and design4all. The principle areas of his research and interest are design thinking, age-friendly design and design management.

Professor Ron A. Nabarro ist Industriedesigner, Stratege, Unternehmer, Forscher und Lehrender. Seit 1970 ist er praktizierender Designer, gestaltete bisher mehr als 750 Produkte für ein breites Branchenspektrum und spielt eine führende Rolle im Bereich des altersfreundlichen Designs und dessen Lehre. Von 1992 bis 2009 war er Professor für Industriedesign am Technologie-Institut Technion Israel, an dem er das Graduierten-programm für fortgeschrittene Designstudien und Designmanagement einführte und leitete. Aktuell unterrichtet Nabarro Designmanagement und Design Thinking an der DeTao Masters Academy in Shanghai, China. Von 1999 bis 2003 war er Vorstandsmitglied des Icsid, für den er aktuell als regionaler Berater tätig ist. Er ist ein gefragter Redner auf Konferenzen, hat bei TEDx-Veranstaltungen präsentiert, hielt Vorträge und Workshops in mehr als 20 Ländern und beriet eine Vielzahl von Organisationen. Zudem ist er Mitbegründer und Geschäftsführer von Senior-touch Ltd. und design4all. Die Hauptbereiche seiner Forschung und seines Interesses sind Design Thinking, altersfreundliches Design und Designmanagement.

01 **AQ Water Bar**
Water purifier, cooler
and heater for domestic
and office use
Wasserreiniger, Kühl- und
Heizgerät für den Einsatz
im Büro und daheim

02 **Jumboard**
Toddler keyboard for
online developmental
computer games
Kinderkeybord für
entwicklungsfördernde
Online-Computerspiele

02

"My work philosophy is to improve the lives of ageing people by gathering leading companies, entrepreneurs, designers and technologists in order to educate and innovate together."

„Meine Arbeitsphilosophie ist die Verbesserung des Lebens älterer Menschen durch Innovation und Erziehung in Zusammenarbeit mit führenden Firmen, Unternehmern, Designern und Technologen."

What impressed you most during the Red Dot judging process?
The most important aspect of the jury process was the way jury members treat the process, in particular the respect for each work presented and the commitments to the high standards of adjudication.

Do you see a correlation between the design quality of a company's products and the economic success of this company?
In most cases this "formula" could work, still it is important to acknowledge the importance of marketing and not put everything on the designers' shoulders. We also can see designs that have been ahead of their time and totally failed, although the design was brilliant.

Which project would you like to realise one day?
At this stage of my life I find more interest in sharing my professional experience and life experience with young designers and design students.

Was hat Sie bei der Red Dot-Jurierung am meisten beeindruckt?
Der wichtigste Aspekt während der Jurierung war die Art, wie die Jurymitglieder dem Prozess begegnet sind, insbesondere der Respekt für jede einzelne Arbeit und die Hingabe an die hohen Jurierungsstandards.

Sehen Sie einen Zusammenhang zwischen der Designqualität, die sich in den Produkten eines Unternehmens äußert, und dem wirtschaftlichen Erfolg dieses Unternehmens?
In den meisten Fällen funktioniert diese „Formel", dennoch ist es wichtig, auch die Rolle des Marketings zu erkennen und nicht alles auf die Schultern der Gestalter zu laden. Wir kennen auch Gestaltungen, die ihrer Zeit voraus waren, aber komplett gescheitert sind, obgleich die Gestaltung brillant war.

Welches Projekt würden Sie gerne einmal realisieren?
In dieser Phase meines Lebens interessiert es mich mehr, meine professionelle Erfahrung und Lebenserfahrung mit jungen Designern und Designstudenten zu teilen.

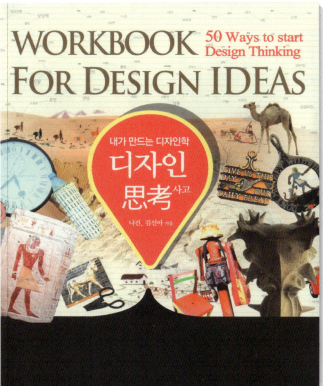

01

Prof. Dr. Ken Nah
Korea
Korea

Professor Dr Ken Nah graduated with a Bachelor of Arts in industrial engineering from Hanyang University, South Korea, in 1983. He deepened his interest in human factors/ergonomics by earning a master's degree from Korea Advanced Institute for Science and Technology (KAIST) in 1985 and he gained a PhD from Tufts University in 1996. In addition, Ken Nah is also a USA Certified Professional Ergonomist (CPE). He was the dean of the International Design School for Advanced Studies (IDAS) and is currently professor of design management as well as director of the Human Experience and Emotion Research (HE.ER) Lab at IDAS, Hongik University, Seoul. From 2002 he was the director of the International Design Trend Center (IDTC). Ken Nah was the director general of "World Design Capital Seoul 2010". Alongside his work as a lecturer he is also the vice president of the Korea Association of Industrial Designers (KAID), the Ergonomics Society of Korea (ESK) and the Korea Institute of Design Management (MIDM), as well as the chairman of the Design and Brand Committee of the Korea Consulting Association (KCA).

Professor Dr. Ken Nah graduierte 1983 an der Hanyang University in Südkorea als Bachelor of Arts in Industrial Engineering. Sein Interesse an Human Factors/Ergonomie vertiefte er 1985 mit einem Master-Abschluss am Korea Advanced Institute for Science and Technology (KAIST) und promovierte 1996 an der Tufts University. Darüber hinaus ist Ken Nah ein in den USA zertifizierter Ergonom (CPE). Er war Dekan der International Design School for Advanced Studies (IDAS) und ist aktuell Professor für Design Management sowie Direktor des „Human Experience and Emotion Research (HE.ER)"-Labors an der IDAS, Hongik University, Seoul. Von 2002 an war er Leiter des International Design Trend Center (IDTC). Ken Nah war Generaldirektor der „World Design Capital Seoul 2010". Neben seiner Lehrtätigkeit ist er Vizepräsident der Korea Association of Industrial Designers (KAID), der Ergonomics Society of Korea (ESK) und des Korea Institute of Design Management (MIDM) sowie Vorsitzender des „Design and Brand"-Komitees der Korea Consulting Association (KCA).

01 **Workbook for Design Ideas**
The Korean and Chinese
version on 50 ways to start
design thinking
Die koreanische und chinesische Version über 50 Wege, mit
Design Thinking zu beginnen

"My work philosophy is to do my best in all areas and every moment of every day, since time never stops and opportunity never waits."

„Meine Arbeitsphilosophie besteht darin, in allen Bereichen und in jedem Moment mein Bestes zu geben, da die Zeit ständig voran-schreitet und sich Gelegenheiten nicht zweimal bieten."

What motivates you to get up in the morning?
My question every morning before getting up is: "What if today were my last day?" This question motivates me to get back to work and use my time and energy in the best possible way.

When did you first think consciously about design?
It was in the winter of 1987, when I had to decide what majors to choose at university. Reading books and articles, I instantly fell in love with the words "Human Factors", defined as designing for people and optimising living and working conditions. Not only physical and physiological characteristics, but also psychological ones are important in "design". Since then, design has been everything to me!

What do you take special notice of when you are assessing a product as a jury member?
I pay attention to the "balance" between form and function. The product should also be well balanced between logic and emotion.

Was motiviert Sie, morgens aufzustehen?
Die Frage, die ich mir jeden Morgen vor dem Aufstehen stelle, ist: „Was wäre, wenn heute mein letzter Tag wäre?" Sie motiviert mich, wieder an die Arbeit zu gehen und meine Zeit und Energie optimal zu nutzen.

Wann haben Sie das erste Mal bewusst über Design nachgedacht?
Es war im Winter 1987, als ich an der Universität meine Hauptfächer wählen musste. Beim Lesen vieler Bücher und Artikel hatte ich mich sofort in die Worte „Human Factors" verliebt. Human Factors bedeutet, etwas für Menschen zu gestalten und die Lebens- und Arbeitsverhältnisse zu optimieren. Nicht nur physische und physiologische, sondern auch psychologische Eigenschaften sind wichtig in der Gestaltung. Seither ist Design mein Ein und Alles!

Worauf achten Sie besonders, wenn Sie ein Produkt als Juror bewerten?
Ich achte auf das Gleichgewicht zwischen Form und Funktion. Zudem sollte das Produkt auch zwischen Logik und Emotion gut ausgewogen sein.

01

Prof. Dr. Yuri Nazarov
Russia
Russland

Professor Dr Yuri Nazarov, born in 1948 in Moscow, teaches at the National Design Institute in Moscow where he is also provost. As an actively involved design expert, he serves on numerous boards, for example as president of the Association of Designers of Russia, as a corresponding member of Russian Academy of Arts, and as a member of the Russian Design Academy. Yuri Nazarov has received a wide range of accolades for his achievements: he is a laureate of the State Award of the Russian Federation in Literature and Art as well as of the Moscow Administration' Award, and he also has received a badge of honour for "Merits in Development of Design".

Professor Dr. Yuri Nazarov, 1948 in Moskau geboren, lehrt am National Design Institute in Moskau, dessen Rektor er auch ist. Als engagierter Designexperte ist er in zahlreichen Gremien des Landes tätig, zum Beispiel als Präsident der Russischen Designervereinigung, als korrespondierendes Mitglied der Russischen Kunstakademie sowie als Mitglied der Russischen Designakademie. Für seine Verdienste wurde Yuri Nazarov mit einer Vielzahl an Auszeichnungen geehrt. So ist er Preisträger des Staatspreises der Russischen Föderation in Literatur und Kunst sowie des Moskauer Regierungspreises und besitzt zudem das Ehrenabzeichen für „Verdienste in der Designentwicklung".

"My work philosophy ranges from functional designs for social regional projects and low-income individuals to engage in joint work with young designers and to integrate Russian design on an international scale."

„Meine Arbeitsphilosophie reicht von betont funktionalen Gestaltungen für soziale regionale Projekte wie für Geringverdienende bis zur Anbindung junger Designer sowie des russischen Designs an internationales Niveau."

What is, in your opinion, the significance of design quality in the product categories you evaluated?
The significance of design quality lies in confirming the usability and safety of the products.

What are the important criteria for you as a juror in the assessment of a product?
Creative ideas and the quality of their realisation.

What impressed you most during the Red Dot judging process?
The most outstanding for me was the mutual understanding of and similarities between our viewpoints.

Do you see a correlation between the design quality of a company's products and the economic success of this company?
It depends on how we interpret economic success. If we talk about profit it may simply be due to a hot commodity. But real design quality always implies taking into account consumer preferences.

Wie schätzen Sie den Stellenwert der Designqualität in den von Ihnen beurteilten Produktkategorien ein?
Der Stellenwert der Designqualität liegt darin, die Bedienbarkeit und Sicherheit der Produkte zu bekräftigen.

Worauf achten Sie als Juror, wenn Sie ein Produkt bewerten?
Auf kreative Ideen und die Qualität ihrer Umsetzung.

Was hat Sie bei der Red Dot-Jurierung am meisten beeindruckt?
Das Hervorstechendste für mich waren das gegenseitige Verständnis und die Ähnlichkeiten unserer Standpunkte.

Sehen Sie einen Zusammenhang zwischen der Designqualität, die sich in den Produkten eines Unternehmens äußert, und dem wirtschaftlichen Erfolg dieses Unternehmens?
Das hängt davon ab, wie man wirtschaftlichen Erfolg interpretiert. Sprechen wir von Profit, mag das schlicht an einem „heißen" Produkt liegen. Aber echte Designqualität impliziert immer die Berücksichtigung der Vorlieben der Verbraucher.

01

Ken Okuyama
Japan
Japan

Ken Kiyoyuki Okuyama, industrial designer and CEO of KEN OKUYAMA DESIGN, was born 1959 in Yamagata, Japan, and studied automobile design at the Art Center College of Design in Pasadena, California. He has worked as a chief designer for General Motors, as a senior designer for Porsche AG, and as design director for Pininfarina S.p.A., being responsible for the design of Ferrari Enzo, Maserati Quattroporte and many other automobiles. He is also known for many different product designs such as motorcycles, furniture, robots and architecture. KEN OKUYAMA DESIGN was founded in 2007 and provides business consultancy services to numerous corporations. Ken Okuyama also produces cars, eyewear and interior products under his original brand. He is currently a visiting professor at several universities and also frequently publishes books.

Ken Kiyoyuki Okuyama, Industriedesigner und CEO von KEN OKUYAMA DESIGN, wurde 1959 in Yamagata, Japan, geboren und studierte Automobildesign am Art Center College of Design in Pasadena, Kalifornien. Er war als Chief Designer bei General Motors, als Senior Designer bei der Porsche AG und als Design Director bei Pininfarina S.p.A. tätig und zeichnete verantwortlich für den Ferrari Enzo, den Maserati Quattroporte und viele weitere Automobile. Zudem ist er für viele unterschiedliche Produktgestaltungen wie Motorräder, Möbel, Roboter und Architektur bekannt. KEN OKUYAMA DESIGN wurde 2007 als Beratungsunternehmen gegründet und arbeitet für zahlreiche Unternehmen. Ken Okuyama produziert unter seiner originären Marke auch Autos, Brillen und Inneneinrichtungsgegenstände. Derzeit lehrt er als Gastprofessor an verschiedenen Universitäten und publiziert zudem Bücher.

02

"My design philosophy reads:
modern, simple, timeless."
„Meine Designphilosophie lautet:
Modern, schlicht, zeitlos."

**What trends have you noticed in the field of
"Vehicles" in recent years?**
Driving performance is no longer a sales point.
Transport design should propose not only mobility but
also a new lifestyle. A car's design reflects its owner's
character and lifestyle more than ever.

**Do you see a correlation between the design
quality of a company's products and the economic
success of this company?**
The correlation is the result of a clear vision and the
teamwork that made it happen, plus the personalities
of individual team members.

**What are the important criteria for you as a juror
in the assessment of a product?**
A juror has to determine a product's value to society
and the market. Therefore, an objective view and wide
ranging knowledge of technology, materials, manufacturing, etc. are necessary.

Welche Trends konnten Sie im Bereich „Fahrzeuge"
in den letzten Jahren ausmachen?
Fahr-Performance ist kein Verkaufsargument mehr. Im
Segment „Transport" sollte Gestaltung nicht nur auf
Mobilität abzielen, sondern auch auf einen neuen
Lebensstil. Das Design eines Autos spiegelt mehr denn
je den Charakter und Lebensstil seines Besitzers wider.

Sehen Sie einen Zusammenhang zwischen der
Designqualität, die sich in den Produkten eines
Unternehmens äußert, und dem wirtschaftlichen
Erfolg dieses Unternehmens?
Diese Wechselwirkung ist das Ergebnis einer klaren
Vision und der ihr zugrunde liegenden Teamarbeit –
plus der Persönlichkeiten der einzelnen Teammitglieder.

Worauf achten Sie als Juror, wenn Sie ein Produkt
bewerten?
Ein Juror muss den Wert bestimmen, den ein Produkt
für die Gesellschaft und den Markt hat. Daher sind
eine objektive Sichtweise und eine große Bandbreite
an Wissen über Technik, Werkstoffe, Herstellung etc.
notwendig.

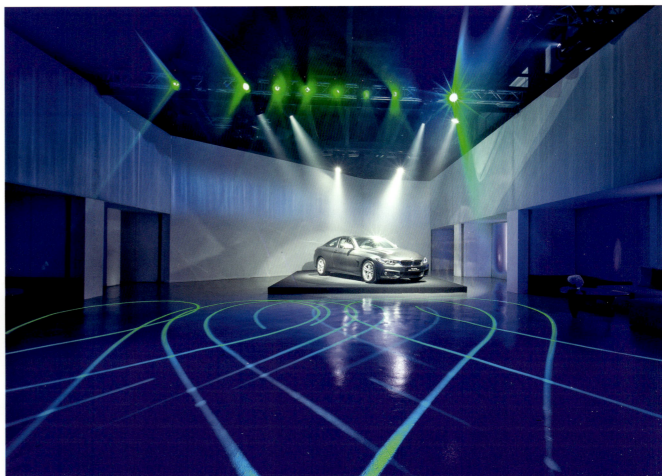

01

Simon Ong
Singapore
Singapur

Simon Ong, born in Singapore in 1953, graduated with a master's degree in design from the University of New South Wales and an MBA from the University of South Australia. He is the group managing director and co-founder of Kingsmen Creatives Ltd., a leading communication design and production group with 18 offices across the Asia Pacific region and the Middle East. Kingsmen has won several awards, such as the President's Design Award, SRA Best Retail Concept Award, SFIA Hall of Fame, Promising Brand Award, A.R.E. Retail Design Award and RDI International Store Design Award USA. Simon Ong is actively involved in the creative industry as chairman of the design group of Manpower, the Skills & Training Council of Singapore Workforce Development Agency. Moreover, he is a member of the advisory board of the Singapore Furniture Industries Council and School of Design & Environment at the National University of Singapore, Design Business Chamber Singapore and Interior Design Confederation of Singapore.

Simon Ong, geboren 1953 in Singapur, erhielt einen Master in Design der University of New South Wales und einen Master of Business Administration der University of South Australia. Er ist Vorstandsvorsitzender und Mitbegründer von Kingsmen Creatives Ltd., eines führenden Unternehmens für Kommunikationsdesign und Produktion mit 18 Geschäftsstellen im asiatisch-pazifischen Raum sowie im Mittleren Osten. Kingsmen wurde vielfach ausgezeichnet, u. a. mit dem President's Design Award, SRA Best Retail Concept Award, SFIA Hall of Fame, Promising Brand Award, A.R.E. Retail Design Award und RDI International Store Design Award USA. Simon Ong ist als Vorsitzender der Designgruppe von Manpower, der „Skills & Training Council of Singapore Workforce Development Agency", aktiv in die Kreativindustrie involviert, ist unter anderem Mitglied des Beirats des Singapore Furniture Industries Council, der School of Design & Environment an der National University of Singapore, des Design Business Chamber Singapore und der Interior Design Confederation of Singapore.

02

"My work philosophy is 'less is more'
in everything we do. Having less
'quantity' or details enables us to
have more time to focus on quality
and what matters."

„Meine Arbeitsphilosophie ist
‚Weniger ist mehr', in allem, was wir
tun. Die Reduktion von Quantität
oder Details gibt uns mehr Zeit, uns
auf die Qualität und das Wesentliche
zu konzentrieren."

What motivates you to get up in the morning?
Humour or having a good laugh. In our rapidly
developing world, we are often so caught up in
our work that we tend to forget the finest thing
in life – that is to laugh.

**In your opinion, which decade brought the greatest
progress in the design sector?**
Affordable computer software in the 1990s opened up
vast opportunities to go beyond traditional processes
in design thinking. Designs could be "tested" through
walk-through imaging – saving time and resources for
prototyping.

What challenges will designers face in the future?
Designers will have to think ahead and look at what
lies beyond sustainable design.

Was motiviert Sie, morgens aufzustehen?
Etwas zum Lachen zu haben. In unserer schnelllebigen
Welt sind wir oft so stark in unsere Arbeit verstrickt,
dass wir dazu neigen, das Beste im Leben zu vergessen –
und zwar zu lachen.

**Welches Jahrzehnt hat Ihrer Meinung nach den
größten Fortschritt im Designbereich gebracht?**
Erschwingliche Computer-Software in den 1990er
Jahren eröffnete riesige Möglichkeiten, über die
Grenzen traditioneller Prozesse im Design Thinking
hinauszugehen. Entwürfe konnten durch Computer-
simulationen „getestet" werden, um so Zeit und Ressourcen bei der Herstellung von Prototypen zu sparen.

**Vor welchen Herausforderungen werden Designer
in der Zukunft stehen?**
Designer werden vorausdenken und ausmachen
müssen, was nach nachhaltiger Gestaltung kommt.

01

Prof. Martin Pärn
Estonia
Estland

Professor Martin Pärn, born in Tallinn in 1971, studied industrial design at the University of Industrial Arts Helsinki (UIAH). After working in the Finnish furniture industry he moved back to Estonia and undertook the role of the ambassadorial leader of design promotion and development in his native country. He was actively involved in the establishment of the Estonian Design Centre and continues directing the organisation as chair of the board. Martin Pärn founded the multidisciplinary design office "iseasi", which creates designs ranging from office furniture to larger instruments and from small architecture to interior designs for the public sector. Having received many awards, Pärn begun in 1995 with the establishment and development of design training in Estonia and is currently head of the Design and Engineering's master's programme, a joint initiative of the Tallinn University of Technology and the Estonian Academy of Arts, which aims, among other things, to create synergies between engineers and designers.

Professor Martin Pärn, geboren 1971 in Tallinn, studierte Industriedesign an der University of Industrial Arts Helsinki (UIAH). Nachdem er in der finnischen Möbelindustrie gearbeitet hatte, ging er zurück nach Estland und übernahm die Funktion des leitenden Botschafters für die Designförderung und -entwicklung des Landes. Er war aktiv am Aufbau des Estonian Design Centres beteiligt und leitet seither die Organisation als Vorstandsvorsitzender. Martin Pärn gründete das multidisziplinäre Designbüro „iseasi", das ebenso Büromöblierung wie größere Instrumente, „kleine Architektur" oder Interior Designs im öffentlichen Sektor gestaltet. Vielfach ausgezeichnet, startete Pärn 1995 mit der Entwicklung und dem Ausbau der Designlehre in Estland und ist heute Leiter des Masterprogramms Design und Engineering, einer gemeinsamen Initiative der Tallinn University of Technology und der Estonian Academy of Arts, u. a. mit dem Ziel, durch den Zusammenschluss Synergien von Ingenieuren und Designern zu erreichen.

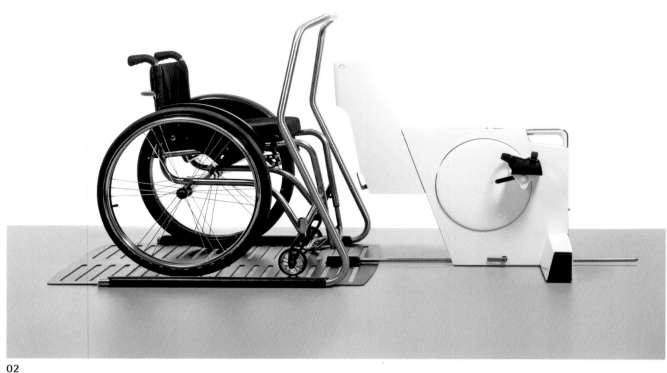

02

"My work philosophy is to search for something that is obvious, but yet unnoticed."

„Meine Arbeitsphilosophie besteht darin, nach etwas zu suchen, das offensichtlich, aber bisher unbemerkt geblieben ist."

What motivates you to get up in the morning?
Sun or, in short, life itself. It is full of new and unseen miracles I do not want to miss.

What challenges will designers face in the future?
The challenge of shifting the focus from fast consumer success towards long-lasting effectiveness and sustainability.

What do you take special notice of when you are assessing a product as a jury member?
I am looking more for the new "Why?" than the new "How?".

What significance does winning an award in a design competition have for the designer?
It means they have gained the respect of their colleagues, and thus it is a matter of honour. The real credits have to be earned in the field.

Was motiviert Sie, morgens aufzustehen?
Die Sonne oder, kurz gesagt, das Leben an sich. Es ist voller neuer und unbemerkter Wunder, die ich nicht verpassen möchte.

Vor welchen Herausforderungen werden Designer in der Zukunft stehen?
Vor der Herausforderung, das Augenmerk vom schnellen Markterfolg auf anglebige Leistungsfähigkeit und Nachhaltigkeit zu verschieben.

Worauf achten Sie besonders, wenn Sie ein Produkt als Juror bewerten?
Ich suche mehr nach einem neuen „Warum?" als nach einem neuen „Wie?".

Welche Bedeutung hat die Auszeichnung in einem Designwettbewerb für den Designer?
Es bedeutet, dass der Designer den Respekt seiner Kollegen gewonnen hat, was daher eine Frage der Ehre ist. Die echten Auszeichnungen müssen in der Praxis verdient werden.

01

02/03

Dr Sascha Peters
Germany
Deutschland

Dr Sascha Peters is founder and owner of the agency for material and technology HAUTE INNOVATION in Berlin. He studied mechanical engineering at the RWTH Aachen, Germany, and product design at the ABK Maastricht, Netherlands. He wrote his doctoral thesis at the University of Duisburg-Essen, Germany, on the complex of problems in communication between engineering and design. From 1997 to 2003, he led research projects and product developments at the Fraunhofer Institute for Production Technology IPT in Aachen and subsequently became head of the Design Zentrum Bremen. Sascha Peters is author of various specialised books on sustainable raw materials, smart materials, innovative production techniques and energetic technologies. He is a leading material expert and trend scout for new technologies. Since 2014, he has been an advisory board member of the funding initiative "Zwanzig20 – Partnerschaft für Innovation" (2020 – Partnership for innovation) by order of the German Federal Ministry of Education and Research.

Dr. Sascha Peters ist Gründer und Inhaber der Material- und Technologieagentur HAUTE INNOVATION in Berlin. Er studierte Maschinenbau an der RWTH Aachen und Produktdesign an der ABK Maastricht. Seine Doktorarbeit schrieb er an der Universität Duisburg-Essen über die Kommunikationsproblematik zwischen Engineering und Design. Von 1997 bis 2003 leitete er Forschungsprojekte und Produktentwicklungen am Fraunhofer-Institut für Produktionstechnologie IPT in Aachen und war anschließend bis 2008 stellvertretender Leiter des Design Zentrums Bremen. Sascha Peters ist Autor zahlreicher Fachbücher zu nachhaltigen Werkstoffen, smarten Materialien, innovativen Fertigungsverfahren und energetischen Technologien und zählt zu den führenden Materialexperten und Trendscouts für neue Technologien. Seit 2014 ist er Mitglied im Beirat der Förderinitiative „Zwanzig20 – Partnerschaft für Innovation" im Auftrag des Bundesministeriums für Bildung und Forschung.

04

"My work philosophy is to constantly discover better solutions
for better products and make
innovative technologies and
sustainable materials marketable."

„Meine Arbeitsphilosophie ist,
beständig bessere Lösungen für
bessere Produkte zu finden und
innovative Technologien und
nachhaltige Materialien marktfähig
zu machen."

**What impressed you most during the Red Dot
judging process?**
The jury with its experts from so many different
countries. This network of opinions emerging from
different cultural backgrounds is the basis for global
acceptance and the success of the Red Dot Award.

**What properties must a new material have to
convince you of its outstanding quality?**
With regards to our resources becoming ever scarcer,
the issues we are faced with concerning waste disposal and the challenges resulting from a growing
world population, I judge the development of a material as outstanding when it opens up the possibility
for sustainable use and leaves a particularly small
ecological footprint.

**Which material development from the last
hundred years has had the biggest influence on
today's world?**
Plastics. They enable the creation of almost limitless properties. However, most of them are not
bio-degradable.

**Was hat Sie bei der Red Dot-Jurierung am
meisten beeindruckt?**
Die Herkunft der Jury mit Experten aus den unterschiedlichsten Ländern. Dieses Netzwerk aus Meinungen verschiedener kultureller Einflüsse ist die
Grundlage für die globale Akzeptanz und den Erfolg
des Red Dot Awards.

**Welche Eigenschaften muss ein neues Material
vorweisen, um Sie von seiner herausragenden
Qualität zu überzeugen?**
Mit Blick auf die knapper werdenden Ressourcen, die
Probleme, die wir mit der Abfallentsorgung haben, und
die Herausforderungen, die sich durch die wachsende
Weltbevölkerung ergeben, bewerte ich Materialentwick-
lungen als herausragend, wenn sie eine Möglichkeit zu
einer nachhaltigen Nutzung aufzeigen und einen besonders geringen ökologischen Fußabdruck offenbaren.

**Welche Materialentwicklung der letzten hundert
Jahre hat den größten Einfluss auf die heutige
Zeit?**
Kunststoffe. Ihre Qualitäten lassen sich nahezu belie-
big einstellen. Ihr Großteil ist jedoch nicht biologisch
abbaubar.

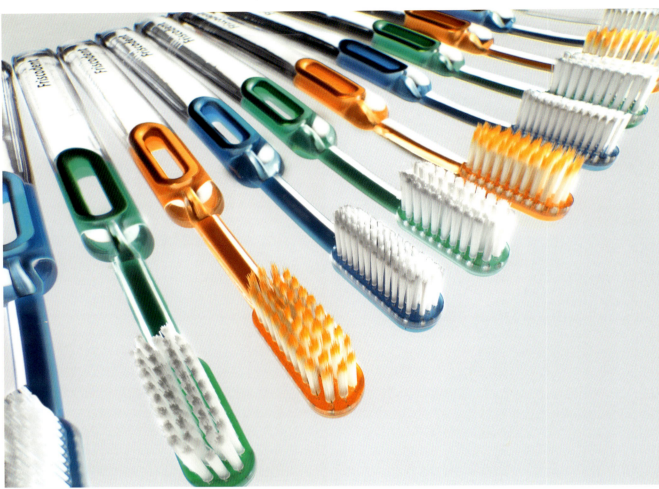

01

Oliver Stotz
Germany
Deutschland

Oliver Stotz, born in 1961 in Stuttgart, Germany, studied industrial design at the University Essen, Germany and at the Royal College of Art in London. As the founder of his own studio "stotz-design.com" in Wuppertal, Germany, he has more than 20 years of experience in the fields of industrial and corporate design. His studio has created established brands such as Proseat in the automotive sector and Blomus in the field of glass, porcelain and ceramic. With his eight-strong team, Stotz advises companies in various industries regarding the implementation and realisation of new design concepts. Since 2010, Oliver Stotz has been a board member of the foundation "Mia Seeger Stiftung", which promotes young designers after graduation with the "Mia Seeger Preis" award. In addition, he is a lecturer at the design department of the University Wuppertal. His design achievements have received several awards and have been on display in numerous exhibitions.

Oliver Stotz, 1961 in Stuttgart geboren, studierte Industriedesign an der Universität Essen und am Royal College of Art in London. Als Gründer des Studios „stotz-design.com" mit Sitz in Wuppertal kann er inzwischen auf mehr als 20 Jahre Berufserfahrung in den Bereichen Industrial und Corporate Design zurückgreifen. In seinem Studio entstanden bereits etablierte Marken wie Proseat im Automotive-Sektor oder auch Blomus im Bereich Glas, Porzellan, Keramik. Mit seinem achtköpfigen Team berät Stotz national wie international Unternehmen aus unterschiedlichsten Branchen bei der Implementierung und Umsetzung von neuen Designkonzepten. Seit 2010 ist Oliver Stotz Vorstandsmitglied der „Mia Seeger Stiftung", die junge Designerinnen und Designer nach ihrem Studienabschluss mit dem „Mia Seeger Preis" fördert. Darüber hinaus ist er als Dozent an der Universität Wuppertal im Fachbereich Design tätig. Seine Designleistungen wurden vielfach ausgezeichnet und in zahlreichen Ausstellungen präsentiert.

02

"My work philosophy is:
a good designer is always
ahead of his time."

„Meine Arbeitsphilosophie ist:
Ein guter Designer ist immer
seiner Zeit voraus."

In which product industry will Germany be most successful in the next ten years?
Germany is currently technology leader in the field of transportation and will achieve more successes in this field. However, networked thinking of the global players is a prerequisite for this to happen.

What are the main challenges in a designer's everyday life?
Keeping it simple is always the main challenge.

The Red Dot Award has been uncovering the best designs for 60 years now. Which product innovation would you like to see in the next 60 years, and why?
Driven by the thought that in future we will have to deal with the digitisation of many processes, as a result of which more and more things will become virtual, I wish to see materiality as product innovation, that haptic qualities become an enrichment of things.

What do you take special notice of when you are assessing a product as a jury member?
Whether it is an original design.

Mit welcher Produktbranche wird Deutschland in den nächsten zehn Jahren am erfolgreichsten sein?
Im Bereich Transportation ist Deutschland Technologieführer und wird Erfolge erzielen können. Voraussetzung ist allerdings, dass die Global Player vernetzt denken.

Was sind große Herausforderungen im Alltag eines Designers?
Keep it simple! Das ist immer wieder die große Aufgabe.

Der Red Dot Award ermittelt seit 60 Jahren die besten Gestaltungen. Welche Produktinnovation würden Sie sich für die nächsten 60 Jahre wünschen, und warum?
Von dem Gedanken getrieben, sich zukünftig mit der Digitalisierung vieler Prozesse auseinandersetzen zu müssen, in deren Folge immer mehr Dinge virtuell werden, wünsche ich mir bei den Produkten als Innovation die Materialität, das Haptische als konkret erlebbare Bereicherung der Dinge.

Worauf achten Sie besonders, wenn Sie ein Produkt als Juror bewerten?
Darauf, ob die Eigenständigkeit gewährleistet ist.

01

Aleks Tatic
Germany / Italy
Deutschland / Italien

Aleks Tatic, born 1969 in Cologne, Germany, is product designer and founder of Tatic Designstudio in Milan, Italy. After his studies at the Art Center College of Design in the USA and Switzerland, he specialised in his focal areas, sports and lifestyle products, in various international agencies in London and Milan. Afterwards, he guided the multiple award-winning Italian design studio Attivo Creative Resource to international success, leading the agency for 12 years. Together with his multicultural team of designers and product specialists, he today designs and develops – amongst others – sailing yachts, sporting goods, power tools, FMCGs and consumer electronics for European and Asian premium brands. Aleks Tatic lectures practice-oriented industrial design and innovation management at various European universities and seminars.

Aleks Tatic, geboren 1969 in Köln, ist Produktdesigner und Gründer der Agentur Tatic Designstudio in Mailand. Nach seinem Studium am Art Center College of Design in den USA und der Schweiz hat er sich zunächst in verschiedenen internationalen Büros in London und Mailand auf sein Schwerpunktgebiet Sport- und Lifestyleprodukte spezialisiert. Danach führte er zwölf Jahre lang das mehrfach ausgezeichnete italienische Designbüro Attivo Creative Resource zu internationalem Erfolg. Heute gestaltet und entwickelt er mit seinem multikulturellen Team von Designern und Produktspezialisten u. a. Segeljachten, Sportgeräte, Hobby- und Profiwerkzeuge, FMCGs und Unterhaltungselektronik für europäische und asiatische Premiummarken. Aleks Tatic unterrichtet an verschiedenen europäischen Hochschulen und Seminaren praxisorientiertes Industriedesign und Innovationsmanagement.

02

"My work philosophy is:
it has to be fun (for my
colleagues and me)!"

„Meine Arbeitsphilosophie ist:
Es muss uns (meinen Kollegen
und mir) Spaß machen!"

Which of the projects in your lifetime are you particularly proud of?
That would be our first sailing yacht, which we designed and then promptly won the "Boat of the Year" award for, even though we were new to the industry. This is the perfect proof that you can design any product even without specialising in the industry.

The Red Dot Award has been uncovering the best designs for 60 years now. Which product innovation would you like to see in the next 60 years, and why?
I would love to have someone launch Scotty's beamer in the market. This would cut travel times to our clients significantly.

What significance does winning an award in a design competition have for the manufacturer?
Many of our clients want to enter the Red Dot Award, because it is often hard for their clients – consumers or buying decision makers – to judge the design quality. Awards such as the Red Dot, as internationally recognised quality seals, are excellent indicators for this purpose.

Auf welches Projekt in Ihrem Leben sind Sie besonders stolz?
Auf unsere erste Segeljacht, die wir als komplett Branchenfremde gestaltet und dann prompt die Auszeichnung „Boat of the Year" gewonnen haben – der perfekte Beweis dafür, dass man jedes Produkt auch ohne Branchenspezialisierung gestalten kann.

Der Red Dot Award ermittelt seit 60 Jahren die besten Gestaltungen. Welche Produktinnovation würden Sie sich für die nächsten 60 Jahre wünschen, und warum?
Ich würde mich sehr freuen, wenn endlich jemand Scottys „Beamer" auf den Markt bringen würde. Die Reisezeiten zu unseren Kunden würden sich endlich deutlich verringern.

Welche Bedeutung hat die Auszeichnung in einem Designwettbewerb für den Hersteller?
Viele unserer Kunden möchten gerne beim Red Dot Award mitmachen. Denn für ihre Kunden – Verbraucher oder Kaufentscheider – ist die Designqualität oft schwierig zu beurteilen. Auszeichnungen wie der Red Dot als international anerkanntes Qualitätssiegel bilden hierfür herausragende Gradmesser.

01

Nils Toft
Denmark
Dänemark

Nils Toft, born in Copenhagen in 1957, graduated as an architect and designer from the Royal Danish Academy of Fine Arts in Copenhagen in 1985. He also holds a Master's degree in Industrial Design and Business Development. Starting his career as an industrial designer, Nils Toft joined the former Christian Bjørn Design in 1987, an internationally active design studio in Copenhagen with branches in Beijing and Ho Chi Minh City. Within a few years, he became a partner of CBD and, as managing director, ran the business. Today, Nils Toft is the founder and managing director of Designidea. With offices in Copenhagen and Beijing, Designidea works in the following key fields: communication, consumer electronics, computing, agriculture, medicine, and graphic arts, as well as projects in design strategy, graphic and exhibition design.

Nils Toft, geboren 1957 in Kopenhagen, machte seinen Abschluss als Architekt und Designer 1985 an der Royal Danish Academy of Fine Arts in Kopenhagen. Er verfügt zudem über einen Master im Bereich Industrial Design und Business Development. Zu Beginn seiner Karriere als Industriedesigner trat Nils Toft 1987 bei dem damaligen Christian Bjørn Design ein, einem international operierenden Designstudio in Kopenhagen, das mit Niederlassungen in Beijing und Ho-Chi-Minh-Stadt vertreten ist. Innerhalb weniger Jahre wurde er Partner bei CBD und leitete das Unternehmen als Managing Director. Heute ist Nils Toft Gründer und Managing Director von Designidea. Mit Büros in Kopenhagen und Beijing operiert Designidea in verschiedenen Hauptbereichen: Kommunikation, Unterhaltungselektronik, Computer, Landwirtschaft, Medizin und Grafikdesign sowie Projekte in den Bereichen Designstrategie, Grafik- und Ausstellungsdesign.

02

"My work philosophy is: design is a language that carries your brand and tells the story of how good your products are."

„Meine Arbeitsphilosophie lautet: Design ist eine Sprache, die deine Marke transportiert und erzählt, wie gut deine Produkte sind."

When did you first think consciously about design?
I was around seven or eight years old when my parents introduced me to arts and design. I remember how I, in my mind, redesigned the interior of my playmates' homes, when I saw how different they looked compared to mine.

What motivates you to get up in the morning?
I have always looked at life as a big apple and I can't wait to get up in the morning to take the next bite.

What impressed you most during the Red Dot judging process?
The vast number of exceptional entries and the fantastic discussions with the other jury members.

What significance does winning an award in a design competition have for the designer?
It is a tribute to the uniqueness of their talent and a reminder not to take talent for granted.

Wann haben Sie das erste Mal bewusst über Design nachgedacht?
Ich war ungefähr sieben oder acht Jahre alt, als mich meine Eltern an Kunst und Design heranführten. Ich erinnere mich daran, wie ich in meinem Kopf die Inneneinrichtung der Häuser meiner Spielkameraden umgestaltete, als ich sah, wie stark sie sich von meiner unterschied.

Was motiviert Sie, morgens aufzustehen?
Ich habe das Leben schon immer als einen großen Apfel betrachtet und kann es kaum erwarten, morgens aufzustehen und den nächsten Bissen zu nehmen.

Was hat Sie bei der Red Dot-Jurierung am meisten beeindruckt?
Die riesige Anzahl bemerkenswerter Einreichungen und die fantastischen Diskussionen mit den anderen Jurymitgliedern.

Welche Bedeutung hat die Auszeichnung in einem Designwettbewerb für den Designer?
Sie zollt der Einzigartigkeit seines Talentes Tribut und erinnert daran, Talent nicht als selbstverständlich anzusehen.

01

Prof. Danny Venlet
Belgium
Belgien

Professor Danny Venlet was born in 1958 in Victoria, Australia and studied interior design at Sint-Lukas, the Institute for Architecture and Arts in Brussels. Back in Australia in 1991, Venlet started to attract international attention with large-scale interior projects such as the Burdekin hotel in Sydney, and Q-bar, an Australian chain of nightclubs. His design projects range from private mansions, lofts, bars and restaurants all the way to showrooms and offices of large companies. The interior projects and the furniture designs of Danny Venlet are characterised by their contemporary international style. He says that the objects arise from an interaction between art, sculpture and function. These objects give a new description to the space in which they are placed – with respect, but also with relative humour. Today, Danny Venlet teaches his knowledge to students at the Royal College of the Arts in Ghent.

Professor Danny Venlet wurde 1958 in Victoria, Australien, geboren und studierte Interior Design am Sint-Lukas Institut für Architektur und Kunst in Brüssel. Nachdem er 1991 wieder nach Australien zurückgekehrt war, begann er, mit der Innenausstattung großer Projekte wie dem Burdekin Hotel in Sydney und der Q-Bar, einer australischen Nachtclub-Kette, internationale Aufmerksamkeit zu erregen. Seine Designprojekte reichen von privaten Wohnhäusern über Lofts, Bars und Restaurants bis hin zu Ausstellungsräumen und Büros großer Unternehmen. Die Innenausstattungen und Möbeldesigns von Danny Venlet sind durch einen zeitgenössischen, internationalen Stil ausgezeichnet und entspringen, wie er sagt, der Interaktion zwischen Kunst, Skulptur und Funktion. Seine Objekte geben den Räumen, in denen sie sich befinden, eine neue Identität – mit Respekt, aber auch mit einer Portion Humor. Heute vermittelt Danny Venlet sein Wissen als Professor an Studenten des Royal College of the Arts in Gent.

02

"My work philosophy is to create objects that, while embodying the theory of relativity, leave us no other choice than to act differently."

„Meine Arbeitsphilosophie besteht darin, Objekte zu kreieren, die – obwohl sie die Relativitätstheorie verkörpern – uns keine andere Wahl lassen, als anders zu handeln."

Which design project was the greatest intellectual challenge for you?
The Easyrider for Bulo, because my briefing consisted of the two opposing concepts of "relaxation/office", which gave birth to, if I may say so, a great innovation.

Which country do you consider to be a pioneer in product design?
Of course Belgium, due to its excellent training, its history and companies which are willing to take the design path.

What is the advantage of commissioning an external designer compared to having an in-house design team?
The biggest advantage is that an external designer is free-spirited and not conditioned by the constraints that an in-house designer might have encountered working in a company. In order to innovate you need to be able to look beyond!

Welches Ihrer bisherigen Designprojekte war die größte Herausforderung für Sie?
Der Easyrider für Bulo, da mein Briefing aus den zwei gegensätzlichen Konzepten „Entspannung/Büro" bestand, aus denen eine, wenn ich das einmal so sagen darf, großartige Innovation hervorging.

Welche Nation ist für Sie Vorreiter im Produktdesign?
Natürlich Belgien, aufgrund seiner hervorragenden Ausbildungsmöglichkeiten, seiner geschichtlichen Vergangenheit und seiner Unternehmen, die gewillt sind, auf Design zu setzen.

Worin liegt der Vorteil, einen externen Designer zu beauftragen, im Vergleich zu einem Inhouse-Designteam?
Der größte Vorteil ist, dass ein externer Designer über mehr geistige Freiheit verfügt und nicht von den Beschränkungen konditioniert ist, denen ein Inhouse-Designer in einer Firma unterworfen ist. Um Innovationen zu kreieren, muss man die Möglichkeit haben, über den Tellerrand hinauszuschauen!

01

Günter Wermekes
Germany
Deutschland

Günter Wermekes, born in Kierspe, Germany in 1955, is a goldsmith and designer. After many years of practice as an assistant and head of the studio of Professor F. Becker in Düsseldorf, he founded his own studio in 1990. He attracted great attention with his jewellery collection "Stainless steel and brilliant", which he has been presenting at national and international fairs since 1990. His designs are also appreciated by renowned manufacturers such as BMW, Rodenstock, Niessing or Tecnolumen, for which he designed, among other things, accessories, glasses, watches and door openers. In exhibitions and lectures around the world he has illustrated his personal design philosophy, which is: "Minimalism means reducing things more and more to their actual essence, and thus making it visible." He has also implemented this motto in his design of the Red Dot Trophy. In 2014, Günter Wermekes was one of ten finalists in the design competition for the medal of the Tang Prize, the "Asian Nobel Prize". His works have won numerous prizes and are part of renowned collections.

Günter Wermekes, 1955 geboren in Kierspe, ist Goldschmied und Designer. Nach langjähriger Tätigkeit als Assistent und Werkstattleiter von Professor F. Becker in Düsseldorf gründete er 1990 sein eigenes Studio. Große Aufmerksamkeit erregte er mit seiner Schmuckkollektion „Edelstahl und Brillant", die er ab 1990 auf nationalen und internationalen Messen präsentierte. Sein Design schätzen auch namhafte Hersteller wie BMW, Rodenstock, Niessing oder Tecnolumen, für die er u. a. Accessoires, Brillen, Uhren und Türdrücker entwarf. In Ausstellungen und Vorträgen weltweit verdeutlicht er seine persönliche Gestaltungsphilosophie, dass „Minimalismus bedeutet, Dinge so auf ihr Wesen zu reduzieren, dass es dadurch sichtbar wird". Diesen Leitspruch hat er auch in seinem Entwurf der Red Dot Trophy umgesetzt. 2014 war Günter Wermekes einer von zehn Finalisten des Gestaltungswettbewerbs für die Preismedaille des „Tang Prize", des „Asiatischen Nobelpreises". Seine Arbeiten wurden mehrfach prämiert und befinden sich in bedeutenden Sammlungen.

01 **Der Halbkaräter**
(The Half-Carat-Piece)
Ring model R0011 made
from stainless steel and
a 0.5 ct tw/vsi brilliant-
cut diamond for the
"Stainless steel and brilliant"
collection
Ringmodell R0011 aus Edelstahl und einem Brillanten von
0,5 ct tw/vsi für die Kollektion
„Edelstahl und Brillant"

02 **Red Dot Trophy**
Trophy made from stainless
steel and acrylic glass for
the Red Dot Design Award,
Essen, Germany, 2013
Preistrophäe aus Edelstahl
und Plexiglas für den Red Dot
Design Award, Essen, 2013

02

"My work philosophy is based on the quotation 'Less is more' by Mies van der Rohe, with regards to design as well as ecology."

„Meine Arbeitsphilosophie fußt auf dem Zitat von Mies van der Rohe ‚Weniger ist mehr', und zwar in gestalterischer Hinsicht wie auch im Hinblick auf Ökologie."

When did you first consciously think about design?
Around 1970/72, when my art teacher at grammar school gave me the following assignment: Design and produce a wooden chair. The chair has never been completed, but my interest in designing objects of everyday life was awakened.

What are your sources of inspiration?
Contrasts, borders, irritations. Actually, anything you can discover outside the box. And the music of Johann Sebastian Bach as a common thread.

The Red Dot Award has been uncovering the best designs for 60 years now. Which product innovations would you like to see in the next 60 years?
I would like to see innovations that are dealing responsibly with our living environment, planet Earth.

What impressed you most during the Red Dot judging process?
The fact that an awareness of good and useful design is attracting increasing attention.

Wann haben Sie das erste Mal bewusst über Design nachgedacht?
So um 1970/72, als mein Kunstlehrer am Gymnasium mir die Aufgabe stellte: Entwirf und fertige einen Holzstuhl. Der Stuhl ist zwar nie fertig geworden, aber das Interesse für die Gestaltung von Alltags-gegenständen war geweckt.

Was sind Ihre Inspirationsquellen?
Gegensätze, Grenzbereiche, Irritationen. Eigentlich alles, was man jenseits des „Tellerrands" entdecken kann. Und als „Roter Faden" die Musik von Johann Sebastian Bach.

Der Red Dot Award ermittelt seit 60 Jahren die besten Gestaltungen. Welche Produktinnovationen würden Sie sich für die nächsten 60 Jahre wünschen?
Ich wünsche mir Innovationen, die verantwortungsvoll mit unserem Lebensraum Erde umgehen.

Was hat Sie bei der Red Dot-Jurierung am meisten beeindruckt?
Dass das Bewusstsein für gutes und sinnvolles Design immer größere Kreise zieht.

Alphabetical index manufacturers and distributors
Alphabetisches Hersteller- und Vertriebs-Register

Alphabetical index manufacturers and distributors
Alphabetisches Hersteller- und Vertriebs-Register

Alphabetical index designers
Alphabetisches Designer-Register

Alphabetical index designers
Alphabetisches Designer-Register

Alphabetical index designers
Alphabetisches Designer-Register

Alphabetical index designers
Alphabetisches Designer-Register

reddot edition

Editor | Herausgeber
Peter Zec

Project management | Projektleitung
Jennifer Bürling

Project assistance | Projektassistenz
Sophie Angerer
Tatjana Axt
Theresa Falkenberg
Constanze Halsband
Danièle Huberty
Estelle Limbah
Judith Lindner
Melanie Masino
Anamaria Sumic

Editorial work | Redaktion
Bettina Derksen, Simmern, Germany
Eva Hembach, Vienna, Austria
Catharina Hesse
Burkhard Jacob
Karin Kirch, Essen, Germany
Karoline Laarmann, Dortmund, Germany
Sarah Latussek, Cologne, Germany
Bettina Laustroer, Wuppertal, Germany
Kirsten Müller, Essen, Germany
Astrid Ruta, Essen, Germany
Martina Stein, Otterberg, Germany
Achim Zolke

Proofreading | Lektorat
Klaus Dimmler (supervision), Essen, Germany
Mareike Ahlborn, Essen, Germany
Jörg Arnke, Essen, Germany
Wolfgang Astelbauer, Vienna, Austria
Sabine Beeres, Leverkusen, Germany
Dawn Michelle d'Atri, Kirchhundem, Germany
Annette Gillich-Beltz, Essen, Germany
Eva Hembach, Vienna, Austria
Karin Kirch, Essen, Germany
Norbert Knyhala, Castrop-Rauxel, Germany
Laura Lothian, Vienna, Austria
Regina Schier, Essen, Germany
Anja Schrade, Stuttgart, Germany

Translation | Übersetzung
Heike Bors-Eberlein, Tokyo, Japan
Patrick Conroy, Lanarca, Cyprus
Stanislaw Eberlein, Tokyo, Japan
William Kings, Wuppertal, Germany
Cathleen Poehler, Montreal, Canada
Tara Russell, Dublin, Ireland
Jan Stachel-Williamson, Christchurch, New Zealand
Philippa Watts, Exeter, Great Britain
Andreas Zantop, Berlin, Germany
Christiane Zschunke, Frankfurt am Main, Germany

Layout | Gestaltung
Lockstoff Design GmbH, Grevenbroich, Germany
Susanne Coenen
Katja Kleefeld
Judith Maasmann
Stephanie Marniok
Lena Overkamp
Nicole Slink

Photographs | Fotos
Dragan Arrigler ("Alpina ESK Pro", juror Jure Miklavc)
Kaido Haagen ("Fleximoover", juror Martin Pärn)
Jäger & Jäger ("Lodelei", juror Martin Pärn)
Kompan marketing department ("elements", juror Nils Toft)
MBD Design ("Optifuel Trailer", juror Vincent Créance)
Masaki Ogawa ("Black+white", juror Herman Hermsen)
Christophe Recoura ("Alstom – Paris Tram T7"/
"Bombardier Regio 2N", juror Vincent Créance)
studio aisslinger ("Upcycling & tuning"/"Luminaire",
juror Werner Aisslinger)
Roberto Turci ("Helmet for Bimota SA", juror Jure Miklavc)
Wagner Ziegelmeyer ("Fabrimar Lucca", juror
Guto Indio da Costa)

Page | Seite
504
Name | Name
Tokyo_Institute_of_Technology_Centennial_Hall_2009
Copyright | Urheber
Wiiii
Source | Quelle
http://commons.wikimedia.org/wiki/File:Tokyo_
Institute_of_Technology_Centennial_Hall_2009.jpg

Distribution of this photograph is protected by copyright
law. A license is required for distribution. More information
is available at
Diese Fotografie unterliegt hinsichtlich der Verbreitung
dem Urheberrechtsschutz. Die Verbreitung bedarf einer
Lizenz. Nähere Angaben dazu finden Sie unter
http://creativecommons.org/licenses/by-sa/3.0/deed.en

Jury photographs | Jurorenfotos
Simon Bierwald, Dortmund, Germany

In-company photos | Werkfotos der Firmen

Production and litography |
Produktion und Lithografie
tarcom GmbH, Gelsenkirchen, Germany
Gregor Baals
Jonas Mühlenweg
Bernd Reinkens

Printing | Druck
Dr. Cantz'sche Druckerei Medien GmbH,
Ostfildern, Germany

Bookbindery | Buchbinderei
BELTZ Bad Langensalza GmbH
Bad Langensalza, Germany

Red Dot Design Yearbook 2015/2016
Living: 978-3-89939-174-9
Doing: 978-3-89939-175-6
Working: 978-3-89939-176-3
Set (Living, Doing & Working): 978-3-89939-173-2

©2015 Red Dot GmbH & Co. KG, Essen, Germany

The Red Dot Award: Product Design
competition is the continuation of the
Design Innovations competition.
Der Wettbewerb „Red Dot Award: Product Design"
gilt als Fortsetzung des Wettbewerbs
„Design Innovationen".

All rights reserved, especially those of translation.
Alle Rechte vorbehalten, besonders die der Übersetzung
in fremde Sprachen.

No liability is accepted for the completeness
of the information in the appendix.
Für die Vollständigkeit der Angaben im Anhang
wird keine Gewähr übernommen.

Publisher + worldwide distribution |
Verlag + Vertrieb weltweit
Red Dot Edition
Design Publisher | Fachverlag für Design
Contact | Kontakt
Sabine Wöll
Gelsenkirchener Str. 181
45309 Essen, Germany
Phone +49 201 81418-22
Fax +49 201 81418-10
E-mail edition@red-dot.de
www.red-dot-edition.com
www.red-dot-shop.com
Book publisher ID no. | Verkehrsnummer
13674 (Börsenverein Frankfurt)

Bibliographic information published
by the Deutsche Nationalbibliothek
The Deutsche Nationalbibliothek
lists this publication in the Deutsche
Nationalbibliografie; detailed bibliographic
data are available on the Internet at
http://dnb.ddb.de
Bibliografische Information
der Deutschen Nationalbibliothek
Die Deutsche Nationalbibliothek verzeichnet
diese Publikation in der Deutschen
Nationalbibliografie; detaillierte
bibliografische Daten sind im Internet über
http://dnb.ddb.de abrufbar.